Biotechnology and its Applications

Biotechnology and its Applications

Using Cells to Change the World

Second edition

W T. Godbey

Tulane University
New Orleans, Louisiana

Academic Press is an imprint of Elsevier
125 London Wall, London EC2Y 5AS, United Kingdom
525 B Street, Suite 1650, San Diego, CA 92101, United States
50 Hampshire Street, 5th Floor, Cambridge, MA 02139, United States
The Boulevard, Langford Lane, Kidlington, Oxford OX5 1GB, United Kingdom

Notices
Knowledge and best practice in this field are constantly changing. As new research and experience broaden our understanding, changes in research methods, professional practices, or medical treatment may become necessary.

Practitioners and researchers must always rely on their own experience and knowledge in evaluating and using any information, methods, compounds, or experiments described herein. In using such information or methods they should be mindful of their own safety and the safety of others, including parties for whom they have a professional responsibility.

To the fullest extent of the law, neither nor the Publisher, nor the authors, contributors, or editors, assume any liability for any injury and/or damage to persons or property as a matter of products liability, negligence or otherwise, or from any use or operation of any methods, products, instructions, or ideas contained in the material herein.

Library of Congress Cataloging-in-Publication Data
A catalog record for this book is available from the Library of Congress

British Library Cataloguing-in-Publication Data
A catalogue record for this book is available from the British Library

ISBN: 978-0-12-817726-6

For information on all Academic Press publications visit our website at https://www.elsevier.com/books-and-journals

Publisher: Katey Birtcher
Senior Acquisition Editor: Stephen Merken
Editorial Project Manager: Chris Hockaday
Production Project Manager: Manikandan Chandrasekaran
Cover Designer: Mark Rogers

Typeset by TNQ Technologies

Printed in United Kingdom

Last digit is the print number: 9 8 7 6 5 4 3 2 1

Contents

List of Figures

List of Tables

Preface

I wrote this book for students who have little to no biology background, with the goal of introducing them to the world of biotechnology in a way that runs deeper than a mere survey. There are myriad biotechnologies in the world today, and the number continues to grow. I happen to find the world of biotechnology an incredibly exciting one, and I want to share that excitement with you, the reader. However, to fully appreciate just how cool some of these technologies are, we must understand some of the science underlying the glamour.

This is not intended to be a comprehensive book on all biotechnologies; one could spend the better part of a decade becoming an expert in only one area. Likewise, although I state that one must understand the science behind a technology to appreciate it fully—and a healthy portion of that science is covered in this book—the text is not intended to be a complete, rigorous reference book for the biological or chemical sciences. Instead, it was my aim to produce a book that would serve as a mix of both basic science and biotechnological applications, so that readers might become energized about biotechnology at a level deep enough to allow further pursuit of a particular area, with the knowledge that a solid foundation has been laid. If you still have a passion for one of the subjects contained in this book after studying it, that passion will probably be quite real. I want you to get enough foundation to appreciate what is out there in the world of biotechnology, so that you will understand new developments in greater depth than you will get from the news and so that you will not be fooled by unsubstantiated claims you might read on the internet.

The book is divided into three units. In the first, basic science is covered to introduce the reader to the cell: how it behaves and what it is made of. For instance, if you want to design a drug that will enter the cell, you must know and understand the barrier that separates the inside of the cell from its exterior environment. If you want that drug to affect the cell's behavior, you should understand how the cell functions so that you can pick a cellular target upon which your drug can act. Perhaps you want the cell to make a product that you can isolate and sell, such as recombinant insulin, ethanol, or even a novel protein you have designed yourself. To be successful in this endeavor, you must understand how the cell would go about generating the product, in addition to knowing exactly what your product would be. Although we do not cover every possible product that could be made by a cell, we do cover some of the building blocks such as amino acids, proteins, and nucleic acids (DNA and RNA).

The second unit is aimed at the biotechnological application of scientific principles in the laboratory. Unit I shows us how things work, but in Unit II, we see how we can employ such knowledge to design, produce, and analyze products on a laboratory scale. These applications have allowed the identification of pathogens so that appropriate antibiotic selections can be made, the production of cell cultures to assist in the study of investigational drugs or engineered tissue constructs, and the amplification of DNA or RNA to permit the engineering and construction of entire genes.

The third unit presents biotechnologies "in the real world," which is not intended to imply that the laboratory is not the real world! Certainly, the laboratory makes up the primary world of many biotechnologists. The unit title refers to technologies that are used for practical purposes to aid nonscientists. Examples include recombinant proteins that are available to millions of patients, plants that have been engineered to produce food for people around the world, and regenerative medicine that may someday allow patients to receive organs that have been grown from their own cells. From fighting crime to removing fingernail polish to powering automobiles without the use of fossil fuels, biotechnology is being used in a wide range of applications that affect the lives of millions of people every day.

When I first undertook the writing of this book, I was under the impression that biotechnology was a new and cutting-edge field. The cutting-edge part is correct, but upon citing important references for the book, I came to realize that the information is not necessarily "new." Sure, there are famous experiments from the 1950s that we learned about in school, but I was a little surprised to find myself citing a paper from 1948 and then heavily relying on a referenced paper from 1852. However, it was when I came across texts that predated the Bible that my opinion of biotechnology really changed. This is not a new field; it is a timeless field. It is true that progress in biotechnology perhaps occurs at a greater rate these days, but it is entirely incorrect to think that the only "real" science has occurred in recent times. We may utilize more sophisticated equipment today, but the science that took place in, for example, G.G. Stokes's laboratory in the mid-1800s was just as real (and the mathematics just as complex) as anything performed today.

Finally, this book was written primarily for the student, not the teacher. The style is informal throughout, which will undoubtedly irritate some professors. Not to worry; they simply will not adopt the book for their course. However, the text is written in a style that is easy to follow and perhaps a little light-hearted in places, with numerous figures to reinforce the information. The material is solid, however. The student and teacher can both feel confident that, despite the material's being presented in a way that is (I hope) entertaining and easy to understand, it is based on decades of personal experience, education, and consultation with experts on the individual topics.

UNIT I

The cell

CHAPTER

An introduction to biotechnology

1

Chapter outline

1.1 An agreement with the student

A good place to begin in any text on biotechnology is with a definition of the term. Technology is easy enough to define—it is the application of scientific principles for a practical purpose. "Bio" is a prefix that refers to life. Unfortunately, a sound definition for "life" is not available; however, we can go back to our basic biology books from middle or high school and find lists of characteristics that describe living things. Those lists might include attributes like:

- an ability to grow
- an ability to reproduce
- an ability to maintain homeostasis
- an ability to respond to stimuli
- the performance of metabolism

Unfortunately, there are examples of living things that do not match up with these characteristics. A mule, the offspring of a horse and a donkey, can pull a plow, but it cannot reproduce. Many grandmothers also cannot reproduce, and some get smaller as they progress deeper into old age (because of bone loss). Nevertheless, it is generally easy to tell when we run across a living mule or grandmother.

What about crystals? They are able to grow, reach equilibrium, and even move in response to stimuli, but we do not call them alive. Rivers move, can grow after rains or snow melts, can spawn babies in the form of tributaries, respond to stimuli like dams or rockslides, and even metabolize by converting some molecules into other forms via hydrolysis. They have energy and perform work, but most people (except maybe riverboat captains) do not consider them to be alive.

Biotechnology and its Applications. https://doi.org/10.1016/B978-0-12-817726-6.00001-0

Then there are tardigrades, also known as water bears (Figure 1.1). Different members of the phylum Tardigrada have been shown experimentally to survive extreme conditions such as exposure to radiation, temperatures ranging from -273°C (near absolute zero) to almost 125°C, high pressures of over 1 million PSI, ultra-low pressure (such as the vacuum of space), and complete dehydration. Some biotechnologists believe that tardigrades hold the secret to successful cryopreservation of humans in the future.

The dehydration issue is especially interesting. If we were to remove all the water from two samples of tardigrades, grind up the second sample into a powder, then rehydrate both samples, the first one would contain living, swimming tardigrades while the second would contain muck. This implies that life has something to do with structure. Does this mean that, if we were able to impart that structure to a chemical compound, it would suddenly come to life? Dehydrated tardigrades come to life when we add water, but what if we don't? Is the dehydrated form alive? It cannot grow, reproduce, respond to stimuli, or metabolize in the dehydrated state. It seems to be devoid of life until we reintroduce water. So at what point is it alive? It depends on whom you ask.

I hope we have just established that a firm definition of life is not available. By extension, there is not a suitable definition for biotechnology because any definition involving the prefix "bio" will reference back to the word "life." If we cannot define the term that is central to this book, then is there any reason to go on reading it? This is the very reason why an agreement must be made with the student right now: without this text providing a definition of life, can we agree that it exists? If so, then we are prepared to learn about many exciting things that can be done for and with living things. We will discuss many phenomena at the organism level, the subcellular level, and the molecular level. We may have trouble with a couple of fundamental issues related to life and death, but even without a definition for biotechnology, I hope we can agree that it exists.

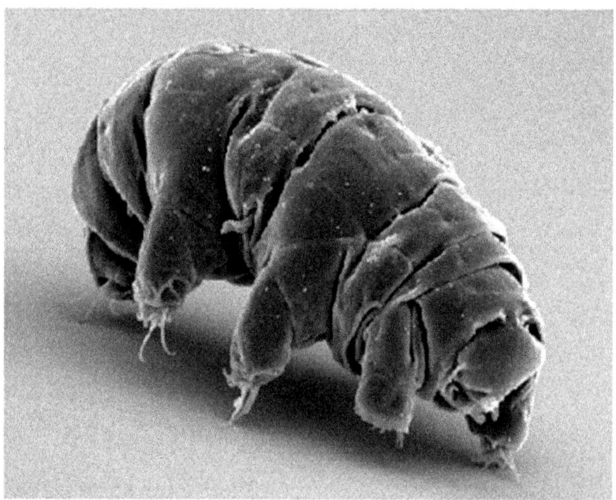

FIGURE 1.1

A tardigrade (water bear).

From: Schokraie, E., Warnken, U., Hotz-Wagenblatt, A., Grohme, M.A., Hengherr, S., Förster, F., Schill, R.O., Frohme, M., Dandekar, T., Schnölzer, M. Comparative proteome analysis of Milnesium tardigradum in early embryonic state versus adults in active and anhydrobiotic state. PLoS One. 2012;7(9):e45682.

1.2 Misconceptions about biotechnology

1.2.1 Biotechnology is a new field

Some of the most exciting advances in the world today have come from the field of biotechnology. Certainly, the prospect of being able to cure certain heritable diseases before a fertilized egg even reaches the womb spells exciting (if not ominous) changes for humans as a species. However, biotechnological advancements did not begin only in the past 20 years, or the past 200, or even the past 2000. The production of beer, which uses microbes such as yeast to alter the composition of a slurry of mashed-up seeds into an alcoholic brew, is a biotechnology that has been mentioned in texts pre-dating the Bible.

You probably know the story of how penicillin was discovered in 1928 when Alexander Fleming observed how a certain mold (*Penicillium notatum*) restricted the growth of bacteria (*Staphylococcus aureus*). Maybe you have heard the next part of the story that details how the drug was developed and scaled up in the late 1930s and early 1940s by Howard Florey, Ernst Chain, and Norman Heatley. While penicillin is often touted as the first antibiotic, antibiotics have actually been used by humans for thousands of years. Chinese folk remedies used moldy soybean curds to treat boils as early as 600 BCE; whether or not the users were aware of its existence, an antibiotic was probably being produced by the mold. Similarly, the ancient Egyptians used honey to prevent or treat infections in wounds. While the concept of using honey to treat wounds is ancient, "medical honey" is in use even today (Figure 1.2). It is not considered an antibiotic or an antiseptic, but it does have antimicrobial properties. The modern product is still just honey, although it has been treated by gamma irradiation to kill spores.

FIGURE 1.2

An example of a medical honey, a modern version of an ancient biotechnology.

FIGURE 1.3

Friendship bread, is an example of biotechnology from the kitchen.

From: https://www.publicdomainpictures.net/pictures/110000/velka/homemade-bread.jpg

1.2.2 Biotechnology is only developed in laboratories

When envisioning biotechnology and how it is developed, one might tend to think of a sterile laboratory with scientists dressed in white coats or Tyvek suits, surrounded by scales, microscopes, and biosafety cabinets. While such environments do exist, they are not required for the development of biotechnology. Ancient biotechnologies were certainly not developed that way.

Biotechnology is occasionally developed in an environment as innocent as somebody's kitchen. Yeast cultures are a good example. Sourdough is the product of yeast digestion of components of flour (and extra sugar), and sourdough bread and pancakes are delicious when done correctly. The trouble is that at first it is not easy to grow the cultures just right; one must do a little research to determine what works best in one's own environment. Culture times; the amount of flour, water, and sugar to add; and even the vessel used to let the culture mature all affect the final product. Friendship bread is another example of a biotechnology from the kitchen (Figure 1.3). For this biotechnology, a starter yeast culture is passed from friend to friend, and sweet, tasty desserts are the result … if the cultures develop properly.

Even the requirement for a clean environment is questionable when one considers pruno, also known as prison wine. This disgusting concoction is a testament to the lengths to which some humans will go to obtain alcohol for consumption. It is a mixture of oranges and fruit cocktail, sugar, and yeast (obtained from bread), mixed together and allowed to ferment in a closed container (such as a plastic bag) in a warm environment such as a prison cell toilet (Figure 1.4). After about a week, the rotten fruit and other solids are strained out to yield the final product. It is not recommended that you try this at home (or anywhere else, for that matter). However, the use of microbes to convert a feed stock into a product is indeed an application of biotechnology.

1.2.3 Biotechnology yields only unsafe results

Humans are wary of things they do not understand. This is a natural phenomenon that could explain the evolutionary success of the species; although our natural curiosity has led to discoveries such as what

FIGURE 1.4

Prison toilets have been used to make pruno, a questionable application of biotechnological practices.

From: https://www.publicdomainpictures.net/pictures/120000/velka/exhibit-of-toilet-in-prison-cell.jpg

foods are safe to eat versus what will make us sick, as a general rule, not every individual in a group stumbling upon some unknown mushroom in the woods will rush to be the first to scarf down handfuls of the fungus. Most will watch one person eat and observe him to see whether he gets sick later. Likewise, when a new biotechnology is made available to the public, the first response is usually one of skepticism, fear, or even outrage. Skepticism is natural and even healthy, but that does not mean we should automatically shun all biotechnologies.

For example, some biotechnologies yield information as their product. A good example is DNA fingerprinting. Because of advances in molecular biology, specific features have been identified in the human genome that can be used to trace the original owner of hair, skin, blood, or even smudged fingerprints left at a crime scene. Use of the data yielded by this technology in the courts was vehemently opposed by defendants' attorneys at first, but now the technology is considered a strong form of evidence for prosecuting wrongdoers and exonerating the innocent.

Recombinant DNA technologies are also monitored by watchful eyes, and that is not a bad thing. However, the fact that a molecule was made by a cell expressing an engineered gene does not necessarily make that molecule harmful. An example is insulin, which has been produced on a commercial scale (Figure 1.5) by delivering genes encoding human insulin into bacteria. The recombinant bacteria are then allowed to grow and multiply while at the same time expressing the newly delivered gene. The insulin is later harvested and purified for use in patients with diabetes. The insulin molecules produced

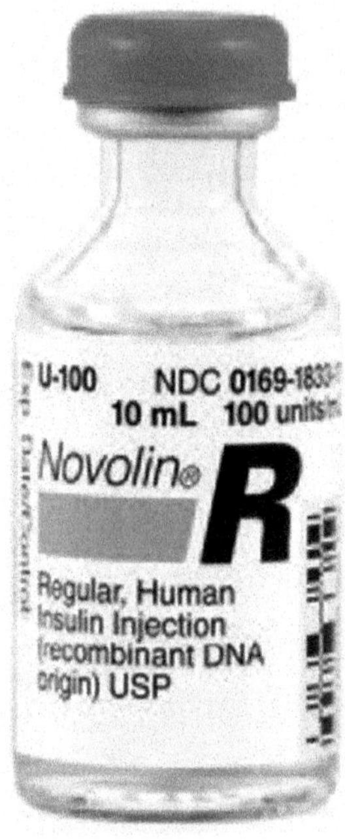

FIGURE 1.5

An example of recombinant insulin.

in this fashion are identical to the insulin produced by the β cells of the normal human pancreas. The recombinant product is no more or less dangerous than its human-produced counterpart.

Other recombinant protein drugs certainly exist, including Aranesp (recombinant erythropoietin, used to treat anemia), Avonex (recombinant interferon β1a, used to treat multiple sclerosis), and Gardasil (recombinant forms of four proteins made by the human papillomavirus, used as the first vaccine against a cancer). Time will tell as to the safety of the medications, which have already passed extensive safety trials. Recombinant proteins are not inherently dangerous, and any adverse reactions will be due to the molecules themselves, not the method by which they were produced.

1.3 Biotechnology is broad and still expanding

As we learn more and more about how cells work, biotechnologists gain a greater refinement in what can be produced through cellular manipulation or the harnessing of biological components. It has been possible for several decades to engineer genes and put them into cells for expression; more recently, the genomes of host cells have been directly manipulated to introduce or remove cellular behaviors. Cellular components can be isolated to allow for the manipulation of molecules such as DNA or proteins in cell-free environments. Cellular processes can be mimicked in a factory to produce cellular products without cells, and cells can be used as factories to produce industrial products. Creative minds are needed to discover new ways in which various components, behaviors, and attributes of cells and cellular products can be applied to improve lives.

Biotechnology is expanding at an unprecedented rate, and a student interested in biotechnology is not restricted to working solely in a biology laboratory. Chemists, engineers, mathematicians, lawyers, and businesspeople are all involved in the field of biotechnology.

Various fields that utilize biotechnology and examples of where biotechnologists may contribute include:

- the medical industry (drugs and vaccines)
- the beverage industry (beer and wine)
- farming (pesticides, insect-resistant plants, high-yield crops)
- food processing (recombinant enzymes for food processing)
- chemical plants (large-scale cell cultures to degrade waste products of chemical synthesis)
- environmental (oil spill remediation)
- alternative fuels (biobutanol, algae-derived oils)

Not all members of the biotechnology field are biologists. Many different talents, qualifications, and backgrounds are needed throughout the industry. While it is clear that bio-related disciplines such as medicine, farming, and ecology all have hands in biotechnology, disciplines not directly involved with cells, plants, and animals are also involved. Examples of non-biologists involved with biotechnology and ways in which they contribute include:

- chemical engineers (scale-up, process optimization)
- chemists (design of molecules to interact with cells)
- marketers (public acceptance of biotechnology-derived products)
- business managers (streamlining of operations for optimal availability of the final product)
- lawyers (intellectual property, patents, and licenses)

It is exciting to think of all the ways biotechnology has touched the world and how many different disciplines are involved. Before we delve into the applications of the field, it would behoove us to have a fair understanding of how cells work. After all, the whole of biotechnology has something to do with cells in some way at some point. In the next chapter, we will take our first look at cells and their initial interactions with bioactive agents, and we will continue to build our understanding of cells throughout the first unit.

Related reading

Boothby, T.C., Tapia, H., Brozena, A.H., Piszkiewicz, S., Smith, A.E., Giovannini, I., et al., 2017. Tardigrades use intrinsically disordered proteins to survive desiccation. Mol. Cell 65, 975–984 e5.

Horikawa, D.D., 2012. Survival of tardigrades in extreme environments: a model animal for astrobiology. In: Altenbach, A., Bernhard, J., Seckbach, J. (Eds.), Anoxia. Cellular Origin, Life in Extreme Habitats and Astrobiology, vol. 21. Springer, Dordrecht, the Netherlands.

Markel, H., September 27, 2013. The real story behind penicillin. PBS NewsHour. https://www.pbs.org/newshour/health/the-real-story-behind-the-worlds-first-antibiotic. (Accessed September 30, 2019).

Olempska-Beer, Z.S., Merker, R.I., Ditto, M.D., DiNovi, M.J., 2006. Food-processing enzymes from recombinant microorganisms—a review. Regul. Toxicol. Pharmacol. 45, 144–158.

Ono, F., Saigusa, M., Uozumi, T., Matsushima, Y., Ikeda, H., Saini, N.L., et al., 2008. Effect of high hydrostatic pressure on a life of a tiny animal tardigrade. J. Phys. Chem. Solids 69, 2297–2300.

Simon, A., Traynor, K., Santos, K., Blaser, G., Bode, U., Molan, P., 2009. Medical honey for wound care—still the "latest resort"? Evid. Based Complement. Alternat. Med. 6, 165–173.

CHAPTER

Voyage into the cell

2

Chapter outline

All biotechnologists will have to deal with cells at some point. Whether they're working on a drug designed to help a patient, a device built to respond to environmental stimuli, or a polymer that serves as a scaffold for tissue regeneration, if "bio" is involved in the process, cells will be involved in some way during the research and development process (if not the final application of the technology itself). After all, the prefix "bio" comes from the Greek word *bios*, which means "life," and the cell is the fundamental unit of life. If we are going to be dealing with cells, we should know something about them.

Biotechnology and its Applications. https://doi.org/10.1016/B978-0-12-817726-6.00002-2

In this chapter we will look at a few structures and properties of cells. What makes their interiors different from what lies outside? If we are to deliver a drug to a cell, how will the drug get in to do its thing, if it gets in at all? First, we must understand the barrier that separates the cell interior from the exterior environment. Next, we must address how things can cross this barrier. More specifically, we will discuss:

- the plasma membrane
- phospholipids
- cholesterol
- membrane proteins
- channels
- active and passive transporters
- endocytosis
- vesicles, endosomes, and lysosomes
- receptors

2.1 Membranes

We will begin our survey of the cell with (one of) the exterior layer(s) of the cell – the plasma membrane. At the cellular level, this is the first point of contact for technologies such as immunotherapy, gene/drug delivery, and patterned cell attachment involving eukaryotic cells. For any biotechnologist designing a particle that is to be taken up by a cell, a thorough knowledge of the plasma membrane is advisable.

2.1.1 Membrane lipids

Both **eukaryotic** cells (which have membrane-bound nucleus and organelles) and **prokaryotic** cells (single-celled organisms with no nucleus or membrane-bound organelles) are surrounded by a plasma membrane (Figure 2.1). The plasma membrane is not the outermost layer of a prokaryotic cell; this point will be addressed in detail later in the text. The plasma membrane is made up of lipids, carbohydrates, and proteins (and various combinations thereof), but the primary constituent is the phospholipid. There are many different types of phospholipids, but those present in the plasma membrane all follow some basic principles.

Membrane phospholipids contain a **hydrophilic** head and two **hydrophobic** tails and as such are said to be **amphipathic**. Phospholipids can be considered as having three constituents linked to a glycerol backbone via reactions with each of the three hydroxyls. The structure of glycerol is given in Figure 2.2.

Of course, there are exceptions, but the following rules of thumb help to describe the structure of a phospholipid built on a glycerol backbone:

1. The first hydroxyl will have been used to react with the carboxyl terminus of a fatty acid to form an ester linkage. The fatty acid is typically 14 to 24 carbons long, with an even number of carbons. The fatty acid will be saturated.
2. The second hydroxyl will also be attached to a fatty acid with an even number (14–24) of carbons in an ester linkage. However, this fatty acid is typically unsaturated. The unsaturated fatty acid

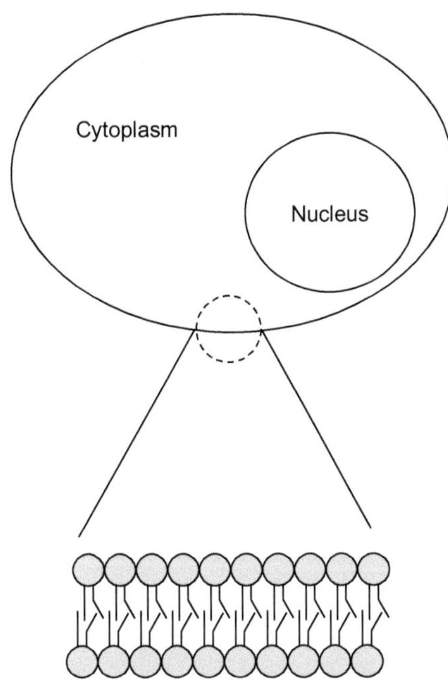

FIGURE 2.1

The eukaryotic plasma membrane.

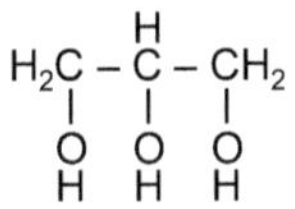

```
       H
H2C – C – CH2
  |    |    |
  O    O    O
  H    H    H
```

FIGURE 2.2

The structure of glycerol, the backbone of the phospholipid structure. Note the three hydroxyl (–OH) groups.

will be kinked, which results in less dense packing in the membrane and therefore a lower freezing temperature.

3. The third hydroxyl will be linked to a phosphate group to form a phosphoester bond. The arm of the molecule containing the phosphate group will constitute the hydrophilic portion of the phospholipid. If this arm has only a phosphate group, the lipid is phosphatidic acid (Figure 2.3). However, additional moieties can be attached to the phosphoryl group to give rise to the range of phospholipids found in the plasma membrane.

Three of the most common phospholipids of the plasma membrane are **phosphatidylethanolamine (PE)**, **phosphatidylserine (PS)**, and **phosphatidylcholine (PC)**. Figure 2.4 shows the general structures of these phospholipids, but it can also be seen that there is not one single formula for a given phospholipid because the compositions of the two fatty acids attached to the glycerol backbone are not fixed. This means there can be several different phosphatidylcholines, for example.

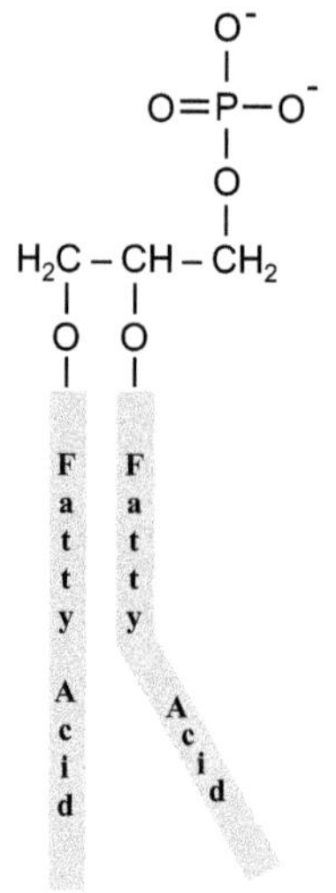

O^-–P(=O)(–O^-)–O–CH_2–CH(–O–Fatty Acid)–H_2C–O–Fatty Acid

FIGURE 2.3

Phosphatidic acid.

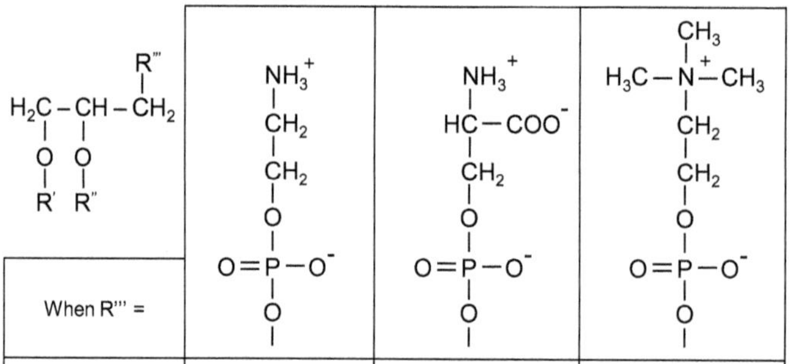

$H_2C-CH-CH_2$ with O–R' on H_2C, O–R'' on CH, and R''' on CH_2

When R''' =	$NH_3^+-CH_2-CH_2-O-P(=O)(O^-)-O-$	$NH_3^+-HC(COO^-)-CH_2-O-P(=O)(O^-)-O-$	$(CH_3)_3N^+-CH_2-CH_2-O-P(=O)(O^-)-O-$
Name	Phosphatidyl-ethanolamine	Phosphatidyl-serine	Phosphatidyl-choline
Abbreviation	PE	PS	PC
Overall charge	Neutral	Negative	Neutral
Found in which layer of PM	Inside	Inside	Outside

FIGURE 2.4

Structures, charges, and general locations of three of the most common phospholipids in the plasma membrane.

Two additional types of lipids that are common constituents of the plasma membrane also bear mention: sphingomyelin and glycolipids. Sphingomyelin is a phospholipid that has a choline head group but differs from phosphatidylcholine in that, instead of a glycerol backbone holding two fatty acid tails, the hydrophobic portion of the molecule is ceramide (Figure 2.5). As the name implies, sphingomyelin is found in the myelin sheath that surrounds the axons of certain nerve cells, but it also makes up a significant percentage of the plasma membranes of red blood cells and liver cells.

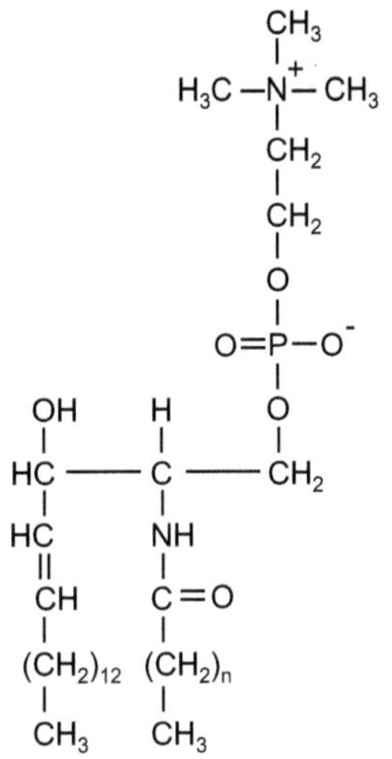

```
                    CH3
                    |+
               H3C -N- CH3
                    |
                    CH2
                    |
                    CH2
                    |
                    O
                    |
                 O=P-O-
                    |
 OH        H        O
 |         |        |
 HC ------ C ------ CH2
 |         |
 HC        NH
 ||        |
 CH        C=O
 |         |
 (CH2)12   (CH2)n
 |         |
 CH3       CH3
```

FIGURE 2.5

Sphingomyelin.

Glycolipids are prevalent in certain plasma membranes. In animal cells, they are found only on the exterior surface, which implies that they may have some function in cell–cell interactions, cellular identification, or cell signaling. Although the glycolipid composition of the plasma membrane differs greatly between cells of different species or even between tissues in the same animal, it is believed that all cells contain at least some glycolipids in the exterior face of their plasma membranes. This is a feature that has yet to be fully capitalized on in fields such as stem cell engineering and tissue engineering, although the medical science and immunology disciplines have recognized the glycolipid signatures of various microbes and have used them to develop therapies.

The plasma membrane is asymmetric due to the preferential locations of the phospholipids at one side or the other. As a general rule, the phospholipids with a terminal amino group (PE, PS) tend to be on the inner (cytoplasmic) face of the membrane, whereas those with a choline in their head group (PC and sphingomyelin) are typically found on the external face. Practically, glycolipids are always on the external face (Figure 2.6A).

These principles do not dictate the absolute positions of membrane constituents, however, because the plasma membrane is fluid in structure. Molecules on the outer face can generally move about like people on the second floor of a crowded mall, whereas molecules on the inner face are like the people on the first floor. Some lipids will stay together in lipid rafts—much like a family might stay together in the mall—whereas others can have their positions changed to the other side of the membrane, like somebody taking the escalator. Such translocation can be seen in specific circumstances, such as when a membrane is being actively formed by the addition of new lipids. Enzymes known as phospholipid translocators are responsible for the phenomenon, which is known as flip-flop. The location of PS at the cytoplasmic face in newly formed plasma membranes is achieved in a specific manner by the action of enzymes called **flippases**.

Another example of phospholipid flipping occurs during the process of programmed cell death known as **apoptosis**. During the early stages of this complex process, the enzyme **scramblase**

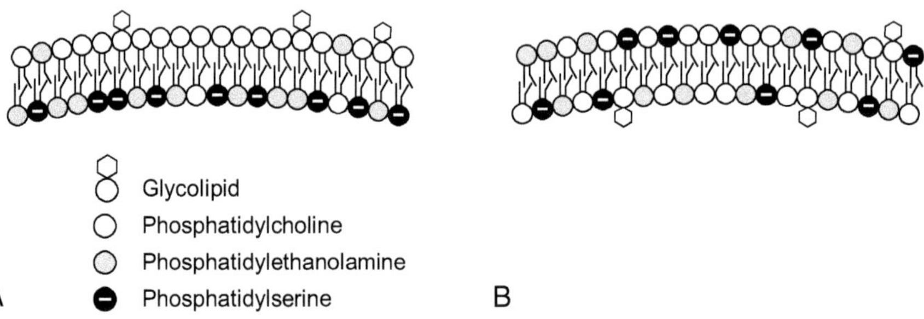

FIGURE 2.6

(A) A representative distribution of membrane phospholipids. (B) During apoptosis, scramblase randomizes phospholipid distributions. The redistribution of phosphatidylserine can be detected by other cells (e.g., macrophages).

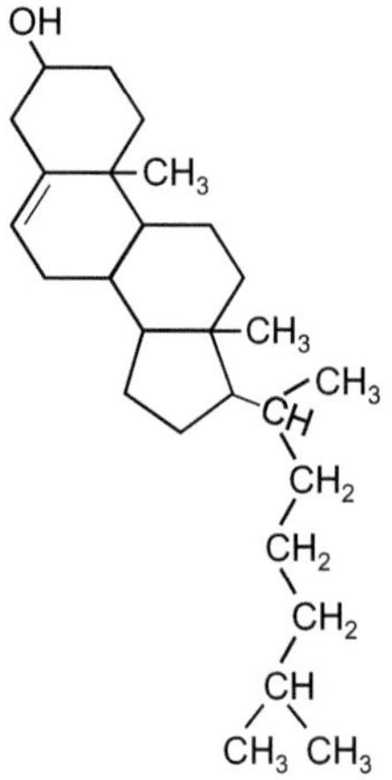

FIGURE 2.7

The structure of cholesterol.

randomizes phospholipid locations relative to the side of the membrane on which they usually reside (Figure 2.6B). One result is that PS becomes translocated to the exterior face of the plasma membrane; the sudden appearance of these negatively charged phospholipids on the cell exterior surface is thought to serve as a signal to macrophages to engulf and degrade the cell.

To help you keep these two translocation enzymes straight, remember that flippases act specifically to create an ordered lipid bilayer with distinctly different faces (negatively charged cytoplasmic and neutral extracellular), whereas scramblases act more randomly. Note that this randomization adds negative charges to the extracellular face of the plasma membrane.

2.1.2 Cholesterol

Cholesterol is another constituent of the plasma membrane. The structure of cholesterol is shown in Figure 2.7. Although it is not a phospholipid, cholesterol is somewhat similar in that it is also

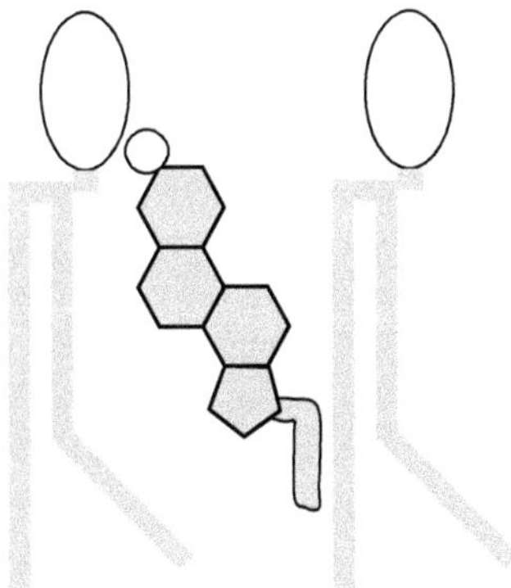

FIGURE 2.8

Cholesterol fits between adjacent phospholipids.

amphipathic: it has a hydrophilic portion (the OH group) and a hydrophobic region (the rest of the molecule). Cholesterol fits between adjacent phospholipids in the plasma membrane, with the hydroxyl being aligned with the polar head groups and the rest of the molecule fitting in among the fatty acid tails (Figure 2.8). Cholesterol can be found in high abundance in animal cells, sometimes at concentrations of one cholesterol molecule for every phospholipid molecule. Although high cholesterol content in the blood is considered a bad thing, cholesterol is nevertheless a necessary component of the membrane of every cell, performing several functions. First, the four steroid rings are relatively rigid, which serves to decrease the fluidity of the membrane and increase its stability. With the hydroxyl group being positioned as it is, the rings will be firmly placed in the area of the glycerol backbones and first few carbons of the fatty acid tails of adjacent phospholipids. This effectively immobilizes these carbons and renders the membrane less fluid. Second, while making the membrane less fluid, cholesterol also serves to make the membrane more flexible. Whereas the lower-numbered carbons of the phospholipid fatty acid tails are held in relative rigidity, the higher-numbered carbons (farthest from the glycerol backbone) are free to gyrate.

With cholesterol serving as a spacer, there is less chance for the carbons at the end of the tails of one phospholipid to associate with the tails of adjacent phospholipids. The fatty acid tails are thus prevented from coming together to form ordered, crystalline structures, as happens during phase transitions at lower temperatures (freezing).

2.1.3 Membrane proteins

A discussion of the plasma membrane would not be complete without a presentation of membrane proteins. However, membrane proteins cannot be adequately described without an introduction to proteins in general. Entire careers have been devoted to proteins, and many wonderful books have been devoted to their structure and function. Although an exhaustive introduction to proteins is not appropriate for an introduction to biotechnology, a rudimentary knowledge is indeed required when learning about the structure and function of the cell. We will begin our survey in the next chapter with the building blocks of proteins: amino acids.

2.2 Cellular transport

Although we do not have an adequate definition of life, we do know that part of a cell's being considered alive has to do with the maintenance of a different environment inside versus outside the plasma membrane. As we have already learned, the plasma membrane serves as an excellent barrier for most molecules. Some molecules can cross more freely than others, as has been shown experimentally with synthetic membranes (Figure 2.9). The hydrophobic tails of phospholipids provide a phase that does not mix well with charged molecules, so the phospholipid bilayer is very effective at separating the ionic content of the cytoplasm from that of the extracellular environment. Conversely, although students occasionally consider oxygen or carbon dioxide to be polar because of the high electronegativity of oxygen atoms, these molecules are in fact nonpolar because the dipole moments are symmetric. Nonpolar molecules have a greater ability to cross the plasma membrane.

2.2.1 Membrane transporters

Although it might seem apparent that certain molecules can leave and enter the cell freely, the question remains as to how a cell can acquire polar molecules or ions, especially at concentrations that differ from those in the extracellular environment. The answer is that the cell contains transporters—membrane proteins that are specific for the molecules they carry. There are different classes of transporters based on the direction of molecule travel. Transporters that carry a single molecule in only one direction (either into or out of the cell) are called **uniporters**. Transporters that carry two or more molecules

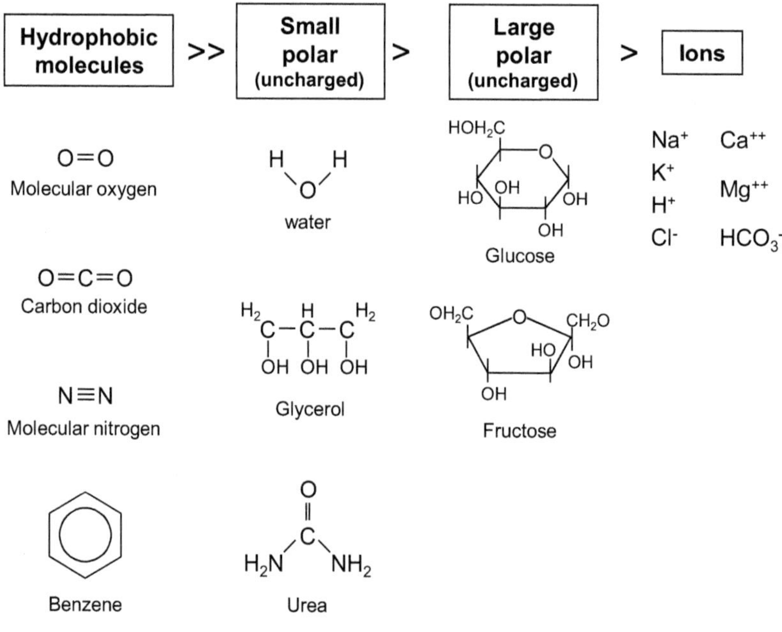

FIGURE 2.9

Relative abilities of types of molecules to cross a synthetic lipid bilayer membrane, with examples.

across the cell membrane at the same time and in the same direction are **symporters**. Transporters that carry two or more molecules across the cell membrane simultaneously, but in opposite directions, are **antiporters** (Figure 2.10). If energy (typically in the form of ATP hydrolysis) is directly required for the transporter to carry molecules across the membrane, the carrier is said to be an **active transporter**. If no ATP hydrolysis is required, the carrier is a **passive transporter**.

A **channel** is another means by which molecules can enter a cell. Ion channels are proteins that span the plasma membrane and open to allow the diffusion of specific ions. Some channels are always open (**leak channels**), and some are opened in response to a stimulus (**gated channels**; e.g., voltage-gated channels). Ion channels have specificity because of their size. Consider for a moment a potassium channel. As a K^+ ion moves through the channel, it will interact with specific amino acids in the channel, commonly via electronegative carbonyl oxygens on the amino acids. The space is so atomically precise that the water molecules that usually surround the ion will be stripped off, being replaced spatially by the amino acids in the channel's selectivity filter as the ion is positioned within the channel. We can see from the periodic table that a potassium ion is smaller than a rubidium ion but larger than a sodium ion. The channel is sized so that the larger Rb^+ cannot fit. Although a Na^+ can fit through the opening of the channel, it is too small to interact with the amino acids in the selectivity filter in an energetically favorable way, so the water molecules surrounding the ion are not stripped off and it cannot interact properly with the amino acids; it will not be able to proceed through the channel. This is how ion channels achieve their high specificity.

We will now look at several examples of transporters, starting with the sodium/glucose symporter. As you read the descriptions, remind yourself of the types of molecules being discussed. For example, the sodium/glucose symporter is not only a symporter but also a passive transporter and a transmembrane protein.

2.2.1.1 The sodium/glucose symporter

Figure 2.11 is a sketch of the sodium/glucose symporter. Initially, sodium ions and glucose molecules that will be transported are on the **extracellular** side but are free to diffuse into the active sites on the symporter. Outside of the cell, there is a relatively high sodium concentration as compared with the

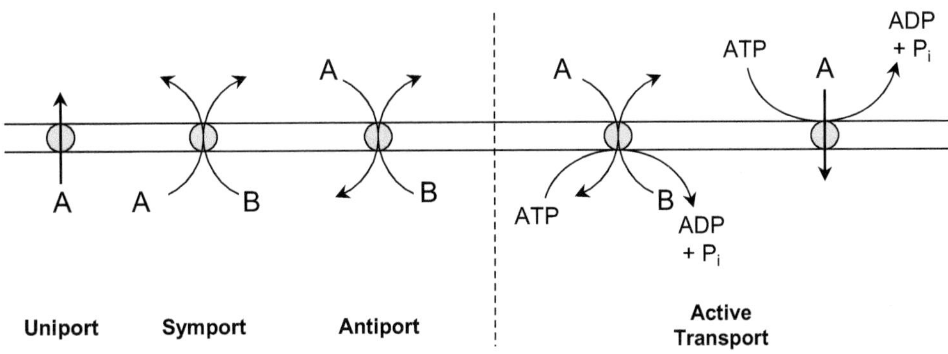

FIGURE 2.10

Examples of transport of substances (A) and (B) across a cell membrane. Note that uniport and antiport are shown twice, as passive and active transporters.

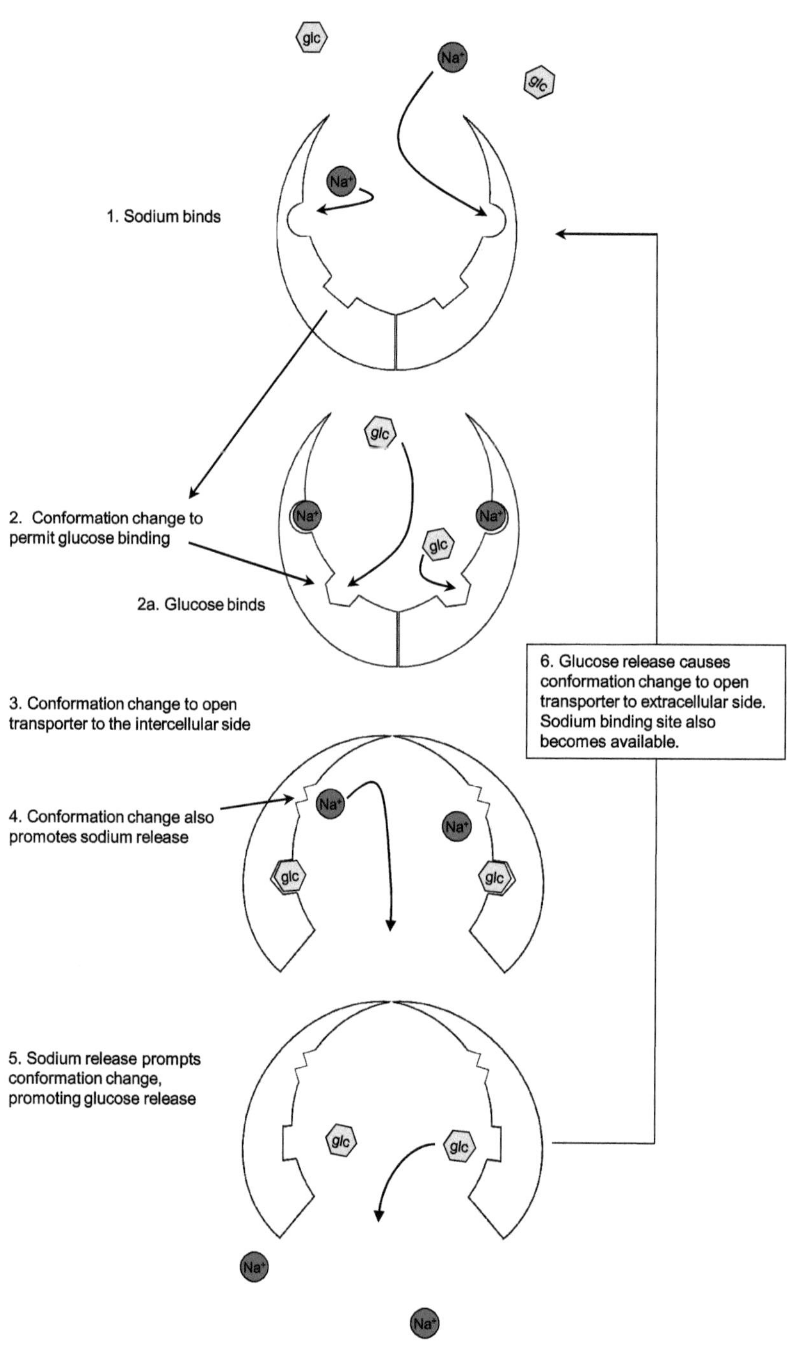
glc
Na+
glc
1. Sodium binds
Na+
2. Conformation change to
permit glucose binding
glc
Na+
Na+
glc
2a. Glucose binds
6. Glucose release causes
conformation change to open
transporter to extracellular side.
Sodium binding site also
becomes available.
3. Conformation change to open
transporter to the intercellular side
4. Conformation change also
promotes sodium release
Na+
Na+
glc
glc
5. Sodium release prompts
conformation change,
promoting glucose release
glc
glc
Na+
Na+

FIGURE 2.11

Mechanism of the sodium/glucose symporter.

inside of the cell (the reason for this difference will be covered later). As a result, there will be numerous sodium ions that could bind to the sodium binding sites in this transporter. After sodium is bound, the conformation of the transporter changes to create active glucose binding sites. When glucose is bound to these acceptor sites, there is a second conformation change in the symporter, effectively resulting in a closing of the external side and the creation of an opening on the **intracellular** side. At the same time, the conformation of the sodium binding sites changes, causing the release of the sodium ions. This release results in another conformation change, this time releasing the glucose molecules. After the glucose has been released, there is a final conformation change that returns the transporter to the original state.

2.2.1.2 Transporters that control pH

Because pH helps to determine fundamentally important things such as protein folding, it is important that we examine the ways in which cells regulate their own pH. There are several mechanisms to achieve this regulation, involving both active and passive transport of ions.

2.2.1.2.1 Examples of passive transport to control pH

$$\mathbf{HOH + CO_2 \Leftrightarrow H_2CO_3 \Leftrightarrow HCO_3^- + H^+}$$

In this reaction series, carbon dioxide and water combine to yield carbonic acid (H_2CO_3). In solution at physiological pH, carbonic acid dissociates from its acid form into a bicarbonate ion and a proton. This dissociation is important for a couple of reasons. Carbon dioxide is nonpolar due to its symmetry (see Figure 2.9), which means it can freely cross the plasma membrane and enter the aqueous solutions of the extracellular fluid or bloodstream. Thus carbon dioxide produced within the cell can end up in the blood as a bicarbonate ion plus a proton. This does two things: it gives the body two methods for transporting the waste carbon and oxygen through the blood (i.e., as CO_2 and as HCO_3^-), and it buffers the pH of the blood.

Suppose you were to hold your breath and run up the stairs. This would create a buildup of carbon dioxide in your body because carbon dioxide is a product of aerobic metabolism via the citric acid cycle. The carbon dioxide created in your muscle cells will diffuse into your blood and dissociate. Usually, when you are breathing normally and your cells are respiring, some of the newly created carbon dioxide combines with water to produce bicarbonate ions and protons, as shown in the reaction series. When the bicarbonate ions and H^+ reach the lungs, they will encounter a lower partial pressure of carbon dioxide due to freshly inspired air, and the reaction series will be driven to the left, re-forming carbon dioxide. This carbon dioxide is usually removed from the body upon exhalation. However, if you do not exhale, you will have a buildup of HCO_3^- and H^+ in the blood. The increase in proton concentration means you will have a lower blood pH, a situation that is termed acidosis; in this specific case, it is **respiratory acidosis**.

Using similar logic, you can intentionally hyperventilate as you read this, which will reduce the partial pressure of carbon dioxide in your lungs to lower than normal. This will pull the reaction series further to the left because the relative concentrations of the reactants and products—the left and right sides of the reaction series—must be held constant for equilibrium. In other words, the ratio of (CO_2 + HOH) to (HCO_3^- + H^+) is a constant at equilibrium. If you get rid of carbon dioxide by hyperventilating, this reaction series will be pulled to the left to reestablish that constant ratio. As a result, protons will be removed from the blood and the blood pH will increase. This is **respiratory alkalosis**. When somebody hyperventilates and passes out, they will have a higher blood pH. The pH change is not so drastic that it reaches values of 10, or even 7.8, because that would lead to protein denaturation (which would be catastrophic), but it will rise above the normal blood pH of 7.4.

Because pH helps to determine fundamentally important things such as protein folding, it is important to understand ways that cells regulate their own pH. There is a difference in pH between the inside and outside of the human cell, with both environments being slightly alkaline (pH = 7.2 and 7.4, respectively). Figure 2.12 shows three common transporters used by cells to regulate pH.

The first transporter shown in the figure is the sodium/proton antiporter. The net effect of this transporter's action is the removal of one proton from the cell. This transporter is useful when the intracellular pH is too low (acidic), meaning there are too many protons inside the cytoplasm. Notice that there is no change in charge: the electrical gradient from the outside to the inside of the cell is preserved as one positive charge comes in and one positive charge comes out. Keep in mind, however, that only protons contribute to pH.

The second transporter shown in Figure 2.12 is the sodium-driven chloride/bicarbonate exchanger. This antiporter transports a sodium ion into the cell along with a bicarbonate ion, while at the same time a proton and a chloride ion are removed from the cytoplasm. Notice again that the electrical gradient is not affected by this transporter, but the pH inside the cell is. The net effect on pH is that one proton is removed from the cytoplasm, and an additional proton can now be buffered ($HCO_3^- + H^+ \leftrightarrow H_2CO_3$) in the cytoplasm. This transporter could be considered a more efficient buffering mechanism than the previous one because, for every transport event, two cytoplasmic protons will effectively be removed.

A variation on the above theme is the sodium-independent chloride/bicarbonate exchanger (Figure 2.12). As with the sodium-driven exchanger, a chloride ion and a bicarbonate ion are being exchanged, but the net result is the loss of one buffering molecule from the cytoplasm. Note that cytoplasmic pH is being affected without the transport of any protons: one buffering molecule is lost, which effectively results in one extra (unbuffered) proton in the cytoplasm, so the pH will fall ever so slightly. This transporter is used when cytoplasmic pH is too high (meaning the concentration of H^+ must be increased).

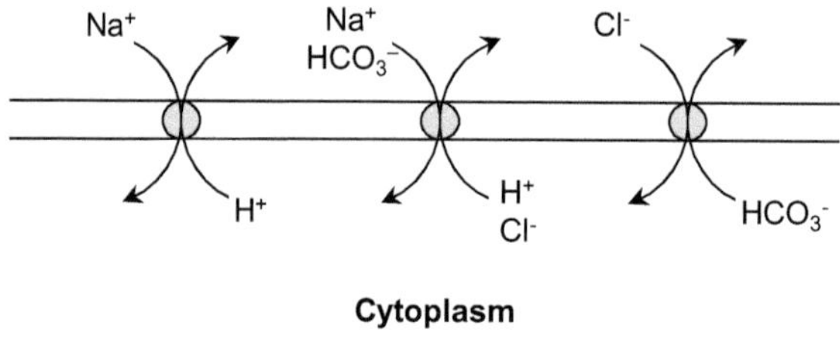

FIGURE 2.12

Three passive antiporters that are used by cells to control cytosolic pH. *Left:* sodium/proton antiporter. *Middle:* sodium-driven chloride/bicarbonate exchanger. *Right:* sodium-independent chloride/bicarbonate exchanger.

2.2.1.2.2 Examples of active transport to control pH: the proton ATPases

Proton ATPases are used by cells to transport protons against the electrochemical gradient by harnessing the energy of ATP hydrolysis. Because ATP is hydrolyzed, this type of transport is referred to as active transport. One use of a proton ATPase is to acidify a cellular compartment, such as a lysosome (covered in the next section). As shown in Figure 2.13A, a proton is transported into the vesicle at the expense of one ATP. This particular transporter is known as a **V-ATPase**. The "V" stands for "vesicular."

There is another type of proton ATPase—the **F-ATPase**, found on the inner mitochondrial membrane—which may appear to be set up to pump protons out of the mitochondrion at the expense of ATP in a similar way to a V-ATPase. F-ATPases are membrane-bound transporters that typically function in the reverse direction (Figure 2.13B). There is a relatively strong proton gradient just outside of the inner mitochondrial

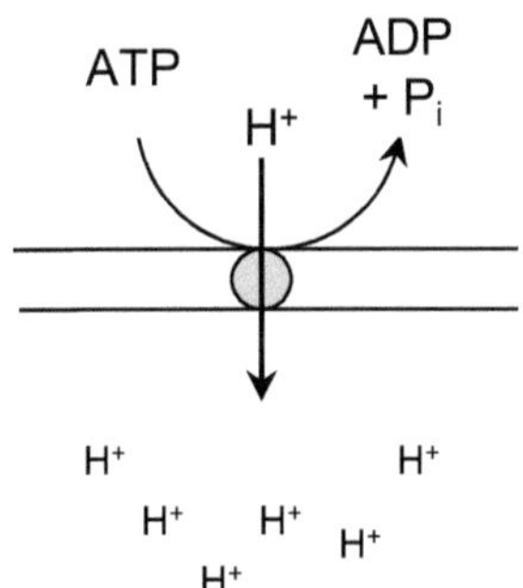

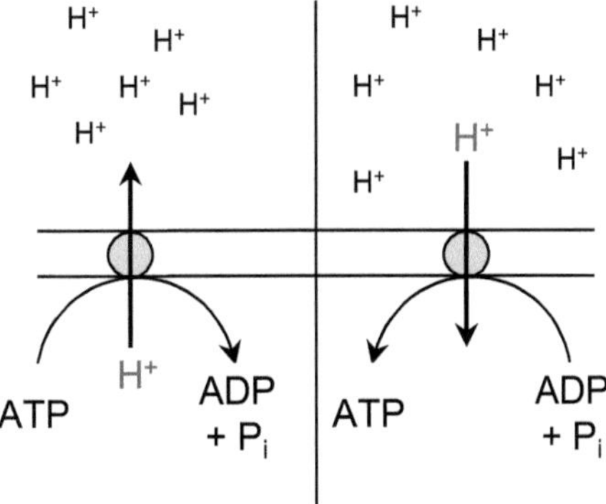

FIGURE 2.13

Actions of two proton ATPases. (A) V-ATPases drive protons against their gradient to acidify vesicles such as lysosomes. (B) F-ATPases are set up similarly to V-ATPases *(left)*, except that the proton gradient is used to drive the production of ATP by mitochondria *(right)*.

membrane in what is known as the intermembrane space of the mitochondrion. This proton gradient is high enough that it drives protons back into the mitochondrion and the ATPase reaction is reversed; the energy from the proton gradient is harnessed by the transporter to form ATP. The proton gradient is established via the electron transport chain, which is used by the mitochondrion to harness energy from certain energy-rich molecules via redox reactions to ultimately reduce oxygen to water plus energy in the form of ATP.

2.2.1.2.3 Lysosomes

There are good reasons why a biotechnologist should be aware of how cells maintain or alter pH levels. First, if one is going to work with cells, it is important to understand how they function and how various drugs might affect the ability of a cell to maintain its own pH. Second, biotechnologists interested in drug delivery or gene therapy should be aware that intracellular vesicles with low pH, known as **lysosomes**, present a significant barrier to successful drug or gene delivery. There will be more on this specific point in a moment, but first let us examine what lysosomes are.

Lysosomes are cytoplasmic organelles that degrade materials transported into the cell. If a cell ingests something, the ingested material may very well be broken down inside a lysosome: proteases break down proteins (think of them as "protein-ases," or enzymes ["ases"] that act upon proteins), lipases degrade lipids ("lipid-ases"), and nucleases break down nucleic acids (DNA and RNA) (think "**nucleic** acid-**ases**"). These different degradative enzymes require a low pH to be active, and this is because they are proteins that require a low pH for proper folding. In this case, "low pH" means 4.5–5.5. Compare this with the normal cytoplasmic pH of 7.2, and you can get a feel for the disparity. This disproportion serves a purpose. Sometimes vesicles burst. If a lysosome were to burst and release its proteases, lipases, and nucleases into the cytoplasm, then the cell would be destroyed if these enzymes remained active. However, at pH of 7.2, these proteins will be folded incorrectly, and therefore they will be inactive. A lysosome is relatively small relative to the entire cell (Figure 2.14), so the bursting of a single lysosome will have virtually no effect on cytoplasmic pH. One can therefore consider the requirement of low pH as a fail-safe mechanism for the cell.

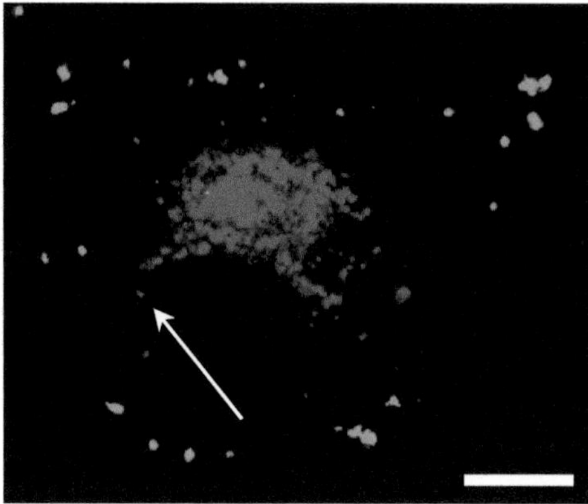

FIGURE 2.14

Image of a cell undergoing gene delivery using green-labeled gene delivery complexes. Lysosomes have been labeled with a red dye. The *green spots* delineate the cell exterior in this micrograph. Note the size of one lysosome (denoted by the *arrow*) relative to the size of the cell. Scale bar = 10 μm.

The reason that lysosomes might be of interest to biotechnologists is that they are important in the areas of gene delivery and drug delivery. We mentioned earlier that the plasma membrane is the first point of contact (the first physical barrier at the cellular level) for technologies such as gene and drug delivery to eukaryotic cells. The lysosome, or endolysosome (which is the fusion of a lysosome with an endosome), serves as the second barrier. Consider gene delivery for a moment. Nonviral gene delivery complexes typically enter cells via endocytosis and are subjected to the harsh interiors of late endosomes and especially lysosomes. There has been a valid concern in the field of gene therapy about how to cause the gene delivery complexes to escape from or completely avoid these organelles. All of this relates directly back to how cells regulate their pH.

The proton pump responsible for acidifying the lysosomal interior is the proton V-ATPase, discussed earlier. The antimalaria drug chloroquine is **lysosomotropic**, meaning it acts on lysosomes; it achieves its effect by shutting down proton V-ATPases. This drug was once thought to hold value for the field of gene therapy, if only in a laboratory setting: by preventing the acidification of (endo)lysosomes, chloroquine prevented activation of the associated degradative enzymes and thus prevented the destruction of gene delivery complexes in the lysosome.

2.2.1.3 Another active transporter: the sodium/potassium ATPase

The regulation of pH is not the only reason to transport molecules into and out of the cell. The sodium/potassium exchanger (Figure 2.15) is perhaps the most famous of the active transporters. This antiporter belongs to a class of transporters known as the P-ATPases, which bind an inorganic phosphate as part of their mechanism (hence the P in the name). By using ATP hydrolysis for energy, the sodium/potassium exchanger is able to pump three sodium ions out of the cell while pumping two potassium ions in. A couple of things are in play here. (1) The cell is establishing an electrical gradient (or voltage gradient) by pumping more positive charges to the outside than are being transported to the inside. For every ATP used, there will be a net change of +1 in the charge on the outside of the cell. (Do not confuse a change in net charge with a change in pH. Only protons are considered in pH.) (2) The cell is building a chemical gradient because with every ATP used, three sodium ions are pumped to the outside for every two potassium ions transported back in. The cell uses the established sodium gradient to drive other transporters. Taken together, (1) and (2) contribute to the **electrochemical gradient**.

Cell exterior

$2K^+$

ATP

$3Na^+$

ADP $+ P_i$

Cytoplasm

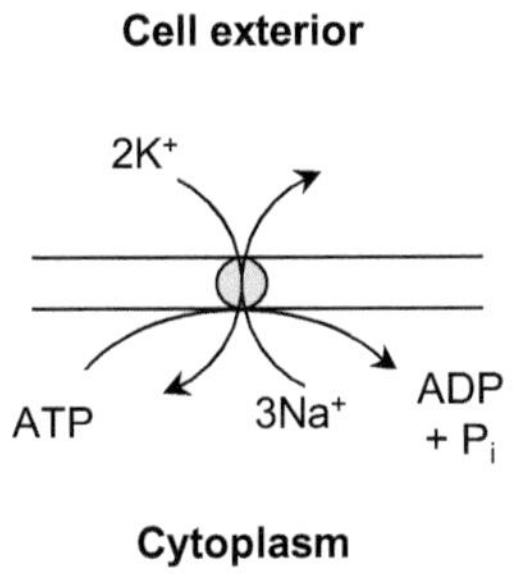

FIGURE 2.15

The sodium/potassium exchanger. This transporter establishes an electrochemical gradient for the cell by pumping three sodium ions out of the cell while moving two potassium ions in. Energy to drive the transporter is supplied by the hydrolysis of ATP.

2.2.1.4 Transporters can be coupled: the sodium-driven calcium exchanger

The sodium-driven calcium exchanger transports three sodium ions into the cell while removing one calcium ion (Figure 2.16). Ca^{2+} concentration outside the cell is about 1 mM (10^{-3} M), whereas inside the cell, it is about 10^{-7} M. The very low concentration inside the cell means that even a small influx of calcium ions will make a marked difference to their concentration in the cytoplasm, permitting the cell to use Ca^{2+} as a signaling molecule. Therefore the cell needs a way to maintain a low cytoplasmic Ca^{2+} concentration. There is a straightforward active transporter that pumps Ca^{2+} out of the cell at the expense of one ATP, but a perhaps more interesting Ca^{2+} transporter is the sodium-driven calcium exchanger. This exchanger pumps calcium outside the cell against its concentration gradient without direct use of ATP. With about 10,000 times more calcium ions outside the cell than inside, energy is required to push the calcium out; this energy comes from the sodium chemical gradient outside of the cell, established by the sodium/potassium exchanger described earlier. Because of the sodium gradient, there is a force that drives reentry of sodium ions into the cytoplasm. This force is harnessed to drive the export of calcium ions against their gradient. In chemistry, when the energy from one reaction is used to drive another reaction, we refer to the reactions as being **coupled**. As shown by this example, transporters can also be coupled.

2.2.1.5 ABC transporters

The ATP-binding cassette (ABC) transporters are a class of active transporters used by both prokaryotes and eukaryotes. Whereas prokaryotes use these transporters to import hydrophilic molecules, both

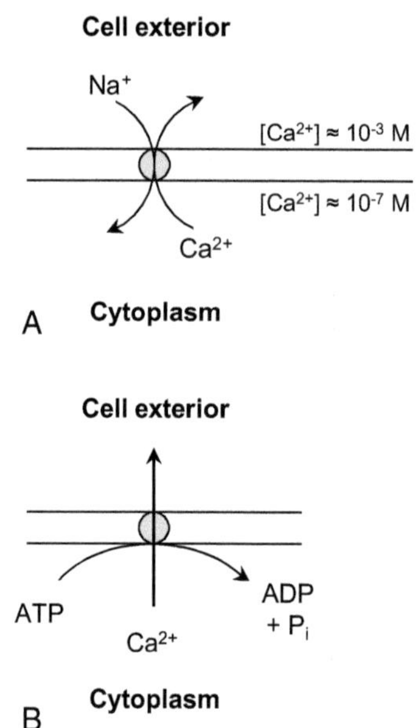

FIGURE 2.16

(A), The sodium-driven calcium exchanger. Using the Na^+ gradient that was established via the Na^+-K^+ ATPase, Ca^{2+} is transported up a steep concentration gradient. This is an example of coupled transporters. (B), Ca^{2+} can also be transported out of the cell by directly using the energy of ATP hydrolysis.

prokaryotes and eukaryotes use them to export molecules such as lipids, steroids, and toxins. ABC transporters are fairly complex proteins that span the plasma membrane multiple times. The transporter consists of two integral membrane proteins that span the membrane six times each, two peripheral membrane proteins that bind and hydrolyze ATP (the ATP-binding cassette), and a substrate-binding protein.

Within the ABC transporter class of proteins, some of the exporters are responsible for pumping toxins or other hydrophobic molecules out of a cell. Cancer cells often utilize ABC transporters to pump drugs out before the cells are negatively affected. In this case, from the perspective of the cancer cell, the drugs are toxins that must be removed. The drugs are hydrophobic, so they can cross the plasma membrane to gain access to the cell interior. However, the cell may be able to recognize the drug via the substrate-binding region of certain ABC transporters and then transport the drug back out using energy from ATP hydrolysis. Such transporters confer **multidrug resistance** to the cancer cell, and the specific ABC transporter in this case is known as a multidrug resistance protein.

As another example of ABC transporter use in nature, consider that four species in the genus *Plasmodium* are able to cause malaria. Malaria is often treated with the drug chloroquine, which works as a lysosomotropic agent (discussed earlier in this chapter). However, some drug-resistant strains of *Plasmodium* have ABC transporters that recognize and remove chloroquine from their cytoplasms, making chloroquine an ineffective treatment for malaria when it has been caused by any such strain.

2.2.1.6 Hydrophilic molecule transport and electrochemical gradients

A general point to keep in mind is that whereas hydrophobic molecules diffuse across the plasma membrane, charged molecules are largely prevented from doing so. But the cell *must* move charged and polar molecules across the plasma membrane if it is to survive. One example of a polar molecule required by the cell is glucose, which is used for energy. (Certain cells, such as those of the brain, use glucose preferentially and almost exclusively as their primary energy source.) An example of the need for charged molecules to cross cellular membranes is that of proton transport to regulate the pH inside the cell or across the membranes of cellular organelles such as mitochondria for ATP production. In addition, keep in mind that while sodium is not used directly to drive certain reactions, the sodium ion gradient established by the sodium/potassium ATPase is often used to drive the transport of other molecules across the plasma membrane; an example of this, again, is glucose. Recall the sodium/glucose cotransporter discussed earlier. The transporter gets its energy from the electrochemical gradient of sodium ions, generated in large part by the sodium/potassium ATPase, which hydrolyzes ATP to establish the gradient.

Earlier, we discussed the specificity of ion channels such as potassium channels—but channel specificity alone does not dictate the direction of ion movement; electrochemical gradients play a defining role. Consider for a moment the chemical gradient for potassium ions: using the transporters discussed so far, which way would K^+ be pushed—into or out of the cell? The sodium/potassium ATPase will pump K^+ into the cell, establishing a chemical gradient pointing toward the outside of the cell for potassium ions. (The direction of a gradient is in the direction of decreasing concentration.) However, this antiporter also pumps more positive charges out of the cell than into the cell and thus establishes an electrical gradient. Considering potassium ions specifically, although the chemical gradient is pushing them toward the outside of the cell, the electrical gradient pushes back in the other direction. For sodium ions, the chemical and electrical gradients push in the same direction to drive sodium ions back into the cell, which is how the cell can harness the transport of sodium to drive things like the sodium-driven calcium pump. But for potassium leak channels, the situation is different. Whereas the chemical gradient forces potassium ions out of the cell, the electrical gradient serves to force them in. Both must be considered together as a singular electrochemical gradient. When the overall chemical gradient of a resting cell balances the electrical gradient across the plasma membrane, the value of the membrane potential (the electrical gradient) is termed the **resting membrane potential**. Note that although in balance with the chemical gradient, the resting membrane potential is not equal to zero unless the cell is dead.

2.2.1.6.1 The Nernst equation

The Nernst equation is used to relate the electrical and chemical gradients to each other to determine at what membrane potential they are in balance. Although the chemical and electrical gradients can be considered separate entities, this equation shows how they can work with or against each other. The equation is given by:

$$V = \frac{RT}{z\mathcal{F}} \ln\left(\frac{C_o}{C_i}\right),$$

in which R = the gas constant (= 8.315 j/mol K = 2 cal/mol K); T = temperature (in K); z = charge of the ion in question; $\mathcal{F}$ = Faraday's constant (= 96,480 j/V/mol = 230,000 cal/V/mol); C_o = concentration outside the cell; C_i = concentration inside the cell.

For potassium ions, suppose the concentration inside the cells is 139 mM, and the concentration outside the cell is 4 mM. At normal body temperature (37°C), the equilibrium potential for K^+ is given by:

$$V_{K+} = \frac{(2.00)\,(310.)}{(1.00)\,(2.30x10^4)} ln\left(\frac{4}{139}\right) = -0.0956 = -95.6mV$$

This value represents the membrane potential required to keep potassium ions from flowing out of the cell. You might think of this in terms of excess positive charges inside pushing K^+ out of the cell, or it might be easier to think of the equilibrium potential as indicating the amount of negative charge (perhaps from chlorine ions, phosphatidyl serine, or negatively charged proteins) needed to pull on K^+ ions to keep them in the cytoplasm. Notice that if the extracellular concentration of K^+ were increased to 139 mM, equal to the intracellular concentration, the ion would be at chemical equilibrium, and no negative charges would be needed to keep K^+ in the cell, a situation that is also reflected by an equilibrium potential = 0 (verify this with the above equation).

2.2.2 Vesicular transporters: endocytosis

2.2.2.1 Phagocytosis

Endocytosis is a process through which cells take up exogenous material: macromolecules, particulates, or even other cells. **Phagocytosis** is a special form of endocytosis whereby large items such as bacterial cells or cell debris are taken up into phagosomes. Certain cells are known as **professional phagocytes**; their main function is to phagocytose particles, debris, or bacterial cells within tissues. Examples of professional phagocytes include macrophages, neutrophils, and dendritic cells. When one of these cells phagocytoses a foreign cell or particle, it bumps into the foreign body and recognizes something that marks it as a target for phagocytosis.

Phagocytosis is a triggered event. Four different triggers are the recognition of:

- antibodies
- apoptotic cells

- complement
- oligosaccharides or glycoproteins

We will now take a closer look at each of these initiators.

2.2.2.1.1 Recognition of antibodies

Let's say your body is host to a bacterium, perhaps group A *Streptococcus* (the microorganism responsible for strep throat), and you are relying on your body to get rid of it. Your body will secrete immunoglobulins, specific protein molecules that are used by the immune system. Immunoglobulins such as the common immunoglobulin G (IgG) have the general structure shown in Figure 2.17. The sketched IgG molecule has two regions: an antigen-binding region ($\mathbf{F_{ab}}$) and a constant region ($\mathbf{F_c}$). These two regions can be separated by enzymatic digestion followed by fractionation, hence the "F" in the region names. F_c is short for "fraction crystallizable": to make a crystal, one must have a pure substance. The

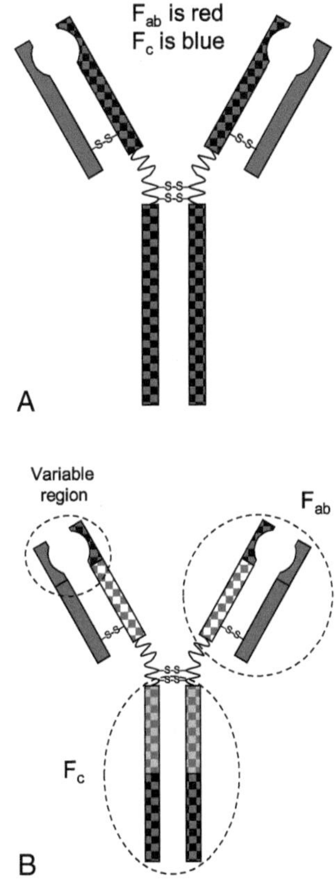

FIGURE 2.17

Schematics of an immunglobulin G molecule. In each panel, the two heavy chains are denoted by *checkered patterns* and the two light chains are *solid*. The chains are held together by disulfide bonds (-S-S-). (A) Note how the heavy and light chains interact to create the antigen-binding region. (B) Each heavy chain has one variable region *(red, checked)* and three constant regions (all with *blue checks*). The light chains also have a variable *(red, solid)* region and a constant region *(blue, solid)*. Different fractions and regions are shown by *dotted lines*.

F_c region is crystallizable because it is constant for all members of this family of immunoglobulins. The F_c region is constant across all members of a species.

Within the antigen-binding region are a constant portion and a highly variable portion that gives the antibody its specificity. The variable region for a given IgG is specific to whatever **antigen** the cell is displaying. IgG antibodies secreted by the immune system are always bumping around and "looking" for things to stick to. If the variable region (F_v) can bind to an antigen displayed by a bacterial cell, the antibody will stick to the bacterium, and eventually, the bacterial cell will be coated with antibodies. Professional phagocytes, such as macrophages, are always wandering around "looking" for something to engulf. They will recognize the exposed F_c regions of the bound antibodies and begin to adhere to them, gradually enveloping the entire antibody-coated bacterium (Figure 2.18).

2.2.2.1.2 Recognition of apoptotic cells

Apoptosis is also known as "programmed cell death." When cells undergo irreparable damage, they may commence a complex series of events involving a molecular cascade. Apoptosis culminates in the cell destroying its own cytoskeleton and genome and packaging the degraded material into apoptotic bodies that will be engulfed and carried off by phagocytes. Although the details of the apoptosis cascade will not be covered here, one important aspect that is relevant to the biotechnologist is the change in the plasma membrane that happens early in the process. Most notably, phosphatidylserine, which typically resides on the inner face of the plasma membrane, is flipped to the outside of the cell. This greatly changes the charge of the exterior cell surface, signaling that the cell is undergoing apoptosis and marking it for destruction by professional phagocytes.

2.2.2.1.3 Recognition of complement

The complement cascade is a complex series of molecular events that results in the permeabilization of microbial membranes. More than 20 proteins are involved in the process, but no cells are in the cascade itself. The complement cascade can be activated via *innate* or *specific* immune responses. The *innate* response involves neither antibodies nor T-cell receptors; a molecule on the surface of a pathogen (e.g., a set of sugars) activates the cascade directly. In the *specific* immune response, antibodies recognize and bind to specific molecules on the microbial membrane, which in turn activates cascade protein C1. Activated C1 then serves to activate other members of the cascade, and so on. (The complement cascade, with its two pathways, is beyond the scope of this book and will not be presented in greater detail.) Chemoattractants are released, causing professional phagocytes to migrate into the area. In addition to recruiting phagocytes, the cascade can result in the drilling of holes through the plasma membranes of targeted microorganisms. These holes allow a free flow of ions, causing the loss of any chemical or electrical gradients and leading to death of the microbe via apoptosis. Notice that the complement cascade involves both antibodies and apoptosis, both of which can lead to phagocytosis by the recruited macrophages.

2.2.2.1.4 Recognition of oligosaccharides or glycoproteins

Recall that the sugars displayed by cells on the extracellular surface are used for cell–cell communication. Bacteria display different patterns of sugars than mammalian cells. These patterns can be recognized by professional phagocytes to help keep the body free of microbial invaders. Cells such as dendritic cells phagocytose and degrade microbes, then attach portions of the degraded cells (e.g., membrane liposaccharides) to **major histocompatibility complexes** (MHCs) for display to other immune cells

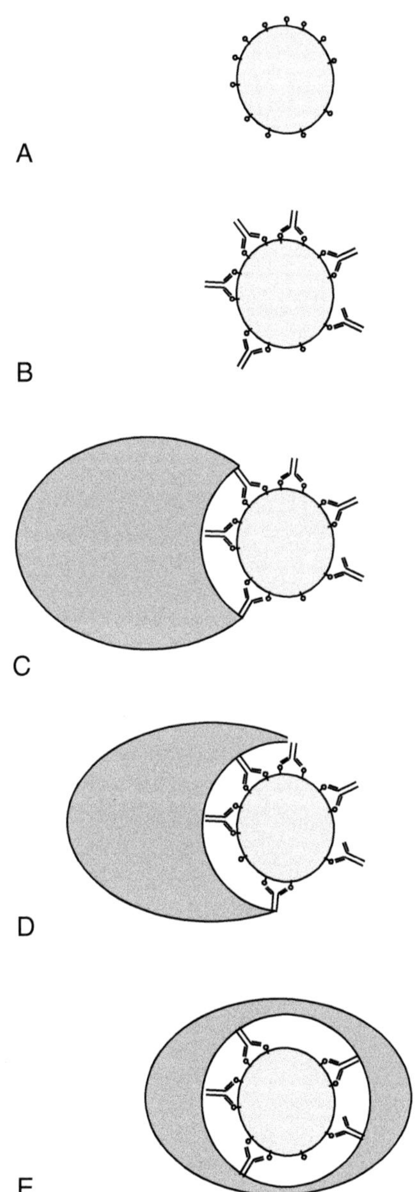

FIGURE 2.18

A macrophage endocytosing a foreign cell. (A) The cell has antigens such as glycoproteins on its exterior. (B) These particles are recognized as foreign through the binding of antibodies such as immunoglobulin G. (C)–(E) The macrophage binds to the Fc regions of the antibodies, gradually engulfing the foreign cell.

(helper T and cytotoxic T cells). Think of the act of display like that of a warrior advertising a victory by making a flag out of parts of his victim (or his clothing). The flag comes from the vanquished microbe, but the flagpole is an MHC. The MHC causes helper T or cytotoxic T cells to notice the displayed antigen on the dendritic cell, and on other cells they will interact with in the future (e.g., B cells, infected macrophages, or virus-infected cells). Interestingly, the flagpole turns out to be an important marker of self. If the wrong MHC is displayed—as generally happens if a foreign tissue is transplanted into the body—the presenting cell is seen as foreign and is destroyed by the host's immune cells. This is what happens when implants such as donated organs or tissue engineered constructs are rejected by the host.

2.2.2.2 Pinocytosis

Endocytosis is the uptake of molecules from outside of the cell by vesicle formation. Phagocytosis typically has to do with the endocytosis of particles, whereas **pinocytosis** is typically associated with the uptake of solutes. More generally, phagocytosis has been termed "cellular eating" and is associated with larger particles, whereas pinocytosis has been termed "cellular drinking" and is associated with smaller particles, generally less than 100 nm in diameter. There is no definitive particle size for distinguishing between phagocytosis and pinocytosis. For example, Percoll particles with 30-nm diameters have been shown to enter rat macrophages via both pinocytosis and phagocytosis; notice that *particles* were pinocytosed, but also something under 100 nm was phagocytosed. Thus a hard and fast rule of size cutoff or architecture to define phagocytosis and pinocytosis is not appropriate (at least not in rats). However, the chance of entry via phagocytosis has been shown to increase with increasing particle diameter.

2.2.2.2.1 Gene delivery is associated with pinocytosis

Recall that endocytosis encompasses both phagocytosis and pinocytosis. In the body, phagocytosis is most often associated with specific cells that wander around looking for things to engulf. In gene therapy, however, gene delivery complexes can be endocytosed by most cell types via the numerous pinocytotic events continually occurring within the plasma membrane. Gene delivery complexes have sizes of the order of 100 nm, so their uptake is generally associated with pinocytosis. There are two types of pinocytotic vesicles: those utilizing **caveolin** and those utilizing **clathrin**. Caveoli ("little cavities") are bulb-shaped indentations in the plasma membrane that utilize the protein caveolin. Caveolin is a multipass integral membrane protein. (We have now seen two specific examples of multipass transmembrane proteins in this book: first ABC transporters and now caveolin.) The other type of pinocytotic vesicle, involving clathrin and clathrin-coated pits, is presented in the next section.

2.2.2.3 Endocytosis via clathrin-coated pits

Individual clathrin molecules have an interesting shape: a **triskelion** (Figure 2.19). Inside the cell, triskelions will self-assemble to form a cage similar to a soccer ball. When bound to the plasma membrane, the three-dimensional geometry of clathrin puts force on the membrane to make it concave. Imagine several clathrin molecules floating around the cytoplasm and one triskelion eventually sticking to the cell membrane. Another clathrin molecule attaches to the first, then another, and so on until a fairly sturdy cage is formed. As the clathrin molecules continue to form this cage, they will pull in on the membrane until it is shaped like a ball (Figure 2.20). There is a constant assembly of clathrin that helps shape the membrane into a ball, which is eventually pinched off by the protein **dynamin**.

Sometimes clathrin molecules seem to be specific for certain receptors. This specificity is achieved through the use of a molecule called **adaptin**. The cytoplasmic side of individual receptors may bind to

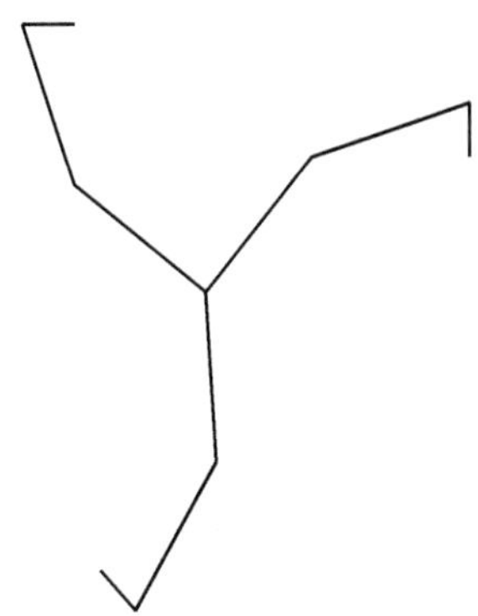

FIGURE 2.19

Clathrin molecules have the shape of a triskelion.

specific adaptin molecules, which are bound to clathrin. This chain of associations can give specificity to clathrin-coated pits, meaning specific receptors can be endocytosed into one pit. In some cases, receptors are endocytosed whether they have bound a ligand or not. In other cases, the receptor only takes on a conformation that is amenable to binding adaptin after the receptor has bound its specific ligand.

If the total number of a specific type of receptor on (and in) the cell were held constant, with the rate of formation of clathrin-coated vesicles also constant, the cell would end up with a variable amount of endocytosed ligand if the amount of substrate outside the cell were varied. If the concentration of ligand outside the cell were increased, there would be a greater chance that one of those ligand molecules would bind to a receptor and be internalized by a clathrin-coated pit. Of course, this is not the only way that a cell can internalize greater numbers of ligand molecules. If the extracellular concentration of ligand molecules were held constant but the number of membrane receptors for the ligand were increased, then the probability of a receptor–ligand interaction would increase, ultimately resulting in greater numbers of ligand molecules being endocytosed by the cell.

If one were to reach blindly into a tank containing only one oyster, then the probability of grabbing an oyster would be relatively low. However, reaching into a tank that contains dozens of oysters provides a greater probability of grabbing one because there are simply more of them around. The same principle works for cells taking up specific nutrients. If the blood contains a nutrient used by the cell, increasing the concentration of that nutrient will increase the probability of one such molecule binding to a receptor, and the cell will ultimately ingest more of that nutrient. Similarly, the number of receptors has an effect. Increasing the number is like reaching into the tank with two hands instead of one; it will be much easier to find a yummy oyster that way. In some instances, when cells need to acquire more of a certain type of molecule, they will increase the number of receptors displayed on the plasma membrane and thereby increase the chances of pulling in the needed molecules when the receptors are endocytosed.

Sometimes there is a conformation change after the receptor binds the ligand, which permits association with clathrin via adaptin (Figure 2.21). The clathrin-coated vesicle forms and pinches off via dynamin. Once endocytosed, the clathrin will come off of the vesicle and migrate back to the cell surface. The uncoated vesicle is then able to fuse with an **endosome**, a membrane-bound structure that is involved with processing endocytotic and pinocytotic vesicles.

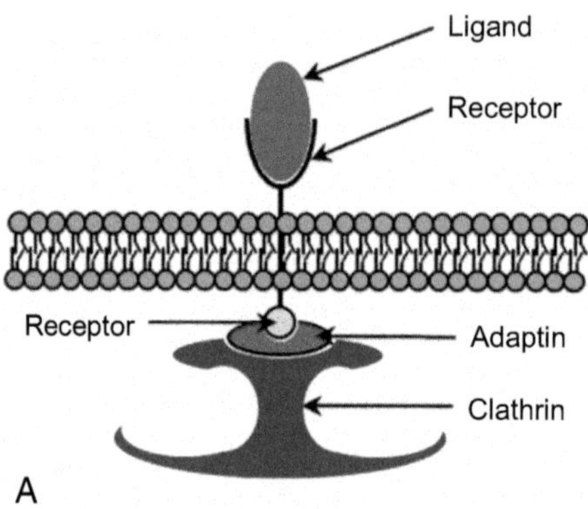

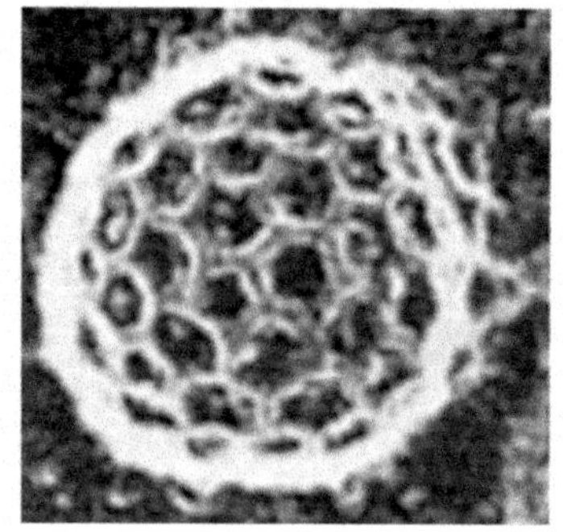

FIGURE 2.20

(A) Clathrin binds to a receptor molecule via adaptin, which gives specificity to a forming pit. Adaptin recognizes the cytoplasmic face of certain receptors. (B) A clathrin-coated pit following pinching off from the plasma membrane, a feat achieved with the protein dynamin (not shown). (C) Electron micrograph of a clathrin-coated pit.

Image in (C) from Heuser, J., Anderson, R. 1989.

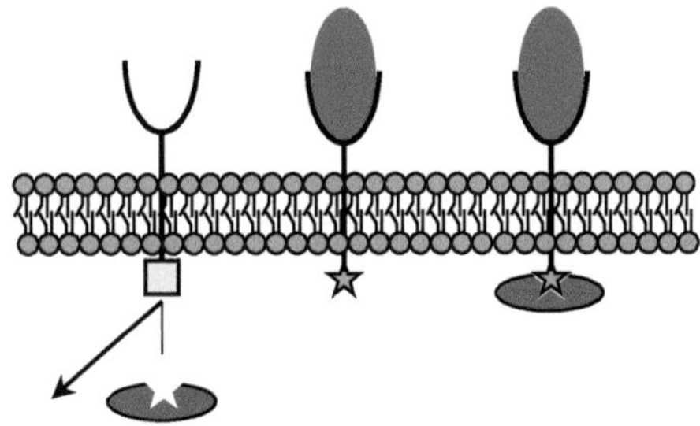

FIGURE 2.21

Sometimes a receptor will not bind its corresponding adaptin molecule unless ligand is bound. Without ligand bound, note that the cytoplasmic side of the receptor on the left cannot bind adaptin. After the ligand binds, there is a conformational change on the cytoplasmic side of the receptor (*middle receptor*), which allows adaptin to bind (as shown by the *receptor on the right*). Once adaptin has been bound, the receptor-ligand complex is ready for clathrin-mediated endocytosis.

2.2.3 Receptor fates

Receptors are proteins, which are encoded by genes. The cell puts in a fair amount of work to create a receptor, so recycling receptors is advantageous. Not all receptors are treated the same within a given cell, though. To illustrate this variation in processing, three examples of receptor fates are presented in this section: one in which the receptor is recycled, one in which both the receptor and ligand are recycled, and one in which neither the receptor nor the ligand is recycled.

2.2.3.1 Receptor recycling: the LDL receptor

The low-density lipoprotein (LDL) receptor provides us with a good example of receptor recycling. As we have seen, not all of the molecules in the cell are made purely of protein, nucleotides, or lipids; hybrids exist. Examples include glycolipids, glycoproteins, and lipoproteins. There exist lipoproteins in the blood, such as apolipoprotein B-100 (Apo B), which can carry various forms of lipids ("apo" is a prefix that means "away from" and is used in biology to denote something that is incomplete). It can encircle triglycerides and cholesterol to form a particle known as LDL. Apo B is the name of the protein by itself.

Apo B can carry certain passengers, including cholesterol. When it is loaded up with triglycerides, fatty acids, and cholesterol (or cholesterol esters), we call the conglomeration very-low-density lipoprotein (VLDL). When VLDL loses fatty acids and triglycerides (e.g., when it delivers them to fat cells (adipocytes) for storage), the density of the lipoprotein particle goes up, and the particle may be referred to as LDL. LDL—which consists of a hydrophobic core of polyunsaturated fatty acids and esterified cholesterols surrounded by Apo B, phospholipids, and unesterified cholesterol—delivers cholesterol to cells. As we learned earlier, cells need cholesterol for fluidity and stability of the plasma membrane. Certain cells also use it as a precursor for steroid synthesis.

High-density lipoprotein (HDL) is used for a process known as "reverse cholesterol transport," in which cholesterol is taken from cells back to the liver for catabolism (breakdown into bile acids that are used for digestion). HDL can also grab cholesterol from LDL and is thought to help combat atherosclerotic plaques in the vasculature via cholesterol scavenging. It might interest you to know that HDL acquires cholesterol from cells via ABC transporters.

Box 2.1 LDL is not cholesterol!

The reason people might think LDL is bad is that if an individual has an excess of this form of the lipoprotein—a form that is carrying cholesterol to cells—it probably means they have either taken in or made too much cholesterol. A large amount of LDL versus HDL indicates that apolipoprotein molecules in the blood have been loaded with an excess of cholesterol. (Be aware that Apo B is not the only apolipoprotein in play here.) The ratio of LDL to HDL serves as an indicator of the relative amount of cholesterol being transported in the blood and the direction of that transport.

High cholesterol levels can lead to the development of atherosclerotic plaques, which are plaques that build up in the vasculature and restrict blood flow and oxygen delivery; this can lead to unwanted events such as myocardial infarctions (heart attacks). Note that despite the terms used in common culture, there is no "good" cholesterol, nor is there any "bad" cholesterol; cholesterol is a molecule, and there are neither good nor bad subtypes. So LDL is not bad cholesterol—in fact, LDL is not cholesterol at all but a lipoprotein. Excessively high levels of cholesterol are undesirable, however, and such levels can be detected by measuring LDL values and HDL/LDL ratios.

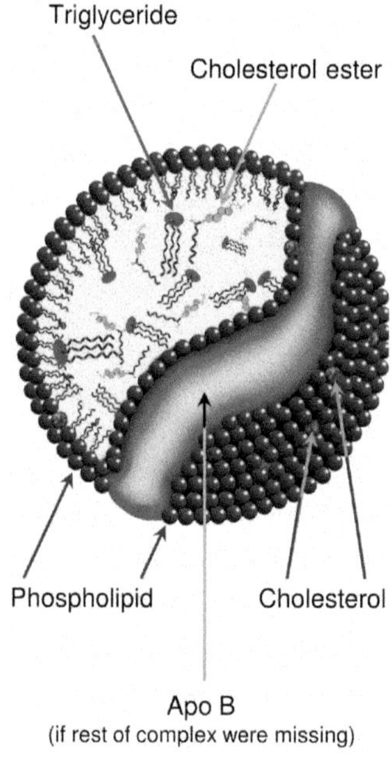

FIGURE 2.22

Structure of a low-density lipoprotein complex.

Simply delivering a cholesterol molecule to a cell does not satisfy the cell's cholesterol needs. The cholesterol must be taken up into the cell, an act that is accomplished by LDL and LDL receptors. Study of LDL receptors and their relation to blood cholesterol levels won the Nobel Prize in Physiology or Medicine for Michael Brown and Joseph Goldstein in 1985, and the work performed to this end played a major role in launching the field of receptor biology.

LDL receptors bind LDL to deliver cholesterol to cells. LDL contains cholesterol bound to fatty acid tails, as well as free cholesterol molecules. To give you an idea of the scale involved, consider that one LDL complex contains around 1500 cholesterol or other sterol molecules bound to fatty acid tails (e.g., cholesterol esters). It also contains about 500 free cholesterol molecules, phospholipids forming a monolayer, and a fairly large protein (~500,000 Da) (Figure 2.22).

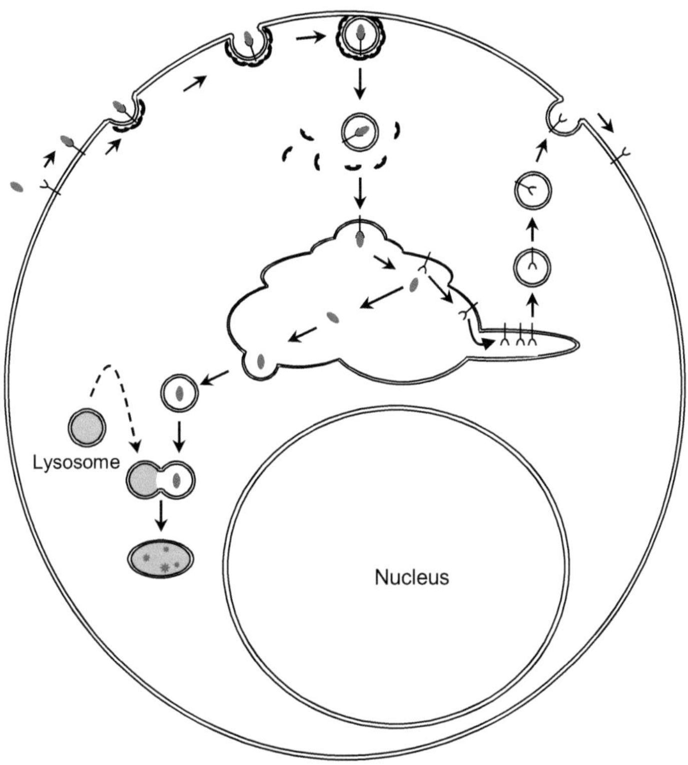

FIGURE 2.23

Endocytosis of low-density lipoprotein bound to its receptor, with the receptor being recycled.

Suppose the cell needs more cholesterol; perhaps you make it in your liver cells, and it is then transported through the blood to cells that need it. When a cell needs cholesterol, there will be an increase in transcription of the gene that codes for LDL receptors, leading to an increase in the amount of the associated messenger RNA which, in turn, leads to an increase in the amount of translation of LDL receptor proteins. More translation of these proteins leads to a greater number of LDL receptors displayed on the surface of the cell, and—remember the oyster tank?—a greater probability that a circulating LDL complex will bind to one of these receptors, so more LDL will enter the cell via endocytosis. This endocytosis will take place via clathrin-coated pits.

LDL receptors are recycled. The recycling of LDL receptors follows the scheme presented in Figure 2.23. Consider a cell with a receptor that binds extracellular LDL. The receptors will diffuse in the membrane, with some of them traveling to the vicinity of a forming clathrin-coated pit. (Notice that the receptors do not necessarily have to be filled to be endocytosed in this case.) Clathrin is shed following endocytosis, leaving the individual triskelions free to participate in additional endocytotic events. After shedding the clathrin coat, the vesicle fuses with an early endosome. Keep in mind that a distinguishing feature of endosomes is a relatively low pH—created by the proton ATPase discussed earlier—and that when the pH drops, the charge of some amino acid side groups is changed, and some polypeptide folding is altered. In this case, when the pH drops, the receptor–ligand interaction is disrupted, allowing

LDL to float free of its receptor. The early endosome forms chutes that contain unbound receptors. The chutes are pinched off the endosome in a process called budding, yielding vesicles that contain unbound LDL receptors. These vesicles are then shuttled back to the plasma membrane where the receptors are redisplayed on the cell exterior. Meanwhile, other vesicles that contain endocytosed ligands, including LDL, pinch off to form late endosomes, which deliver the ligands to lysosomes via fusion to form endolysosomes. Cholesterol esters in the LDL are degraded via lysosomal enzymes to release free cholesterol.

Note: The formation of vesicles is always going to happen, like waves on the beach; even when the beach is deserted, it's safe to assume that the waves still wash up on the shore. The formation of pinocytotic vesicles by the cell is similar. For example, whether or not LDL is bound to LDL receptors (or is even present), clathrin-coated pits will be forming and pinching off. Vesicle formation happens all the time, whether or not receptors have been filled with ligands. The pinching off of these vesicles is going on at all times as well, as is the merging of vesicles into larger units and the formation of chutes with subsequent budding from these larger vesicles. Some of these budded vesicles may return to the plasma membrane, whereas others may interact with lysosomes.

Box 2.2 Treatments for high cholesterol

There are two common causes for a person to have LDL levels. First, they could ingest too many saturated or *trans* fats. If you wake up every morning and have an eight-egg-yolk omelet, chased down with whipping cream and avocados, then you will probably have high LDL levels because your liver can convert molecules from these foods into cholesterol, which will then be carried through the blood via LDL. The good news is that such high LDL levels can be changed by modifying the diet.

A second cause is a condition called *familial hypercholesterolemia*. This is a problem rooted at the genetic level; it is an autosomal dominant condition passed down through families. Patients with familial hypercholesterolemia are unable to remove LDL from the blood due to a problem with their LDL receptors (Figure 2.24). This may be that the receptors are unable to bind LDL, or perhaps they can bind LDL, but the cytoplasmic portions of the receptors are unable to bind clathrin. In other cases, not enough receptors are made in the first place.

The first treatment for this condition is simply a modification of the diet, preferably along with an increase in exercise. However, it is quite common that a dietary fix alone is insufficient to lower blood cholesterol levels adequately, so biotechnologists in the pharmaceutical industry created a class of drugs known as statins. Examples of statin drugs include atorvastatin (brand name Lipitor), simvastatin (Zocor), lovastatin (Mevacor), fluvastatin (Lescol), and pravastatin (Pravachol). Cholesterol is made in the liver through a complex series of chemical reactions that are made more efficient by enzymes. The statin drugs act as competitive inhibitors to *HMG CoA reductase*, the enzyme that catalyzes the rate-limiting step in the cholesterol synthesis pathway. The result is a decrease in the amount of cholesterol being synthesized.

A secondary effect of the statins is that sterol regulatory element binding protein (SREBP) will be activated in cells. This protein is a transcription factor that causes an increase in the expression of the gene that encodes the LDL receptor. An increase in LDL receptor expression results in more LDL being taken up by the cells, thereby lowering LDL concentration in the blood (assuming that functioning receptors can be made).

2.2.3.2 Receptor and ligand recycling: the transferrin receptor

There is another class of receptors in which the cell recycles both the receptor and its ligand. For example, consider **transferrin** and its receptor (Figure 2.25). Iron is taken up into cells via transferrin receptors, meaning the receptors are specific for transferrin, an iron-transporting glycoprotein found in the blood. When the protein is not bound with iron, it is called **apotransferrin**. Apotransferrin will not bind with a transferrin receptor at physiological pH. Once iron is bound, however, apotransferrin

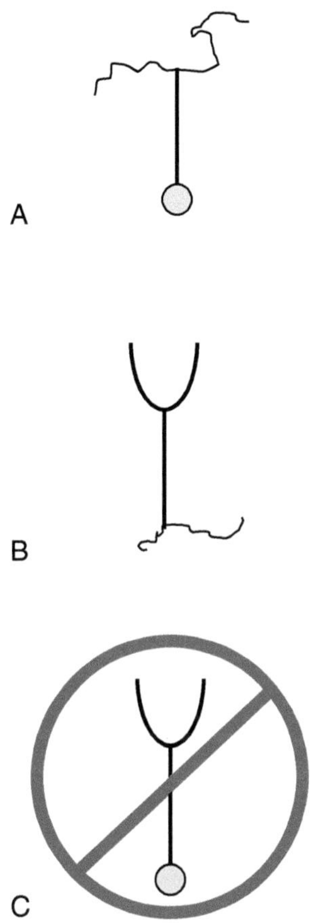

FIGURE 2.24

Problems with low-density lipoprotein (LDL) receptors include (A) a defective LDL binding site, (B) a defective binding site for clathrin or adaptin, or (C) too few receptors made at all.

undergoes a conformational change, becoming transferrin and gaining the ability to bind with a transferrin receptor. The receptor–ligand complex is endocytosed, the vesicle undergoes a drop in pH (as discussed earlier), and the iron is released from the protein, with the protein still bound to its receptor. As in the previous example, chutes that contain the receptor (still bound to the protein ligand this time) will form and bud, returning the receptor–protein complex to the cell surface. The pH of the exocytotic vesicle is still low prior to fusion with the plasma membrane.

After the phospholipid bilayers of the vesicle and the plasma membrane fuse, what was the interior of the vesicle is exposed to the extracellular environment, which has a relatively higher pH (~7.4). At the higher pH, the apotransferrin goes back to its initial conformation, which does not fit the active site of the receptor, and is released from the receptor. The result of this process is that both the receptor and the ligand are recycled.

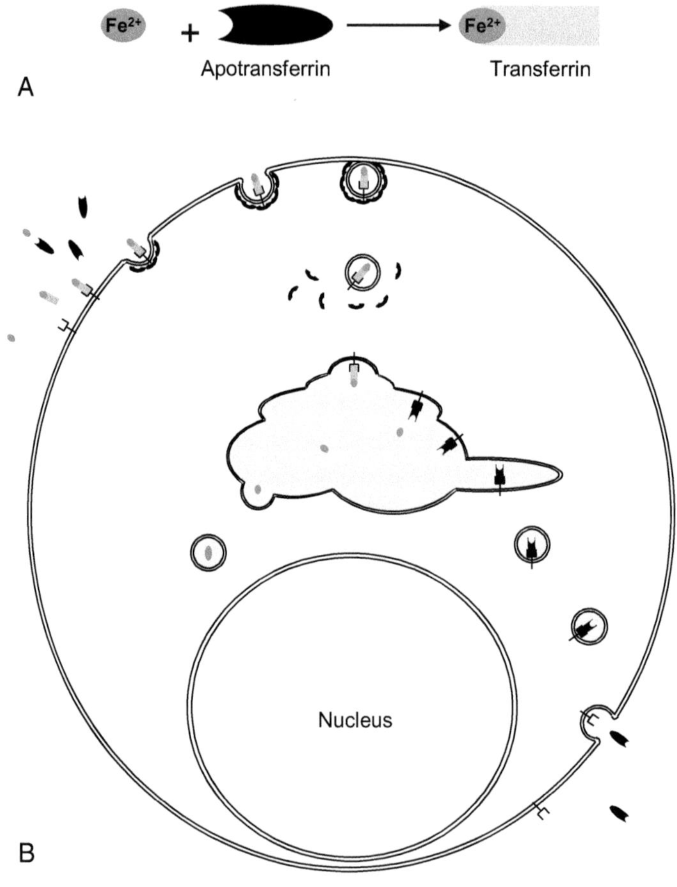

FIGURE 2.25

(A) Apotransferrin binds Fe^{2+} to create a transferrin molecule. (B) Endocytosis of transferrin bound to its receptor, with both the receptor and (apo) ligand being recycled.

2.2.3.3 Neither receptor nor ligand are recycled: the opioid receptor

Let us now consider another scenario in receptor processing, which will be represented by the opioid receptor. The opioids are a class of drugs used to relieve pain and include morphine, heroin, and hydrocodone (Vicodin). Opioid receptors bind their ligands *before* attaching to a forming clathrin-coated pit (unlike LDL receptors, which can be endocytosed whether or not they have ligand bound to them). After the receptor–ligand complex is endocytosed, the complex can follow the same pathway as LDL receptors, where the receptor undergoes recycling, or the entire complex can be directed to a lysosome for destruction. Endocytosing opioid receptors reduces their plasma membrane concentration, thereby rendering the cell less sensitive to additional stimulation by the opioid. Although the cell can transport fresh receptors (although not necessarily identical ones, in the case of morphine receptors) to the surface, such an event requires a stimulus, so replacement is not guaranteed. This leads to patient desensitization to the drug—sometimes a long-term condition. The result is that patients build up a tolerance

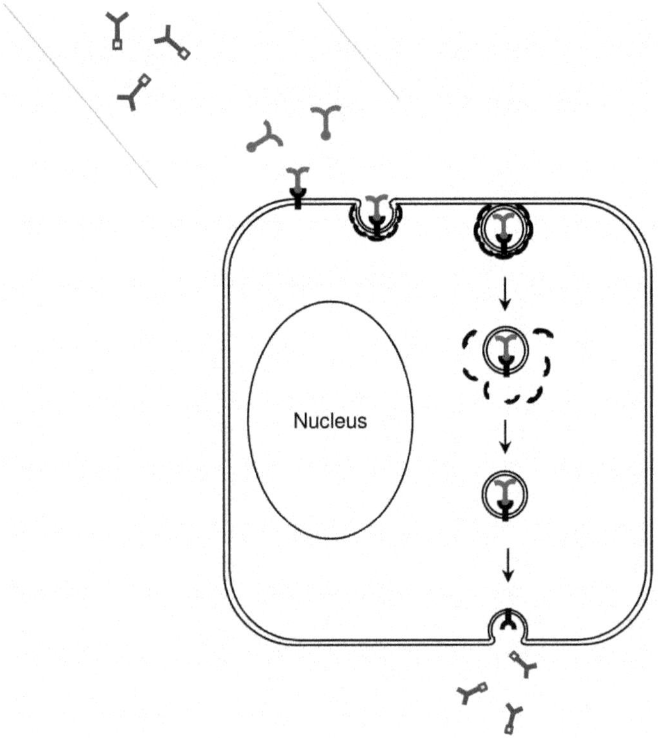

FIGURE 2.26

Transcytosis. In this example, secretory immunoglobulin A (IgA) molecules from the mother's milk *(top left, in blue)* change conformation when they hit the acidic pH of the baby's stomach *(red)*. The altered folding allows the IgA molecules to be recognized, endocytosed, and transported across the cell for exocytosis. The pH on this side of the cell is closer to neutral, so the IgA molecules will revert to their original conformations *(blue)* and be released from the receptors into the infant's body.

to the drug, requiring greater and greater doses to achieve the same analgesic effect. In the case of heroin, this explains why addicts commonly overdose, given enough time.

2.2.3.4 Transcytosis

In addition to the recycling mechanisms just described, some receptors participate in transcytosis, whereby the receptor and ligand are endocytosed, transported to the other side of the cell, and the cargo released outside of the cell (Figure 2.26). Transcytosis is what happens when a baby suckles her or his mother. Early breast milk, called *colostrum*, is secreted by the mammary glands during the first few days following childbirth. It has a higher concentration of antibodies than breast milk, although both provide antibodies for the baby's adaptive immune system. When the baby drinks, the milk enters the acidic environment of its stomach and small intestine. The folding of the maternal antibodies (in this case, IgA) changes in the low pH, and the altered conformation is recognized by specific receptors on the **lumenal** side of the baby's gut. The antibodies are pulled out of the digestive tract and into gut epithelial cells.

Then, rather than destroy these antibodies, the cells relocate them to the plasma membrane on the other (**basolateral**) side of the cell, where they are exocytosed, making them available to the baby's lymphatic system. The pH of the extracellular fluid on this side of the gut epithelia is near-neutral, so the antibodies regain their original conformation; this causes them to dissociate from the receptors. In other words, the antibodies are pulled from the intestinal lumen, transferred through the cell, and released into the baby's body.

2.2.4 Lysosomes are for degradation, but are they safe?

The proton ATPases are always working in the lysosomal membrane, so lysosomal pH may drop to as low as 4.5. Just as pH affects the conformation of receptor/ligand complexes, it also affects the conformation of lysosomal proteins. Specifically, the acid hydrolases of the lysosome, which include proteases, nucleases, and lipases, begin to take on an active conformation at around a pH of 5 or 6. This dependence on acidic conditions is a safety feature for the cell. Suppose several lysosomes rupture, perhaps for osmotic reasons. This will release proteases, nucleases, and lipases into the cytoplasm. However, because the volume of the cytoplasm is very large relative to a lysosome and because it is buffered, the pH of the cytoplasm (7.2) will be enough to change the conformation of these hydrolases and render them inactive. This is a fail-safe feature for the cell.

Acid hydrolases are proteins, and as such are composed of amino acids. Earlier we mentioned (and will discuss further in Chapter 3) that pH can alter the charge of certain amino acids. Altering the ionic state of amino acids in the primary structure of a protein can have a profound effect on the tertiary structure of the protein. It can mean the difference between an active hydrolase and an inactive polypeptide.

2.2.4.1 Identification of intracellular vesicles

Although all the smaller vesicles inside the cell are interrelated, there are guidelines for defining the identities of each type. Early endosomes have a slightly acidic pH (~6.5) and do not contain active degradative enzymes. They serve as a sorting station for freshly endocytosed vesicles. As already discussed, some vesicles bud off of the early endosome to return to the plasma membrane, and some are transported to late endosomes. Late endosomes have a more acidic pH (~6), and several types of **acid hydrolases**, pH-sensitive degradative enzymes. The enzymes and lower pH are the result of the late endosome serving as another sorting station, this time not only for the processing of endocytosed materials but also for the final portion of lysosome construction. One of the later steps in this process involves the delivery of the acid hydrolases from an organelle known as the Golgi apparatus to late endosomes. Final lysosome maturation is a gradual process, with endosomal membrane proteins being trafficked away from the developing lysosome. Lysosomes have a loose definition because they vary so much in morphology, even within the same cell. Generally, however, lysosomes are characterized by a low pH (~5) and the presence of *active* acid hydrolases.

2.3 Summary

We have seen that cells maintain interior environments that differ from what lies outside. The difference is preserved by the plasma membrane, which has a hydrophobic portion that prevents free travel of many molecules. The plasma membrane is primarily constructed of phospholipids but also contains molecules such as cholesterol, proteins, or protein hybrids.

Membrane proteins are used for cell signaling, identification, and transport of molecules across the plasma membrane barrier. For a large, polar, or charged molecule to cross the plasma membrane, some form of membrane protein must be involved. Channels provide a means by which specific ions can cross via facilitated diffusion. Transporters, be they active or passive, bind to the molecules to be transported and undergo a conformation change that allows release of the bound molecules into (or out of) the cytoplasm. Receptors bind to specific ligands and either signal to the cell via a conformation change that the ligand has been bound or are internalized into the cell via endocytosis.

When inside the cell, the endocytosed vesicle may meet up with an endosome, where receptor and ligand are separated via a change in pH. The receptors may be recycled back to the cell surface, or may be degraded. The ligands may undergo further processing by leaving the endosome via another vesicle and interacting with lysosomes. Lysosomes use pH-dependent enzymes (acid hydrolases) to break down lipids, nucleic acids, and proteins.

The bottom line is that, for most molecules, getting into the cell interior is not such a simple process. Even when a molecule is not being delivered, one must consider how cells will interact with the external substances they might touch, such as signaling molecules or surfaces meant for attachment (or the prevention thereof). Membrane proteins are involved in these situations, too.

In the next chapter, we will take a step back to focus upon proteins themselves: what they are, how they are constructed, their structures, and how biotechnologists might isolate them for study.

Related reading

Bie, B., Pan, Z.Z., 2007. Trafficking of central opioid receptors and descending pain inhibition. Mol. Pain 3, 37.

Cahill, C.M., Holdridge, S.V., Morinville, A., 2007. Trafficking of delta-opioid receptors and other G-protein-coupled receptors: implications for pain and analgesia. Trends Pharmacol. Sci. 28, 23–31.

Ejendal, K.F., Hrycyna, C.A., 2002. Multidrug resistance and cancer: the role of the human ABC transporter ABCG2. Curr. Protein Pept. Sci. 3, 503–511.

Hankins, H.M., Baldridge, R.D., Xu, P., Graham, T.R., 2015. Role of flippases, scramblases, and transfer proteins in phosphatidylserine subcellular distribution. Traffic 16, 35–47.

Hellgren, M., Sandberg, L., Edholm, O., 2006. A comparison between two prokaryotic potassium channels (KirBac1.1 and KcsA) in a molecular dynamics (MD) simulation study. Biophys. Chem. 120, 1–9.

Heuser, J., 1980. Three-dimensional visualization of coated vesicle formation in fibroblasts. J. Cell Biol. 84, 560–583.

Heuser, J.E., Anderson, R.G., 1989. Hypertonic media inhibit receptor-mediated endocytosis by blocking clathrin-coated pit formation. J. Cell Biol. 108(2): 389–400.

Lever, J.E., 1984. A two sodium ion/D-glucose symport mechanism: membrane potential effects on phlorizin binding. Biochemistry 23, 4697–4702.

Mindell, J.A., 2012. Lysosomal acidification mechanisms. Annu. Rev. Physiol. 74, 69–86.

Pratten, M.K., Lloyd, J.B., 1986. Pinocytosis and phagocytosis: the effect of size of a particulate substrate on its mode of capture by rat peritoneal macrophages cultured in vitro. Biochim. Biophys. Acta 881, 307–313.

Questions

1. Sometimes phospholipids are denoted as shown here:
   What is responsible for the bend in the second tail?

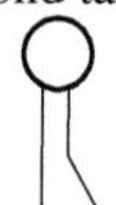

2. Describe the usual structure of a phospholipid built on a glycerol backbone.
3. What is the most common molecule found in the plasma membrane?
4. Which of the phospholipids below might be a common constituent in the cell membrane? For the ones that are not expected to be common, give a reason why not.

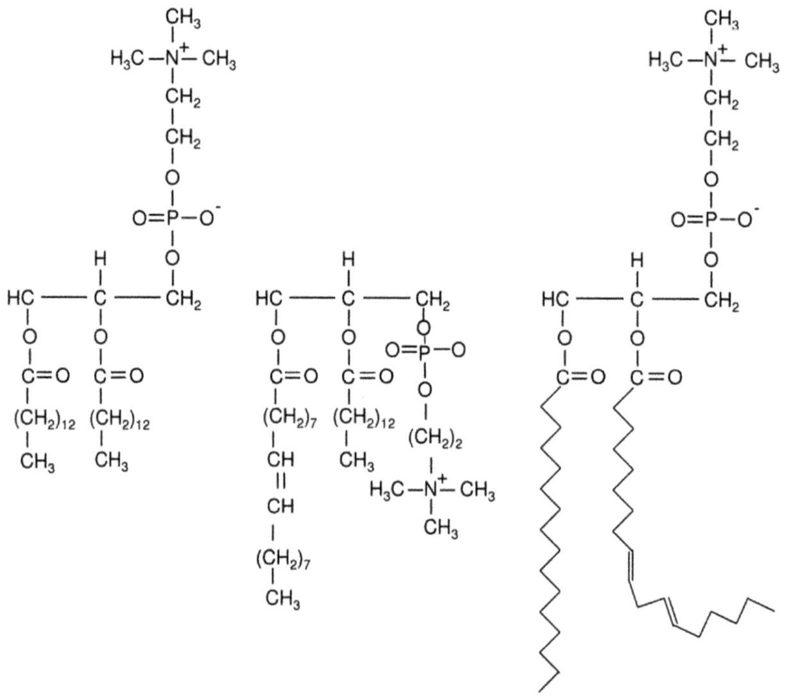

5. A phospholipid head has a nitrogen group attached to four carbons. Can it be found on the inside or outside layer of the plasma membrane or both? Explain your answer.
6. Why is the charge on phosphatidylserine useful?
7. Are phospholipids able to move within the plasma membrane? If so, describe the movement.
8. What does cholesterol do for the cell membrane?
9. What is the difference between flippases and scramblases?

10.

a. Given the following chart, what is the equilibrium potential for sodium, chloride, and calcium ions?

	Concentration in cell (mM)	Concentration in blood (mM)
K^+	139	4
Na^+	12	145
Cl^-	4	116
Ca^{2+}	<0.0002	1.8

b. Consider the ion X^z, which has an equilibrium potential of +43 mV. If it were normally more concentrated inside the cell than outside, what can you deduce about the charge of this ion?

c. Consider the ion Y^z, which has an equilibrium potential of -57 mV. If it were normally more concentrated outside the cell than inside, what can you deduce about the charge of this ion?

11. Consider an endocytosed molecule bound to a receptor. In the early endosome, it will be released from its receptor due to a conformation change in the receptor, which is a protein.

a. Which amino acids might contribute to the change in conformation in the early endosome because of an ionization state change?

b. Which amino acids might contribute to a change in conformation because of an ionization state change if the protein is eventually transported to a lysosome?

12. Which would you expect to cross a synthetic membrane the easiest: a lipid or sucrose? Why?

13. Can transporters denature just like other folded proteins? Why or why not?

14. Do ions with the same charge have the same resting membrane potential? Why or why not?

15. What does the Nernst equation calculate?

16. Assuming the membrane potential of a cell is only affected by potassium ions, and given an extracellular concentration of potassium ions of 5 mM and a membrane potential of -92mV, what is the concentration of potassium ions inside the cell?

17. Can potassium ions move through Na^+ channels even if $V_{Na+} = 0$? What if $V_{cell} = 0$?

18. Does a chemical gradient produce a measurable force? (Consider this in the absence of an electrical gradient.)

19. a. What are ABC transporters? What do they do? What does "ABC" mean?

b. Name two instances when ABC transporters are utilized. Decide whether each would be good or bad for a human patient.

20. Draw the Na^+/Ca^{++} exchanger and describe how the energy of ATP is involved.

21. Describe how a potassium leak channel is specific for potassium ions.

22. What is the difference between a channel and transporter?

23. Is it accurate to call LDL "bad cholesterol"? Why or why not?

24. What would happen if the cell did not make dynamin?

25. Sarah is in charge of giving a presentation about lysosomes to her classmates. During the presentation, she says that if the lysosome ruptures, the cell will die because the degradative enzymes of the lysosome will be released into the cytoplasm, where they will break down healthy components of the cell. Is this a correct statement? Why or why not?

26. If a lysosome pumped HCO_3^- into its interior, would it still function correctly?
27. Why should we care that IgG antibodies have separate Fc and Fv regions?
28. a. What are four triggers of phagocytosis?
    b. Is pinocytosis a triggered event?
29. Heroin is an opioid drug. Using facts about receptor processing presented in the text, explain why heroin addicts take ever-increasing dosages of the drug over time. What will be the eventual end result of taking more and more of the drug?

CHAPTER

Proteins

3

Chapter outline

At this point, we should pause to understand more about proteins. Although we mentioned them in the previous chapter, the flow of our voyage into the cell did not permit an in-depth look into just what proteins are. The membrane proteins, channel proteins, receptors, and degradative enzymes of the lysosome all share properties that are common to all proteins. In addition to understanding the roles of proteins as cellular structures, the biotechnologist should be aware of their roles as **bioactive** substances. Some drugs are proteins, and most toxins are proteins. Proteins can be manufactured synthetically, or the biotechnologist might harness cells to manufacture specific proteins—even novel proteins designed by the scientist.

In this chapter we will discuss protein structure, characteristics, and methods of isolation. More specifically, we will discuss:

- amino acids
- amino acid charge, pK_a, and isoelectric point
- primary protein structure
- secondary protein structure (such as the α-helix or α-pleated sheet)
- tertiary and quaternary protein structure
- the hydrophobic effect

Biotechnology and its Applications. https://doi.org/10.1016/B978-0-12-817726-6.00003-4

Then we will return to the cell and look more closely at membrane proteins with respect to:

- how proteins are embedded in the membrane
- movement within the membrane
- isolation for study:
  - Detergents, soaps, micelles
  - Electrophoresis: SDS-PAGE.

3.1 Amino acids

Amino acids are so named because they contain both an amino group and a carboxylic acid. The structure of an amino acid at neutral pH (pH 7.0) is given in Figure 3.1. Disregarding the side group R for a moment, the figure shows that the molecule is a **zwitterion**—a molecule that has both positive and negative regions of charge. With respect to amino acids, the carbon that connects the amino and carboxylate groups is said to be a **chiral** center if the R group is anything other than a hydrogen atom. *A chiral carbon has four different groups attached to it.* Paraphrasing Lord Kelvin, if the mirror image of a molecule is not superimposable upon the molecule itself, then the molecule is said to be chiral. Chiral molecules will rotate polarized light, with L-forms rotating polarized light to the left (counterclockwise) and D-forms rotating it to the right. Examples of L- and D- forms of an amino acid are given in Figure 3.2; note that the two forms are mirror images of one another. Although chirality may be thought of as interesting only to the organic chemist, it is important to biotechnologists as well. In the case of amino acids, it is the L-form that is used in constructing proteins such as enzymes, receptors, and signaling peptides. D-form amino acids are sometimes used, with varying success, to produce biologically inert molecules.

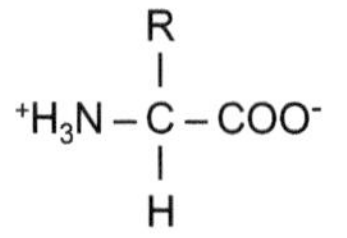

R
|
^+H_3N – C – COO^-
|
H

FIGURE 3.1

The structure of an amino acid at neutral pH.

COO^-
^+H_3N — C — H
R

COO^-
H — C — NH_3^+
R

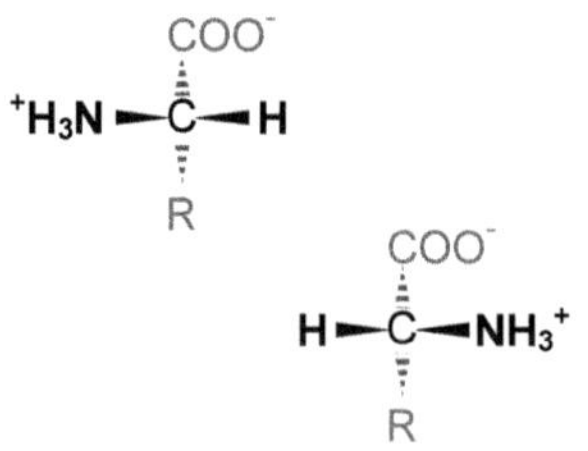

FIGURE 3.2

L- *(top)* and D- *(bottom)* forms of the same amino acid. *Heavier fonts* indicate atoms coming toward the reader out of the plane of the page, and *lighter fonts* indicate atoms oriented away from the reader below the plane of the page.

Although all amino acids have the basic structure shown in Figure 3.1, the identity of the R group defines each specific molecule. There are 20 amino acids commonly found in nature to form proteins, each with a defined R group. These amino acids are shown in Figure 3.3.

After you have examined Figure 3.3, a couple of points may need further clarification. First, notice that aspartate and asparagine are grouped together, as are glutamate and glutamine. The difference between each pair of side groups is that the negatively charged oxygen has been replaced by an amine group (NH_2), which is hydrophilic. These amino acids are paired together because (1) aspartate and asparagine are very similar chemically, as are glutamate and glutamine, and (2) they are all hydrophilic. Appearing in the second half of the figure are the positively charged amino acids, which are hydrophilic, followed by the aromatics, which tend to be hydrophobic. A case can be made that tyrosine can be considered hydrophilic because of the hydroxyl group on the aromatic ring; this is why the one-letter code "Y" is not circled for tyrosine. Note that the other hydroxyl-containing amino acids are also hydrophilic.

Cysteine is a special amino acid in that its side group contains a sulfhydryl group. In a protein, cysteines often occur in pairs. The paired cysteines do not have to be next to each other (or even close to each other) in terms of the order of amino acids, but when cysteines pair in three dimensions as a part of protein folding, they will be in close proximity and form a disulfide bond. The pair of bonded cysteines will be known as a cystine (note the spellings). The disulfide bonds can be broken via reduction, which means acidic environments or proton donors can convert cystines back into a pair of cysteines. As we will see, this is a property one must contend with in the laboratory, especially in terms of identifying the primary sequence of a protein.

Even though biotechnology students might at first be overwhelmed with the prospect of learning the structures of all 20 common amino acids, they should nevertheless strive to acquaint themselves with the structures and functions of these molecules. At the very least, they should have a working knowledge of the different classes of amino acids as they delve into different biotechnical applications. For instance, if one were interested in finding specific amino acids with a propensity to bind with DNA or perhaps the potential to serve as delivery vehicles for siRNA, negatively charged side groups (D, E) should be avoided because they would be repelled by the negative charges in the DNA or RNA molecules. Similarly, using a polypeptide made up entirely of nonpolar and uncharged amino acids (G, A, V, L, I, P, F, W, M) as a signaling molecule to be delivered via the blood would not be advisable because the molecules would be hydrophobic and therefore not soluble in the aqueous environment of the blood. Any sort of protein engineering should be accompanied by knowledge of which amino acids are small, large, inflexible, charged, polar, hydrophobic, or cross-linkable.

3.1.1 pK_a

Consider a pH scale and the side group for aspartate (—CH_2-COO^-). This amino acid is often referred to as aspartic acid, which has the side group —CH_2-COOH. The naming convention has the "acid" form as the form with its complete set of protons. When a proton is lost and the species carries a negative charge, we refer to this as the "ate" form; hence, "aspartate," as opposed to "aspartic acid."

Not all acids ionize at the same pH. A way to characterize different acids with respect to the pH at which they will tend to lose a proton is the **pK_a value**. Without deriving it here, know that the pK_a value is related to the dissociation constant. The definition of pK_a is the pH at which an ionizable species is 50% ionized. Considering the side group for aspartate/aspartic acid, the pKa is 3.65. This implies that

Nonpolar, aliphatic R groups

Glycine
Gly
G

Alanine
Ala
A

Valine
Val
V

Leucine
Leu
L

Isoleucine
Ile
I

Proline
Pro
P

Negatively charged R groups and their amine-containing counterparts

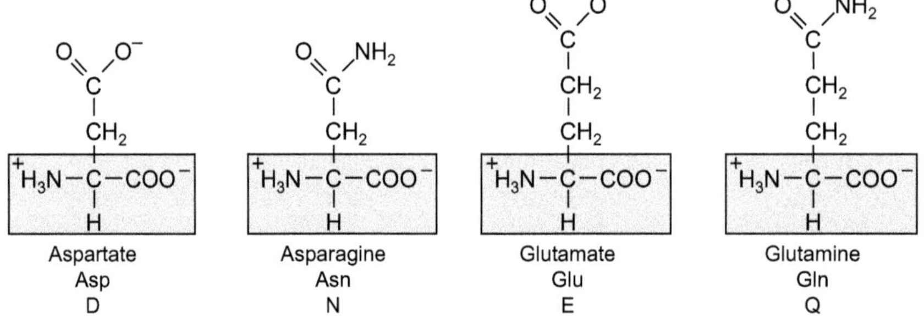

FIGURE 3.3

Structures, names, three-letter abbreviations, and one-letter codes of the 20 common amino acids used to form proteins. These formulae reflect the prevailing ionization states at a pH of 7.0. The *shaded portions* of the structure are the entities common to each amino acid and would form the backbone in a polypeptide; the R group of each amino acid is *unshaded*. Because the pK_a of the histidine R group is relatively close to 7.0, both ionization states are shown (with a hydrogen and positive charge in parentheses). The nonionized form of the histidine R group will predominate at a pH of 7.0. *Circles* around single-letter codes indicate nonpolar side groups. The *circle* around G is dashed because, although the small hydrogen side group is most easily classified as nonpolar, it contributes very little to hydrophobic interactions. Also note that C is considered polar here, although the side group is commonly bound to another C via a disulfide linkage, which is strongly nonpolar.

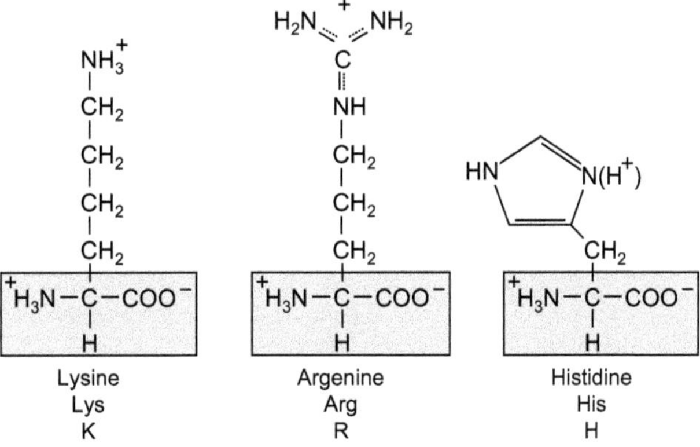
Positively charged R groups
Lysine
Lys
K
Argenine
Arg
R
Histidine
His
H

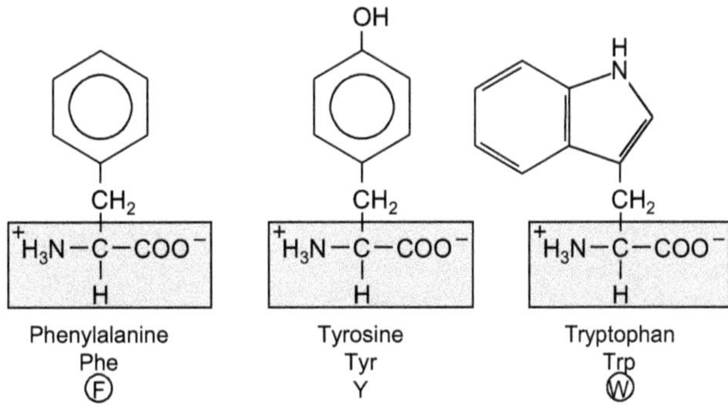
Aromatic R groups
Phenylalanine
Phe
F
Tyrosine
Tyr
Y
Tryptophan
Trp
W

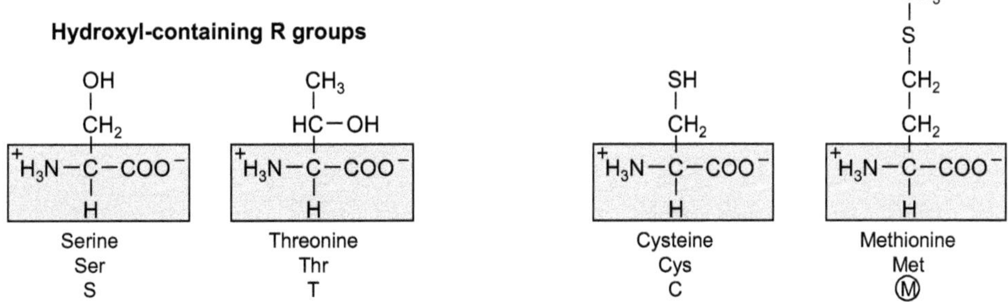
Hydroxyl-containing R groups
Serine
Ser
S
Threonine
Thr
T
Sulfur-containing R groups
Cysteine
Cys
C
Methionine
Met
M

FIGURE 3.3—CONT'D

if one has a beaker of aspartate/aspartic acid in an aqueous, buffered solution held at pH 3.65, half of the amino acid molecules will be aspartate, and half will be aspartic acid. What this tells us is that for aspartate/aspartic acid at physiological pH (7.2 inside the cell), the pH is a good distance from the pK_a value, so virtually all of the molecules will be in the form of aspartate. As a general rule, a "good distance" in these cases is at least one pH unit away from the pK_a. The pK_a value of aspartic acid being 3.65 does not imply that at pH 3.66 all of the molecules will be in the form of aspartate; one would expect that slightly more than 50% would be. As the pH is increased further from the pKa value, a greater percentage of the side groups would be in the "ate" form rather than the acid form. In the body, changes in pH by 0.2 units can have profound effects on protein folding.

When deciding on the ionization state of a species, a handy rule to keep in mind is that at pH values below the pK_a for that species there will be a relative abundance of protons compared with the species, so the species will carry as many protons as possible. At pH values above the pK_a for the species, there will be a relative shortage of protons, so the species will tend to release protons into solution. In the case of carboxylic acids, low pH will mean the species is in its acid (—COOH) form. At a pH of 1, the amino acid Asp would be in the form of aspartic acid. The same is true for Glu: the side group has a pK_a of 4.25, so the amino acid would be in the form of glutamic acid.

The same rules apply to species that ionize to carry a positive charge, such as the side group for lysine (—CH_2-CH_2-CH_2-CH_2-NH_3^+). In this case, the side group has an ionizable amine, with a pK_a value of 10.53. This means that at a pH of 10.53, half of the amines will be NH_2, and half of them will be NH_3^+. At physiological pH, which is a good distance below the pK_a value for the Lys side group, the amine will carry as many protons as possible and be in the form —NH_3^+.

Carrying the general ionization rule one step further, at "low" pH, where there is a relative excess of protons and the ionizable groups are carrying as many protons as possible, the ionizable species will carry the highest charge they can. For an amino acid such as aspartate, low pH means the side chain will carry a charge of 0 instead of -1. For an amino acid such as lysine, low pH will cause the side chain to carry a charge of +1 instead of 0.

Using the pK_a values for every ionizable species in the molecule, one can determine the **isoelectric point** of the molecule—the pH at which the predominant net charge of the molecule in solution is zero. This is a term that is typically applied to proteins and polypeptides.

Consider glycine, the amino acid for which the side group is a hydrogen. At physiological pH, it can be described with the formula H_3^+N-CH_2-COO^-. Note that the N-terminus and the C-terminus can each be ionized; this implies that each terminus has its own pK_a value. In fact, this is true, and the pK_a values are 9.60 and 2.34 for the N- and C-termini, respectively. From these two values, the ionization state of the entire molecule can be written for any given pH value (Figure 3.4). At a low pH such as 1.0, each of the termini will carry as many positive charges as possible, so the structure of glycine will be H_3^+N-CH_2-COOH, and the overall charge of the molecule will be +1. If the pH were to be gradually increased, after the lowest pK_a value is crossed the corresponding molecular group will change to the ionization state carrying a charge one lower than before. For glycine, the lowest pK_a value (2.34) corresponds to the C-terminus, which will lose a proton to yield H_3^+N-CH_2-COO^-. The overall charge of the molecule is now 0 (neutral), and the molecule will tend to retain the same state until the pH reaches the next pK_a value (9.60). At this point, the N-terminus will lose a proton and the molecule will have the structure H_2N-CH_2-COO^-, with a net charge of -1. To determine the isoelectric point of a molecule, simply take the average of the pK_a values that surround the pH range where the charge of the molecule is zero. For glycine, we would take the average of 2.34 and 9.60 to obtain an isoelectric point of 5.97.

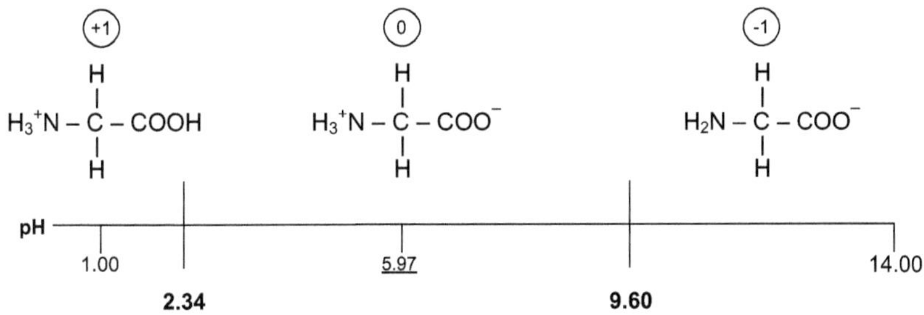

FIGURE 3.4

Ionization states of glycine at different pH values. Each ionizable species has a distinct pK_a value, shown on the number line in *bold*. The isoelectric point for glycine—the average of the two pK_a values—is *underlined*. *Circled numbers* above the structures indicate the predominant net charge of the molecule in that pH range.

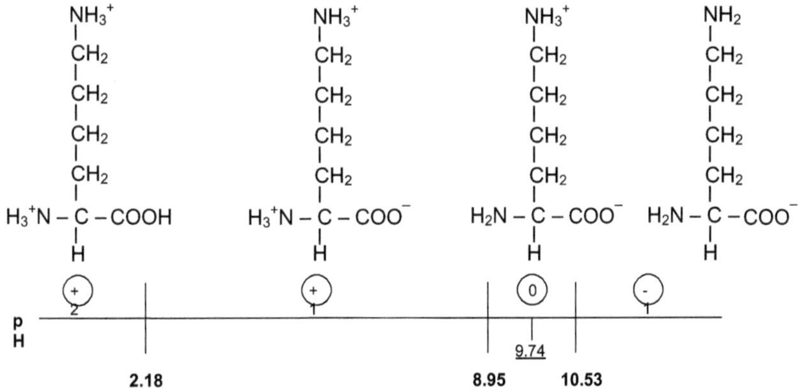

FIGURE 3.5

Ionization states of lysine at different pH values. To determine the isoelectric point (pI), take the average of the two pK_a values (shown in *bold*) that bound the pH range where the net charge = 0. *Circled numbers* indicate the predominant net charge of the molecule in the given pH range. In this case, pI = 9.74 *(underlined)*.

As another example, consider lysine. For this amino acid, in addition to N- and C-termini, there is a side group that can be ionized. The pK_a value for the N-terminus is 8.95, whereas for the C-terminus, it is 2.18. (The pK_a value for any ionizable group depends upon the structure of the entire molecule, which explains why the N- and C-termini have pK_a values that differ from those of glycine.) The side group of lysine has a pK_a value of 10.53. The ionization states of this molecule with respect to pH are shown in Figure 3.5. Note that, again, the molecule will hold as many protons as possible at low pH. When the pH is gradually raised, as each pK_a value is crossed one proton will be lost from the species that corresponds with the pK_a value just crossed, and the overall charge of the molecule will go down by 1. Even though there are more than two pK_a values for lysine, determining the isoelectric point is straightforward as long as one is aware of the different ionization states of the molecule for each pH. Applying the same rule we used for glycine, we can calculate that the isoelectric point is the average of the two pK_a values that surround the pH region for which the net charge of the molecule is neutral. For lysine, the isoelectric point is $(8.95+10.53)/2=9.74$.

After you figure out what the overall charge of the molecule is at a very low pH, every time you cross a pK_a value, it reduces the overall charge by 1. Using this rule, it should be fairly straightforward to determine the isoelectric point of a polypeptide with hundreds of ionizable side groups. Figure out how many pK_a values must be crossed to obtain an overall charge of zero, then take the average of the two bounding pK_a values. To find the overall charge of the molecule at low pH, it's simply a matter of counting the number of groups that ionize to a positive value. Lysine has two groups that ionize to a positive value, meaning the overall charge of the molecule at pH 1 is equal to +2, which means two pK_a values must be crossed to bring the molecule to a neutral state, so the isoelectric point will be the average of the second and third pK_a values (as ordered by increasing value on the number line).

3.2 Protein structure

3.2.1 Primary structure

Amino acids can be polymerized to form a polypeptide, which gets its name from the peptide bonds that are formed by the loss of water molecules during polymerization (Figure 3.6). Once formed, the peptide bond is rigid and planar, but it can be broken by a water molecule (hydrolyzed) in the reverse direction of the reaction shown. Notice that when peptide bonds are formed, there is still a primary amine on one side of the polymer backbone and a carboxyl group on the other—termed the N-terminus and C-terminus, respectively. It is customary to draw a polypeptide in the N → C direction.

Rather than draw a polypeptide using a complete molecular representation, one can imply its structure through proper naming. Using the same convention of beginning with the N-terminus, a pentapeptide made of phenylalanine, histidine, lysine, isoleucine, and threonine could be called phenylalanylhistidinyllysinylisoleucinylthreonine. What a great way to win bets over who knows the longest word! However, naming polypeptide sequences in this manner is not very practical. Another way to write the ordered amino acid sequence, or **primary sequence**, of a polypeptide is to use three-letter code of each amino acid. Using this system, our pentapeptide is now Phe-His-Lys-Ile-Thr. An even more efficient convention, used for longer sequences, is to use the one-letter codes for the amino acids, still adhering to the principle of the N-terminus always appearing first. One would write the pentapeptide in this example as FHKIT.

3.2.2 Secondary structure

The **secondary structure** of a protein describes the spatial arrangement of the atoms in the protein backbone. The repeating pattern of –HN–Cα–C=O can exist in a disorganized array called a random coil, or in

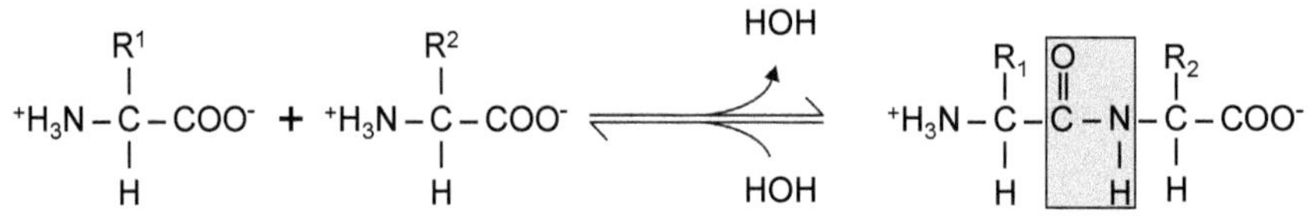

$$^{+}H_3N-\underset{H}{\overset{R^1}{C}}-COO^- + \ ^{+}H_3N-\underset{H}{\overset{R^2}{C}}-COO^- \underset{HOH}{\overset{HOH}{\rightleftharpoons}} \ ^{+}H_3N-\underset{H}{\overset{R_1}{C}}-\overset{O}{\overset{\|}{C}}-\underset{H}{N}-\underset{H}{\overset{R_2}{C}}-COO^-$$

FIGURE 3.6

The polymerization of two amino acids results in the loss of a water molecule and the formation of a peptide bond, shown by the *shaded box*.

a distinctly well-defined manner, with the angles of the two planar peptide bonds attached to each α carbon repeating in a regular fashion. The most common secondary structures are α-helices, β-pleated sheets, and β-turns. The α-helix is a right-handed helix with amino acid residues spaced at 3.6 residues per turn and a rise of 0.54 nm per turn. (A "right-handed helix" is wound such that if one were to curl the fingers of the right hand around the backbone of the helix with the fingers pointing in the N → C direction, the thumb would point along the helical axis in the direction of N → C helix progression.) Figure 3.7 illustrates approximate amino acid placement and rise for an α-helix. Notice that amino acid 0 is not directly below the third or fourth subsequent residue. If the coil were a number line, amino acid 0 would be beneath 3.6, which is where "3.6 residues per turn" comes from. Because of this, amino acid 0 can form hydrogen bonds with the third or fourth subsequent residue, as shown in the second panel of the figure. The side groups of each amino acid would point outward from the helix (not shown).

The β-pleated sheet is comprised of multiple relatively straight segments within one polypeptide, with the strands lining up adjacently (Figure 3.8). It is the zigzag nature of the N-C_α-C backbone that gives the sheet its pleats. The sheet is stabilized by hydrogen bonding between peptide oxygens and amide hydrogens

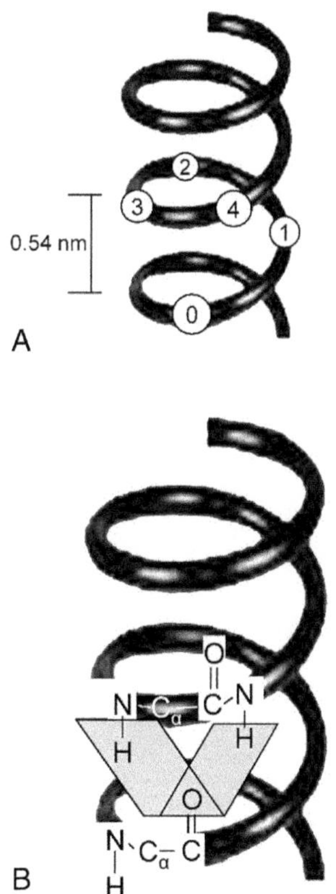

FIGURE 3.7

Placement and interactions of amino acids in an α-helix. (A) The helix has 3.6 residues per turn and a pitch of 0.54 nm per turn. (B) The structure is made stable by internal hydrogen bonds: a peptide oxygen can interact with a peptide hydrogen three or four residues away.

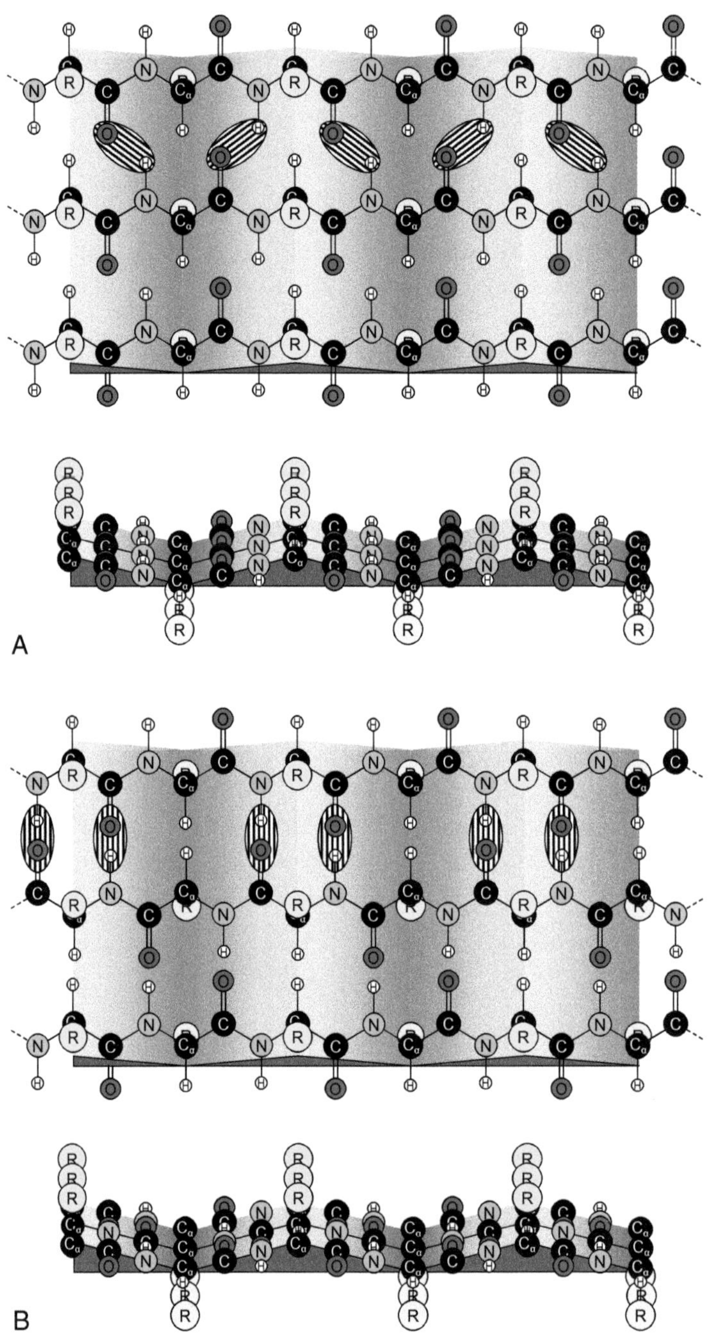

FIGURE 3.8

Placement and interactions of amino acids in β-pleated sheets. Hydrogen bonds between the first two strands are circled. (A) Parallel β-sheets. (B) Antiparallel β-sheets.

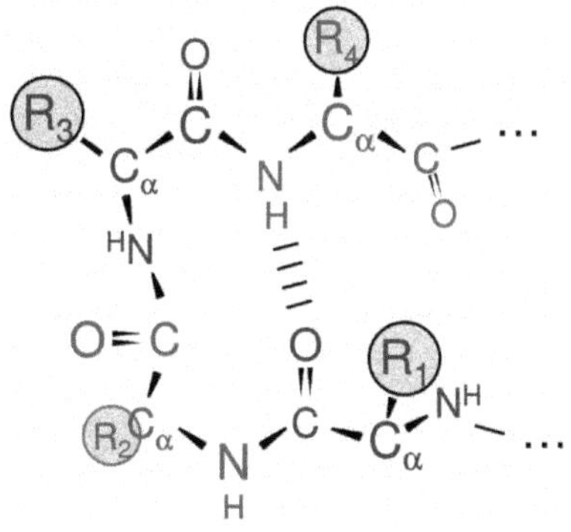

FIGURE 3.9

Four amino acids make up a β-turn. Amino acids 1 and 4 lend stability to the turn through a hydrogen bond, which is shown. The majority of β-turns are type 1, which means they have a proline residue in position 2. Another common turn, type 2, is defined by residue 3, which is a glycine.

on a neighboring segment. The straight segments can be parallel—meaning the N → C directions of all β-sheet segments point the same way—or antiparallel, meaning the segments alternate in direction. Antiparallel β-sheets have hydrogen bonds perpendicular to strand directions, whereas parallel sheets have hydrogen bonds occurring at alternating angles. Both forms of the β-pleated sheet have amino acid side groups occurring parallel and on the same side of the sheet. In the figure, take note of the N → C directions, hydrogen bonding, and alignment of the side groups for both parallel and antiparallel configurations.

A β-turn (also known as a β-bend) is a structure that allows for sharp turns in the middle of a polypeptide. One way in which β-turns appear in proteins is during the formation of β-sheets. A β-turn is made up of four amino acid residues, with residues 1 and 4 often interacting as illustrated in Figure 3.9. The amino acids glycine and proline are often present in β-turns. Glycine works well in a β-turn because its small side group (–H) is sterically favorable for the compact structure, rendering the residue relatively flexible. (Keeping this in mind, one might also correctly predict that alanine would be a relatively common amino acid in β-turns.) Proline also works well because of some interesting properties that merit deeper discussion.

Proline is a unique amino acid in that its side group is attached to its backbone atoms in two places: the chiral carbon and the amino nitrogen. This renders proline a rigid amino acid. As such, it is not often found in α-helices but is found in β-turns. There are two different conformations for proline (Figure 3.10). In the *trans* conformation, the polypeptide backbone is somewhat straight, whereas in the *cis* conformation, it makes a sharp turn. Because the β-turn itself is a sharp turn, the *cis* conformation is overwhelmingly favored for prolines involved in β-turns.

In summary, for β-turns: amino acids 1 and 4 interact with each other, and glycine and proline are commonly present—glycine because it is small and proline because it is rigid and can naturally form part of the sharp turn of the β-bend.

3.2.2.1 Super-secondary structure

Super-secondary structure is simply a combination of secondary structures. Consider a piece of paper; if it were rolled into a tube, it would still be a piece of paper, but it would have been modified into a

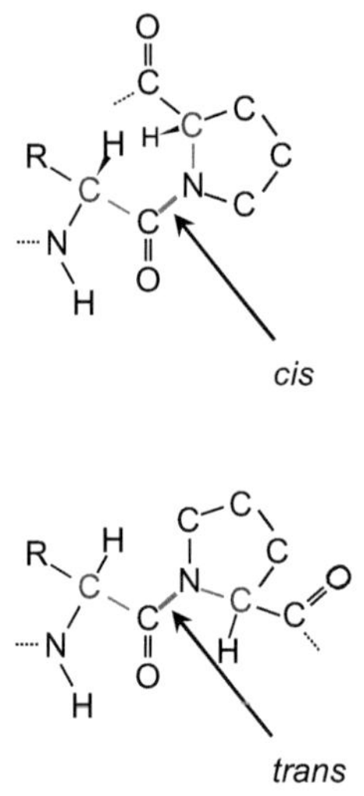

cis

trans

FIGURE 3.10

Proline in *cis* and *trans* conformations. α carbons are shown in blue; arrows point to peptide bonds. The arrangement of the three bonds in red (the positions of the two α carbons relative to the peptide bond) determines the conformation.

more complex structure. The same can be done to a β-sheet, and in this case the resulting structure would be a β-barrel. Any combination of secondary structures can be used to produce super-secondary structure. For example, there is the helix-loop-helix, where two α-helices are connected by a β-turn. This structural motif can be found in many transcription factors. (Transcription factors are used in gene processing and will be discussed in a later chapter.)

3.2.3 Tertiary structure

Tertiary structure—the three-dimensional structure of a protein—is the next level of complexity in protein folding. Whereas individual amino acids in the primary sequence can interact with one another to form secondary structures such as helices and sheets, and individual amino acids from distant parts of the primary sequence can intermingle via charge-charge, hydrophobic, disulfide, or other interactions, the formation of these bonds and interactions serve to change the shape of the overall protein. The folding that we end up with for a given polypeptide is the tertiary structure.

3.2.4 Quaternary structure

Quaternary structure is the interaction of two or more folded polypeptides. Many proteins require the assembly of several polypeptide subunits before they become active. If the final protein is made of two subunits, the protein is said to be a *dimer*. If three subunits must come together, the protein is said to be a trimer; four subunits make up a tetramer, and so on. If the subunits are identical, the prefix "homo" is used, as in "homodimer." If the subunits are different, we use "hetero," as in "heterodimer."

Hemoglobin is the protein responsible for carrying oxygen in the blood. It is made up of four polypeptides: two α and two β subunits. One α and one β subunit come together to form a heterodimer, and two of these heterodimers interact to form one hemoglobin molecule. Hemoglobin can therefore be thought of as a dimer of dimers, which come together to give the final protein its quaternary structure.

3.3 The hydrophobic effect

We mentioned earlier that some individual amino acids carry charges, but others do not. At physiological pH, the amino acids that can ionize negative (Asp and Glu) will carry charges on their side groups, meaning the forms of these two amino acids will be aspartate and glutamate. The positively charged amino acids lysine and arginine (and to a lesser extent, histidine) have protonated side chains at this pH.

The interaction of amino acid side groups plays a pivotal role in protein folding. Charge-charge interactions are important, as is the interplay of polar side groups with each other and with water molecules. The nonpolar (hydrophobic) amino acids are also very important to protein folding. These amino acids (which include alanine, valine, leucine, isoleucine, and phenylalanine) tend to fold toward the interiors of proteins because of the **hydrophobic effect**. Like oil and water, hydrophobic and hydrophilic molecules do not mix well. Nevertheless, proteins within the cell must exist in the hydrophilic environment of the cytoplasm, even though the hydrophobic amino acids within these proteins do not mix well in an aqueous environment.

The hydrophobic effect is the primary driving force behind protein folding. It is directly related to the thermodynamic property of a system known as Gibbs free energy. Interactions and reactions occur in ways that tend to minimize Gibbs free energy, and processes that decrease the value of Gibbs free energy are said to be **spontaneous**. Changes in free energy are given by the formula:

$\Delta G = \Delta H - T\Delta S$, where

Δ = change,
G = Gibbs free energy,
H = enthalpy,
T = temperature (in Kelvin), and
S = entropy (disorder).

Note that increases in entropy (disorder) contribute to negative free energies. This implies that increasing the order in a system is not favored. To understand the hydrophobic effect and its relation to Gibbs free energy, let us consider an example of hydrophobic molecules (in this case, methane) in an aqueous solution (Figure 3.11A). In the two shaded beakers, the one on the left appears to have more entropy because the methane molecules are dispersed throughout the beaker with great disorder. However, you probably know from your own experience that if we place oil into a beaker of water, the oil coalesces more like the figure on the right: it floats to the top. How can this possibly be, when this scenario appears to have less entropy? To resolve the issue, one must consider the entropy of the entire system, not just the hydrophobic molecules. Water molecules form a cage around a single hydrophobic molecule, and in doing so they form an ordered structure. If additional hydrophobic molecules were added and kept separate from each other, separate cages would form around each one (Figure 3.11B). This is different from the situation in which the hydrophobic molecules come together and become surrounded by a larger, single cage (Figure 3.11C). The larger cage incorporates more water molecules

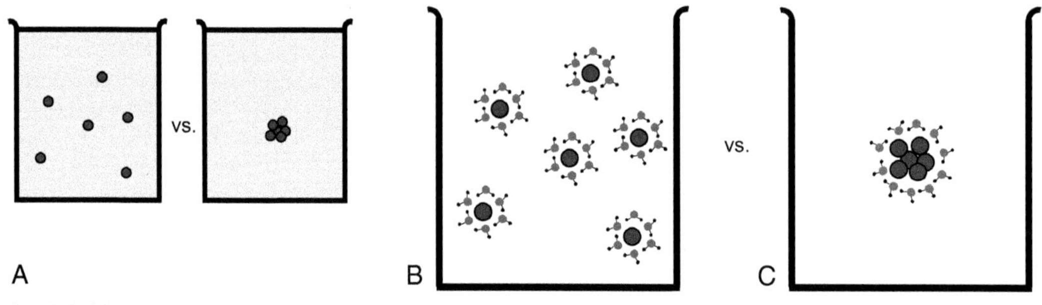

FIGURE 3.11

(A), Hydrophobic molecules in an aqueous solution might disperse throughout the solution, or they might coalesce. (B), When the hydrophobic molecules are dispersed, water molecules (represented by) will form a cage around each one. (C), When the hydrophobic molecules coalesce, a cage of water molecules will still form around them. Although this cage will be larger than the cages in (B), the total number of water molecules in the cage will be less than the sum of all those in the cages formed around individual hydrophobic molecules. This will leave a greater total number of molecules in a disordered state, which will increase entropy and decrease Gibbs free energy. The formation in (C), is therefore favored.

than an individual small cage but involves fewer water molecules overall compared with forming separate cages around each individual hydrophobic molecule. Water molecules that are not involved with cages are free to move virtually anywhere, which contributes to the overall entropy of the system.

The above discussion illustrates the hydrophobic effect, and the hydrophobic effect explains why hydrophobic amino acids tend to fold into the interiors of proteins. As a corollary, because hydrophobic amino acids tend to be on the inside of a folded protein, the exterior portion of the protein tends to be made up of charged and polar amino acids.

Protein folding is one reason why the charge of individual amino acids in the primary sequence is important; another reason is molecular recognition. Consider the protein shown in Figure 3.12. In this example, as is the case for many proteins, there is an indentation that might appear as a hollowed-out spot. This could be because two or more polar, or charged, amino acids are interacting to help hold the shape of the cavity and thereby preserve the shape (conformation) of the protein. The resulting cavity could serve as a point of interaction with another molecule. Suppose the protein is a receptor. The molecule that the receptor is intended to receive (its **ligand**) may have positive, negative, or polar regions that line up with negative, positive, or polar regions of the receptor. The receptor and ligand will fit with some degree of precision—the greater the precision, the higher the **specificity** of the receptor for its ligand.

A similar example is that of **enzymes**, which are proteins that catalyze chemical reactions. An enzyme uses its **active site** to interact with its **substrate** or substrates. There are over 4000 enzymes within a human cell, and each enzyme catalyzes a specific chemical reaction or type of reaction. Even the polymerization of amino acids into proteins (Figure 3.6) is catalyzed by enzymes which also happen to be proteins).

3.4 A return to membranes

The early material in this book focuses on cells and the basic entities a bioactive agent will first encounter as it attempts to gain entry into them. As already mentioned, cells are surrounded by a plasma

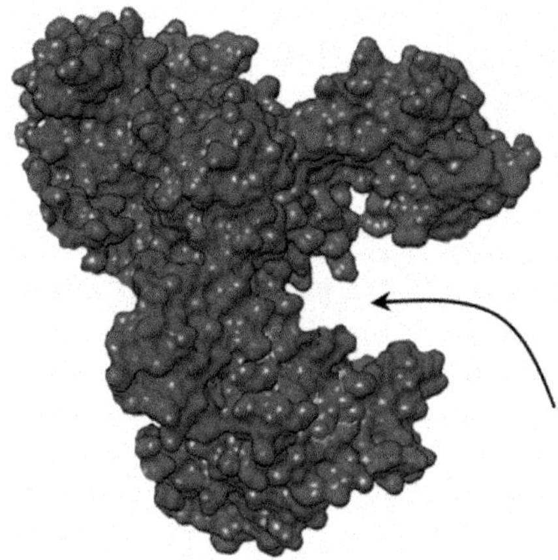

FIGURE 3.12

A hypothetical protein, with an active site in a cleft *(arrow)*.

membrane that is a lipid bilayer, and the bilayer is primarily made up of phospholipids. We also discovered that mixed in among these phospholipids are other molecules such as cholesterol and transmembrane proteins. Now that proteins have been more thoroughly introduced, let's discuss them in the context of the plasma membrane.

Membrane proteins can be embedded partially in the membrane, may cross the entire membrane once, or may cross it multiple times (Figure 3.13). Proteins that completely span the membrane are known as transmembrane proteins. The proteins can be anchored in the phospholipid bilayer by, for example, fatty acids; however, when a transmembrane protein spans the entire membrane, the portion that interacts with the lipid tails is typically in the form of an α-helix.

The number of times that a transmembrane protein spans the lipid bilayer gives rise to the terms **single-pass** and **multipass** transmembrane proteins. One important class of multipass transmembrane proteins is the seven-pass transmembrane proteins. These are often receptors, and they are used to transmit a signal from outside of the cell to the inside of the cell, thus initiating a chemical cascade inside the cell without transporting a signaling molecule across the membrane. Other transmembrane proteins make up **channels** that allow certain molecules to cross the plasma membrane into and out of the cell (e.g., sodium channels). Some membrane proteins that do not completely span both layers of the plasma membrane are anchored within the hydrophobic portion of the lipid bilayer. Remember that the middle of the lipid bilayer contains fatty acid tails, so the anchoring moieties will also be hydrophobic, as is the case with common anchoring groups such as fatty acids and prenyl groups ($-CH_2$-CH_2-CH=C-$(CH_3)_2$).

3.4.1 Protein movement within the plasma membrane

As mentioned in Chapter 2, components of the plasma membrane can be likened to people walking around in a crowded mall. Individual components—phospholipids, transmembrane proteins, lipid rafts, and so on—are able to move around in the membrane. Although two individual phosphatidylcholine molecules might be next to each other one minute, they are able to move around and may become separated in the next minute. Let's look at an example to illustrate this point.

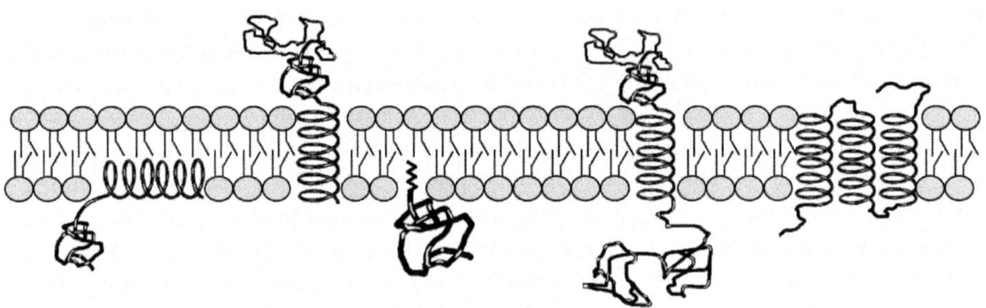

FIGURE 3.13

Membrane proteins can be exposed on one or both sides of the membrane, can cross the membrane once or several times, or can be anchored in the membrane via amino acids or other molecules (e.g., fatty acids).

Consider a mouse cell, which has the same phospholipids we discussed in Chapter 2. The cell also has transmembrane proteins, but some are unique to mouse cells. At the same time, let us also consider a human cell, which has the same phospholipids and many transmembrane proteins that are unique to the human cell. These two cells can be fused together. Specific proteins can be labeled using antibodies that carry a **fluorescent tag**—a molecule that will glow with a specific color when exposed to the right light conditions (this will be discussed in more detail in a later chapter). For this example, let us suppose certain mouse proteins are labeled with antibodies carrying a green tag, and certain human proteins are labeled using a red tag. If the hybrid cell were to be examined immediately after the fusion event, the green and red labels would be separate, with green on one side of the cell and red on the other. However, after only 40 minutes, the labels would appear to be randomly dispersed (Figure 3.14). This experiment demonstrates that membranes are dynamic, with components such as transmembrane proteins being able to travel about like people in a mall.

Box 3.1 The creation of fusion cells

Fusing cells together is a two-step process. First, the cells must be touching each other, and then an electrical current is delivered, which will cause a perturbation of the phospholipid bilayer. To get the cells to touch each other, one might use polyethylene glycol. The electrical pulse will be in the order of 1 to 20 kV/cm, delivered for approximately 1 to 20 μs ($1–20 \times 10^{-6}$ s). When such an electrical pulse is delivered to cells, pores will spontaneously form in the plasma membrane. (The delivery of a current to produce pores is also a technique used for gene delivery, which will be discussed later.) After the current is removed, these pores and other effects of the membrane disruption will resolve, but individual phospholipids from one cell membrane may flip into the phospholipid bilayer of the other. The result will be that where once there were two distinct phospholipid bilayers, there will now be one. When two cells are touching during the membrane perturbation, fusion cells may result.

3.4.2 Restriction of protein movement within the plasma membrane

Although individual components of the plasma membrane are often able to migrate around the entire cell, this is not always the case. There are several instances in which multiple membrane components must be held together as a unit. For some cells—such as those comprising the gut epithelium, cells of the gastric lining, or sperm cells—some proteins are displayed in only one region of the plasma

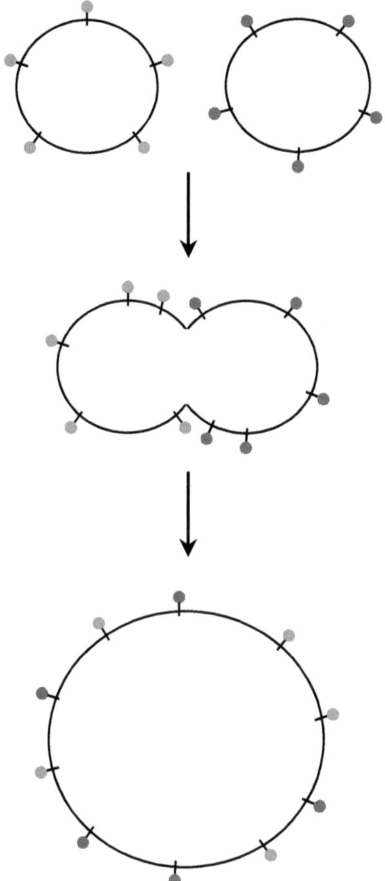

FIGURE 3.14

Proteins are able to migrate in the plasma membrane. Here, individual cells with labeled membrane proteins have been fused, with redistribution of labeled molecules occurring in a matter of minutes.

membrane because of the spatial location required for certain functions. Consider gut epithelium and the gastric lining, where absorption of nutrients occurs. There are proteins that serve as receptors and transporters for certain nutrients (such as glucose) that the body needs to pull out of the stomach or intestine for transport into the body. If these proteins were displayed on both sides of the cell, then between meals the body might pull glucose out of the bloodstream and dump it back into the alimentary tract, which would not be a good thing. There are different ways that such membrane proteins can be sequestered on one area of the plasma membrane: aggregation, tethering, and blocking via intercellular junctions (Figure 3.15).

Aggregation is where proteins of the cell membrane stick together. Tethering can occur by attaching the membrane protein to an extracellular protein of another cell, some other extracellular structure, or an internal structure such as the cytoskeleton. For an example of the final sequestering mechanism,

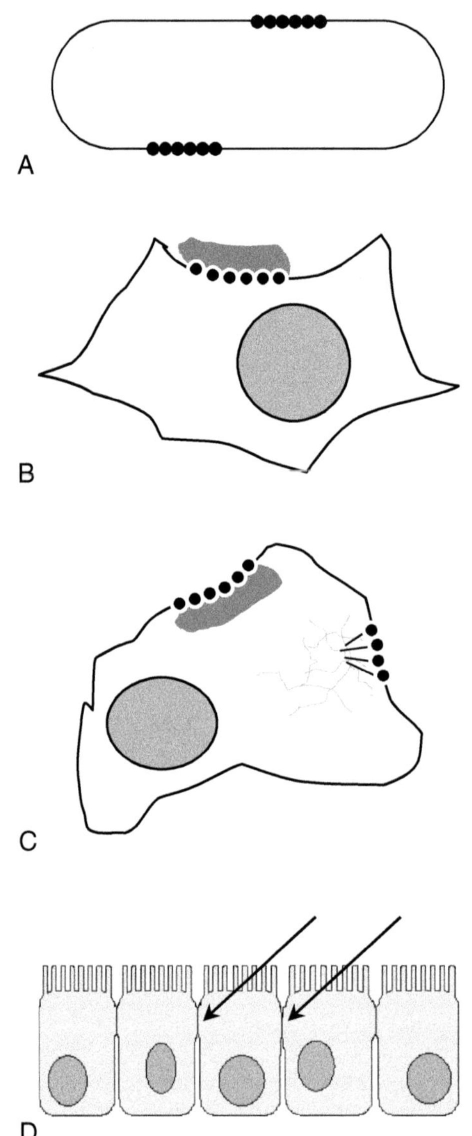

FIGURE 3.15

Not all constituents are free to diffuse throughout the membrane. Some membrane proteins are held together via aggregation (A), whereas others are held in place via tethering to macromolecules outside (B) or inside (C) the cell. In many instances, proteins are tethered to the cytoskeleton (C). Tight junctions (D, *arrows)* can prevent proteins in the outer layer of the plasma membrane from diffusing to the opposite face of the cell and vice versa.

consider the cuboidal cells of the alimentary tract: these cells have to stick together very tightly, and they do so via intercellular junctions. Intercellular junctions, while holding two cells together, can also serve to prevent the movement of membrane proteins from one side of the cell to the other.

So now we know our cell membranes contain phospholipids, cholesterol, and proteins. We have special proteins that can act as identifiers of self and others that serve as receptors. Membrane proteins are also used to help transport molecules into and out of the cytoplasm. Before we launch into that topic, however, let us discuss how proteins can be isolated for further study. The discussion is of interest at this point because the procedure involves molecules very similar to the phospholipids we have already studied.

3.4.3 Protein isolation often involves detergents

To isolate transmembrane proteins, the membranes must be dissolved. This can be accomplished with surface active agents (**surfactants**) such as detergents. Detergents have structures similar to fatty acids and soaps, in that there are a polar head group and a hydrophobic tail.

One might wonder why detergents are used instead of soaps to disrupt plasma membranes for the isolation of transmembrane proteins. To answer this question, let's begin with the difference in chemical structure between the two classes of molecules. Although both detergents and soaps are surfactants, they have different polar head groups. Soaps are sodium salts of fatty acids. One common soap is sodium stearate, an 18-carbon molecule made of a long hydrocarbon tail that terminates in a carboxylic acid (Figure 3.16, left). For comparison, consider the detergent sodium lauryl sulfate, which is a long hydrocarbon tail that terminates with a sulfate group (Figure 3.16, right). These different head groups make a practical difference. Soaps, by virtue of their carboxylate head groups, form precipitates in the presence of calcium or magnesium, two ions found in cells. If the soap precipitates out of solution, it will not be available to form the structures needed to isolate membrane proteins. Detergents, on the other hand, do not precipitate out of solution in the presence of these ions, so they are more effective for protein isolation.

Box 3.2 Aside: don't pour hand soap into the dishwasher!

Soaps and detergents are found in the home as well. The calcium in hard water is responsible for some of the soap scum that accumulates in bathtubs and sinks. When we wash our clothes, calcium inactivates soaps by removing them from the cleaning solution via precipitation. Such precipitates can harm appliances and are a main reason why dishwashers and washing machines require detergents instead of soaps as cleaning agents.

The amphipathic structure of surfactants allows them to form micelles in aqueous solutions. Consider what happens when clothes are washed. One might put water into a bucket, add a little detergent, then throw a dirty, greasy shirt into the bucket. The shirt is agitated in the water and detergent mixture, rinsed, and pulled out sparkling clean. Where did the oil go? The oil interacts with the detergents to form micelles. As shown in Figure 3.16, detergents have structures similar to those of the phospholipids in the plasma membrane, but detergents have only one fatty acid tail. The hydrophobic portion of a detergent does not have to be a long hydrocarbon tail, but the detergent does have to contain a polar head group and a hydrophobic portion. If several detergent molecules are added to water, they spontaneously form structures known as micelles (Figure 3.17). In this scenario, the hydrophobic

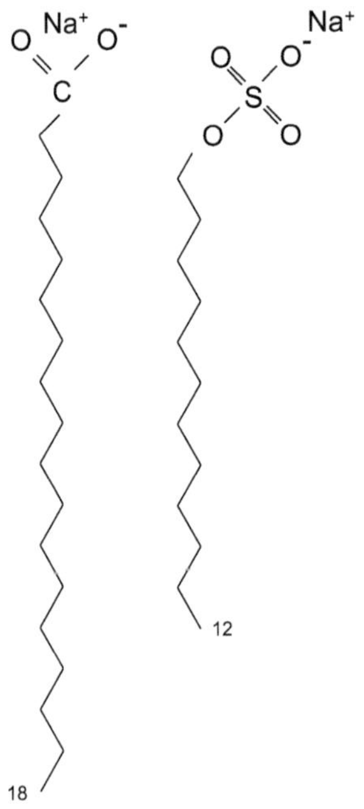

FIGURE 3.16

The structures of sodium stearate, a soap *(left)*, and sodium lauryl sulfate, a detergent *(right)*.

portion faces the interior of the micelle. Hydrophobic particles, such as oils from the skin, can also be carried in the micelle interior as **cargo molecules**. That is what happens in the washing machine when body oils are separated from articles of clothing.

The same principle is also used for certain types of drug delivery. Hydrophobic drugs are often delivered using surfactants. Micelles form, with the drugs being carried as cargo. Hydrophilic head groups make up the exteriors of the micelles, making them free to interact with the water content of blood, the cell cytoplasm, or any other aqueous environment.

As an example of how this is used in practice, think of a cancer cell as being a good cell that has gone bad. If the cancer cell were a foreign cell, your immune system would sense it as foreign and destroy it almost immediately. But a cancer cell is derived from one of your own cells, so it expresses all the normal proteins your cells typically express. Cells of the immune system might find it and examine it, but if the cancer cell is displaying a signal that identifies it as one of your own cells, it will not be destroyed. That is great if you are the cancer cell but not if you are the person with the tumor.

Now suppose you are a researcher trying to come up with a biotechnology to recognize cancer cells based on the proteins they display. You might wish to isolate these proteins for further study, and this is achieved by disrupting the phospholipid bilayers of the isolated cells. As with the example of throwing a greasy shirt into a bucket with some detergent, in this case you are essentially going to throw some detergent onto the cells. The detergent will disrupt the plasma membranes and allow you to isolate the embedded proteins, so that you can identify (through more testing) the unique proteins displayed by this class of cancer.

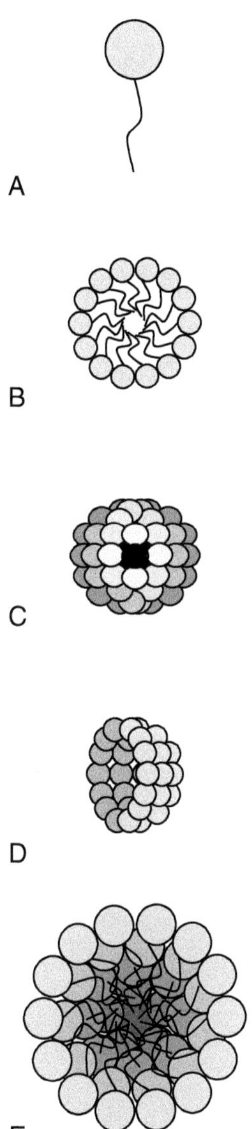

FIGURE 3.17

(A), General structure of a surfactant, such as a detergent. (B), Amphipathic molecules can come together to form a micelle. (C–E), Three-dimensional view of a spherical micelle from the outside (C), cut open to show its hollow interior (D), and turned to reveal a hydrophobic interior that can carry cargo molecules (E).

What detergent might you use for the above example? One common detergent for laboratory applications is sodium dodecyl sulfate, which is the chemical name for sodium lauryl sulfate, the detergent shown in Figure 3.16. The term "dodecyl" means 12, and here it refers to the hydrophobic tail, which has 12 carbons in it. The sulfate portion is an SO_4^-. Because of the charged head group, sodium dodecyl sulfate is classified as an **ionic detergent**.

Triton is another detergent used for cellular disruption and, being a detergent, has a polar head group and a hydrophobic tail group (Figure 3.18). However, Triton does not carry a charge, so it is known as a **nonionic detergent**. Nonionic detergents are often used in protein research because they typically do not **denature** (cause the unfolding of) proteins. They can surround membrane proteins embedded within membranes without changing their native conformations, so the proteins will retain their functions and folding patterns for molecular recognition.

To isolate individual membrane proteins according to size, one would use sodium dodecyl sulfate poly(acrylamide) gel electrophoresis (SDS-PAGE). The detergent is mixed with the cells and energy is added by vortexing, heating, or sonication of the mixture. Micelle-like structures are formed via the phospholipids and detergent molecules, and the membrane proteins will be unfolded. The SDS molecules interact not only with the plasma membrane (forming micelles) but also with membrane proteins. Recall that membrane proteins typically must have at least one hydrophobic section because they must either cross or be anchored in the hydrophobic region of the plasma membrane. Because of this, the hydrophobic tail of SDS can interact with that section of the protein as it also disrupts the phospholipid bilayer. At the same time, interactions between the hydrophobic tails of several SDS molecules and the protein denature the protein.

Because the SDS is an ionic detergent, the now-unfolded proteins can be separated with the aid of an electrical field. The negatively charged SDS molecules are pulled toward the anode, carrying the unfolded proteins with them. Note that it does not matter whether a given protein is large or small; the concentration of SDS per unit length of protein is going to be roughly the same regardless of protein size. Larger proteins will complex with more SDS molecules, but the number of SDS molecules per unit length will remain

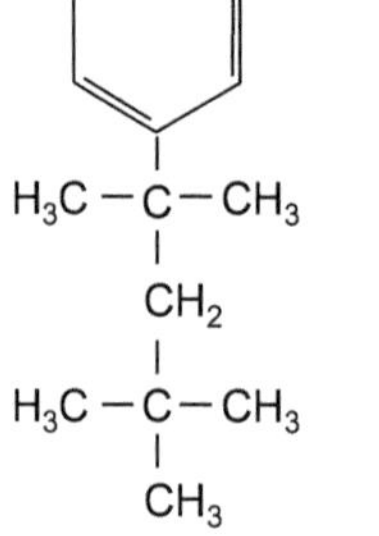

FIGURE 3.18

Triton X-100, a nonionic detergent.

roughly the same for large and small proteins alike. This implies that large and small proteins will be pulled with the same relative amount of force toward the anode. Separation is achieved because of the differences in physical size between the unfolded proteins. Large polypeptides have a more difficult time traversing the twists and turns of the poly(acrylamide) gel than smaller polypeptides, so smaller polypeptides migrate further in a given amount of time. This is the principle behind protein separation via SDS-PAGE.

Box 3.3 Counter-ions

In the laboratory, sodium dodecyl sulfate is used, as opposed to dodecyl sulfuric acid (which would have a hydrogen in place of the sodium). The problem with the acid is that when it is placed in water, it gives up the hydrogen and thereby lowers the pH of the surrounding solution. Rather than having a hydrogen atom associated with the dodecyl sulfate, a sodium counter-ion is used, thus forming the sodium salt. "Sodium dodecyl sulfate salt" is another name for the molecule, but its common name in the laboratory is SDS. When you look at the chemical names of many pharmaceuticals, you will see the counter-ion listed at either the beginning or the end of the name. Examples include atorvastatin calcium (Lipitor), alendronate sodium (Fosamax), and doxacurium chloride (Nuromax).

3.5 Summary

We have just taken a closer look at proteins and how the biotechnologist can think of them as folded, functional polypeptides. We have seen the formation of polypeptides in a chemical sense as a dehydration reaction between amino acids, and we have looked at amino acids in terms of what they have in common: N- and C- termini, a (typically chiral) central carbon, and the presence of a side chain. We have also seen that these side chains are what gives each specific amino acid its identity.

Protein folding is the result of interactions within a polypeptide itself and is the spontaneous result of hydrogen bonding, ionic interactions (via the side chains), and the hydrophobic effect. The three-dimensional structure of a protein is fundamentally important to its function. Alteration of the structure via hydrolysis or changes in conformation (perhaps via pH change or the binding of a ligand) can alter the function of a polypeptide, including whether it has a function at all.

Knowledge of protein structure is important to biotechnologists because the structure–function relationship is key to designing novel drugs, counteracting toxins, or coating surfaces with proteins for cell recruitment or attachment.

In the next chapter, we will look at where proteins come from, from the viewpoint of the cell. We will see that proteins arise from RNA and that the RNA arises from a DNA blueprint. We will start at the molecular level to understand just what nucleotides are and how they can be polymerized to form DNA or RNA. We will see how these polymerization processes are controlled and then move on to learn how to translate the genetic code into polypeptides. The biotechnologist should know about DNA because cellular function can be controlled via DNA manipulation. This powerful approach is being used in attempts to cure disease, restore function to damaged tissues, and create commercial products such as polymers and biofuels.

Related reading

Alberts, B., Johnson, A., Lewis, J., Raff, M., Roberts, K., Walter, P., 2008. Molecular Biology of the Cell, fifth ed. Garland Science, New York.

Benz, R., Beckers, F., Zimmermann, U., 1979. Reversible electrical breakdown of lipid bilayer membranes: a charge-pulse relaxation study. J. Membr. Biol. 48, 181–204.

Neumann, E., Rosenheck, K., 1972. Permeability changes induced by electric impulses in vesicular membranes. J. Membr. Biol. 10, 279–290.

Sugar, I.P., Förster, W., Neumann, E., 1987. Model of cell electrofusion. Membrane electroporation, pore coalescence and percolation. Biophys. Chem. 26, 321–335.

Teissie, J., Tsong, T.Y., 1981. Electric field induced transient pores in phospholipid bilayer vesicles. Biochemistry 20, 1548–1554.

Zimmermann, U., Pilwat, G., Riemann, F., 1974. Dielectric breakdown of cell membranes. Biophys. J. 14, 881–899.

Questions

1. Identify the level of structure (primary, secondary, super-secondary, tertiary, or quaternary) for the following:
   a. β-barrel
   b. interactions between polypeptides
   c. sequence of amino acids
   d. 3-D structure of polypeptide
   e. α-helix
   f. heterodimer (protein made of two different subunits)
   g. helix-loop-helix
   h. asp-leu-tyr.
2. a. Which is more likely to be found in a β turn, *cis* or *trans* proline? Why?
   b. Which is more likely to be found in a membrane phospholipid, a *cis* or *trans* double bond in an unsaturated fatty acid? Why?
3. Could histidine act in place of proline in a β-bend structure? Why or why not?
4. Identify the bonds or interactions that lend to the stability of secondary structures. Which portion of an amino acid is responsible for these interactions?
5. What is the difference between a polypeptide and a protein?
6. Why are glycine and proline usually involved in β-turns?
7. Why is the charge of a given protein not important when SDS-PAGE is used to separate proteins by size?
8. Why do we use detergents instead of soaps to dissolve membranes?
9. Why don't we use Triton X-100 for gel electrophoresis?
10. Compare and contrast ionic and nonionic detergents.
11. Does it make a difference for the SDS-PAGE analysis if a protein is denatured before beginning the procedure? Why or why not?
12. What are the general structures of soaps and detergents? How do the structural differences between the two affect their functionality?
13. If a small, neutral protein and a large, negatively charged protein were to be separated via an SDS-PAGE, which protein would be expected to move faster through the gel?

14. Below is the result of an SDS-PAGE experiment:

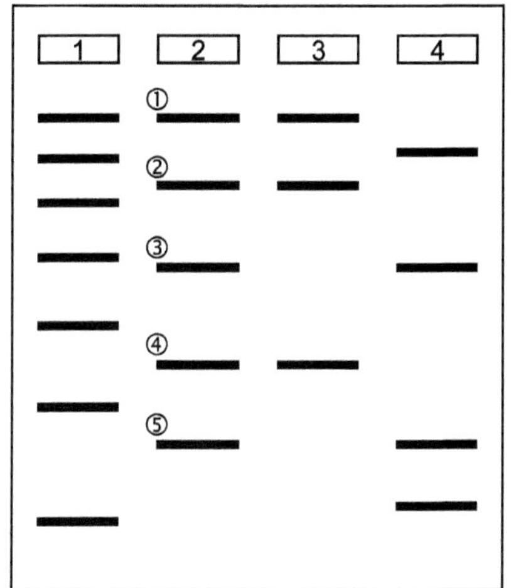

Lane 1: MW marker
Lane 2: Lyse then label
Lane 3: Label then lyse
Lane 4: Expose to proteinase K, wash, lyse, label

a. Do the bands in lane 3 represent intracellular, extracellular, or transmembrane proteins?
b. To which of the numbered bands in lane 2 does each band in lane 4 correspond?
c. Which is most likely to be the classification of protein ①: adhesion molecule, signaling protein, nuclear protein, phospholipid, or glycolytic enzyme? Explain your answer.

15. Of the three amino acids with positively charged R groups, which one would be least likely to aid in transporting a negatively charged molecule across a cell membrane? Explain your answer.
16. In the laboratory, when might one use an anionic detergent? When might one use a nonionic detergent?
17. What is the hydrophobic effect? Is it important at all?
18. When a surfactant such as a detergent is added to an aqueous solution, organized micelles form. However, this phenomenon is said to increase the entropy of the solution compared with having the surfactant molecules randomly scattered throughout the solution. Justify this apparent contradiction.
19. Given the following sequence of amino acids, which region will most likely be in the middle of the folded polypeptide?

Region:	I	II	III	IV
	Lys – Arg –	His – Ser – Cys – Thr –	Phe – Trp – Pro –	Glu – Asp – Gln

20. Would valine tend to be found on the interior or exterior of a cytosolic protein? What phenomenon drives the amino acid to this location? To what level of structure does this relate?
21. Consider a cellular transmembrane protein such as a receptor. Suppose that another molecule, a ligand, binds to the extracellular side of the protein. How could the receptor signal to the cell that a ligand has been bound, without having the ligand internalized?

22. Why is sodium, as opposed to a proton, used as a counterion with dodecyl sulfate when isolating polypeptides?

23. a. In the presence of SDS, both large and small proteins denature, and the large proteins have more SDS molecules associated with them. Despite having a greater number of negative charges, the larger proteins will migrate more slowly during PAGE. Why?

    b. Suppose the SDS–protein solution from part a were placed in the middle of a solution of electrophoresis buffer and a voltage applied. Which protein would migrate faster, the large or small? Why?

CHAPTER

Genes: the blueprints for proteins

4

Chapter outline

Proteins, proteins, proteins! We have spent so much time discussing them and with good reason: most of the bioactive molecules produced by a cell are proteins. Biotechnologists should be interested in proteins because they could be the final product of a process (e.g., recombinant insulin), or they could be a key enzyme used in producing the final product (e.g., the use of the enzyme zymase in the production of ethanol).

Cells produce their own proteins based on a set of plans—blueprints, as it were—that contain not only the sequence of amino acids that will produce the primary sequence of the protein but also instructions on how much of the protein to make and when. The sequence of events that goes into protein production is often referred to as the *central dogma* of molecular biology (Figure 4.1). In this scheme, DNA is used to produce an intermediate poly nucleic acid, RNA, via a process called transcription, and the

Biotechnology and its Applications. https://doi.org/10.1016/B978-0-12-817726-6.00004-6

FIGURE 4.1

The central dogma of molecular biology.

RNA is then converted into a protein sequence via a process known as translation. However, the entire journey from DNA to protein is far more intricate than this sequence suggests. Presenting the details and subtleties of the pathway is the aim of this chapter. Specifically, we will discuss the following topics:

- the phosphoribose backbone;
- nucleotide bases, nucleosides, and nucleotides;
- DNA as the genetic material;
- the double-stranded nature of genomic DNA;
- the semiconservative nature of DNA replication;
- the genetic code;
- genes;
- transcription;
- the production of mRNA; and
- translation.

4.1 Nucleotides and nucleic acids

4.1.1 The phosphoribose backbone

In Chapter 2 we learned about membranes and their compositions, which include not only phospholipids but also cholesterol, proteins, glycolipids, and glycoproteins. Membrane proteins serve as channels, pores, identifiers, or receptors. In this chapter we will take a detailed look at nucleic acids, which carry the codes for protein production and thereby play a central role in cellular growth, replication, function, and survival.

There exist two types of nucleic acids in the cell: ribonucleic acid (RNA) and deoxyribonucleic acid (DNA). The structure of RNA contains the sugar β-D-ribose (Figure 4.2A). In nucleic acids, the hydroxyl group (–OH, indicated by an arrow) on carbon 1 of the ribose is substituted by a nitrogenous base; we will get to that in a moment. Figure 4.2B illustrates the molecule phosphoribose. The only difference between ribose and phosphoribose is that a hydroxyl has been replaced by a phosphate on carbon 5. Phosphoribose could serve as a monomer in the formation of a poly(phosphoribose), the polymer that makes up the backbone of RNA. Polyphosphoribose is a succession of phosphoribose molecules that are attached between carbon 3 on one phosphoribose molecule and the phosphate group attached to carbon 5 on the other molecule (Figure 4.2C), Depending on the R groups, the molecule shown in Figure 4.2C might be the phosphoribose backbone for an RNA dinucleotide. If each of the ribose units were missing a hydroxyl on carbon 2, the molecule would be DNA (the "deoxy" part of its name refers to the loss of the oxygen atom from carbon 2, as seen in Figure 4.3A).

A closer look at the carbons of the ribose in a nucleotide reveals that carbon 1 holds the R group, carbon 2 determines whether the molecule is DNA or RNA, carbon 3 is used for polymerization, carbon 4 is part of the ribose ring structure, and carbon 5 is involved with polymerization just like carbon 3.

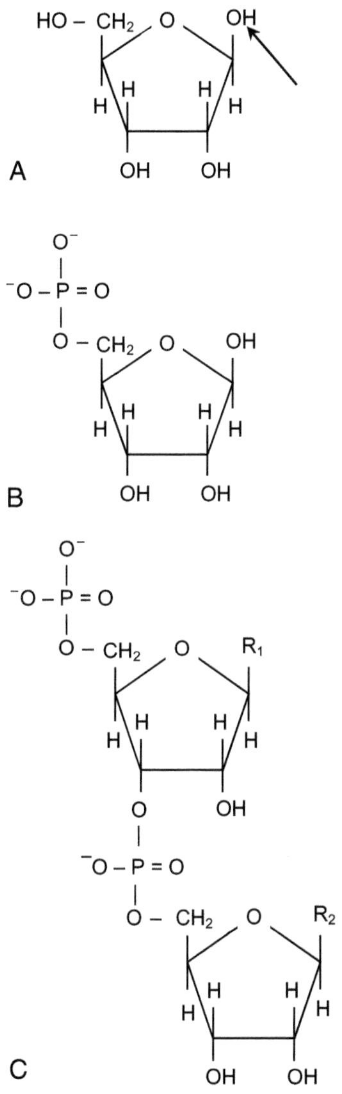

A

B

C

FIGURE 4.2

(A) Ribose, a sugar. The *arrow* indicates the hydroxyl attached to carbon 1. (B) Phosphoribose. (C) An RNA dinucleotide. Replacement of the carbon 2 hydroxyls with hydrogen atoms would make this a DNA dinucleotide.

Note that the phosphate group carries a negative charge. There will be one negative charge for every nucleotide in a DNA or RNA polymer (except for the one on the end with an unpolymerized phosphate, which typically carries two negative charges).

The repeated negative charges of DNA phosphates are key for gene delivery techniques that utilize synthetic delivery vectors. Most of the lipids used for liposome-mediated gene delivery are cationic, as are the polymers used for gene delivery; even electroporation works, in part, because of the numerous negative charges on the delivered DNA (DNA flows in the same direction as the electrons in an applied current because of the negative charges on the phosphates in the polynucleotides).

A linear DNA or RNA polymer has two ends that are chemically distinct. On one end there is an exposed phosphate group because it is only attached to one (deoxy) ribose. The other end has an unpolymerized hydroxyl on carbon 3 of the terminal (deoxy)ribose. These two distinct ends give directionality to the polynucleotide and give rise to the terms 5′ phosphate and 3′ hydroxyl, or the 5′ and 3′ termini of a linear DNA or RNA molecule.

Box 4.1 Ethers, esters, and phosphodiesters

An ether is two alkyl or aryl groups connected by an oxygen atom. For example, $H_3C\text{-}CH_2\text{-}O\text{-}CH_2\text{-}CH_3$ is diethyl ether. An ester involves two oxygen atoms, one that links two carbons in the same fashion as in an ether and one that serves as a carbonyl oxygen on one of the two carbons attached to the linking oxygen. Another way to think of an ester is as a carboxylic acid with the hydroxyl hydrogen replaced by a hydrocarbon group.

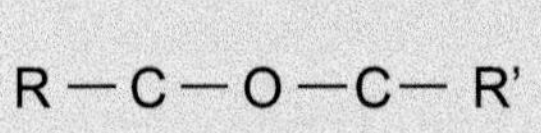

R—C—O—C—R'

Ether

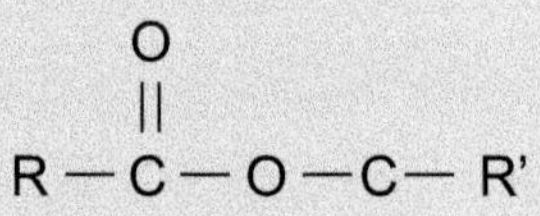

O
||
R—C—O—C—R'

Ester

A phosphodiester bond involves two ester-like configurations *(circled)*, with one of the carbons being replaced by a phosphorus atom. When nucleotides are polymerized, phosphodiester linkages are formed.

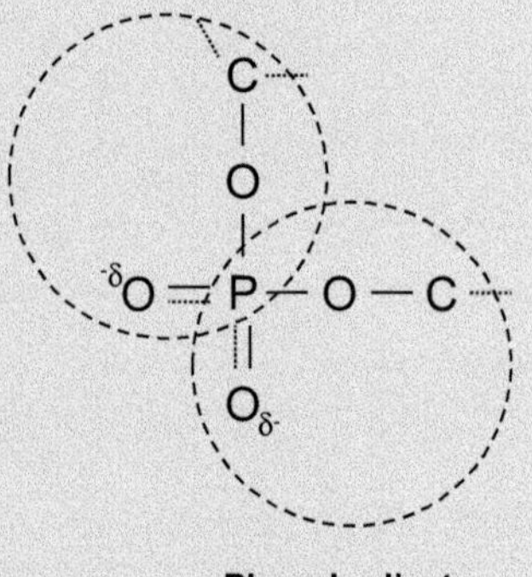

C
|
O
|
$^{-\delta}$O=P—O—C
||
$O_{\delta-}$

Phosphodiester

The polymerization of two nucleotides (the joining of two DNA molecules to make a single dinucleotide) requires +25 kJ of free energy per mole. The same reaction could proceed in the reverse direction to *yield* 25 kJ of free energy per mole (the change in Gibbs free energy is -25 kJ/mol). Because the reverse direction yields free energy, it is the favored direction of the reaction. In other words, hydrolysis of polynucleotides (the breaking of polynucleotides via water) is favored. If this is the case, then how can we even be alive? How can our cells synthesize DNA if the reaction is energetically unfavorable? The answer is that we do not polymerize using nucleotides with only one phosphate attached; rather, we use nucleotides with three phosphates attached and couple the cleavage of two of those phosphates with the formation of a phosphodiester bond. Let NMP stand for "nucleotide monophosphate" and let NTP stand for

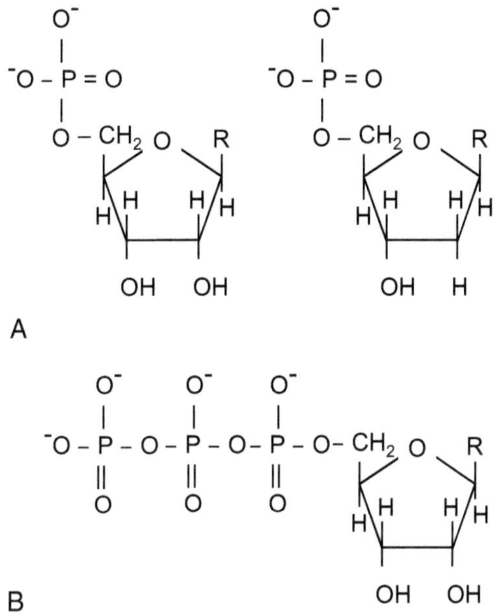

FIGURE 4.3

(A) *Left*: a nucleotide, also known as a nucleotide monophosphate (NMP). Right: a deoxynucleotide, or dNMP. (B) A nucleotide triphosphate (NTP).

"nucleotide triphosphate" (Figure 4.3). Cells polymerize RNA using NTPs and polymerize DNA using deoxy-NTPs, or dNTPs. The three phosphates in tandem in NTP make it a high-energy molecule: when one or two of the phosphates are cleaved, a relatively large amount of energy is released. The reaction ATP → AMP + PP_i (used as a model for NTP → NMP + PP_i) yields around 43.1 to 45.6 kJ/mol of energy under standardized conditions. (An additional 19.2–21.3 kJ/mol potentially could be produced by subsequent hydrolysis of the PP_i into two molecules of organic phosphate.) When using NTPs for nucleotide polymerization, the formation of a new phosphodiester bond is coupled with NTP hydrolysis, yielding an energy balance of $[^{+}25 + (-45.6)] = -20.6$ kJ/mol, meaning 20.6 kJ/mol is released (Figure 4.4).

4.1.2 Nucleotide bases, nucleosides, and nucleotides

The R group attached to carbon 1 of (deoxy)ribose could be a hydroxyl, although it seldom is. When a nucleotide is constructed from scratch by the cell, the R group at some point will be a pyrophosphoryl ($–O–PO_3^-–PO_3^{2-}$). However, for nucleotides, the R group will be one of the DNA or RNA bases: guanine, adenine, cytosine, thymine, or uracil. The structures of these nitrogenous bases are given in Figure 4.5. DNA normally utilizes guanine, adenine, cytosine, and thymine (G, A, C, and T). The bases are the same in RNA, except that thymine is replaced by uracil (U). Note that there are two types of ring structures for these bases: a bicyclic structure containing a five-member and a six-member ring, and a monocyclic structure with a six-member ring. The bicyclic structure is characteristic of the bases known as purines (guanine and adenine), whereas the monocyclic structure is characteristic of the pyrimidines cytosine, thymine, and uracil.

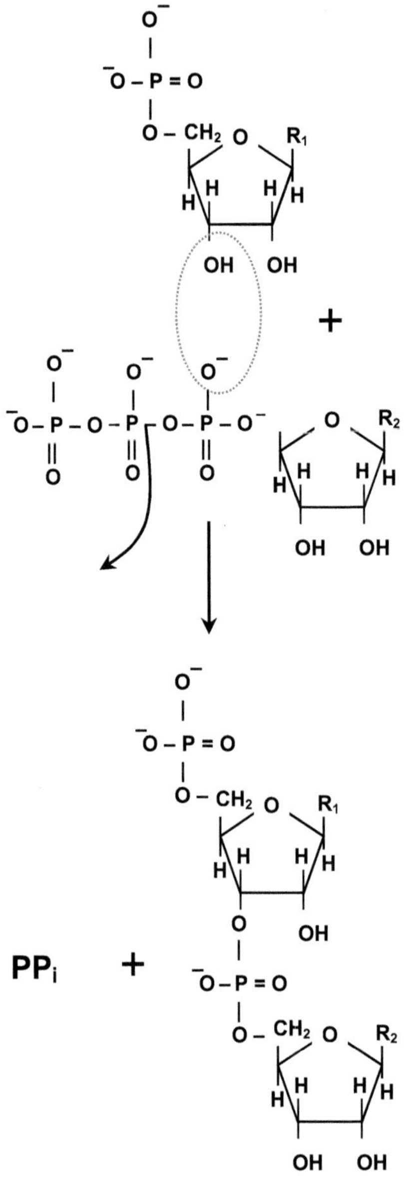

FIGURE 4.4

Nucleotide triphosphates (NTPs) are used in growing a poly(nucleotide) chain. Whereas the reaction of a 3′ hydroxyl with a 5′ phosphate requires energy (25 kJ/mol), the cleavage of the high-energy phosphate bond in the NTP to produce a pyrophosphate group (shown as PP_i) yields 43.1–45.6 kJ/mol. The two reactions are coupled to produce a net energy yield of approximately 20.6 kJ/mol, making the polymerization reaction energetically favorable.

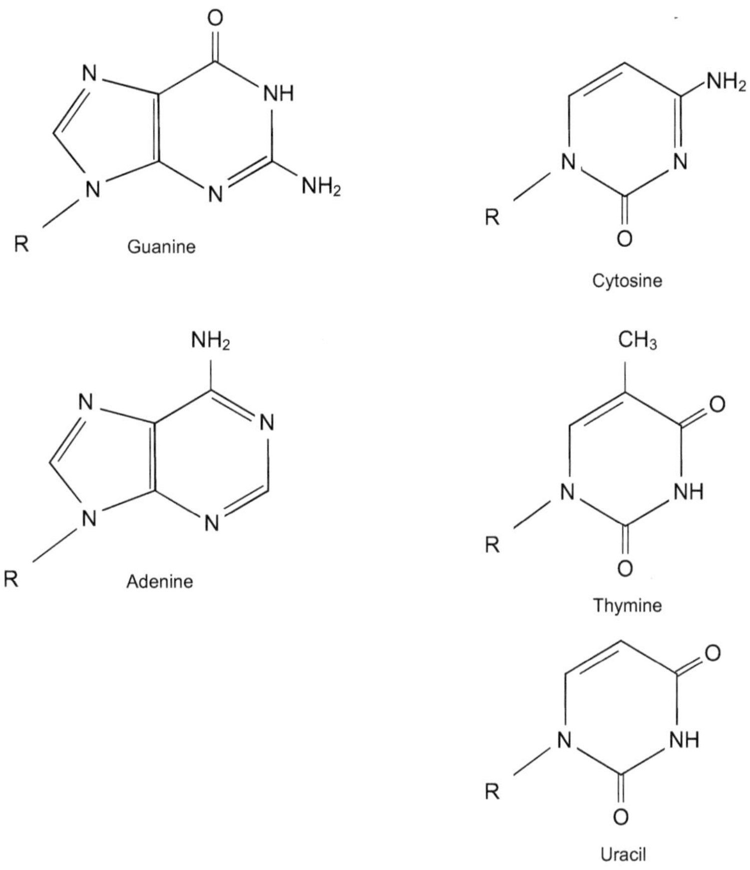

FIGURE 4.5

Structures of the nucleotide bases. Purines are on the *left,* and pyrimidines on the *right.* The makeup of the R group will determine whether the molecule is a base, a nucleoside, or a nucleotide (see also Table 4.1).

Again referring to Figure 4.5, note the R group attached to each base. When the R group is a hydrogen atom, the molecule is a purine or pyrimidine base. If the R group is a ribose, then the molecule is a **nucleoside**. If the R group is a phosphoribose, then the molecule is a **nucleotide**. The names of the most common of these molecules are given in Table 4.1.

We have seen that RNA is polymerized using NTPs. This means that for DNA, polymerization is carried out using dGTP, dATP, dCTP, and dTTP (the lower-case d indicates "deoxy.") For RNA, UTP is used instead of TTP. Note that the ATP used in RNA polymerization is the same molecule we saw in earlier chapters as an energy donor for other reactions and processes, such as active transport. It is also possible for the other NTPs to serve as energy donors in specific cases. Knowledge of the structures of the five nitrogenous bases is important for understanding DNA base pairing, as well as knowing why certain transcription factors are specific for certain sequences. Note that Figure 4.5 is drawn so that the bottom-left nitrogen of each of the bases will be the atom that attaches to carbon 1 of the ribose sugar.

Table 4.1 Names of purine and pyrimidine bases, nucleosides, and nucleotides

Base [Base]	Nucleoside [(deoxy)Ribose–Base]	Nucleotide [Phosphate–(deoxy)Ribose–Base]	
Guanine	Guanosine Deoxyguanosine	Guanylate Deoxyguanylate	(GMP) (dGMP)
Adenine	Adenosine Deoxyadenosine	Adenylate Deoxyadenylate	(AMP) (dAMP)
Cytosine	Cytidine Deoxycytidine	Cytidylate Deoxycytidylate	(CMP) (dCMP)
Thymine	Deoxythymidine	Deoxythymidylate	(dTMP)
Uracil	Uridine	Uridylate	(UMP)

When a DNA sequence is written out, it is always written in the 5′→3′ direction, meaning the nucleotide with the unpolymerized phosphate on carbon 5 is on the left and the nucleotide with the unpolymerized hydroxyl on carbon 3 is on the right. The structure of a trideoxynucleotide is shown in Figure 4.6. For simplicity, however, DNA sequences are typically written out with just the one-letter abbreviations of the bases. The three-base sequence shown in the figure can therefore be written as CTG. From this nomenclature, not only is the directionality of the sequence implied, but we can also tell whether the molecule is DNA or RNA based on the presence of thymine versus uracil.

4.1.3 DNA is the genetic material

In the early days of molecular biology, it was not readily apparent what the genetic material was made of; in fact, for many years, proteins were thought to contain the genetic blueprints for heredity. In 1952, however, Alfred Hershey and his research assistant Martha Chase published the results of the experiments that proved DNA serves as the genetic material.

Hershey and Chase capitalized on the fact that DNA and proteins can be distinguished by the presence of distinct atoms: a phosphorus atom can be found in DNA but not in polypeptides. Conversely, sulfur atoms can be found in most proteins (by virtue of the amino acids cysteine and methionine) but not in DNA. Taking advantage of this difference, the team used the radioactive tracers ^{32}P and ^{35}S to distinctly label DNA and proteins (Figure 4.7). Viruses, which have proteins in their coats and DNA in their payloads, were labeled with ^{32}P or ^{35}S. *Escherichia coli* cultures were exposed to the labeled viruses, which bound to the bacterial cells and injected their payloads. A blender was used to detach the viruses from the cells, and the researchers found that the shear force from the blender caused ~75% of the ^{35}S but only ~15% of the ^{32}P to be released into the (extracellular) solution; most of the ^{32}P had been transferred into the cells. The viruses were injecting DNA, not proteins, into the cells.

After a period of incubation, new viruses were formed in the cells and the viral progeny were released. About 30% of the original parental ^{32}P, but less than 1% of the ^{35}S, was recovered in the progeny. We now interpret this to mean that the transfer of DNA was responsible for the viral coats in the progeny: while the newly formed viruses all had viral coats, the coat proteins were not constructed from the labeled proteins of the parent viruses. Instead, the DNA delivered by the parent viruses had served as a blueprint that led to the formation of protein products. DNA had been shown to be the genetic material.

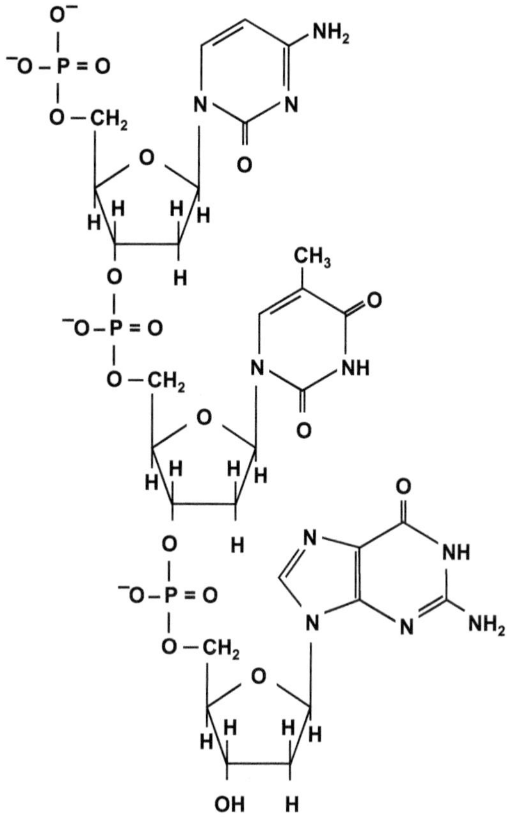

FIGURE 4.6

The trideoxynucleotide CTG. Note that the abbreviations for DNA sequences can be written in the same manner as for RNA (e.g., CTG instead of dCdTdG), especially for longer sequences. Also note the phosphate on the 5′ terminus of the structure and that the DNA sequence CTG is written with the 5′ terminus on the left.

4.1.4 Genomic DNA is double stranded

After it was established that DNA was the genetic material, the question arose as to how it could ever be replicated in such a way that a mother could pass on a copy of her genes to her offspring; there must be some way of conserving the information from one generation to the next. In the 1950s a scientist named Erwin Chargaff reported finding that the composition of DNA varied from one species to another. However, he also noted that although the percentage of DNA nucleotides containing adenine might vary between species, the percentage of DNA bases containing thymine was always roughly equal to the percentage of bases containing adenine. The same was true for guanine versus cytosine. These relations, now referred to as **Chargaff's rules**, were early indications of the double-stranded nature of DNA and the regular pairing of bases between the two strands. Combined with the x-ray crystallographic data of Rosalind Franklin, which indicated that DNA has a helical structure, Chargaff's

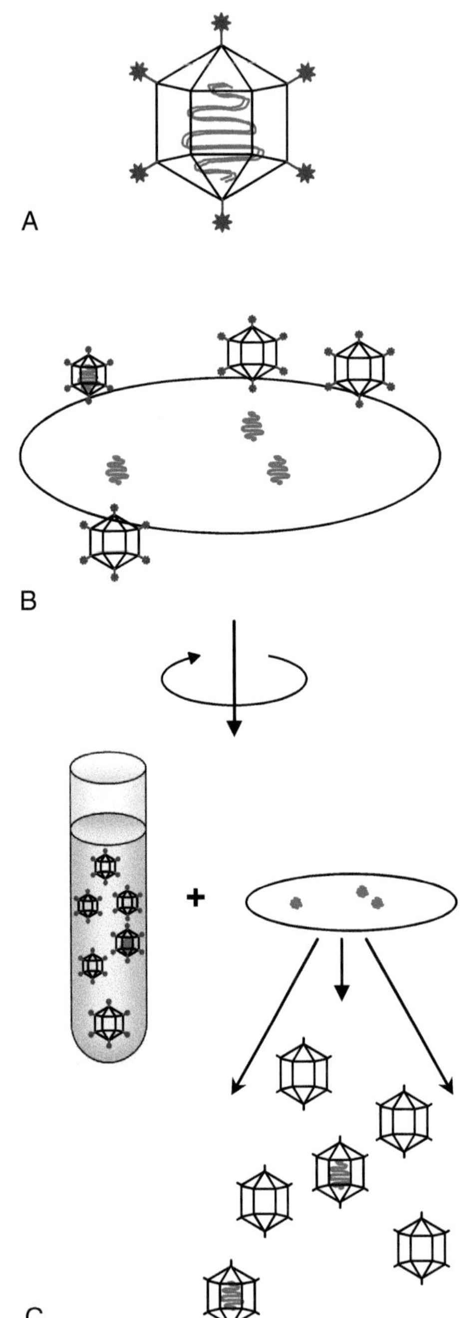

FIGURE 4.7

Summary of the Hershey Chase experiments. (A) Hypothetical virus with ^{32}P *(red)*-labeled DNA and ^{35}S *(blue)*-labeled proteins. Double labeling is shown, although the experiments used the labels separately. (B) When exposed to cells, labeled viruses adhere via their coat proteins and inject their DNA. A blender was used to knock viral particles off the cells to assess the locations of isotopes. Labeled proteins remained outside the cells, with viral ghosts. DNA, however, was found inside the cells. (C) Infected cells produced viral progeny containing some labeled DNA but no labeled proteins.

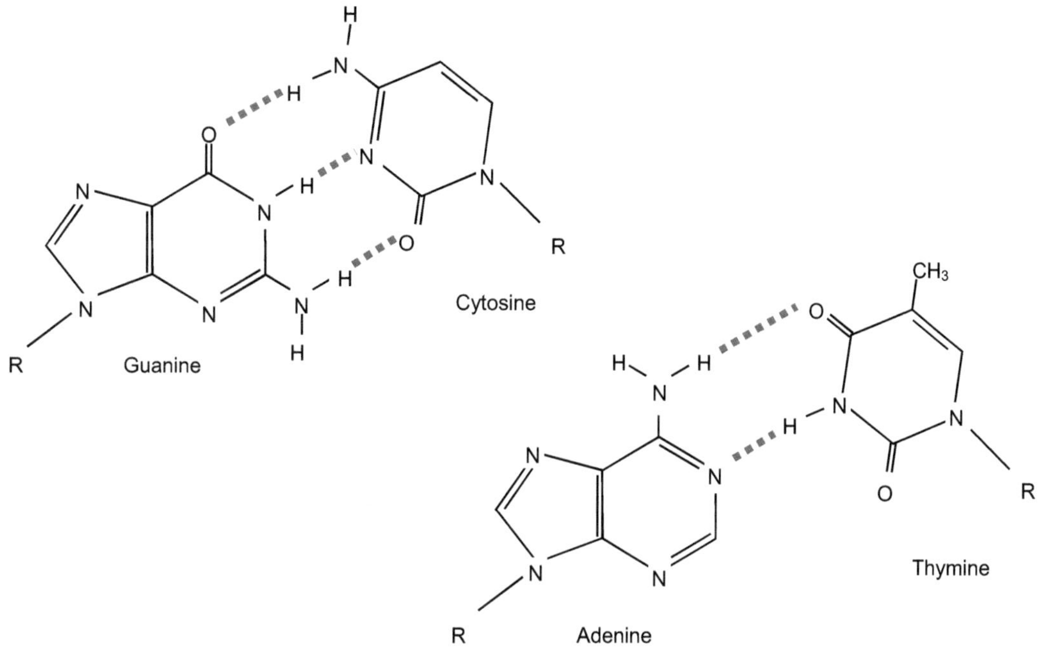

FIGURE 4.8

Hydrogen bonding, as shown by the *dotted lines*, can occur between the DNA base pairs.

data led to the notion of a double-stranded DNA helix with complementary strands due to A-T and G-C pairing.

Using the structures presented in Figure 4.8, we can see that hydrogen bonding can occur simultaneously at three sites within guanine–cytosine pairs. Similarly, two hydrogen bonds can form in A-T pairs. Because the bases in double–stranded DNA generally adhere to these pairing rules, knowing the sequence of one strand in a double-stranded polynucleotide provides enough information to deduce the sequence of the complementary strand. Thus, although a sequence could be written as ATCGT, it is understood that this written sequence actually refers to:

5′–ATCGT–3′

3′–TAGCA–5′

Consider two 100-base-pair stretches of DNA, one composed solely of G-C base pairs and the other consisting of only A-T base pairs. Because each G-C pair is held together by three hydrogen bonds, the bases adhere to one another more stably than those of the A-T base pairs, which only have two hydrogen bonds. Extending this idea to our two hypothetical DNA fragments, we could guess would be harder to separate the double-stranded DNA (dsDNA) fragment made completely of G-C base pairs into two single-stranded DNA (ssDNA) fragments; because of the 100 extra hydrogen bonds, it would take a greater amount of heat energy to melt the poly(G-C) strand into its single-stranded components. The temperature at which a dsDNA fragment is melted into two ssDNA fragments is known as the melting temperature (T_m) and is

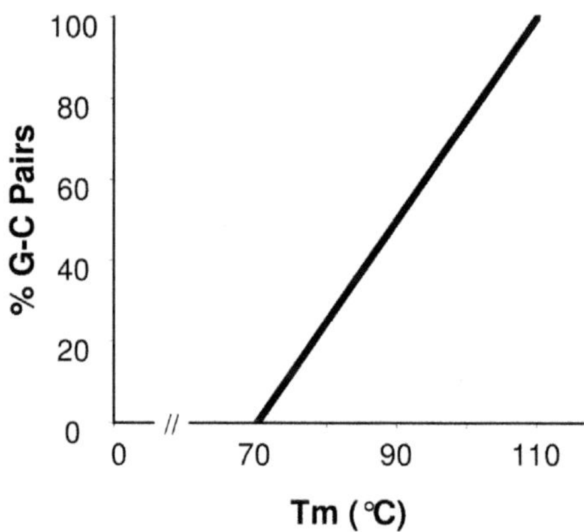

FIGURE 4.9

Because of increased hydrogen bonding, G-C base pairs require more energy to separate than do A-T base pairs. As a result, the greater the percentage of G-C pairs in a piece of dsDNA, the higher the melting temperature (T_m).

directly dependent on the base composition of the dsDNA in question. Figure 4.9 illustrates this correlation. If we took a solution of several identical dsDNA fragments, then added heat to the system and monitored when they separated into single-stranded fragments, we would be able to determine their (G+C) content. Notice that the percentage (G+C) would be the same for the dsDNA and each separate ssDNA fragment.

4.1.5 DNA replication is semiconservative

The fact that DNA replication takes place in a semiconservative fashion was shown in 1957 by equilibrium sedimentation experiments performed by Matthew Meselson and Franklin Stahl. Suppose we have a bacterial culture grown in a controlled environment. The bacteria, of course, contain DNA. Before a given bacterium can divide into two daughter cells, new DNA must be manufactured, and an entire genome must be constructed. The materials for this DNA will ultimately come from the cells' environment; in the laboratory, this will be growth medium. If we were to grow the bacterial culture for multiple generations in an environment that contained only the heavier (^{15}N) form of nitrogen atoms, then (through subculturing techniques) we could obtain a culture in which virtually every nitrogen atom of the bacterial DNA was ^{15}N. If we were to extract the DNA from these cells and centrifuge it in a cesium chloride gradient, the genomic DNA would migrate a distance of x through the gradient (Figure 4.10). Because of the ^{15}N, the DNA from these bacteria would have a higher density than the DNA from a bacterial culture grown in normal (^{14}N-containing) medium, which would migrate a shorter distance (say, z) under the same conditions.

Next, suppose that some of the bacteria grown in the heavy medium are transferred to an environment containing normal (^{14}N-containing) medium and allowed to proliferate for one generation. The results of DNA extraction and centrifugation in the CsCl gradient would produce a band at y, a distance somewhere between x and z. Even more interesting, if this newly transferred bacterial culture were allowed to proliferate for one additional generation, then two bands would be seen upon centrifugation. The bands would correspond to the distances y and z (Figure 4.10).

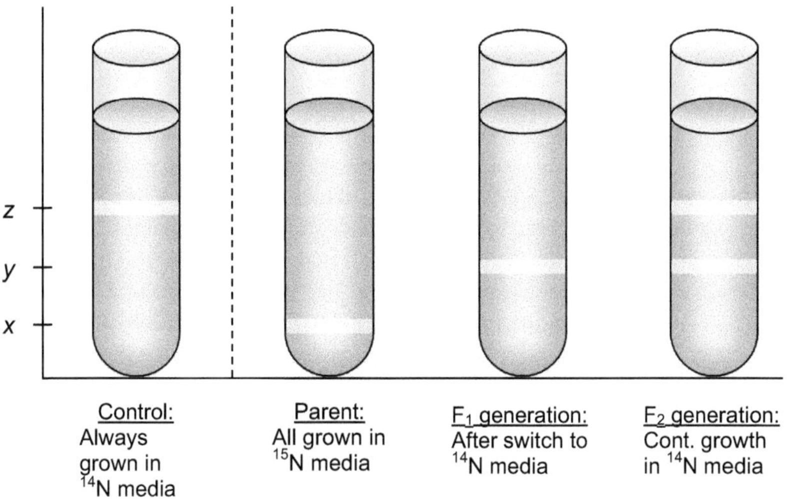

FIGURE 4.10

Cells grown in ^{15}N-containing media have DNA of a higher density than untreated controls, as seen following centrifugation in a CsCl gradient. After moving the culture to ^{14}N-containing media, the first generation of daughter cells has genomes that migrate a lesser distance (*y*) than did the genomes from the parent culture (*x*). The second generation of cells grown in the ^{14}N-containing media has genomes that migrate upon centrifugation either the same distance as the F_1 generation or the same distance as untreated controls (*z*).

The reasoning offered by Meselson and Stahl to explain the above results was that during DNA replication, the parental dsDNA strand was separated into two single-stranded templates, and a single strand of newly synthesized DNA was created to match up with each of them. Each daughter cell would then contain dsDNA composed of one parental strand and one newly synthesized strand that contained ^{14}N. This would explain the single, hybrid band of DNA that migrated the intermediate distance of *y* in the above centrifugation experiments. For the second generation of replication, each hybrid strand would separate in the same fashion into two ssDNA templates. This time, however, one of the single-stranded templates would contain the heavy nitrogen, and the other template would contain ^{14}N. New DNA strands would be created to pair with the template strands, yielding two daughter dsDNAs composed of ^{15}N-^{14}N and ^{14}N-^{14}N. Half of the parent DNA would be conserved in each of the daughter cells, hence the term **semiconservative replication**.

4.2 From genes to proteins

4.2.1 Introduction to the genetic code

The act of **transcription**, whereby RNA polymerase traverses genomic DNA to produce an RNA transcript that will be further processed to yield **messenger RNA (mRNA)**, proceeds in the 5′→3′ direction, and the mRNA is produced in the same direction. This may at first seem contradictory, until one considers that the genomic DNA is double-stranded. We may write a code like ATGATGATGTGA, but recall that such a code refers to:

5′– ATGATGATGTGA –3′

3′– TACTACTACACT –5′.

The strand that is written on top is referred to as the **coding strand**; think of it as the strand you would talk about if you were discussing a gene. The other strand, written in the 3′→5′ direction above, is called the **template strand**. RNA polymerase will bind both strands and they will be separated. The template (bottom) strand will be used by the enzyme to create a primary RNA transcript, and the enzyme will move from the 5′ end to the 3′ end (which is the 3′→5′ direction on the template strand). We refer to the movement of the enzyme as being in the 5′→3′ direction because that is the direction the enzyme is moving in terms of the coding strand and it is the direction in which the enzyme makes an RNA sequence. The new RNA sequence will be complementary to the template strand and identical to the coding strand (with the exception that uracils are used in place of thymines). The messenger RNA sequence corresponding to the dsDNA above would thus be 5′ -AUGAUGAUGUGA -3′.

Transcription takes place in the nucleus of eukaryotic cells (and in the cytoplasm of prokaryotic cells) because that is where the genomic DNA is. **Translation**, or the conversion of an mRNA sequence into a polypeptide sequence, takes place in the cytoplasm. In eukaryotes, this necessitates the transport of the mRNA from the nucleus to the cytoplasm through nuclear pores. Translation is then carried out by molecular machines called **ribosomes**.

When a ribosome processes a polynucleotide, it handles the bases in groups of three, referred to as **codons**. The mRNA sequence produced in our example above can now be thought of as AUG AUG AUG UGA. Refer to Figure 4.11 to determine the amino acids that are encoded for a given codon sequence. Polypeptides are produced beginning with the N-terminus and ending with the C-terminus, or the N→C direction.

4.2.1.1 Degeneracy and wobble

Refer again to the genetic code in Figure 4.11 and note that, for example, valine is pretty easy to decode after the first two bases have been located: GUU, GUA, GUC, and GUG all code for valine. The fact that the genetic code often has more than one codon for a particular amino acid is termed **degeneracy**. Also notice that when an amino acid is represented by more than one code—which is the case for every amino acid except methionine and tryptophan—the first two bases are identical if there are four or fewer redundant codes for the given amino acid. This has been hypothesized to be due to a condition known as "wobble," which is a straightforward concept, but to appreciate it we must first discuss ribosomes and their interactions with **transfer RNA (tRNA)**.

4.2.1.1.1 Ribosomes and translation

Ribosomes attach to messenger RNA and travel along it in the 5′→3′ direction, interpreting the codons and building a primary amino acid sequence with the help of tRNAs. A tRNA molecule is itself a polynucleotide, an RNA molecule that is typically 73 to 93 nucleotides in length. The 3′ end of the tRNA molecule can be covalently bonded to a specific amino acid. A tRNA molecule pairs with a codon via three bases in the tRNA structure known as the **anticodon** (Figure 4.12). If the mRNA codon is CUG, the tRNA used in translation of this codon will have the anticodon CAG (verify this for yourself, remembering that the sequence of the anticodon is written in the 5′→3′ direction).

The large ribosomal subunit has three places that can bind tRNA: the A site, the P site, and the E site. The A site typically holds an aminoacyl-tRNA, meaning a tRNA carrying an amino acid. The P site

First	Second: U		C		A		G		Third
U	UUU	Phe	UCU	Ser	UAU	Tyr	UGU	Cys	U
U	UUC	Phe	UCC	Ser	UAC	Tyr	UGC	Cys	C
U	UUA	Leu	UCA	Ser	UAA	Stop	UGA	Stop	A
U	UUG	Leu	UCG	Ser	UAG	Stop	UGG	Trp	G
C	CUU	Leu	CCU	Pro	CAU	His	CGU	Arg	U
C	CUC	Leu	CCC	Pro	CAC	His	CGC	Arg	C
C	CUA	Leu	CCA	Pro	CAA	Gln	CGA	Arg	A
C	CUG	Leu	CCG	Pro	CAG	Gln	CGG	Arg	G
A	AUU	Ile	ACU	Thr	AAU	Asn	AGU	Ser	U
A	AUC	Ile	ACC	Thr	AAC	Asn	AGC	Ser	C
A	AUA	Ile	ACA	Thr	AAA	Lys	AGA	Arg	A
A	AUG	Met (Start)	ACG	Thr	AAG	Lys	AGG	Arg	G
G	GUU	Val	GCU	Ala	GAU	Asp	GGU	Gly	U
G	GUC	Val	GCC	Ala	GAC	Asp	GGC	Gly	C
G	GUA	Val	GCA	Ala	GAA	Glu	GGA	Gly	A
G	GUG	Val	GCG	Ala	GAG	Glu	GGG	Gly	G

FIGURE 4.11

The genetic code. To read this chart, begin with the left-hand column to find the first base of your codon. This will determine the four rows of interest. Next, use the column headings to find the second base of the codon. This will determine the column of interest. The intersection of the four rows and column will yield four entries, a set that can be narrowed to a single entry by using the base indicated on the right side of the chart that corresponds to the third position in your codon.
Degeneracy is shown by colors: amino acids with two codes are highlighted in *orange* or *pink*, the amino acid Ile has three codes highlighted in *purple*, amino acids with four codes are highlighted in *yellow*, and amino acids with six codes are highlighted in *light green* or *blue*. The three stop codons are highlighted in *red*. The single codons of Trp and Met (which is also the Start codon) have a *white background*.

typically holds the peptidyl-tRNA, meaning the tRNA that is holding the growing peptide chain. The E site can be thought of as a temporary holding site for empty tRNAs (Figure 4.13).

Translation of mRNA via ribosomes can be thought of as a three-step process. Figure 4.14 (top) shows a ribosome having the P and A sites already filled with tRNAs linked to the growing polypeptide or a single amino acid, respectively. The first step of the process has an aminoacyl-tRNA associating with the ribosomal A site. The ribosome may process several different tRNAs until a proper match is found between the anticodon of the tRNA and the mRNA codon lined up with the A site. Whether the proper tRNA is held in the A site is determined via base pairing (through hydrogen bonding) between the anticodon and the corresponding mRNA codon. In step 2 of the process (Figure 4.14, middle), the

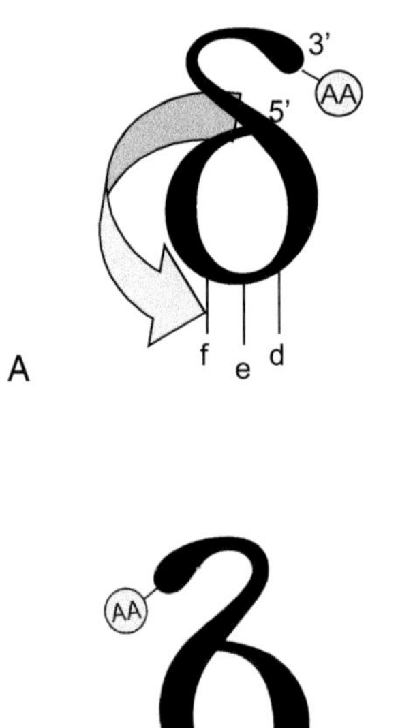

FIGURE 4.12

(A) The structure of a tRNA molecule is similar to the shape of the lower-case Greek letter delta. Here, an amino acid specific to the particular tRNA is attached to the 3′ terminus. The anticodon, which will pair with an mRNA codon, is shown at the bottom of the panel. The arrow denotes the 5′→3′ direction. (B) The same tRNA interacting with an mRNA molecule (ribosome not shown). This panel illustrates the direction of the tRNA (3′ to 5′) relative to the mRNA molecule. The letters in the mRNA and anticodon represent a fictitious base-pairing system where A pairs with a, B pairs with b, and so on.

growing polypeptide chain is cleaved at its C-terminus from the tRNA, then moved over to the α-amino group of the amino acid on the adjacent tRNA. A new peptide bond is formed to make the growing chain one residue longer. Note that due to a conformation change in the ribosome, the growing peptide chain remains associated with the P site in the ribosome, even though the chain is now held by a different tRNA. In step 3, the ribosome regains its original conformation but is now three nucleotides closer to the 3′ terminus of the mRNA (Figure 4.14, bottom). The now-empty tRNA is released from the E site in this step, rendering the ribosome ready to start the process again. Note that while the ribosome is progressing in the 5′→3′ direction, the polypeptide chain is growing in the N→C direction.

4.2.1.1.2 Back to wobble

The first two bases of the codon/anticodon pair must match up very well, but there is a little bit of play in the tRNA as it is held by the ribosome, so it can physically wobble into and out of contact with the third base of the codon. This implies that the third base is perhaps not as important as the first two in

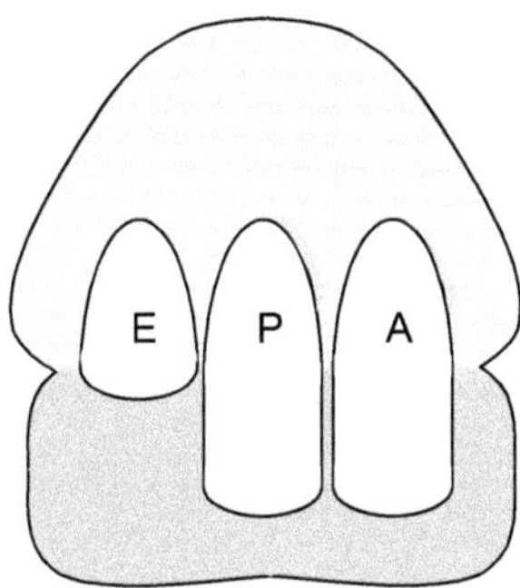

FIGURE 4.13

Schematic of the ribosome, with the large subunit shown in *light gray* and the small subunit shown in *darker gray*. The three binding sites for tRNA are shown by the letters E (empty tRNA), P (peptidyl-tRNA), and A (aminoacyl-tRNA).

terms of coding. Refer again to Figure 4.11 and note that GUU, GUC, GUA, and GUG all code for the same amino acid, valine. The **wobble effect** helps to explain this degeneracy of the genetic code.

4.2.1.2 Mutations and their effect on translation

Changing the third base in the codon CUG to an A (i.e., CUG→CUA) would be a mutation. However, it would be a **silent mutation** because the base change would not have any effect on the identity of the amino acid added to the growing polypeptide; the final polypeptide would be unaffected. A more serious type of mutation is the **frameshift mutation**. Such a mutation involves a shift in the groupings that define codons. For instance, while the sequence

$$\text{ACG|AGC|AAC|GAA|UGA|AAA}\ldots \overset{\text{Codes for}}{\rightarrow} \text{Thr-Ser-Asn-Glu(-Stop)},$$

shifting the reading frame to the right by one base has this mRNA sequence code for something completely different:

$$\text{A |CGA|GCA|ACG|AAU|GAA} \rightarrow \text{Arg-Ala-Thr-Asn-Glu-}\ldots$$

Not only are the encoded amino acids different, but the stop codon has been lost, so the mRNA will continue to be translated into a larger polypeptide. A single frameshift mutation can therefore have a profound effect on the cell. These mutations can be brought about by **insertions** or **deletions**: adding an extra nucleotide or removing one nucleotide, respectively.

4.2.2 Genes

We have now seen the basics of converting DNA into a polypeptide. However, DNA contains more than just coding regions. There are also control regions: regions that serve to activate or inhibit transcription. Some regions serve as filler, not coding for anything, not necessarily activating or inhibiting transcription but allowing the cell to gain different polypeptides from a single unit of DNA. There may also be regions where we don't know the function (yet) but whose presence is important for the survival of the cell. Let us therefore use the following definition for a **gene**: a stretch of DNA that functions as a unit to give rise to a polypeptide product via an RNA intermediate.

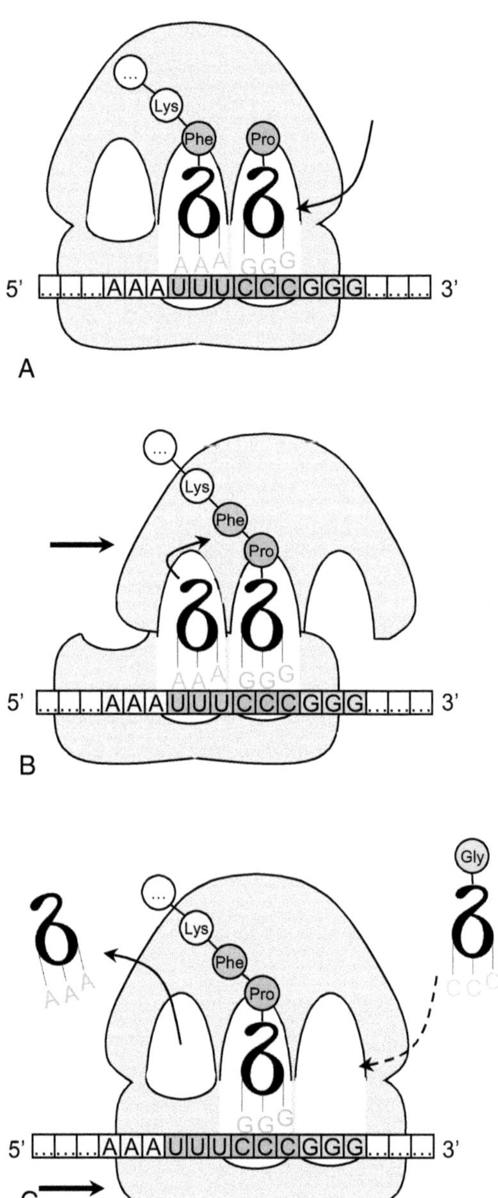

FIGURE 4.14

For the ribosome, translation can be thought of as a three-step process. (A) An aminoacyl-tRNA is loaded into the A site and checked for accuracy via hydrogen bonding of the anticodon with the mRNA codon. (B) The growing polypeptide chain is transferred to the aminoacyl-tRNA acquired in step 1. A conformation change keeps the peptidyl-tRNA in the ribosomal P site, and the now-empty tRNA is located in the ribosomal E site. (C) Another conformation change causes release of the old tRNA while the small ribosomal subunit realigns with the large subunit and is positioned three nucleotides (1 codon) downstream, ready to begin the process again.

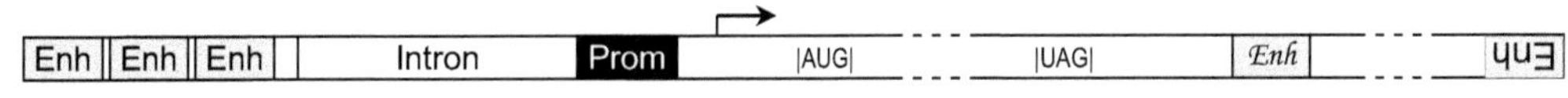

FIGURE 4.15

Hypothetical setup of a gene, showing enhancers (Enh), a promoter (Prom), the transcriptional start site *(arrow)*, the translational Start codon (AUG), and a translational stop codon (UAG in this example). Note that enhancers can occur as a string of repeats, that a single gene can have different enhancers (denoted here by different fonts), and that some enhancer sequences can appear in either direction.

Consider a gene that codes for the green fluorescent protein. The part of the gene that specifically codes for the amino acids in the green fluorescent protein is known as the **exon**. An exon is a stretch of DNA that will be represented in a mature RNA transcript. The transcript could be mRNA, ribosomal RNA (rRNA), or tRNA. Before the gene is transcribed, RNA polymerase must first bind to the DNA. The spot at which it binds is referred to as a **promoter**. Other bases in the DNA sequence may serve to enhance the binding of the transcriptional machinery and are referred to as **enhancers**. Enhancers can be located upstream (on the 5′ side of the coding region) or downstream (on the 3′ side) of the exon. Enhancer sequences can also be backward and still serve to enhance transcription. Another noncoding DNA sequence that is considered part of a gene is the eukaryotic **silencer** (known as an **operator** in prokaryotes). Figure 4.15 shows a general layout of some important genetic elements.

At one point in time, research showed that each step in a metabolic pathway was controlled by the product of a gene; if one could somehow destroy that gene, the function of the enzyme at that particular step in the pathway would be knocked out. This led to the "one gene, one enzyme" hypothesis, which we now know is not entirely correct. Since the completion of the project to sequence the human genome, it may seem that we could now analyze the database and elucidate every gene in the human genome. After the sequencing was completed, some hoped that for any malfunctioning protein an individual might produce, any corresponding mutated genes could be located and treatments be developed. But the problem was not that simple because there is not a linear pathway from protein back to gene. This fact is illustrated by the example of Gregor Mendel and his peas. Mendel published his famous work in 1866 (although it went largely unrecognized until 1900), but the gene responsible for producing wrinkled versus smooth peas was not discovered until 1990. (The gene codes for a starch-branching enzyme. A defect in that gene causes the peas to have wrinkled coats because they lack appropriate branching to push out on the coat of the pea.) The moral is that it is easy to discover a phenotype and relatively easy to discover the protein responsible for it, but it can be far more difficult to identify the gene or genes responsible for the phenotypic change.

Rather than say a gene is producing a protein, it is more robust to say that it is coding for a polypeptide. Think back to polypeptide structure (Chapter 3) and recall that quaternary structure is the interaction of two or more tertiary structures. For example, recall the structure of hemoglobin: it is a dimer of identical dimers and can also be called a tetramer. Each of the four subunits of hemoglobin is made from polypeptides, and the polypeptides come together to make the functioning protein subunits. Each alpha subunit in hemoglobin happens to be a protein that is encoded by four genes, and each beta subunit is made from polypeptides encoded by two genes. Because the functioning protein subunits are made up of multiple polypeptides, the "one gene, one protein" hypothesis is inaccurate. The hypothesis must be reframed as "one gene, one polypeptide."

Having the primary sequence of a polypeptide does not mean we can backtrack directly to the genomic sequence; recall that more than one codon can code for a given amino acid in all but two cases. To make matters more complicated, having the primary sequence of a functioning polypeptide does not account for any amino acids that were cut out during folding and processing by the cell to produce the functioning unit. Bases can even be cut out during RNA processing in producing a functioning mRNA. The point is that although a gene codes for a polypeptide sequence, so many modifications happen between gene and final polypeptide (which may only be one piece of a functioning protein) that the link between a phenotype and the associated gene can be difficult to establish, even with a sequenced genome.

4.2.2.1 How many genes are in the human genome?

There are approximately 3.08×10^9 bases in the human genome. For many years, it was believed that there were to about 100,000 genes, based on the average number of amino acids in a protein. However, the completion of the Human Genome Project revised that estimate down to around 33,000, and after further scrutiny of the results, the number was further revised down to about 25,000. At the end of 2010, four different counts put the total number of human genes at between 18,877 and 38,621. In August, 2020 the RefSeq database, maintained by the National Institutes of Health in the United States, listed the number of human genes at 37,805 (an increase of 12,241 from 2011). One might think that with the entire genome sequenced, it would be a straightforward matter to determine the exact number of genes in a human genome. The task is not so simple because, in addition to all the noncoding regulatory regions we have discussed, one or more **introns** (which also do not encode any part of a polypeptide) can appear in the middle of a gene. Moreover, if we say that the average gene uses 145,267 nucleotides (the average length of a human gene, according to RefSeq), 37,805 genes would make up only about 56% of the total genome.

One reason why the estimate of the total number of genes in the human genome went down from 100,000 was the discovery that certain polypeptides that are used to make large proteins are reused for other proteins. Because these polypeptides can be used in more than one protein, the product of a gene can be used in more than one place. "One gene, one polypeptide" still holds, but it must be kept in mind that one gene can contribute to many different proteins.

4.2.2.2 Phenotypes

Why is it that one phenotype is dominant over another? It has to do with the proteins that are formed by an organism. Each of our genes has two **alleles** (copies) because we get one set of DNA from our mother and one set from our father. So we have two versions of every (non–sex-linked) gene in our bodies. If the mother and father of an organism each passed on a dominant form of the gene, then the offspring would display the corresponding dominant (also known as **wild-type**) phenotype. If, however, one allele coded for the wild type of the protein (e.g., the brown pigment seen in human eye color) and the other coded for a recessive form (perhaps blue, or *null*), the wild-type gene would be responsible for the cell producing a brown pigment while the other allele failed to yield brown pigment. The offspring would still have the brown pigment in their irises, albeit possibly in a lower quantity. In most cases, having only one functioning version of the polypeptide encoded by a gene is enough to give one the wild-type phenotype, and the only time we can get a mutant phenotype is when we have two mutant copies of the gene.

For an example of how the alteration of one gene can affect an entire organism and how that alteration can be passed to future generations, consider blood types. Most people are aware that there are four main blood types: O, A, B, and AB. (We will not consider Rh factors in this example.) These four types indicate a specific antigen, known as the H antigen, that is attached to certain blood proteins (or a sphingolipid) on red blood cells. The gene responsible for the H antigen codes for a fucosyltransferase and is one gene in a group of seven that constitutes the ABO locus found on chromosome 19 (a **locus** is a location—in other words, where a gene can be found in the genome). The three blood types arise from the presence or absence of a gene encoding an enzyme responsible for transferring galactosyl groups. The A type of the gene codes for an N-galactosyltransferase, which transfers an N-acetylgalactosamine to the H antigen. The B type of the gene codes for a galactosyltransferase, which transfers a galactose to the H antigen (Figure 4.16). The O type of the gene does not code for such an enzyme, so the H antigen will not be further modified at the glucose end. The reason that the O form of the gene is

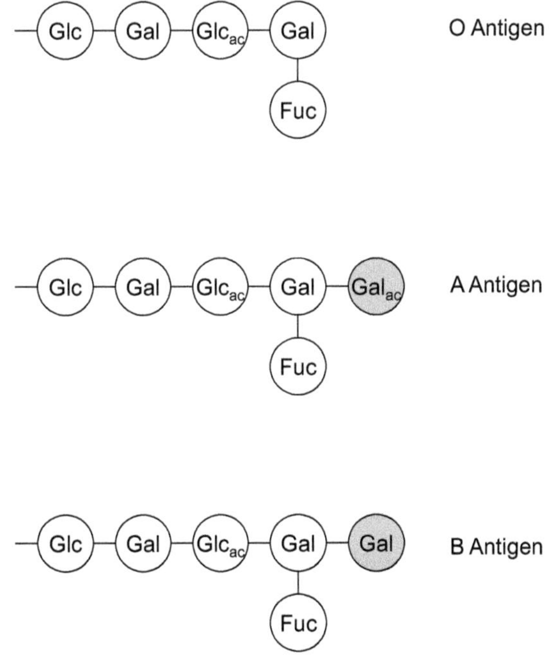

FIGURE 4.16

Structures of the H antigen, which determines ABO blood type. The sole difference between the A and B antigens from the O antigen is shown in *pink*. *Fuc*, Fucose; *Gal*, galactose; *Gal_{ac}*, N-acetylgalactosamine; *Glc*, glucose; *Glc_{ac}*, N-acetylglucosamine.

inactive is that the gene is missing a single guanine in one codon, which causes that codon and all subsequent codons to be misread (a frameshift mutation). The result is that no active transferase enzyme is made from that gene.

The H antigen is not interpreted as foreign, which means type O blood can be transfused into any individual. However, type A or B blood put into an individual with type O blood would cause a problem because the red blood cells would contain glycoproteins that were not produced in the recipient's body and were therefore recognized as foreign. Individuals with type O blood are known as universal donors, but they are certainly not universal recipients; that title goes to the type AB individuals because no matter what blood type they receive, they should not mount an immune response to it because their bodies already make matching glycoproteins.

To link this idea with our discussion of alleles, know that you have two alleles for every ABO locus: one from your mother and one from your father. Each of these alleles could contain the A, B, or O form of the gene; all the possible combinations are shown in Figure 4.17. It is clear that somebody with two type A alleles will have type A blood, but a person with one type A and one type O allele will also be said to have type A blood: if any amount of the type A oligosaccharide is made, the phenotype of the blood will reflect this. Even a small amount of the A oligosaccharide would be recognized as foreign in an individual who did not have a type A allele. The same is true for type B alleles.

Karl Landsteiner discovered the ABO blood types in 1900; 40 years later he, along with Alexander Wiener, also discovered the Rh blood types. The term "Rh" is short for Rhesus; Rhesus macaques were the monkeys initially used in making the antiserum for blood typing. This blood typing system is very complex because it involves 45 different red blood cell antigens. However, the entire system is essentially controlled by two genes, and inheritance has an even simpler, binary form (either you have the antigens, or you don't). A person with type O blood displaying the antigens is said to be O+ ("O-positive"), as opposed to O- ("O-negative").

Box 4.2 Mother–child blood issues

Consider for a moment a type O- mother who is carrying a type O+ fetus. If the blood of the fetus were to mix with hers, the mother's immune system would mount a response against the Rh antigen of the fetus, which would be devastating for both. This event is prevented in utero by a sort of blood–baby barrier. The bloods of the mother and fetus are two separate tissues (yes, blood is a tissue) that are kept separate, although close, in the placenta to ensure they do not mix during gestation. However, during childbirth or the latter months of pregnancy, the placenta tears away from the uterus, and some mixing of the mother's and child's bloods can occur. The amount of mixing is relatively low and does not present a significant problem for either; however, the mother's immune system becomes sensitized to the Rh antigens of that child and will be able to mount a quicker and more intense response against those Rh factors in the future, as is possible if she carries another Rh-positive fetus. This could be disastrous during a subsequent pregnancy: fetal red blood cells would be destroyed, leading to anemia and a rise in unconjugated bilirubin levels, followed by several adverse conditions, including brain damage or even death of the fetus or baby shortly after birth.

To prevent this type of immunologic attack, Rh-negative mothers-to-be may be given the drug RhoGAM, which contains antibodies against Rh antigens. The principle is that Rh antigens will be bound by the inactive antibodies in the drug before they can come into contact with the mother's active antibodies.

A type O mother giving birth to a second type A or B child can also pass some of her antibodies to the child during childbirth. The result will again be red blood cell destruction, which leads to anemia and heightened unconjugated bilirubin levels (jaundice). Such a condition will be treated with phototherapy ("bili-lamps") to break down the bilirubin and blood replacement transfusions to clear out the maternal antibodies if the condition is severe enough.

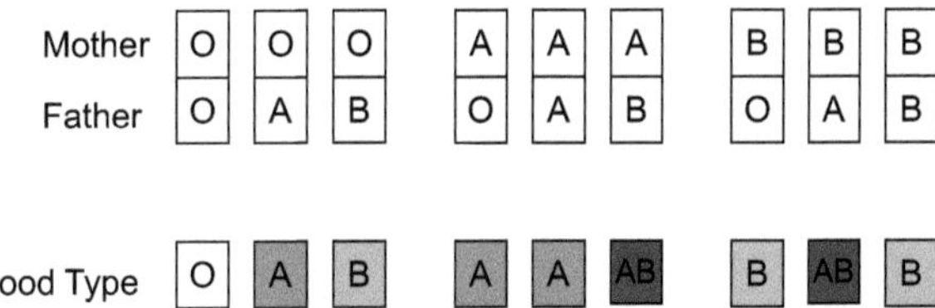

FIGURE 4.17

Blood type is determined by the gene for the H antigen, one allele of which is inherited from each parent. Each column shows one of the nine possibilities for inheritance of the H antigen genes. Note that, for example, a child inheriting one A and one O allele is considered type A because of the immunologic implications.

The concept of typing can be further extended to include bone marrow or organs for transplantation. An in-depth matching must be performed to prevent the recipient's body from rejecting the transplant. Rejection involves the recognition of certain molecules by the immune system; the fact that such recognition occurs qualifies these molecules as antigens. When the antigens are identified as "foreign," cells of the immune system will be recruited to the area of transplantation, and the foreign cells will be destroyed.

4.2.3 Transcription

Our general discussion has taken us from DNA to RNA to protein. Let us now examine transcription—the process of producing RNA from the DNA template—at greater depth.

4.2.3.1 The start of transcription: RNA polymerase binds to DNA

In human cells, with only about 25,000 genes in a genome of over 3 billion base pairs, one might wonder how RNA polymerase determines where to start transcription. Referring back to the genetic code in Figure 4.11, we can see that the codon AUG codes for start. AUG also codes for methionine, which indicates that polypeptides must all begin with a methionine. However, the initial polypeptide will often undergo modification, including the removal of some amino acid residues, as it matures into a functional polypeptide/protein.

4.2.3.1.1 Prokaryotes

The way that RNA polymerase determines where to start differs between eukaryotes and prokaryotes. The prokaryotic method is described here for simplicity to give the reader the general insight needed to appreciate the principles of transcriptional commencement.

Bacterial **RNA polymerase** contains a core enzyme, made from several polypeptide subunits, that performs the task of synthesizing an RNA strand from a DNA template. A separate subunit, known as **sigma (σ) factor**. can attach to the core enzyme, and in so doing forms the RNA polymerase holoenzyme. The RNA polymerase holoenzyme will readily but weakly bind to DNA but is apt to fall off the DNA strand as it quickly slides along the polymer. However, when it slides across a promoter region, the σ factor interacts with the bases of the promoter to add strength to the holoenzyme–DNA interaction and briefly halt the sliding. This strengthened binding causes a conformational change in the holoenzyme that serves to separate the dsDNA into two separate strands without hydrolyzing ATP. The polymerase is then able to synthesize an RNA polymer of ~10 nucleotides that is complementary to the DNA template strand.

At this point, the σ factor that served to halt the RNA polymerase holoenzyme at the promoter is now acting as a sort of anchor, preventing further transcription from taking place because the holoenzyme cannot move farther downstream. The RNA polymerase core enzyme then releases the σ factor and continues its slide down the DNA, this time synthesizing complementary RNA at a rate of about 3000 nucleotides per minute.

At the end of the DNA coding region is a DNA sequence termed the **terminator** sequence, which contains a string of A-T base pairs. When the RNA polymerase traverses this region, the core enzyme dissociates from the DNA template strand; this may be because A-U base pairs (formed as the RNA polymerase synthesizes a strand complimentary to the terminator sequence) are less stable or perhaps because self-pairing of As and Us within the final portion of the primary transcript causes a hairpin structure to form. This dissociation, in turn, causes the newly formed RNA strand to be released from the core enzyme, most likely due to a conformation change in the protein. The core enzyme is then free to associate with another σ factor and start the process again.

4.2.3.1.2 Eukaryotes

Eukaryotes begin transcription in a similar way to prokaryotes. However, they use three different RNA polymerases: RNA polymerase I is involved with transcribing genes encoding rRNA, RNA polymerase II transcribes genes that will become proteins (via mRNA) and some snRNAs (covered later, under "Splicing") and RNA polymerase III mainly transcribes genes encoding tRNAs but also some snRNAs and one particular rRNA gene. The type of transcription that is most interesting to most biotechnologists yields a product that becomes an mRNA that will be translated into a protein, so we will focus on RNA polymerase II in the following sections.

If the prokaryotic RNA polymerase core enzyme can be likened to eukaryotic RNA polymerase II, then the prokaryotic σ factor can be likened to general **transcription factors**, which are proteins that must bind to RNA polymerase II before it can start transcription. Transcription factors are abbreviated "TF," and those that bind to RNA polymerase II are abbreviated "TFII." We shall run into TFIIB and TFIID soon.

In eukaryotes, the product of RNA polymerase is indeed an RNA molecule, but it is not mRNA; three modifications are required to convert it into a mature mRNA ready for translation. Before we delve further into eukaryotic transcription, let us discuss these modifications in greater detail.

4.2.3.1.2.1 The eukaryotic 5′ mRNA cap Even before RNA polymerase II has finished creating the primary transcript, the 5′ terminus is modified through capping. The cap can be thought of as essentially an upside-down GTP molecule that is methylated at the nitrogen in position 7 in the purine base, sometimes with an additional methylation of one or both of the first two bases in the transcript (Figure 4.18). Creation of the cap is more complicated, though, as illustrated in Figure 4.19. The primary transcript will have a triphosphate at the 5′ end, and the first step in the capping process is the removal of one of these phosphates. Next, a GTP molecule is added in a 5′-5′ linkage, with some of the energy for the reaction coming from cleavage of a pyrophosphate group. (Think of a pyrophosphate as two connected phosphates.) The third step entails methylation of the guanosine base at the nitrogen in position 7 in the purine rings. Occasionally, one or both of the ribose rings of the first two nucleotides in the primary transcript are then also methylated, at position 2.

Keep in mind that mRNA is not the only type of RNA in a cell; rRNA and tRNA are transcribed by RNA polymerases I and III, respectively. Other uncapped RNAs exist, too. RNA polymerase II is the

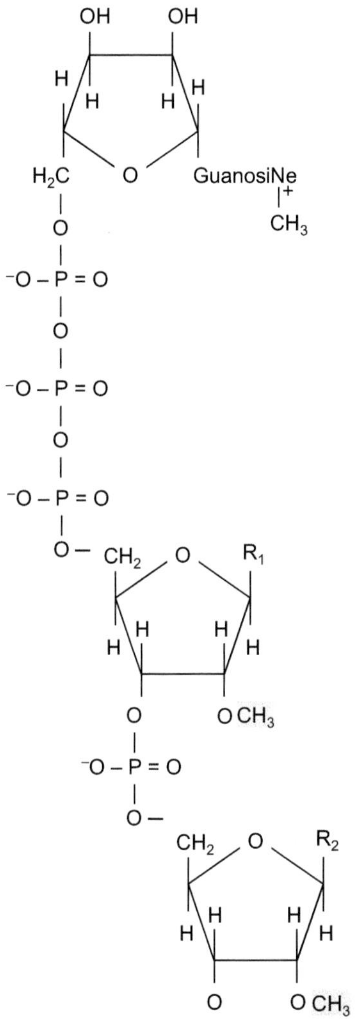

FIGURE 4.18

Structure of the 5′ cap added to primary transcripts created by eukaryotic RNA polymerase II. It consists of the positively charged 7-methylguanosine in a 5′-5′ linkage with the 5′ terminus of the RNA transcript. The first and/or second riboses of the primary transcript chain may also be methylated (shown by *shaded* CH_3 groups). The capital N in "guanosiNe" is meant to show that the methyl group attaches to a positively charged nitrogen atom.

only one of the three eukaryotic RNA polymerases that produces transcripts with a 5′ cap, though, because it has a tail section that binds the three enzymes needed for the three capping steps.

The cap serves as a signal that the molecule is an mRNA molecule. By removing the 5′ phosphate, it prevents degradation of the pre-mRNA by 5′ exonucleases. The cap will also be bound by the cap-binding complex, which marks and mediates export of the mRNA from the nucleus into the cytoplasm.

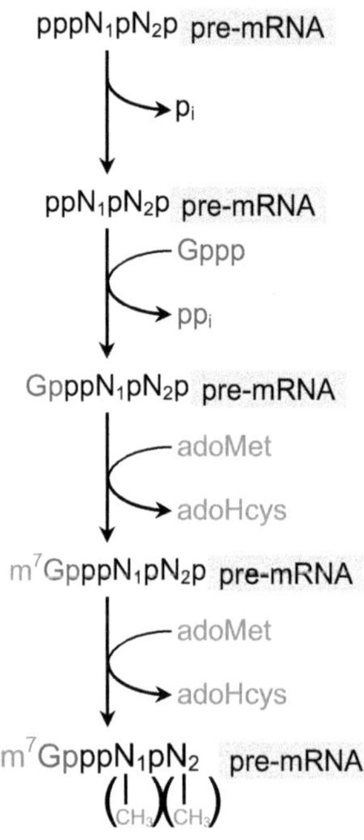

FIGURE 4.19

Order of steps in the creation of the 5′ RNA cap. A different enzyme catalyzes each step. *adoHcys*, S-adenosylhomocysteine; *adoMet*, S-adenosylmethionine, a methyl donor.

As we will soon see, the cap also serves as an initial binding site for part of the ribosome to begin the translation process.

4.2.3.1.2.2 Splicing The primary transcript contains introns as well as exons. In producing a mature mRNA, the cell will remove the introns by a process known as **splicing**. Figure 4.20 shows this process with two levels of detail. On the left, it can be seen that a specific adenylate residue is brought to the 5′ border of the intron. Hydrolysis at the 5′ border is immediately followed by creation of a new phosphodiester bond between the exposed 5′ intron phosphate and the 2′ carbon on the ribose of the adenylate residue. This creates a loop structure in the intron. The newly created 3′ hydroxyl at the end of the previous exon becomes available to form a phosphodiester bond with the 5′ phosphate of the next exon after another hydrolytic reaction frees the intron. The intron is released from the pre-mRNA as a loop structure with a tail, as structure known as a **lariat**.

Splicing out introns as lariats is a complex process mediated by over 50 proteins and 5 additional RNA molecules. These five RNA molecules—U1, U2, U4, U5, and U6—are known as **small nuclear RNAs** (**snRNAs**). Each snRNA is less than 200 nucleotides long and participates in the recognition of intron/exon boundaries as well as remodeling of phosphodiester bonds, but they do not work alone. They associate with several **small nuclear ribonuclearproteins** (**snRNPs**) to create a complex called the **spliceosome**. The right-hand side of Figure 4.20 shows some basic interactions between the snRNPs. First, the branch point adenine nucleotide is recognized by the branch point binding protein and a helper peptide (not shown), which recruit snRNP U2 to the site. At the same time, snRNP U1

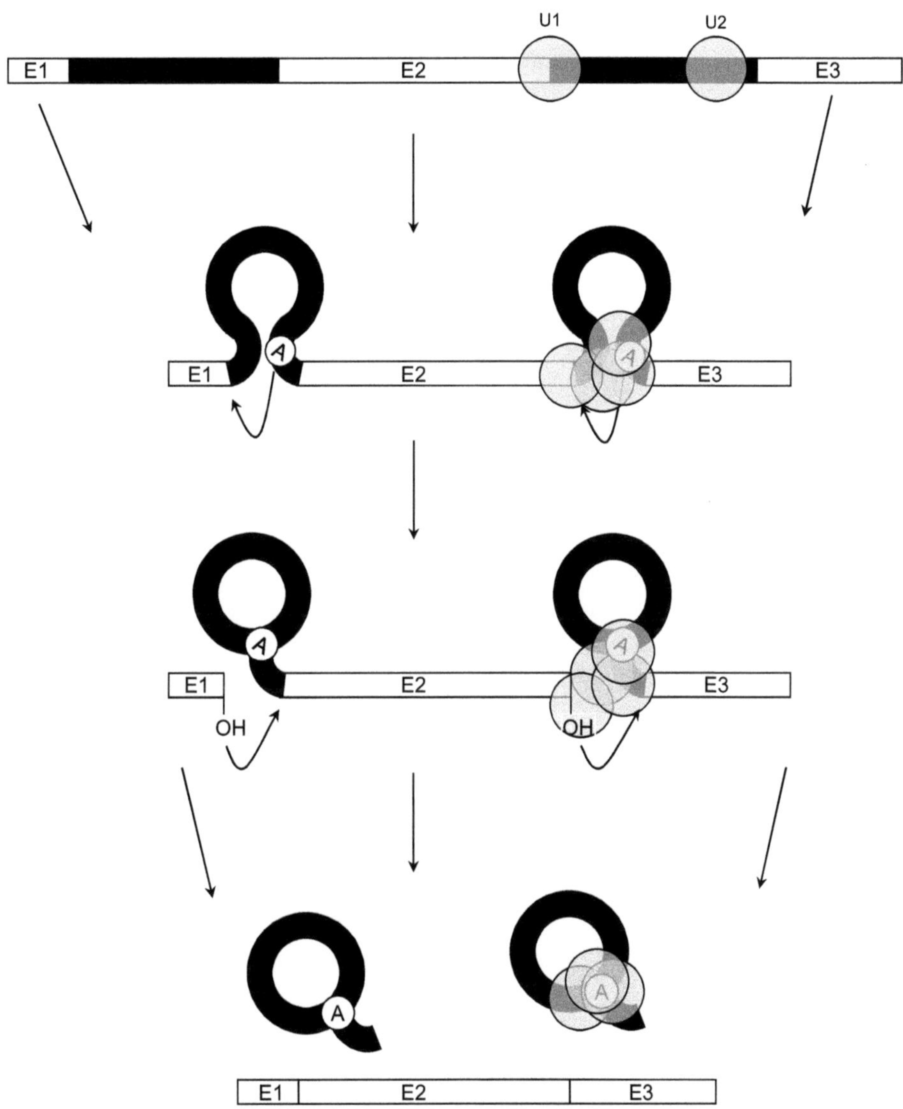

FIGURE 4.20

Splicing: the removal of introns from eukaryotic RNA. The figure shows two introns being removed in different levels of detail. A specific adenylate residue is brought into the vicinity of the 5′ border of the intron, where hydrolysis and recreation of a phosphodiester bond are used to remove the intron as a lariat. The right-hand side of the figure shows the same process but in the presence of snRNPs, which are essential in the formation of the spliceosome.

identifies and associates with the upstream border between the intron and the adjacent exon. The other three snRNPs (U4/U6-U5) then become involved as a unit to help bring the 5′ splice site to the branch point and facilitate hydrolysis with the aid of additional proteins and ATP. During the process, there are rearrangements and dissociations of snRNPs in the spliceosome, resulting in only the U2, U5, and U6 snRNPs remaining with the lariat as it is released.

4.2.3.1.2.3 The eukaryotic poly(A) tail In prokaryotes, transcription of a gene halts when the RNA polymerase passes a termination signal, at which point it releases both the DNA and the RNA transcript. Because prokaryotes lack a nucleus, no further processing is needed to allow the transcript to traverse a nuclear envelope.

Eukaryotic mRNAs must traverse the nuclear envelope, through nuclear pore complexes, to get to the cytosol for translation. The 3′ end of a mature mRNA consists of a string of about 200 adenylate residues, which play a role in exporting the mRNA from the nucleus and lend stability to the molecule once it is in the cytoplasm. These A residues are not encoded in the genome but are added after RNA polymerase II transcribes a cleavage signal found in the gene at the terminus of the final coding region. The signal to add the poly-A tail has the consensus sequence AAUAAA-$[N]_{10\text{-}30}$ –CA in mammals; the cleavage signal immediately following is U- or G/U-rich and is less than 30 nucleotides in length (Figure 4.21). Although the process of adding the poly(A) tail is far more complex than that of adding the 5′ cap, the progression of events boils down to:

1. Polyadenylation and cleavage signals in the DNA are transcribed by RNA polymerase II.
2. Cleavage and polyadenylation specificity factor (CPSF) binds to the AAUAAA signal, and cleavage stimulation factor F (CstF) binds to the G/U-rich signal.
3. RNA is cleaved by an endonuclease. The G/U-rich sequence will be degraded in the nucleus.
4. Poly-A polymerase (PAP) binds to the pre-mRNA and produces the poly-A tail. PAP operates like the conventional RNA polymerases, except that no DNA template is required.
5. Several copies of poly-A-binding protein bind the newly created tail.

4.2.3.2 Regulation of transcription

Proteins bind to DNA, and when they do, it is probably for a specific reason. RNA polymerase binding occurs because transcription is about to start. Still other proteins can bind to DNA and act as "fly paper" to help catch RNA polymerase (through mediators). Sometimes it is important for a cell to reduce or completely shut off the expression of a specific gene, and this is accomplished by another class of regulatory protein binding to a general (or specific) sequence of DNA. The portions of DNA that bind proteins for the purpose of transcriptional regulation are promoters, enhancers, silencers, and operators.

4.2.3.2.1 Promoters and promoter elements

Gene regulatory elements play a significant role in determining how much of a gene is transcribed at any given moment. Perhaps the most important of these elements is the promoter, which serves as a binding site for transcription factors (in eukaryotes) or sigma (σ) factors (in prokaryotes). This binding serves to position RNA polymerase for transcription. We will now look at promoters in more detail.

4.2.3.2.1.1 TATA box Perhaps the best-known promoter element is the **TATA box**, which gets its name from its **consensus sequence**, or the most common order of bases that make up this element. In eukaryotes, the TATA box is typically located 25 to 35 bases upstream of the transcriptional start site.

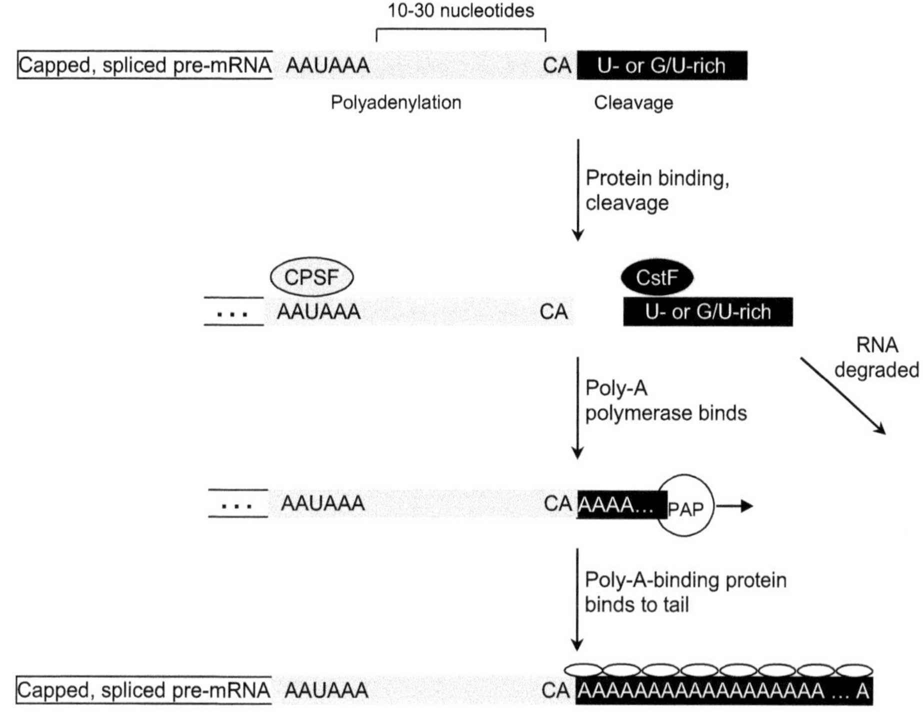

FIGURE 4.21

Eukaryotic mRNA has a string of A residues at its 3′ end, but these are not directly coded for in the genome. Rather, the gene has bases coding for RNA polyadenylation and cleavage signals that are transcribed and bound by three proteins, two of which are cleavage and polyadenylation specificity factor (CPSF) and cleavage stimulation factor (CstF). These proteins help facilitate polyadenylation and cleavage, respectively. Once the cleavage signal is removed via hydrolysis, the poly(A) tail is synthesized by poly-A polymerase (PAP) and coated with multiple copies of poly-A-binding protein.

No single sequence is used universally as the TATA box, but some sequences are used far more than others. Table 4.2 shows the top TATA consensus sequences taken from a ranking of the 1024 possible sequences of (TA[A/T][A/T]NNNN) (where N = any nucleotide) in yeast. From the list, it seems evident that an octamer TATA sequence ought to adhere to four rules: it begins with TATA, the sixth base is an A, the fifth and seventh bases are either As or Ts, and the eighth residue is a purine. So the consensus sequence is not like the spelling of a word, which has only one correct order of letters but rather it is a reflection of the letters that are most commonly used to make up the ordered bases.

In prokaryotes, the promoter situation is a little bit different. *E. coli* has two upstream sequences that often serve as promoter elements, being found in the -35 and -10 regions of the transcriptional start site. As already stated, σ factors bind to these sites. It just so happens that the transcriptional factor σ^{70} has the -10 consensus sequence of TATAAT, very similar to the TATA box of many eukaryotic genes. (For completeness, the -35 region consensus sequence for σ^{70} is TTGACA.) The -10 binding region in prokaryotes is sometimes referred to as the **Pribnow box**.

Table 4.2 Consensus sequence of the TATA box in yeast

Position	1	2	3	4	5	6	7	8
	T	A	T	A	T	A	A	A
	T	A	T	A	T	A	A	G
	T	A	T	A	A	A	T	A
	T	A	T	A	A	A	A	G
	T	A	T	A	T	A	T	A
	T	A	T	A	A	A	A	A

Consensus: T A T (A/T) A (A/T) (A/G).
The sequence was worked out from all possible 8-mers of (TA[A/T][A/T]NNNN). The top six sequence candidates are listed in the table.

For eukaryotic genes that use the TATA box, alterations in the position or composition of this promoter element can have significant effects on gene expression. If we were to change one base in the TATA box sequence, transcription of the corresponding exon(s) would go down; this should be intuitive. If we change the base sequence between the TATA box and the resulting transcriptional start site, not much changes in the transcription rate or product of the gene. However, if we remove bases between TATA and the usual transcriptional start site, RNA polymerase will begin transcription at a new site. How can changing bases in a region not have any effect while removing bases from the same region yields a new transcriptional start site?

The answer lies in the fact that, once assembled, the transcriptional machinery has a fixed size. The transcription factor **TATA-binding protein (TBP)** binds to TATA and aids in the binding of RNA polymerase to the gene. The active site of RNA polymerase is a set distance away from TBP when bound to it; this distance is about the same as that covered by a 25- to 35-base stretch of DNA. The DNA base that is next to the active site of RNA polymerase once the transcriptional machinery has been assembled will be where transcription starts. The identity of the bases between TATA and the active site of RNA polymerase is not of great importance, but the space they occupy helps to define the transcriptional start site.

4.2.3.2.1.2 CpG islands The TATA box is not the only feature that can be used to identify the location of the gene. For instance, there is a class of constantly expressed genes known as **housekeeping genes**, which code for proteins that are used for vital cellular processes. They are necessary to keep the cell alive and must be present in the cell at all times. Virtually all the cells in a given organism express the same housekeeping genes. For example, consider that cells always need energy. One of the most common ways to obtain this energy is through the process of glycolysis, which breaks a glucose molecule into two pyruvate molecules while generating two molecules of NADH plus a net of two ATPs. The ATPs can then be used by the cell for energy. Glycolysis is, for all practical purposes, always occurring in the cell; hence glycolytic enzymes are always required, and the genes for these enzymes must always be expressed.

Very often, housekeeping genes are preceded by a stretch of DNA that is very rich in Cs and Gs. This stretch, often called a CG island, or more commonly a **CpG island** to distinguish it from a C-G base pair, is 1000 to 2000 base pairs long and remains unmethylated in all cell types. (Methylation, or the addition of a $-CH_3$ group to a nucleotide base, is a genomic modification often utilized by a cell to inactivate a gene.) CpG islands often surround the promoters for housekeeping genes. Unmethylated C-G base pairs are not so common in the genome, so having an extended stretch of unmethylated C-G appear in the genome is due to more than mere chance and serves to indicate the presence of an active gene in the vicinity.

Table 4.3 Consensus sequences of several transcriptional regulatory elements

Element	Consensus sequence	Binds protein
TATA box	TATA(A/T)A(A/T)(A/G)	TATA-binding protein (TBP)
GC box	GGGCGG	SP1 transactivator Sp1
CAAT box	GG(T/C)CAATCT	CAAT enhancer binding protein (C/EBP)
BRE^u	(G/C)(G/C)(G/A)CGCC	Transcription factor IIB (TFIIB)
BRE^d	(G/A)T(G/A/C)(G/T)(G/T)(G/T)(G/T)	Transcription factor IIB (TFIIB)
DPE	(A/G)G(A/T)CGTG	Transcription factor IID (TFIID)
INR	(C/T)(C/T)AN(T/A)(C/T)(C/T)	Transcription factor IID (TFIID)

BRE^u = transcription factor IIB upstream response element, found just before the TATA box in eukaryotes
BRE^d = transcription factor IIB downstream response element, found just after the TATA box in eukaryotes
DPE = downstream promoter element, found after transcriptional start site
INR = initiator element, encompasses transcriptional start site

4.2.3.2.1.3 GC box A CpG island is not the same as a **GC box**, which is a transcriptional regulatory element containing the sequence GGGCGG. Assembly of the transcriptional machinery is enhanced by the binding of the protein Sp1, a transcription factor, to a GC box. Sp1 is expressed in all eukaryotic cells, so having a GC box present in a transcriptional regulatory region increases the amount of transcription for that gene. GC boxes are often found within 100 bases upstream of the transcriptional start site, and although they may appear as a single instance, they are commonly repeated 20 to 50 times. Sometimes GC boxes appear in genes that do not use a TATA box.

4.2.3.2.1.4 CAAT box Another transcriptional regulatory element worthy of mention is the CAAT box. It has the consensus sequence of GG(T/C)CAATCT and can be found approximately 75 base pairs upstream of the transcriptional start site. Keep in mind that the CAAT box, GC boxes, and even the TATA box are **promoter elements**; they are sequences that have been found within various promoters. Other common consensus sequences found in the vicinity of the transcriptional start site are shown in Table 4.3.

The TATA box is a promoter element. Some people say there is only one promoter, which contains the TATA box, and anything else that acts like a promoter should be termed a promoter proximal element. Regardless of semantics, it is generally agreed that promoters are always upstream of the transcriptional start site, they tend to occur within 200 base pairs of this start site, and they must appear in the correct orientation (meaning the backward version of the sequence will not work). In addition, if they occur within 50 base pairs of the transcriptional start site, their location is fixed. As with most rules in life, however, there are exceptions.

4.2.3.2.2 Enhancers

As well as promoters, other sequences in the DNA can also bring about increased transcription levels of certain genes. These sequences are known as enhancers and can be located very far from the transcriptional start site—sometimes over 1000 nucleotides away. They can appear upstream or downstream of the transcriptional start site, and they are active in either the forward or reverse orientation. Often, they are short sequences that are repeated several times.

An enhancer can show activity despite being far away from the transcriptional start site because of the overall conformation of a stretch of DNA. DNA is not a rigid entity; it is flexible and can be folded, perhaps to bring an enhancer closer to where the action is (Figure 4.22).

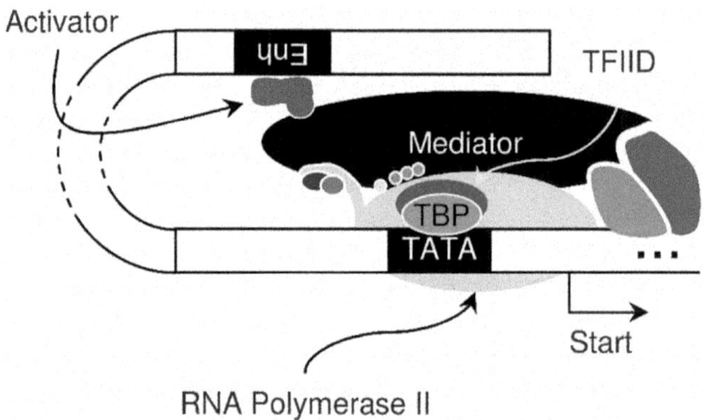

FIGURE 4.22

An example of the eukaryotic transcription machinery (not all proteins are shown). Note that the DNA enhancer is brought close to the transcriptional start site by an activator protein binding to a mediator that also interacts with RNA polymerase II. A gene may have many activator proteins and enhancers. The TATA box is bound by TBP, which is in turn bound by TFIID, one of many transcription factors involved with initiating eukaryotic transcription. Proteins involved with chromatin remodeling appear to the right, ahead of the polymerase. *TBP*, TATA-binding protein.

Enhancers work by binding **activators**—proteins that help to build the transcriptional machinery. One could think of the transcriptional machinery as being like a car built out of Lego blocks. When you get your kit, there might be a big flat piece, some wheels, some longer red blocks, some short white blocks, and perhaps some intermediate-sized blue blocks. You could use these blocks to build a house, a car, a train, or even a city-destroying monster. By assembling the blocks in different ways, we can build all kinds of different houses, monsters, or vehicles. We can think of the transcriptional machinery as a vehicle that will bind to DNA and roll down it, polymerizing an mRNA molecule as it travels.

In some ways, a Lego train and the transcriptional machinery are similar. The enhancer sequence will bind some of the "Lego pieces" and the cell will start to build the locomotive. Because enhancers can be repeated, several of the same type of activator protein can be bound. These pieces help to bind other subunits of the transcriptional machinery—perhaps RNA polymerase II, the core of the locomotive. Other sequences in the DNA bind further proteins, such as TBP. Still other proteins can collect these individual DNA-bound proteins into a functioning unit: the **mediator** protein complex attaches to activator proteins, general transcription factors, and RNA polymerase II. More proteins, such as chromatin remodeling complexes and histone modifying enzymes, bind during the assembly of a eukaryotic transcriptional machine and act as a sort of cow catcher on the front of the train to clear the track—unwinding the DNA and translocating the histone proteins used for DNA organization—to allow the transcriptional train to pass and do its job.

Enhancers, promoters, and promoter proximal elements all serve to increase the amount of transcription of a given gene, but there are differences between them; for example, promoters and promoter proximal elements are relatively close to the transcriptional start site (within 100 or 200 bases) and have a specific orientation, whereas enhancers can be found either upstream or downstream, perhaps at great distances from the transcriptional start site, and in either the forward or backward orientation.

Enhancers serve to enhance transcription levels (Table 4.4), but an enhancer cannot take the place of a promoter. Keep in mind that an enhancer binds to activator proteins that are to be bound by the

Table 4.4 Hypothetical plasmids and the amount of gene expression attainable with various gene elements

Plasmid contains:	Gene expression level
Exon only	-
Promoter + exon	+++
Enhancer + promoter + exon	+++++
Enhancer + exon	-

mediator complex. A promoter, on the other hand, serves as the site for building the transcriptional machine, which includes RNA polymerase II. If RNA polymerase II is not bound, no mRNA polymer will be generated.

An example of an enhancer is the simian virus 40 (SV40) enhancer, which was the first enhancer to be adequately characterized. Its sequence is about 100 base pairs long and appears about 100 base pairs upstream of a viral early transcription start site. The SV40 enhancer has turned out to be a great tool for a genetic engineer constructing a gene because it has been found to stimulate transcription from all mammalian promoters. It can be inserted into a plasmid in either orientation, even thousands of base pairs from the transcriptional start site. Detailed research has shown that this enhancer is composed of several individual regions that act as protein binding sites, each of which contributes to the total activity of the enhancer.

As with promoter elements, there are also enhancer elements. The **E-box** is a transcriptional element found within many enhancers (hence the name E-box); it has the sequence CANNTG, where "N" stands for any nucleotide. Notice that a stem-loop structure can be formed via the C pairing with the G and the A pairing with the T. This three-dimensional structure of E-boxes can be recognized by transcription factors that contain a basic helix-loop-helix structural motif (recall the discussion of supersecondary structure in proteins) in their folded conformation. The two "N" bases in the middle of the E-box sequence can determine specificity for certain transcription factors. Several E-boxes can appear in succession in a transcriptional regulatory region.

4.2.3.2.3 Silencers and operators

The functional converse of an enhancer is the silencer. In eukaryotes, a silencer sequence binds a **repressor** protein. A silencer shares most of the same properties with an enhancer in that it can be very far from the transcriptional start site, can appear in either orientation, and can bind a protein transcription factor; the most notable difference is that it induces a *lowering* of the transcription level of a particular gene after the appropriate transcription factor is bound.

In prokaryotes, however, gene regulation is a little different: prokaryotes do use promoters, but they do not use enhancers. There are, however, operator sequences, which are bound by repressor proteins. (Note that the name of the protein here–a repressor–is the same as the name used for the eukaryotic protein that binds to a silencer.) Once a repressor protein has bound to its operator, the binding of RNA polymerase to the promoter is blocked. If an operator is bound, there is no transcription. If the operator is free, transcription is free to proceed. It's like an on/off switch.

4.2.3.3 Locating transcriptional control regions: deletion analysis

Deletion analysis is a relatively straightforward way to detect and locate where transcriptional control regions such as promoters and enhancers lie within a stretch of DNA. Knowing a precise location is very economical for a biotechnologist who delivers genes into cells. If, for example, we are constructing a gene that is influenced by a specific enhancer and we know the enhancer lies within 4000 base pairs of the transcriptional start site, we could use the entire 4000-base-pair upstream region in constructing an engineered gene. However, if the enhancer region only spans 500 bases and the precise location of the enhancer is known, it is far more efficient to include only the 500 base pairs in the engineered gene because more copies of the gene could be delivered for a given mass of DNA.

Suppose the location of a particular gene's transcriptional start site is known. A large region of DNA upstream of transcriptional start can be isolated and inserted into a circular piece of DNA called a **plasmid** (exactly how this is done will be discussed in Chapter 11). The plasmid will typically contain, among other things, an exon for a **reporter gene**. A reporter gene produces a product that is easily detectable, such as a fluorescent protein. As a part of the deletion analysis procedure, the fragment that contains the suspected control region is inserted upstream of the exon for the reporter. Many copies of this engineered plasmid are created and delivered into cells to see how much of the reporter is expressed.

In a parallel set of steps, the control region is shortened and inserted into another reporter plasmid, many copies of the plasmid are created, and the smaller plasmids are delivered into cells to assess how much reporter is expressed. The process of shortening the suspected control region is performed several times to generate many plasmids with control regions of decreasing sizes. (It may seem odd right now, but the shortening of the suspected control region to produce many DNA fragments of varying lengths is often performed as a single step.) Figure 4.23 gives a pictorial overview of the process.

4.2.3.3.1 An example of deletion analysis

Suppose we performed a deletion analysis and generated the following data following delivery of our plasmids:

Plasmid	Number of bases in inserted region	Bases of original gene represented	Reporter expressed (RFU)
1	2500	-2500 to -1	10,473
2	2000	-2000 to -1	10,672
3	1500	-1500 to -1	4873
4	1000	-1000 to -1	4688
5	500	-500 to -1	4907
6	0	NA	–

RFU = relative fluorescence units

In analyzing the data, look for where there is a change in reporter expression; in this case, there was a decrease in reporter expression between plasmids 2 and 3. This implies that a transcriptional control region was lost when fragment 2 was shortened, so a control region must lie within the sequence that was lost. In this example, there must exist a control region somewhere between 2000 and 1500 bases upstream of transcriptional start (bases -2000 to -1500; the negative numbers indicate we are in regions upstream of transcriptional start, which is counted as base number +1; there is no base number 0). Further analysis of the table shows there is a second control region somewhere between bases -500 and -1 (Figure 4.24).

Also note that RFU values appeared to increase when plasmid 1 was shortened to plasmid 2. In this example, the increase is due to noise as opposed to the elimination of a silencer. Noise will yield small

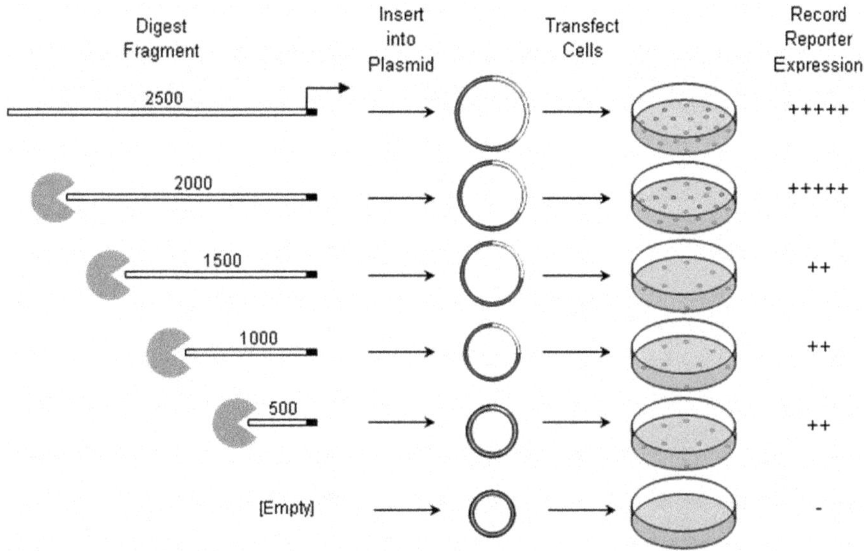

FIGURE 4.23

Overview of deletion analysis. DNA fragments are digested with an exonuclease to produce fragments of varying sizes. The fragments are put into plasmids which are then replicated, purified, and delivered into cells. Promoter/enhancer/repressor activity is implied by reporter gene expression in the transfected cells. Spots in the culture plates shown represent reporter expression, not total cells.

variations in detection, whereas the loss of a transcriptional control element should yield a significant change in RFU value. Notice the noise reflected between RFU values associated with plasmids 3, 4, and 5. In Figure 4.24, RFU values have been converted to relative activity levels (+).

To solve deletion analysis problems, both in theory and in the laboratory, it might help to make a master map of the region being tested. Every time a control region is elucidated by a change in reporter expression, mark it as illustrated at the bottom of Figure 4.24. As can be seen from the results in the figure, the 2500-bp upstream area has control regions somewhere in the ranges -2000 to -1500 and -500 to -1.

In practice, it is difficult to control exactly how long each digested fragment will be. An exonuclease—an enzyme that chews up a polynucleotide from the end—is used to digest the DNA region of interest to yield many fragments with a distribution of sizes. Plasmids are made and amplified, and the reporter expression experiments are run to see which fragments produce the greatest amount of reporter expression. DNA sequencing is performed separately, perhaps later, to determine which DNA regions were actually being tested when the effects were seen. We will not get the nice, controlled fragment sizes shown in the example, but we will still be able to locate regions of interest in the gene.

4.2.4 Translation

4.2.4.1 Initiation of translation in eukaryotes

The transcriptional start site is not the same as the translational start site. Finding a transcriptional start site is not as easy as looking for a specific DNA sequence; the location is often determined by the

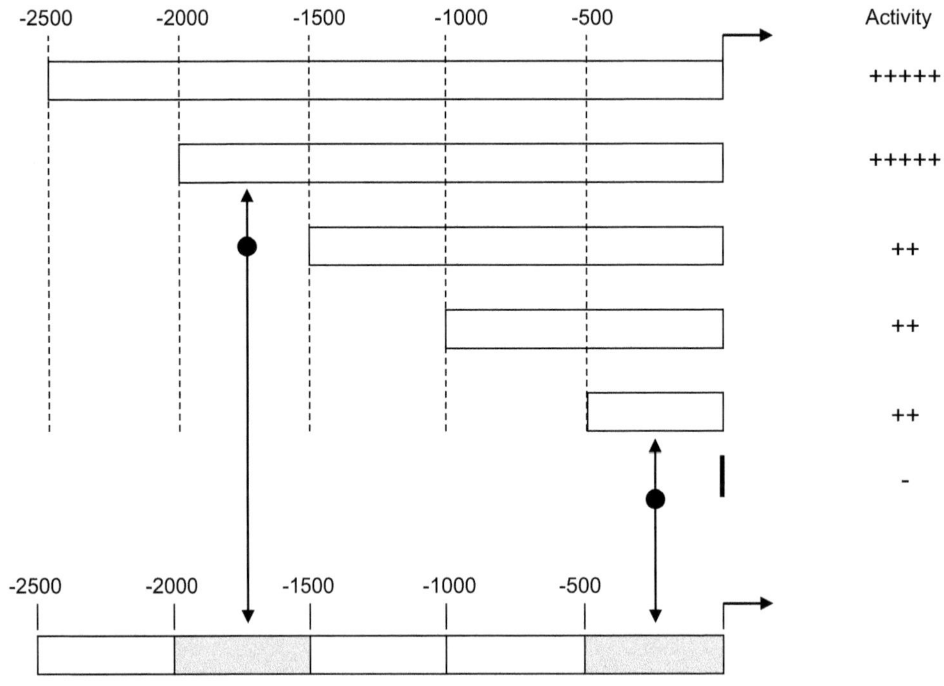

FIGURE 4.24

A straightforward way to map out deletion analysis data. Fragments are first ordered from largest to smallest. Writing the corresponding reporter expression levels to the right, note significant drops or increases in expression activity. When a significant drop or increase is detected, it is purportedly due to the segment of DNA that is present in the next-largest fragment but missing from the present fragment (noted by dotted lines).

physical size of the transcriptional machinery, and the specific place where it is assembled on the DNA. We have discussed how the transcriptional start site might exist approximately 25 to 35 bp after a TATA box in a eukaryotic cell, but it is such a complex problem that computer programs have been developed to assist with predicting where the transcription of genes will commence.

Locating the translational start site, however, is much more straightforward: it will be represented by the codon AUG on the messenger RNA. That makes the job of finding the translational start site very easy unless there is more than one AUG. This problem is eliminated in the eukaryotic cell via the **Kozak sequence**.

There are several possible RNA arrangements that can serve as Kozak sequences. Consider, for example, the strong Kozak sequence GCC**A**CCAUG**G**. Note the underlined AUG; in the Kozak sequence, this AUG is the translational Start codon. The base found three bases upstream of the Start codon must be a purine (an A or a G); usually an A, the base in this position is considered crucial for efficient initiation of translation. In the absence of this purine, a G immediately following the Start codon is essential. (Both of these important bases are shown in bold in our example sequence.) The 10-base sequence GCC(A/G)CCAUGG, for which the next base is not a U, is optimal, and (A/G) NNAUGG(not U) is strong (A at -3 is stronger than G). Anything else is considered weak at best. If you

Table 4.5 A shortened list of initiation factors and their roles in ribosomal assembly

Role	Eukaryotic initiation factors involved
Bind to the small ribosomal subunit	eIF-1, eIF-1A, eIF-3
Binds to initiator tRNA (holding Met)	eIF-2 (plus GTP)
mRNA recognized and brought to the ribosome	eIF-4E (directly recognizes 5′ cap) eIF-4G (binds eIF-4E, poly(A) binding protein, and eIF-3) eIF-4A, eIF-4B

are engineering a gene, knowing that eventually you want it to be translated, it is strongly advisable to have your Start codon appear in the middle of a Kozak sequence.

Ribosomes consist of distinct subunits that must come together for translation of mRNA to occur. In eukaryotic cells, the 5′ cap serves as an assembly point for the beginning of this process, which involves the small ribosomal subunit, the first tRNA, and several initiation factors. Table 4.5 lists the initiation factors (abbreviated eIF for "eukaryotic initiation factor") to give a glimpse into the complexity of the process. These initiation factors help to bring the mRNA to the ribosome for translation. Note that, through eIF-4G, the initiation complex also recognizes the 3′ end of mRNAs. This is believed to be the reason that poly(A) tails have a stimulatory effect on translation.

Recall that the Start codon, AUG, codes for methionine; this means the translation-initiating tRNA will carry a Met residue. In eukaryotes, before being paired with the mRNA molecule, the tRNA-Met molecule is loaded into a small ribosomal subunit together with the eIFs. It is interesting to note that the tRNA-Met molecule is the only one of the aminoacyl tRNAs that can bind tightly to the small ribosomal subunit without the large subunit being present. When the loaded small ribosomal subunit is being assembled on the mRNA at the 5′ cap, it recognizes the mRNA by virtue of initiation factors that have already bound to the cap. After some reorganization, the ribosome apoenzyme (recall that "apo" means "not complete") starts sliding down the mRNA. When it gets to the Kozak sequence, progress will stall, and the GTP bound to eIF-2 will be hydrolyzed, enabling both the release of all of the initiation factors and the binding of the large ribosomal subunit to yield a functioning ribosome that will perform the task of polypeptide production. Cells do not have functioning ribosomes floating around looking for mRNA; complete ribosomes must be assembled, just like the transcriptional machinery must be assembled. At the 5′ cap of an mRNA, the small ribosomal subunit and initiation factors assemble into (to continue a past analogy) a Lego locomotive that doesn't yet have a roof. The "locomotive" can roll down the assembly line until it encounters a Kozak sequence. At this point it will stall, the initiation factors will fall off, and the roof—the large ribosomal subunit—will be put on. The cell will now have a functioning ribosome to translate the mRNA message, starting with the AUG in the Kozak sequence.

The nucleotides immediately surrounding the Start codon in eukaryotic mRNAs influence the efficiency of AUG recognition during the above scanning process. If this recognition site differs substantially from the consensus recognition sequence (Kozak sequence), scanning ribosomal subunits will sometimes ignore the first AUG codon in the mRNA and skip to the second or third AUG codon instead. Cells frequently use this phenomenon, known as **leaky scanning**, to produce two or more proteins, differing in their N-termini, from the same mRNA molecule. For some genes, this allows the production of a single protein that will be directed to two different compartments in the cell. The variable N-termini are removed after the protein is delivered to its destination.

Ribosomes dissociate from the mRNA at the poly(A) tail. This is likely due to the poly(A) tail having only two sites per adenine available for hydrogen bonding (as opposed to three hydrogen bonds with guanine), which provides less force for holding the ribosome to the mRNA. The poly(A) tail also serves to increase the stability of mRNA by inhibiting the assembly of RNases that chew from 3′ to 5′. Poly(A) tails are also involved in the export of mRNA from the nucleus to the cytoplasm.

4.2.4.2 Initiation of translation in prokaryotes

In prokaryotes, a similar but distinct means of determining the translational start site involves the **Shine-Dalgarno sequence**. Simply put, this sequence performs the same function in bacteria as the Kozak sequence in eukaryotes. The Shine-Dalgarno sequence is typically found around position -7 to -4 of the translational Start codon and has the sequence AGGAGG. This sequence is complementary to part of the 3′ end of 16S rRNA: GAUCA<u>CCUCCU</u>UA-3′ (the portion that is complementary to Shine-Dalgarno is underlined).

The way in which bacteria determine the Start codon differs from the method used in eukaryotes. Bacterial mRNAs do not have 5′ caps, so ribosomal assembly in prokaryotes must be different from that in eukaryotes. In prokaryotes, the small ribosomal subunit contains rRNA that is complementary to the Shine-Dalgarno sequence. Hence, while the Shine-Dalgarno sequence is similar to a Kozak sequence in that both help to determine the position of the Start codon, the Shine-Dalgarno sequence is different because it allows the bacterial ribosome to be built at an interior position on the mRNA through direct binding to this sequence.

Because prokaryotic ribosomes can be assembled wherever there is a Shine-Dalgarno sequence and bacterial mRNAs can contain more than one start site, one mRNA can often code for multiple polypeptides. A polynucleotide unit that codes for a polypeptide is called a **cistron**. When an mRNA codes for multiple polypeptides, the mRNA is said to be **polycistronic**. Prokaryotes use polycistronic mRNAs via **operons**. An operon is a contiguous stretch of DNA that codes for more than one polypeptide and whose transcripts are contained in a single mRNA. Just as an opera contains many different songs in a single continuous story, an operon contains many different messages represented in a single mRNA. For example, in *E. coli*, the *lac* operon codes for a single mRNA, but that mRNA encodes three proteins. Each of the protein-coding segments has its own translational start site and Stop codon.

Box 4.3 Polycistronic mRNAs in eukaryotes

In eukaryotes, polycistronic mRNAs are very rare, but they do occur. This can happen because of:

- **Leaky scanning**—This is where ribosomes bypass an AUG codon located close to the 5′ cap or when a given Kozak sequence is weak. This allows the ribosome to reach another AUG that is further downstream.
- **A lack of detachment**—After the translation of one part of the mRNA, a ribosome might continue scanning. In doing so, it might attach a new set of initiation factors and reinitiate at a downstream AUG or Kozak sequence.
- **An internal ribosome entry site**—In very rare instances, a ribosome might enter at an internal ribosome entry site, much like the bacterial model just presented.

4.3 Summary

We have taken an in-depth look at nucleic acids and the molecular differences between DNA and RNA. We discussed these molecules as polymers and paid particular attention to the nucleotide bases, which give each nucleotide its specific identity. We also discussed historically important experiments that led scientists to realize and confirm that DNA is the genetic material.

With regard to the central dogma of molecular biology, we followed the detailed pathway that took us from the genetic blueprints of DNA, to RNA via transcription, on to mRNA messages, and finally the conversion of mRNA to polypeptides via translation. We saw that a gene is more than just a strand of DNA that can be directly converted into the sequence of a functioning protein; it may contain regulatory elements (e.g., promoters, enhancers, silencers/operators) and spacers (introns) that allow the cell to fine-tune when a gene is expressed and, occasionally, in what form.

After transcription, in eukaryotes, the newly created RNA is converted into mRNA by three distinct processes: the addition of a 5′ cap, the splicing of introns, and the addition of a poly(A) tail. The mRNA is transferred to the cytoplasm for translation.

Whether a prokaryote or a eukaryote, a cell performs translation via the attachment of ribosomal proteins to RNA. Differences between prokaryotic and eukaryotic mRNA, ribosomes, and translation start sites were highlighted as we discussed this process, which is one of the most fundamental to all life forms as we know them.

Related reading

Alberts, B., Johnson, A., Lewis, J., Raff, M., Roberts, K., Walter, P., 2008. Molecular Biology of the Cell, fifth ed. Garland Science, New York.

Basehoar, A.D., Zanton, S.J., Pugh, B.F., 2004. Identification and distinct regulation of yeast TATA box-containing genes. Cell 116, 699–709.

Brockmann, C., Soucek, S., Kuhlmann, S.I., Mills-Lujan, K., Kelly, S.M., Yang, J.C., et al., 2012. Structural basis for polyadenosine-RNA binding by Nab2 Zn fingers and its function in mRNA nuclear export. Structure 20, 1007–1018.

Cooper, G.M., 2000. The Cell: A Molecular Approach, second ed. Sinauer Associates, Sunderland, MA. Accessed via: http://www.ncbi.nlm.nih.gov/books/NBK9849/.

Ford, L.P., Bagga, P.S., Wilusz, J., 1997. The poly(A) tail inhibits the assembly of a 3′→5′ exonuclease in an in vitro RNA stability system. Mol. Cell Biol. 17, 398–406.

Hershey, A.D., Chase, M., 1952. Independent functions of viral protein and nucleic acid in growth of bacteriophage. J. Gen. Physiol. 36, 39–56.

Nelson, D.L., Cox, M.M., 2017. Lehninger Principles of Biochemistry, seventh ed. W. H. Freeman and Co.

Quigley, F., Martin, W.F., Cerff, R., 1988. Intron conservation across the prokaryote-eukaryote boundary: structure of the nuclear gene for chloroplast glyceraldehyde-3-phosphate dehydrogenase from maize. Proc. Natl. Acad. Sci. U.S.A. 85, 2672–2676.

Rubin, E., Farber, J.L., 1994. Development and genetic diseases. In: Rubin, E., Farber, J.L., Pathology, second ed. JB Lippincott Company, Philadelphia, PA, p. 256.

Schuegraf, A., Ratner, S., Warner, R.C., 1960. Free energy changes of the argininosuccinate synthetase reaction and of the hydrolysis of the inner pyrophosphate bond of adenosine triphosphate. J. Biol. Chem. 235, 3597–3602.

Shatkin, A.J., Manley, J.L., 2000. The ends of the affair: capping and polyadenylation. Nat. Struct. Biol. 7, 838–842.

Shi, W., Zhou, W., 2006. Frequency distribution of TATA box and extension sequences on human promoters. BMC Bioinf. 7 (Suppl. 4), S2.

Thrall, J.M.H., Goodrich, J.A., 2013. "Promoters", from Brenner's Encyclopedia of Genetics. In: Stanley M, Kelly H., (Ed.), pp. 472–474.

Questions

1. Using single-letter notation, what is the sequence of the following DNA strand? *(Remember to show your sequence in the 5′ to 3′ direction.)*

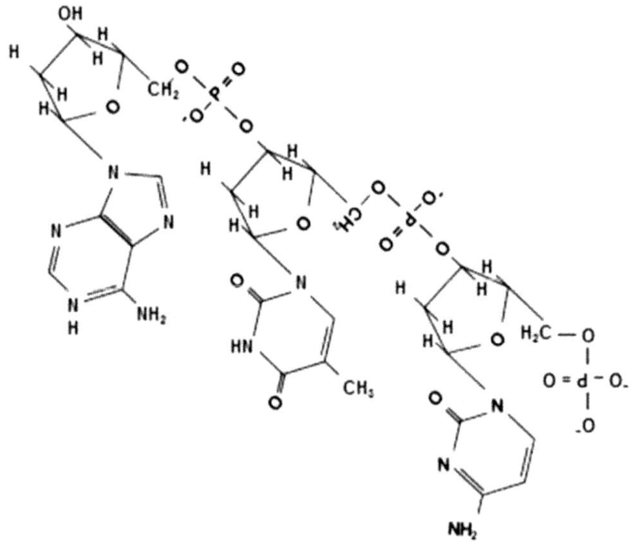

2. Draw a graph to represent T_m versus %(A+T).
3. Suppose a promoter for a gene encoding a nonessential protein experiences a random mutation such that transcription of the gene can no longer occur. Hypothesize how a such a nonlethal mutation might provide an evolutionary benefit to the species.
4. How does the wobble effect explain the degeneracy of the genetic code?
5. What helps position RNA polymerase II?
6. One reason why DNA is thought to be evolutionarily favorable over RNA as genetic material is its inherent stability due to structure. However, it is thermodynamically more favorable to hydrolyze DNA than RNA. Resolve the paradox.
7. **a.** If my genome is 37% A, about what percentage of it is G?
   b. If my genome is 26% A, about what percentage of it is U?
   c. If Koko the gorilla's genome is 21% C, about what percentage of Mighty Joe Young the gorilla's genome is A?
   d. If a raccoon's genome is 21% G, about what percentage of a giraffe's genome is C?
   e. If a wallaby's genome is 50% A, about what percentage of its genome is G?
8. Imagine life evolving on an alien planet in such a way that the genetic code is binary: only guanine and cytosine are used to code for a total of 12 amino acids. What is the minimum number of nucleotides required per codon to code for the 12 amino acids in this system?
9. How much energy does DNA polymerization require? Where does the energy come from?
10. How can an enhancer be thousands of base pairs away from the promoter and still interact with it?

11. Would it be a good or bad idea to make our genomes more efficient by deleting all of the introns? Explain your answer.
12. Explain why the sequence of the TATA box is not always TATA(AAT).
13. Below is an mRNA sequence. Indicate where the promoter is.
CAAUGCGCGCGCGCGCGAGCAUUCAGGCGUAUAAAUAUCGGCAUUCACCGAUG
14. (Make your answers for parts a and b different.)
   a. Why doesn't RNA form double helices in the human cell?
   b. Why doesn't RNA form double helices in the test tube?
15. What does the "deoxy" mean in "deoxyribonucleic acid?"
16. a. Consider two test tubes, each containing a double-stranded segment of RNA in the same concentration. The first tube has dsRNA with one of the strands being comprised of $[U]_{100}$ (meaning a polymer composed of 100 U residues), and the second tube has dsRNA with one of the strands being composed of $[G]_{100}$. Which sample will have the higher RNA melting temperature, and why?
   b. Consider the same situation as in part a, except now the tubes contain dsDNA. The first tube has dsDNA with one strand having the sequence $[A]_{100}$, and the second tube has dsDNA with one strand having the sequence $[AT]_{50}$. Which sample will have the higher DNA melting temperature and why?
17. What did the Hershey Chase experiments show?
18. a. Consider the experiments first described by Meselson and Stahl. Your friend Rosie claims that, even after 100 generations, there is a non-zero chance that there will be ^{15}N-DNA present in the final DNA preparation. Is she correct?
   b. Suppose there is neither cell death nor DNA degradation during the experiments and that 100% of cellular DNA is collected. After 100 generations, what is the ratio of light/light : heavy/light : heavy/heavy DNA?
19. Which blood type is known as the universal donor and why? Which blood type is known as the universal recipient and why?
20. Consider the combinations of alleles for blood type A/O, A/A, and O/A. If a person had any of these three combinations, he would be said to possess type A blood. How can that be?
21. Billy's mother has type O blood, and his father has type B blood. Billy's blood type is unknown, but doctors decide to give him type B blood during his surgery on the basis that Billy received a type O allele from his mother and either a type O or type B allele from his father, so type B blood should be given to cover their bases. Will Billy be okay?
22. Is the transcriptional start site the same as the translational start site? Explain.
23. Why is gene expression nearly nonexistent without a promoter?
24. Does transcription of a gene start at ATG?
25. Is the TATA box a promoter, an enhancer, or something else?
26. If an intron contains the sequence AUGAAACGCUGA, what might it encode?
27. a. What are three differences between promoter and enhancer sequences?
   b. What is the main difference between promoter and enhance function?

28. (Refer to Chapter 9 for more information on agarose gels, if needed.) Suppose we want to locate upstream control regions in a certain gene. We have used an exonuclease to chew the gene from the 5′ end. Below is the result of several runs, in which we have let the exonuclease chew for different periods of time.

Experiment 1: no exonuclease exposure
Experiment 2: 5-min exonuclease exposure
Experiment 3: 10-min exonuclease exposure
Experiment 4: 15-min exonuclease exposure
Experiment 5: 20-min exonuclease exposure

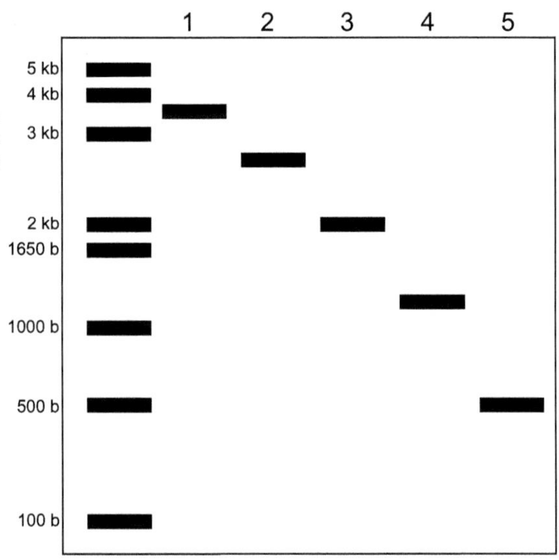

We cloned the fragments into a plasmid containing the coding region for a reporter gene, then transfected cells with copies of the plasmid. The following results for reporter expression were obtained:

	Relative fluorescence units (RFU)
Plasmid 1	1.000
Plasmid 2	0.993
Plasmid 3	0.658
Plasmid 4	0.648
Plasmid 5	-0.003

(Plasmid n *was created from fragment* n, *from experiment* n*)*

a. Estimate the location of each control region.
b. Estimate the nuclease efficiency of the enzyme used in bases/sec.

29. Using deletion analysis, researchers are trying to find the location of transcriptional control elements for a gene. They observe a sharp decrease in polypeptide production when the -1092 to -957 range is removed, and a lack of any reporter expression when the -92 to -1 range is missing. However, there is full expression of the reporter when the reporter plasmid contains the upstream sequence -888 to -1. Explain these results, including any identifiable DNA elements.

30. a. Mathematically, how many amino acids *could* be defined by our genetic code?
    b. Rather than the number you found for part *a*, our genetic code defines 20 amino acids. What one-word term explains the discrepancy?
    c. How is it that we have so many possibilities for unique amino acids provided by the genetic code, but there are only 20 amino acids encoded?
31. In eukaryotic cells, what three modifications are made to a newly transcribed RNA molecule to turn it into mRNA?
32. What is the importance of the Kozak sequence?
33. Why are insertions much more detrimental than the change of a single nucleotide in an exon?
34. a. Does prokaryotic mRNA have a 5′ cap?
    b. Does prokaryotic mRNA have a 3′ poly(A) tail?
35. AUG is the codon for methionine, and it can also serve as the translational start site. Sometimes translation does not start at the first AUG in an mRNA molecule. Give two reasons why this occurs. (It might help to think of eukaryotes and prokaryotes separately.)
36. What is the difference between a CpG island and a GC box?
37. How does an E-box function?
38. What is the function of the 5′ cap on eukaryotic mRNA?
39. a. Name two functions of the poly(A) tail in eukaryotes.
    b. Prokaryotes also use 3′ poly(A) tails in their mRNA. One reason for this is that it allows for the binding of proteins that will be a part of the initiation complex. What does this imply about mRNA structure in prokaryotes?
40. How is it possible that a cell might have 100,000 different proteins but only ~30,000 genes?

CHAPTER

Cell growth

5

Chapter outline

Biotechnology and its Applications. https://doi.org/10.1016/B978-0-12-817726-6.00005-8

As we have already mentioned, the biotechnologist must consider cells, whether directly or indirectly. Cells in culture will grow, but if they are not cared for properly, they will die. If you have a company that uses cells to produce something specific, how will you make sure you always have enough cells for your purposes if they are growing, reproducing, and dying at rates that depend on specific conditions? In this chapter, we will be concerned with cell growth, how to model it, and how to quantify it. More specifically, we will address:

- the eukaryotic cell cycle, phases of mitosis, and control of the cell cycle;
- growth curves and their phases;
- mathematical descriptions of the growth curve;
- counting cell numbers via hemacytometer, agar plates, cell counters, and flow cytometers;
- counting cell mass via packed cell volume, wet and dry weight, and optical density; and
- scale-up.

5.1 The eukaryotic cell cycle

Let's suppose we want to harness some eukaryotic cells to produce a product for us. The cells could be yeast that can produce ethanol under the right conditions, or they may be engineered islet cells that can produce recombinant human insulin. Many different cell types can be harnessed to produce a wide array of products. As a biotechnologist working in the industry, it will be important that you have the cells produce the greatest amount of product for the lowest overall cost. If we have a cell type that makes a known amount of a given product per cell, we need to have as many cells as possible producing it simultaneously. There are, however, limits to the numbers of cells that can be grown in finite cultures. To understand these limits, we must first understand how cells grow, how they divide, and how we can get the maximum numbers of cells in our bioreactors to generate products for us. To gain such understanding, we must return to the cell cycle (Figure 5.1).

Two basic events occur during the cell cycle: cells double their DNA, and they divide. The period of DNA synthesis is known as the **S** (synthesis) **phase**, and the period devoted to cell division is known as the **M phase** (with M standing for **mitosis**, a process whereby a single nucleus is separated into two distinct nuclei, each with its own complete copy of the genome). Cells do not replicate their genomes and then immediately undergo division, nor do they divide and immediately start making more DNA;

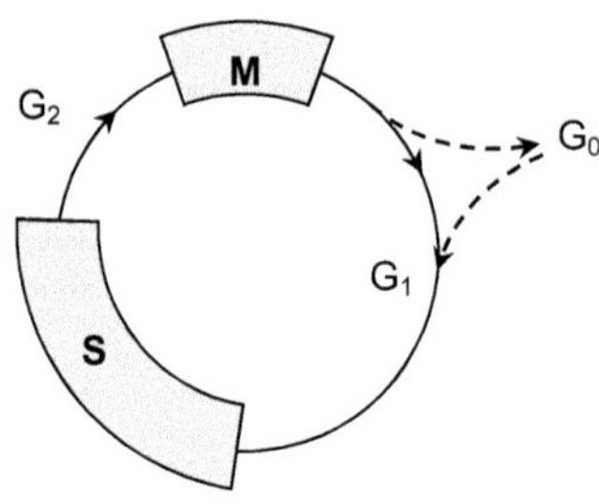

FIGURE 5.1

The cell cycle, including synthesis (S), mitosis (M), and gap (G) phases. G_0 serves as a holding point for the cell, allowing it to function normally without cycle progression.

there are gaps between the S and M phases. For an actively growing cell, the gap preceding S phase is referred to as G_1, whereas the gap between S phase and mitosis is called G_2.

Although many cells in adult humans are growing and dividing for processes such as manufacturing new blood cells, replacing skin or mucosal cells that have sloughed off, or repairing lesions due to cuts or abrasions, most of the cells in the body are not engaged in growth and division. Cells can exist in this state of existence and maintenance for an extended, indefinite amount of time. This special gap phase is conventionally drawn just outside the main cell cycle and is termed G_0 to distinguish it from G_1. Cells can remain in G_0 virtually forever (meaning for the life of the adult organism). A cell could exit M phase and detect that there is no demand for additional cells of its kind or sense that it does not currently possess the necessary energy or nutrients required for the production of a duplicate genome, so it will enter G_0. In G_0, the cell can still perform its normal life functions, such as respiration or the production of a particular substance (depending on the type of cell it is), but it will not double its genome or divide until certain conditions have been met.

5.1.1 Phases of mitosis

Let us now consider mitosis in a little more detail. Looking at the cell cycle solely in terms of mitosis, there are five stages: interphase, prophase, metaphase, anaphase, and telophase. Prophase, metaphase, anaphase, and telophase are used to describe the process of cell division. Interphase refers to the rest of the cell cycle, including S phase and the three gap phases (Figure 5.2).

5.1.1.1 Prophase

This stage of mitosis involves the condensation of DNA into chromatids. Looking at an unstained cell in interphase (Figure 5.3A), it might be easy to discern the location of the nucleus, but it would be difficult to visualize the DNA within. The nucleus of an unstained cell in prophase might look like it contained a lot of dark spaghetti because the DNA would have condensed into **chromatids** (Figure 5.3B).

5.1.1.2 Metaphase

In this stage, the cell dismantles its nuclear envelope, and the chromatids line up along an axis through the middle of the cell by way of a newly formed **mitotic spindle**. Because the cell has already performed S

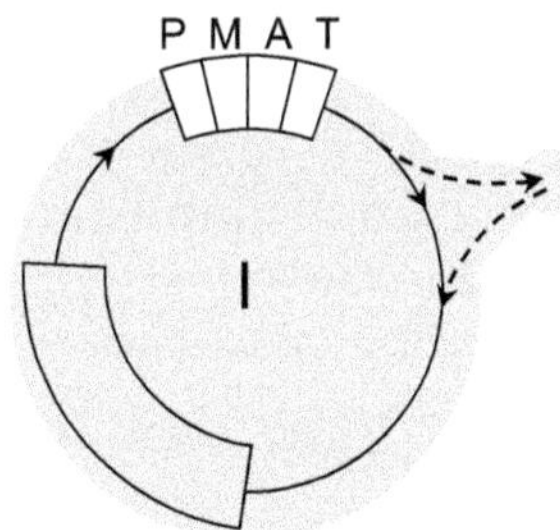

FIGURE 5.2

Another perspective of the cell cycle. Four of the five phases—prophase (P), metaphase (M), anaphase (A), and telophase (T)—are associated with the M phase of Figure 5.1. The remaining phase—interphase (I), shaded—includes everything that remains, including the S and gap phases.

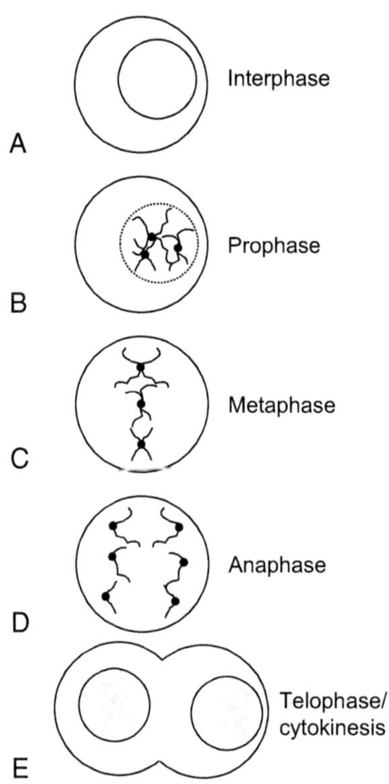

FIGURE 5.3

The stages of mitosis for a hypothetical cell with three chromosomes.

phase, it now contains two complete copies of its genome. That is why the chromosomes look like the letter X (Figure 5.3C): each X represents two copies of a single chromosome (sister chromatids), held together at the **centromere** (Figure 5.4A). The centromere is a special region of chromosomal DNA. Not only is it where the two copies of the chromosome are held together, but it is also where a structure known as the **kinetochore** forms. The kinetochore is made of several layers of protein. There are multiple insertion points for microtubules in the outer plate of the vertebrate kinetochore. These microtubules, also known as spindle fibers, connect the chromosome to **centrioles** (also known as spindle poles), which will anchor the chromosomes as they are pulled apart in the next step of mitosis (Figure 5.4B). The transition from prophase to metaphase is sometimes termed "prometaphase." It is at this stage that we see the mitotic spindle start to form and the chromosomes attaching to it. Metaphase is the point at which they line up.

5.1.1.3 Anaphase

During anaphase, the sister chromatids are pulled apart, thus completely separating the two daughter chromosomes. The pulling is accomplished through the shortening of the kinetochore microtubules. At the same time, the centrioles start to move further apart, and each chromosome is slowly pulled toward one of the two as they separate (Figure 5.3D).

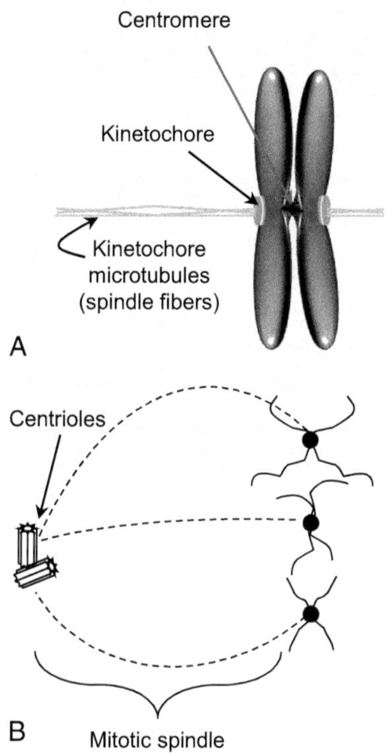

FIGURE 5.4

(A) Location of the centromere, kinetochore, and kinetochore microtubules following metaphase. (B) Attachment of mitotic chromosomes to centrioles (spindle poles).

5.1.1.4 Telophase

This step is marked by the reformation of the nuclear envelope around each set of daughter chromosomes, which decondense after reaching the spindle pole (Figure 5.3E). The cell now contains two separate, self-contained nuclei. This binucleate cell will typically then divide into two daughter cells through a process known as **cytokinesis**.

Cytokinesis involves the division of the cytoplasm through the closing of a contractile ring and is considered to be separate from mitosis. In most animal cells, cytokinesis begins during anaphase and ends slightly after telophase and the completion of mitosis. As the ring becomes smaller, there is a need for additional plasma membrane to supplement the newly forming daughter cells; the total surface area of the two daughter cells' plasma membranes will be greater than that of the parent cell. This new plasma membrane comes from the fusion of intracellular vesicles with the existing plasma membrane.

5.1.2 Control of the cell cycle

Although Figures 5.1 and 5.2 depict the cell cycle as a continuous process, the cycle does not proceed in a steady fashion like the hands on a clock. The lengths of the gap phases are variable; this is partly

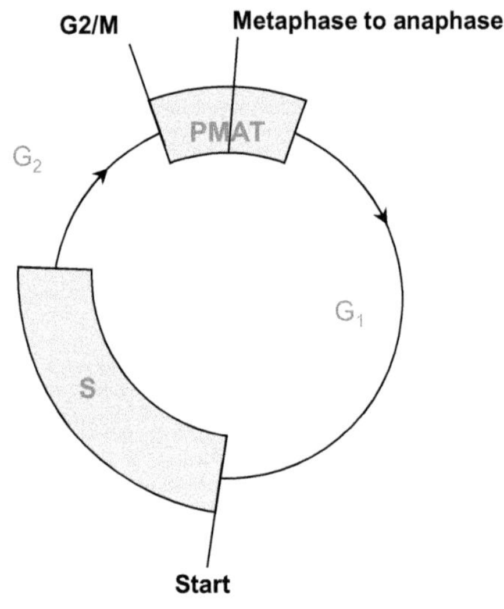

FIGURE 5.5

Cell cycle checkpoints (in *black*): start, G2/M, and M to A transition.

due to control points through which the cell monitors its own status and prevents execution of certain events when the time is not yet right. For example, replication of the entire genome is a very energy-expensive event, requiring at least $2 \times (3.08 \times 10^9)$ ATP equivalents of energy just for polymerization. If the cell does not have sufficient energy stores to complete this task, it will not even begin S phase. This particular control point in the cell cycle is called Start (Figure 5.5).

Some questions that must be favorably satisfied for the cell to proceed beyond Start and into S phase are (1) Is the energy state favorable for replication? (2) Are extracellular conditions favorable? (3) Are there enough dNTPs present to begin replication of the entire genome? (4) Are there enough nutrients? The parent cell must ensure that there will be enough mitochondria and other organelles, proteins and other macromolecular products, and raw materials for forming enzyme proteins, structural proteins, and RNA in the two daughter cells. The synthesis phase is such a taxing process that if the cell is not adequately prepared, it will die. However, if the answer to all the above questions is "yes," the cell will progress past Start and commence the DNA replication of the S phase.

There are two other checkpoints that are vital to successful cell division (Figure 5.5). One occurs at the transition from the G_2 to the M phase, when the cell again determines whether conditions are favorable for progressing into the next phase of the cell cycle: mitosis. The questions are similar to those of the Start checkpoint: (1) Is the environment favorable? Perhaps the cell is facing a large metabolic challenge. Perhaps during S phase, the environment was just fine, but now every other cell in the region is going through S phase or mitosis so the number of cells competing for nutrients has greatly increased. Either of these reasons can render the environment unfavorable because nutrients and other resources have been depleted. (2) Is the replication of the genome complete? It would not be advisable for the cell to split in half if it did not have two full genomes for the daughter cells, so the cell must ensure that S phase has been completed.

The other remaining checkpoint is located within the M phase and occurs at the transition from metaphase to anaphase. Recall that in anaphase, the chromatid pairs are separated; one major thing that could go wrong is that the separation could commence before all the chromatids have been attached to microtubules via the kinetochores. If this were to happen, one or more chromosomes would be lost during the separation, resulting in a daughter cell with an incomplete genome; such a cell would be severely challenged or die. So, this checkpoint asks the question: is every chromatid attached to a spindle fiber?

Consider the amount of DNA per cell with respect to the cell cycle. In G_0/G_1, there is a constant amount of genomic DNA per cell (Figure 5.6A). During S phase, this amount increases until there is double the starting quantity. From the end of S phase through G_2, there will again be a constant amount of DNA in the cell. During M phase, although the cell is forming mitotic spindles and separating chromatid pairs, it still contains the constant, double amount of DNA, even though it will contain two nuclei by the end of telophase. It is only after cytokinesis, at the point when the contractile ring completely closes and the parent cell is cleaved into two daughter cells, that the amount of DNA per cell is reduced back to the original value.

Consider the number of mammalian cells in each phase of the cell cycle at a given time in a culture (Figure 5.6B). The majority of the cells will be in G_0/G_1. (If all of the cells are well fed and actively growing, as should be the case in a culture plated recently, there will be little need for G_0 and the majority of cells will be in G_1.) The figure also illustrates that most of the other cells are in G_2/M. Because we can't distinguish G_2 from M in terms of DNA content per cell, the two phases are often combined and referred to as G_2/M. Everything between G_0/G_1 and G_2/M is S phase. Note that relatively few cells are in S phase. S phase costs so much energetically that the cell is committing a lot to go through it, which is why passing Start is such a key event in the cell cycle.

The relative amount of dsDNA per cell can be determined using **propidium iodide**, a dsDNA-intercalating dye that brightly fluoresces red upon stimulation. The number of cells in a particular stage of the cell cycle can be determined experimentally using this dye, with typical results similar to those shown in Figure 5.6C. If we compare panels (B) and (C) in the figure, the difference between theory and practice should become evident. Whereas panel (B) shows discrete points for the beginning and ending DNA concentrations, panel (C) shows that these values are represented by approximately normal distribution curves.

Using propidium iodide and cell cultures that have undergone synchronization of their cell cycles, the length of the cell cycle can be determined through a time course experiment. Synchronization can be achieved through the restriction of certain nutrients, thereby stalling cells at the Start checkpoint. Human cell types in culture typically have cell cycles of approximately 24 hours. Although relatively complex, mitosis only takes about one hour; S phase, on the other hand, usually lasts for about 8 hours. Other types of cells can have quite different cell cycle lengths. For example, *Escherichia coli* and budding yeasts can make it through the entire process in 20 to 30 minutes.

5.2 Growth curves and their phases

Now that we know how an individual eukaryotic cell grows and divides, we will move on to discuss how cell populations grow. We have mentioned that the length of the cell cycle for microbes is very short compared with that of human cells, which allows them to be used as cellular factories to produce molecular products because their numbers can be amplified greatly in a matter of hours. Although most of this section will focus on prokaryotic cells, the principles for eukaryotic cell cultures are the same.

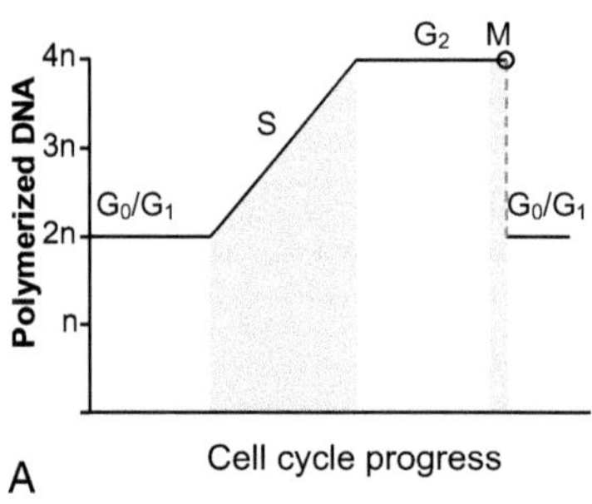

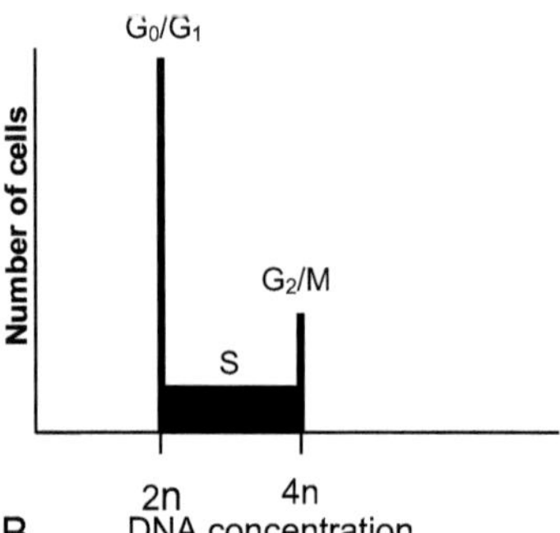

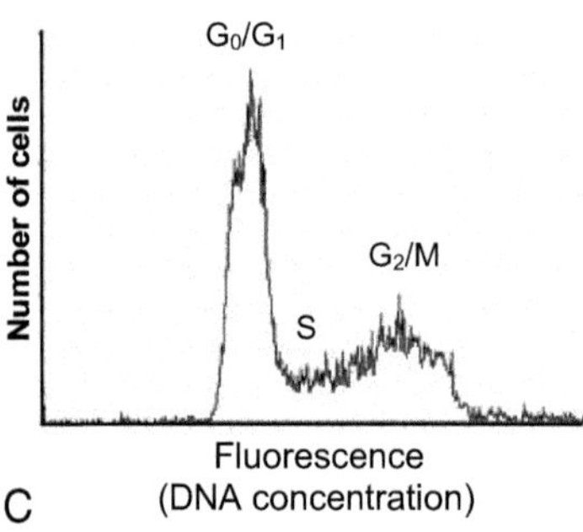

FIGURE 5.6

(A) The amount of polymerized DNA in a cell during the mitotic cell cycle, where n = an arbitrary amount of DNA, depending on the type of organism. For humans, it would be 23 chromosomes' worth of DNA. (B) The theoretical relative number of cells in each phase of the cell cycle. (C) Measured number of cells in each phase of the cell cycle, as determined by propidium iodide fluorescence (indicating DNA concentration).

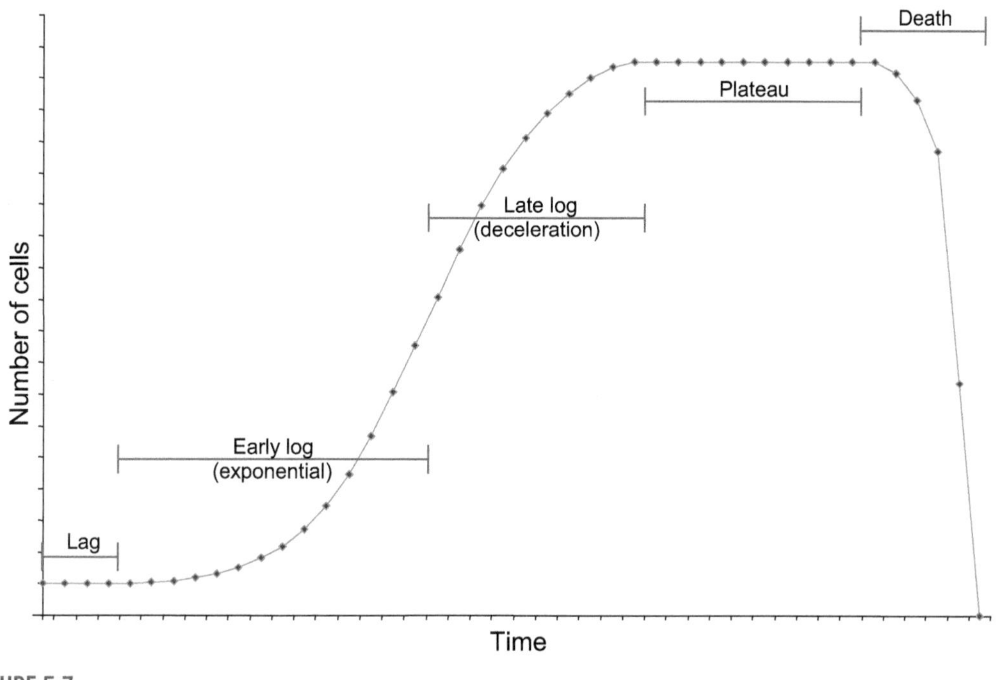

FIGURE 5.7

A hypothetical growth curve showing the progression of cell numbers in a culture over time. Basic phases have been labeled.

A graph of cell number versus time, commonly called the **growth curve**, is shown in Figure 5.7. There are several stages to this curve, both mathematical and physical, that warrant further discussion. These stages are lag, log (early and late), plateau, and death.

5.2.1 Lag phase

Every cell culture begins with the inoculation of a medium with at least one cell, although typically many cells are used. This is true for cultures in the laboratory and cultures growing on the food in your refrigerator. When the cells are first put into the growth medium, they do not begin to replicate immediately because they need time to acclimate to their new surroundings. This is called the lag phase: a period of time that must pass before the cells begin to replicate and populate the medium.

The reason for the lag phase is that after we inoculate our medium with cells, it takes a while for them to acclimate to the new environment from where they originated. Perhaps they were frozen at -80°C, maybe they were taken from another established culture, or maybe they were on the hands of an unsanitary food preparer. Whatever the previous conditions of the inoculate, the new culture environment will be different, and the cells will need time to sense their new surroundings and adjust.

5.2.2 Log phase

This is a period of exponential growth. There are two distinct regions within the log phase:

- *Early log* is a period of accelerated cell growth when cell numbers increase exponentially. Another name for this phase is the *exponential* phase.
- *Late log* is characterized by certain restrictions to cell growth. Although the slope of the curve is positive, its second derivative is negative, meaning the shape is concave down. Another word for the late log phase is the *deceleration phase.*

The log phase is associated with plentiful nutrients and ideal growing conditions. Cells in culture at this stage have warmth, a large amount of space, and plenty of food, so they divide very quickly; hence, the exponential nature of the early portion of this phase. However, in the deceleration phase, nutrients are beginning to become less plentiful. Referring back to the cell cycle, when the cycle comes around to the Start checkpoint, individual cells may sense lower nutrient availability and delay further progression to allow for the necessary buildup of dNTPs and energy stores. The deceleration in the rate of cell division is due to the gradual reduction in nutrient availability, an increase in waste concentration, and a reduction in the amount of free space in the culture container. Most cells (cancer cells being a notable exception) adhere to the principle of contact inhibition, meaning they will not be as likely to divide if they bump into other cells or the sides of a culture container. As a cell culture contains more and more cells, more and more contact will occur between them, leading to greater contact inhibition and a decrease in proliferation rates.

5.2.3 Plateau phase

In the plateau phase, although cells continue to divide, the division rate is greatly hampered. Any increases in cell numbers are offset by cell death, leaving the number of cells roughly constant in this phase.

5.2.4 Death phase

This phase is typical for cells grown in a vessel with a fixed amount of medium (a **batch culture**) for too great a time. Cell division stops as more and more of the cells die.

In the death phase, conditions are horrible. There is no discernible cell growth, and cell numbers drop precipitously because of the environment that is now virtually devoid of nutrients and may contain toxic waste products or secondary metabolites.

5.2.5 Cell mass versus cell number

One can think of growing cells in culture as a chemical reaction that begins with cells and substrates. We can inoculate some growth medium with cells on one day, and by the next day the culture vessel will contain something different. There will (we hope) be a greater number of cells and cell products, plus unused substrates.

$$\text{Original cells} + \text{Substrates} \rightarrow \text{Original cells} + \text{New cells} + \text{Products} + \text{Unused substrates}$$

Although the concept is straightforward, different mathematical formulae are required to describe cell growth in each of the phases.

We can talk about how much *cell mass* is in a culture during exponential cell growth by describing the *specific growth rate*, designated by μ_{net}:

$$\mu_{net} = \frac{1}{x}\frac{dx}{dt}, \text{ where } x = \text{cell mass concentration}, \quad t = \text{time.} \tag{5.1}$$

Note that the above equation is not entirely rigorous. The term μ_{net} is often presented as a function of (the amount of growth) - (the amount of death). Cells are always dying off for one reason or another, even during exponential growth. To keep the discussion straightforward, we have omitted the death parameter here.

Sometimes it's convenient to talk about total cell mass in a culture, but at other times we might be more interested in the *number* of cells. The above equation can be modified to reflect the *rate of replication*, designated by μ_{rep}.

$$\mu_{rep} = \frac{1}{n}\frac{dn}{dt}, \text{ where } n = \text{cell number concentration.} \tag{5.2}$$

Calculations involving cell mass concentration are more convenient to work with because x represents a real number, whereas n is a positive integer. (It is not feasible to have half of a cell.) This implies that the function for μ_{rep} is not a continuous function. Again, μ_{net} represents the change in cell mass while μ_{rep} describes the change in cell number.

Equations 5.1 and 5.2 are not always interchangeable.

We cannot always quantify cell mass and convert it to a number of cells with accuracy because μ_{rep} does not always equal μ_{net}. When cells have a lot of nutrients and a lot of space, as is the case during early log phase, they can grow to a certain size and then divide freely. This will provide a condition where all cells are roughly the same size, so $\mu_{rep} = \mu_{net}$. However, when nutrients are scarce, some cells may become larger because of growth, but they will not replicate; that is to say, they will not pass Start. This is the case in plateau phase. The mass per cell will be greater, meaning *x/n* (cell mass / cell number) will increase. In other words, the cells will tend to be bigger. We must therefore be cognizant of whether we are talking about cell mass or the actual number of cells in a culture.

5.2.6 Growth curve state: a biotech company example

Suppose you have a biotech company that uses engineered cells to produce a specific biological product. The cells only make this product when they are not stressed, meaning the supply of nutrients and space is plentiful. When there aren't enough nutrients, the cells will not have the luxury of making any extra products; they will only make what they need to keep themselves alive. This implies that when the cell culture is in the plateau phase, your company will not be harvesting much (if any) product. Also, when the culture is in the lag phase or in the early part of log phase, although there will be bountiful nutrients and space for growth, there will not be as many cells around to make your company's desired product. As a business person, you will want to maximize production, so you should strive to keep the greatest number of cells as productive as possible. In this case, profits can be maximized by keeping the cells in late log phase.

When this particular culture enters the late log phase, we will have to feed the cells because they will not yield product after entering the plateau phase. The plateau phase can also be considered a warning phase: if the cells do not attain nutrition soon, they will die. Before this happens, the cells could be "passed" to new cultures, meaning the cells in one flask can be divided among several new flasks and permitted to grow with new growth curves. If we divide one flask of cells among eight new flasks, the culture is said to be undergoing **expansion**. However, because any bioreactor facility has a finite amount of space, at some point most of the cells of a culture being **split** must be either frozen to create a **stock** or intentionally destroyed. Instead of undergoing expansion, the culture is now said to be in a **maintenance** situation.

Students are often bothered by the fact that when mammalian cells are passed, over 75% (usually 7/8 or 15/16) of them will be intentionally destroyed. This is a reasonable concern given that these cells may be producing a product that the company sells for profit; more cells would mean more profit. However, every laboratory has physical limitations—there is only so much incubator space, and rooms can house only so many incubators. Other costs must be considered, too, such as reagents, disposables, and personnel. Even if a setup generates profit, doubling its size does not guarantee increased success; one must also consider market size and other economic factors. Having business-savvy professionals in your company's employ is an important part of running a successful biotech company.

5.2.7 Be aware of the lag phase

Consider a case where a batch cell culture has been incubated for too long, and a small aliquot of cells is transferred from this plateau phase culture and placed in a new bioreactor with fresh culture medium. Referring to the graph in Figure 5.8, we see that the new growth curve has the same

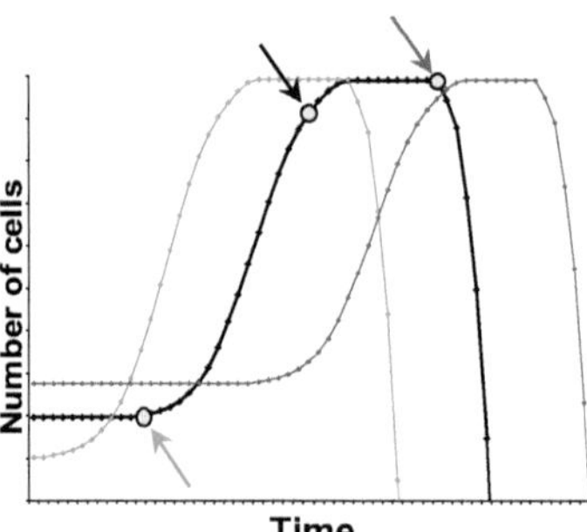

FIGURE 5.8

Growth curves to consider. Starting with a parent culture of *Escherichia coli* cells *(black curve)*, if we were to take a standard aliquot at the point indicated by the *black arrow* and start a new culture, the new culture would have a growth curve like the *black line*. Taking the standard aliquot at an earlier time *(blue arrow)* would yield a new culture with a growth curve like the *blue curve*. Note the lower starting concentration, shorter lag phase, and slightly longer plateau phase. Taking the standard aliquot at a later time *(red arrow)* would yield a new culture with a growth curve like the *red curve*. Note the higher starting concentration, longer lag phase, and slightly shorter plateau phase. The maximum number of cells in each culture should be the same because the results of contact inhibition should be unchanged for these cells. (Note that the curves do not represent actual values but are drawn to illustrate the principles mentioned here.)

elements we've already covered: it has the same general shape and the same phases as before but a new y_0 and a different lag phase length. The length of the lag phase is related to how different the new culture environment is from the old one. For instance, taking cells from the end of lag phase or the very beginning of the log phase and starting a new culture with them should yield a relatively short lag phase for the new culture because the difference between the media should be very small. However, starting a new culture with cells from a plateau phase culture should yield a longer lag phase because the old and new culture media are very different. The cells may have to turn on or off many genes because they are being transferred from a nutrient-depleted environment to one that is rich in resources. The transferred cells first must detect the new surroundings, then begin transcribing genes that are appropriate for the new surroundings. Because the cells will begin growing and dividing again, they will have to produce enzymes that will aid with the production of dNTPs, DNA polymerase, transcription factors, cytoskeletal proteins, and so on. The further away the parent culture was from the lag phase on its own growth curve (every culture has its own growth curve), the longer the next lag phase will be. That's why, instead of taking our culture from late in the plateau phase (which might make sense because there will be a maximal number of cells), we might choose to pass the cells to a new culture during late log phase to shorten the time needed to bring them back into exponential growth. Again, this discussion is for cells grown in batch.

5.2.8 Cryptic growth

There does come a point in the life of some cell cultures when, once a large number of cells begin to die, other cells cannibalize them. This small increase in nutrient availability might allow for a brief period of cell growth at the end of the plateau phase, a period called **cryptic growth**. It is called cryptic because, from a naïve perspective, the medium should be virtually depleted of nutrients, so it is not obvious why there would be an increase in cell numbers right before the death phase. We know now, however, that nutrients are made available by the death and lysis of cells. An example of such a growth curve is given in Figure 5.9.

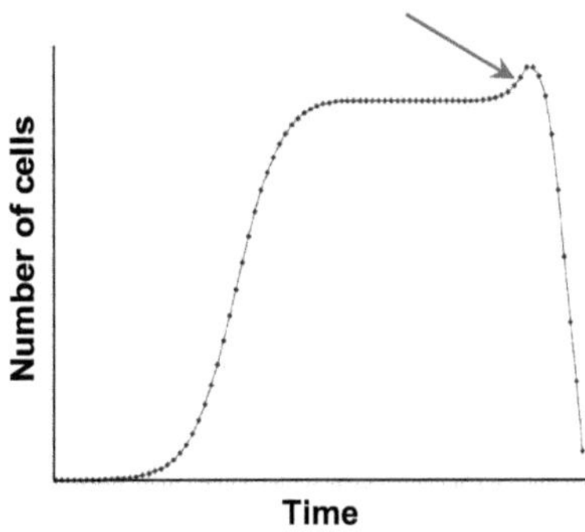

FIGURE 5.9

A hypothetical growth curve with a cryptic growth phase, denoted by the *arrow*.

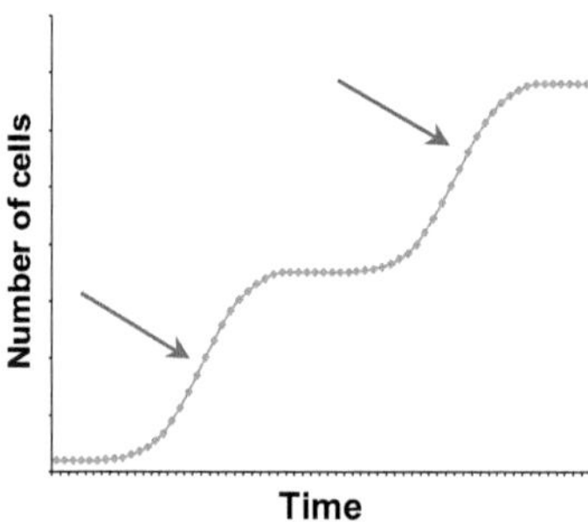

FIGURE 5.10

A hypothetical growth curve representing diauxic growth, in which the culture switches to a secondary carbon source to achieve a second exponential growth phase. The two exponential phases are indicated by *red arrows.*

5.2.9 Diauxic growth

The growth curve shown in Figure 5.10 illustrates an example of **diauxic growth**; note that there are two log phases. Diauxic growth is characterized by cells that utilize two carbon sources, where there is a preference for one over the other. For instance, cells that have the *lac* repressor can use either glucose or lactose for energy. In a medium that contains both sugars, they will preferentially use glucose, but when glucose becomes scarce (corresponding to the first plateau phase in the growth curve), they can alter which genes they transcribe, allowing them to utilize lactose for energy. This switch to a new carbon source permits a second exponential phase of growth.

Returning to the example of a biotech company using cells to produce a commodity, we originally reasoned that passing cells to a fresh culture environment during the deceleration phase was preferable; however, this is not always the case. On many occasions, cells in culture do not produce a product of interest until they are in the plateau phase (e.g., some antibiotic-producing bacteria). By the standards of human society, the microbial world can be very brutal. Some microbial cells will freely kill other cells when times get tough. When resources are limited and microbes are struggling for their own survival, certain microbes will start making antibiotics to kill other cells, especially of other cell types—so if you happen to run a biotech company that produces antibiotics, you might choose to let your cell cultures progress into the plateau phase to maximize product generation. Similarly, if you work with a type of cell that only expresses the gene of interest when the cells are utilizing an alternative carbon source, such as during the second phase of diauxic growth, you might start your cultures with glucose to rapidly increase cell numbers. After an adequate population density has been generated, you might let the cells metabolize the remaining glucose in the culture and then introduce the second carbon source so that the culture begins making the product of interest. We will see in a later chapter that this idea of rapidly growing cells and then changing the culture conditions to obtain a desired bioproduct has been used for thousands of years.

5.3 Mathematics of the growth curve

5.3.1 Exponential phase (early log)

We have already mentioned exponential cell growth and that this phase of the growth curve is made possible because nutrient concentration is reasonably high, so there is relatively little to impede the cells from growing as fast as they can. Cells will progress unimpeded through the cell cycle and divide;

one cell will become two, two will become four, four will become eight, and so on. If we were to model that mathematically, we would get an exponential curve. In the exponential phase, because nothing is being inhibited (the cell is growing and dividing as quickly as possible), $\mu_{net} = \mu_{replication}$, which means the equations that describe cell number and cell mass are interchangeable.

The differential (originally shown as Equation 5.1) can be split up to allow integration of both sides:

$$\mu_{net} dt = \frac{1}{x} dx$$

$$\int_0^t \mu_{net} dt = \int_{x_0}^{x_t} \frac{1}{x} dx$$

$$\mu_{net} t = (\ln x_t - \ln x_0) = \ln\left(\frac{x_t}{x_0}\right)$$

$$e^{\mu_{net} t} = \frac{x_t}{x_0}$$

$\boxed{x_0 e^{\mu_{net} t} = x_t}$ → This means the cell mass at time t is equal to (the initial amount of cell mass) × e raised to [(the rate of change in cell mass concentration) × (the amount of time that has elapsed)].

5.3.1.1 Doubling time

Doubling time, indicated by $t_{1/2}$, is the amount of time it takes for cell mass (or cell number) to increase to twice the starting concentration. Another way to say that is $x_t = 2x_0$.

Plugging this relation into the equation we just derived:

$$x_0 e^{\mu_{net} t} = 2x_0$$
$$\Rightarrow e^{\mu_{net} t} = 2$$
$$\Rightarrow \mu_{net} t = \ln(2)$$

$\boxed{\Rightarrow t = \frac{\ln(2)}{\mu_{net}}}$ → The amount of time needed to double cell mass is equal to 0.6931/(rate of cell mass increase).

Sometimes $t_{1/2}$ is written τ_d. Keep in mind that we are still discussing the exponential phase, and in the exponential phase, we can go back and forth between cell mass and cell numbers, so the time needed to double cell mass is equal to the time needed to double cell number:

$$t_{1/2} = \tau_d = \frac{\ln(2)}{\mu_{net}} = \frac{\ln(2)}{\mu_{replication}} = \tau_d'$$

As a simple example, suppose a certain cell population has a doubling time of 20 minutes, and we start with 200 of these cells in culture. How many cells are in the culture after 1 hour? First, determine the number of doublings: ($t/\tau_d = 60\,\text{min}/20\,\text{min} = 3$ doublings), then plug into:

$$n_0 \times 2^{(\#\ of\ doublings)} = 200 \times 2^3 = 200 \times 8 = 1600 \text{ cells}.$$

5.3.2 Deceleration phase (late log)

Left unattended, the growing culture will undergo deceleration because nutrient supply is no longer seemingly limitless, and waste products are beginning to build up. This is a phase of unbalanced growth, so x and n are no longer interchangeable. As a result, the doubling time is different for cell mass versus cell number. Physically, the cells will be larger before they start to divide. Part of the reason for this is that, when the cell is replicating its DNA, it is using up a lot of dNTPs, which represents a great amount of energy. Keep in mind that every base pair replicated requires two dNTPs—one for each new base being added to the growing double-stranded chains. This energy is in addition to the resources required to manufacture new deoxynucleotides and dNTPs. The ever-decreasing availability of nutrients in the medium is enough to hinder cells from passing the Start checkpoint as easily as before, hence the larger average cell size. Another reason that $\mu_{net} \neq \mu_{replication}$ is that the cells are no longer focused on replication but are starting to secure their own survival.

In Figure 5.7, the amount of time spent in the exponential phase is roughly equal to that spent in the deceleration phase, but this is not necessarily the case for a given bacterial population. As the energy source (carbon source) present in the medium is metabolized by the cells, their growth will eventually be limited. The relation between μ_{net} and the concentration of the carbon source is given by:

$$\mu_{net} = \frac{\mu_{\max} S}{k_S + S}$$, where S = the concentration of the energy source (the substrate) and k_s = a constant describing the affinity of the microorganism for the energy source.

If we set the substrate concentration to a value equal to k_s, the equation becomes:

$$\mu_{net} = \frac{\mu_{\max} k_S}{k_S + k_S} = \frac{\mu_{\max}}{2}$$

This implies that k_s can be defined as the substrate concentration that will produce a growth rate equal to half of the maximum possible rate. This constant serves as a measure of an organism's affinity for an energy source.

During the early log phase, when there is a very large relative concentration of substrate (S), the rate of growth (μ_{net}) will be as fast as it can be (i.e., when $S >> k_s$, μ_{net} will mathematically approach the asymptotic value μ_{max}).

Another use for k_s has to do with where the transition between exponential and deceleration phase appears on the growth curve. A low k_s implies that the microorganisms have a high affinity for the substrate, so they will latch onto it efficiently enough that deceleration will not occur until a greater percentage of the energy source has been used, and the deceleration phase will be relatively short. If k_s is high, affinity for the substrate will be low, meaning the binding efficiency to the substrate is low, and the microorganisms will begin to sense that the energy source is becoming scarce at a higher absolute concentration of substrate. The deceleration phase will last longer (Figure 5.11).

5.3.3 Plateau phase

Recall that the net growth rate in this phase is equal to zero. There are enough nutrients for cell maintenance, but continued large-scale replication is not supported. There may still be growth of some

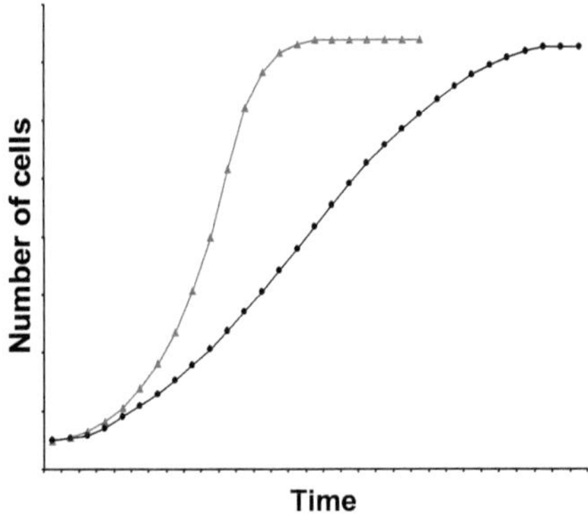

FIGURE 5.11

When cells have a different affinity for their primary energy source, the shape of the growth curve will be altered. When there is high affinity (low k_s), the substrate will be bound efficiently, even at relatively low concentrations. leading to a short deceleration phase *(red curve).* When the binding affinity is lower (high k_s), the cells will sense a lower substrate concentration earlier. The deceleration phase will begin earlier, take longer, and be more gradual.

individual cells but, as pointed out earlier, net growth = (total growth) - (death). The bioreactor and medium cannot support any more population growth.

Two scenarios may occur regarding cell mass concentration in relation to the total number of cells. Either

$\mu_{net} = \mu_{rep} = 0$ (i.e., there are no cells dividing) or

$\mu_{net} = 0$ and $\mu_{rep} < 0$ (cell mass concentration is constant while the number of viable cells is decreasing).

However, there will necessarily be a change in cell mass due to the maintenance of life. This change will be proportional to the initial number of cells:

$$\frac{dx}{dt} = -k_d x$$

$\Rightarrow x_t = x_0 e^{-k_d t}$, in which x_0 = cell mass concentration at the beginning of plateau phase and k_d = a constant describing the rate at which the cell metabolizes its own reserves for the maintenance of life.

Often during the plateau phase, the cells will begin to produce secondary metabolites. These metabolites may have value to the laboratory controlling the culture; the plateau phase is when some companies get their money. In the exponential phase, cells are dividing and are focused on replication. In the plateau phase, cell behavior is more concerned with survival.

Secondary metabolites are different from primary metabolites. Primary metabolites are produced during the exponential phase and are the natural result of the normal processes of the cell. For instance, even as you read this, your cells are converting glucose into pyruvate, and the pyruvate is probably being converted to CO_2 and water. The pyruvate, CO_2, and water are all primary metabolites. Secondary metabolites are not directly involved in the chemical processes that keep an individual cell alive. In the case of the mold *Penicillium chrysogenum*, penicillin is a secondary metabolite that is produced when nutrients become scarce. Under laboratory conditions, even though a culture of *P. chrysogenum* might be pure (meaning the only living things in the culture are *P. chrysogenum*), the cells still produce the antibiotic as nutrients become scarce, regardless of whether they are truly competing against other bacterial strains. This is because the genetics of the bacteria in that culture dictate that the antibiotic genes are expressed in times of limited resources.

It is now appropriate to reintroduce the question "If we have a biotech company that deals with cell cultures, during what stage of the growth curve should cells be passed into a fresh culture?" Earlier, when we were only concerned with cell numbers or the production of primary metabolites, the answer was the deceleration phase. However, if the cells are making a product for us in the plateau phase, we would want to maintain them in the plateau phase. The means by which this can be accomplished will be discussed shortly.

Consider the situation in which a microbe produces something that is toxic to itself, such as yeast producing alcohol. Recall from earlier in this book, from the discussion on sterilization, sanitization, and aseptic conditions, that alcohols can be toxic to cells. Yeast are commonly used to produce the alcohol (ethanol) in fermented products such as beer, even though high concentrations of ethanol will kill the yeast cells themselves. Initially, there will be little to no alcohol in the cell culture, but the concentration will increase slowly as the yeast cells metabolize glucose in the absence of oxygen. When the ethanol concentration reaches a certain level, the yeast will stop net growth; this marks the beginning of the plateau phase. We might ask how we can get around this to obtain a greater concentration of alcohol from the yeast culture. One method could be to extract the alcohol from the culture, perhaps through filtration, distillation, and replacement of the medium.

In general, there are several ways to extract secondary metabolites from a culture. One way is to complex the product with a **nonmetabolite**—a molecule that will not be used by the cell during metabolism. Such a molecule could be used to bind the secondary metabolite, allowing us to pull the complex out of solution. This form of harvest will allow the cells to continue to produce the secondary metabolite without toxic effects to themselves.

5.3.4 Death phase

The death phase is very similar to the plateau phase. The plateau phase is zero order on a logarithmic scale (N is a constant), but it could also be considered a special case of first order where $dN/dt = c = 0$, meaning the graph is linear but the slope is 0. The death phase is also a linear phase on a logarithmic plot, but it has a negative slope, equal to $-k_d$.

$$\frac{1}{x}\frac{dx}{dt} = -k_d$$

$$\int \frac{1}{x}dx = \int -k_d dt$$

$$\ln x = -k_d t + c$$ ← Equation for a line, on a logarithmic scale.

Using definite integrals, $x_t = x_0 e^{-k_d t}$

5.4 Counting cell numbers

To be able to use the above equations, one must have a means of quantifying cell mass and/or cell numbers. Let us first discuss methods for determining cell numbers.

5.4.1 Hemacytometer

One common and relatively inexpensive way of counting cells is with a hemacytometer (Figure 5.12). In Figure 5.12B, note that the main square is divided into nine regions, of which five have been labeled. The etched lines on the hemacytometer are very precise. For instance, the lines inside area E are 0.2 mm apart, and each of the squares A–E denotes a standard area of 1 mm^2. The subdivisions of these areas permit easier counting. The coverslip accompanying the hemacytometer is also specific, such that when 10 μL of cell suspension are loaded into one of the chambers, the height of the liquid will be 0.1 mm. This means the volume of liquid over any one of the areas A–E is

$$\begin{aligned}&1\,\text{mm}\times 1\,\text{mm}\times 0.1\,\text{mm}\\&=0.1\,\text{mm}^3\\&=1\times 10^{-4}\,\text{cm}^3.\end{aligned}$$

A good reason to convert to cm^3 is that $1\,cm^3 = 1\,mL$; we can now say that the volume is 1×10^{-4} mL. This means that if one were to count the number of cells in area E, the total could be multiplied by 1×10^4 to obtain an approximation of the number of cells per milliliter. Counting the cells in areas A through D and dividing by 4 will give an even better representation of the cell concentration in the original suspension.

Counting cells using a hemacytometer is a very straightforward process, with one possible exception. Note the cells denoted by blue arrows in Figure 5.13; each one could be considered to lie either inside or outside the region being counted. When counting, it would be inefficient to also keep track of fractions of cells, so a binary system is used whereby a cell is either counted or not, according to defined criteria. A common practice is to count the cells that touch the left border of the counting area but not those that touch the right border, based on the assumption that the probabilities of a cell touching either border are the same. The same reasoning applies to counting cells that touch the top border but not the bottom border.

Using the above rules and scale-up factor, you should be able to count 43 cells in the region shown in Figure 5.13 (verify this for yourself). This would lead us to approximate the concentration of cells in our original suspension as 430,000 cells/mL. If the total volume of the original suspension were 8 mL, this would imply that we have 3,440,000 cells.

In terms of microbial cell growth, that's fine. You write down the number and decide whether the concentration is good enough to accomplish the next thing you need to do (e.g., isolating a desired mass of DNA). If you are working with a mammalian cell culture, you might need to use a specific number of cells, for example, to seed a porous scaffold for tissue engineering. Determining the volume of cell suspension that will contain a specific number of cells is a common practice for many laboratory applications.

Consider a gene delivery experiment that will involve the transfection of 100,000 canine skeletal muscle cells. We would typically grow such a culture to late log phase and trypsinize it (meaning we use an enzyme, trypsin, to cleave proteins used for cell adhesion so that the cells will no longer stick to the culture flask and will float in the solution). The cells would then be treated to slow the activity of the trypsin, and the resulting suspension collected and centrifuged to force the cells into a solid pellet, thus allowing us to pour off the trypsin-containing supernatant. The cells would then be resuspended in a known volume of medium for quantitation of cell number. Suppose that after performing these steps,

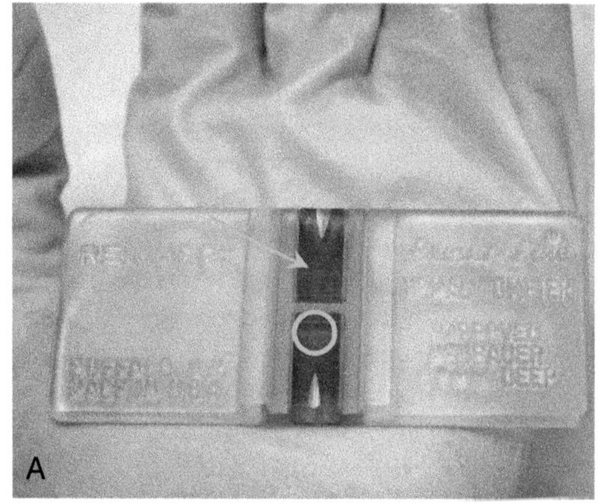

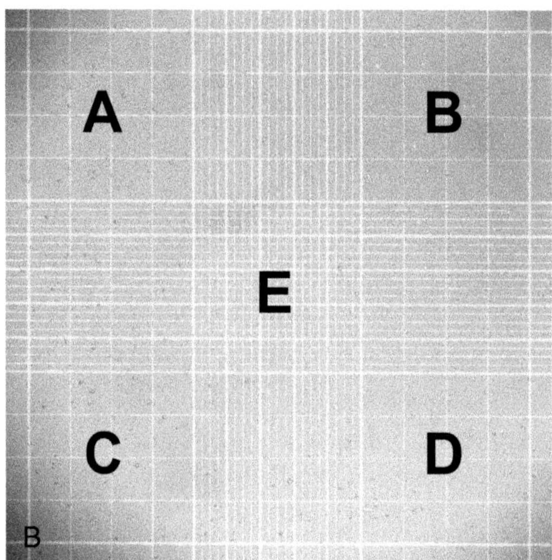

FIGURE 5.12

(A) A standard hemacytometer. The etched lines of the two counting grids are barely visible with the naked eye, being denoted here by a *blue arrow* and *circle*. (B) The area in the *circle* above as viewed under a microscope. With a 4× objective, nine distinct sub-areas are clearly visible. The areas A–E are used to determine the number of cells per milliliter.

10 μL of the cell suspensions loaded into a hemacytometer and subsequent counting yielded an average of 77 cells per counting area. The question is now "How much of this cell suspension should be plated for a 100,000-cell transfection experiment?"

$$77 \text{ cells/counting area} \Rightarrow 770{,}000 \text{ cells / mL}$$

$$\frac{770{,}000 \text{ cells}}{1 \text{ mL}} = \frac{100{,}000 \text{ cells}}{x \text{ mL}}$$

$$x = 0.1298 \text{ mL} = 129.8 \text{ μL}$$

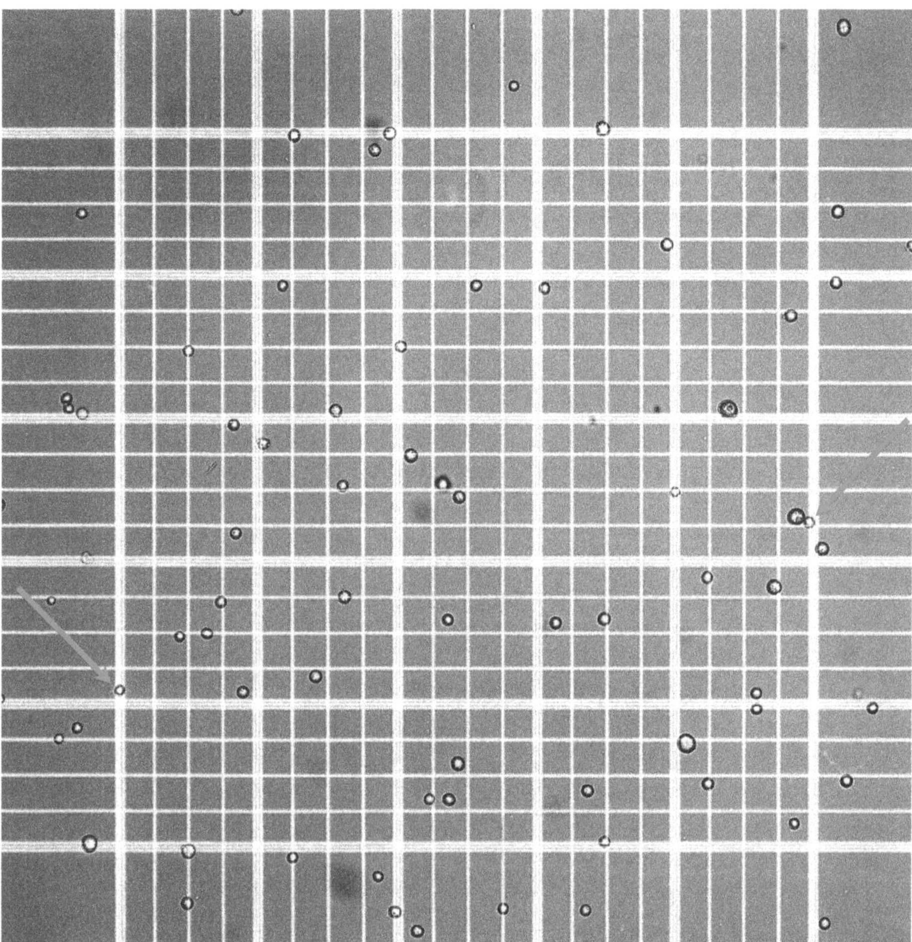

FIGURE 5.13

The region E of a hemocytometer (Figure 5.12), magnified using a 10× objective lens. Note that this level of magnification is needed to be able to count cells; the same cells are present in Figure 5.12B. Also note that sometimes cells will fall on boundaries *(blue arrows)*, making it confusing as to whether a cell is to be counted, not counted, or counted as half a cell. The simple rule of counting cells on left and top borders, but not counting cells on right and bottom borders, easily removes this quandary.

We would plate 129.8 μL into one cell well, let it grow overnight, and the next morning we would be able to transfect roughly 100,000 cells. One might ask why the cells do not multiply during this incubation period; there are three reasons. First, after taking the cells from the flask and transferring them into a new culture environment, there will be a lag phase, as discussed earlier. Second, because the cell cycle for mammalian cells is about 24 hours and the incubation time is overnight, there will not be enough time for a complete population doubling. Third, some cells will die, and some will possibly make it to cell division. Although the actual number of cells transfected the next day will probably not be exactly 100,000, having a standard number of cells plated per well will help ensure that roughly the same number is being transfected no matter how many wells, or hours following plating, we use for our experiments.

5.4.2 Agar plates

For bacterial cultures, one can determine cell number with the aid of agar plates. If we were to evenly plate a given volume from a cell culture onto an agar plate and allow it to grow overnight, the next morning we should be able to count the number of colonies on the plate with the naked eye (Figure 5.14). Each of these circular bacterial colonies originated from a single cell that was plated the night before. So, if we plated 100 μL of a bacterial culture and the next morning, we saw 100 colonies growing on the plate, we could infer that the culture contained one **colony-forming unit** (CFU) per microliter when the cells were originally plated. It may be easy to think of a CFU simply as being a bacterium, but this would not be correct. Dead and most dying bacteria will not form colonies, so the number of CFUs rarely equals the number of cells in an aliquot; it indicates the number of originating cells that were able to grow on the agar to form colonies.

We could take a sample of cells from our culture, dilute the sample with medium, take a known volume (e.g., 100 μL) from it, and evenly distribute it over the surface of an agar plate. After an overnight incubation at an optimal temperature (37° for *E. coli*, 30° for yeast), the number of colonies can be counted and divided by the volume of culture that was initially plated to yield the CFU concentration at the time the cells were plated.

5.4.3 Cell counters and flow cytometers

Consider two electrodes connected to a power source with an ohmmeter in the circuit. If the electrodes are placed into an electrolyte solution, then some current will be able to pass between them, and the current (the lack of resistance) can be noted. If the electrodes are very small and very close and we place a single cell between them, an increase in resistance will be detected because the plasma membrane, made largely of phospholipids (see Chapter 2), is a poor conductor of electricity. Figure 5.15 illustrates how these principles have been applied to produce a cell counter, which is a special type of particle counter.

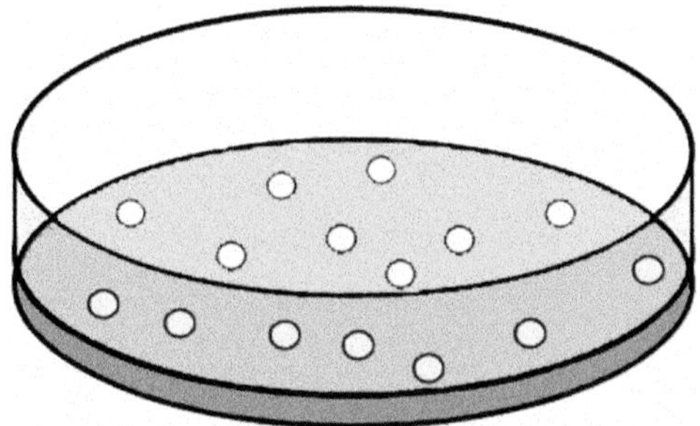

FIGURE 5.14

Agar plates can be streaked with a known volume of bacterial cell suspension to get an idea of cell concentration. Presuming they do not overlap, the resulting circular colonies can each be assumed to have come from individual cells that were able to proliferate. Such cells are known as colony-forming units.

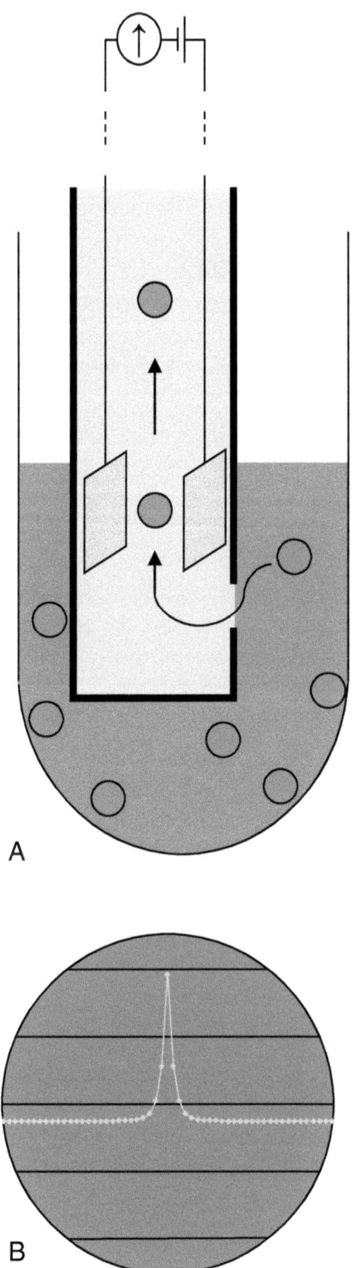

FIGURE 5.15

(A) Schematic of a cell counter. Cells *(green)* and medium are drawn up the collection tube via a vacuum. When a cell passes between the two electrodes *(gray)*, a momentary increase in resistance between the electrodes will be detected (B).

Cell medium, regardless of the type of cell for which it was designed, contains salts, because otherwise the environment would be hypotonic, and the cells would swell/burst. As such, cell media conduct electricity. Negative pressure is used to pull cells into and up a small cylinder that contains an electrode. In the cell counter, the cylinder contains an aperture that should ideally allow no more than one cell to pass at a time. A second electrode resides on the other side of the cylinder. As a cell enters the cylinder and passes between the two electrodes, a momentary increase in resistance is detected on the ohmmeter; the signal is interpreted as one cell. At the same time, the total amount of fluid that has been pulled through the cylinder is monitored so that the number of cells per unit volume can be determined. This calculated concentration is usually output as cells per milliliter.

Flow cytometers work in a similar fashion to cell counters, except that cells are detected optically instead of via changes in resistance. A laser is directed across the inlet tube, and cells are detected when the laser light is scattered away from the detector on the other side.

5.5 Counting cell mass

5.5.1 Packed cell volume

The preceding three methods are means by which one can determine cell numbers, but it can be faster or less expensive to determine cell mass. One straightforward means of obtaining such values is via the **packed cell volume**. Starting with cells in suspension, the culture is centrifuged in a special graduated conical tube. The resulting pellet is measured against the graduations to determine the packed cell volume. If the density of the specific cell type is known, then cell mass can also be calculated.

5.5.2 Wet and dry weight

Perhaps a more precise application of the above method is in determining the **wet weight** of the cells. Once again, cells undergo centrifugation to form a pellet, then the supernatant is poured off and the pellet weighed. The weight of the tube plus pellet is determined, and the tare weight of the empty tube subtracted to yield the wet weight of the cells. There is no need to utilize an average cell density because mass is determined directly via a balance. This method is very straightforward and is routinely used to produce an estimate of cell mass.

Whereas determination of the wet weight is simple, there is error associated with the variable amount of liquid remaining over the pellet and trapped between individual cells after the supernatant has been removed. A more accurate estimate of cell mass can be made using the **dry weight** of the cells. The procedure for determining the dry weight is essentially the same as above, except that after the first centrifugation the cells are washed and undergo a second centrifugation, followed by drying of the pellet. The wash step is very important. Without it, salts, proteins, carbohydrates, and other constituents of the extracellular environment will contribute to the weight of the pellet, even after it has been dried.

5.5.3 Optical density

The spectrophotometer can be used to determine cell mass for Gram-negative bacteria such as *E. coli*. Optical density measurements are taken at 600 nm (red light) in plastic cuvettes, and the reading is compared to a calibration curve. The calibration curve must be generated for the specific spectrophotometer being used because optical density relies in part on light scattering, and readings will vary depending on the distance of the photodiode detector from the sample (which may vary between machines). Generation of the calibration curve is time consuming and often utilizes other methods (e.g., agar plates) to determine the cell counts of the standards. However, once a reliable curve has been generated, it can be repeatedly used to determine cell mass concentration for many experiments to come.

It might be interesting to note that—instead of the quartz cuvettes used to determine the absorbance of solutions containing DNA/RNA—plastic cuvettes are used for determining the optical density of bacterial cultures because (1) quartz cuvettes are over 100 times more expensive than plastic, and (2) at 600 nm, the light wavelength used is of relatively low energy and will not induce autofluorescence, unlike the 260 nm (ultraviolet) light that is used to determine DNA concentration. There is no need to use expensive quartz cuvettes when autofluorescence is not an issue.

5.6 Scale-up

Although a process might work on a very small scale, making it work on a larger scale involves more than just using bigger tubes. This issue of scale-up is very important for the biotechnologist, whether it be used for producing viruses for transduction on an industrial scale, large amounts of bacterial cell culture for production of recombinant protein, or large amounts of herbicide to selectively protect genetically modified plants. To appreciate the significance of scale-up more fully, let us address the problem in terms of increasing the volume of bacterial cell cultures. We'll begin with some basic geometry.

Consider two cylinders, the first with radius and height of 1 (arbitrary units) and the second with radius and height of 2. Using $v=\pi r^2 h$, the volumes of the two cylinders are found to be equal to:

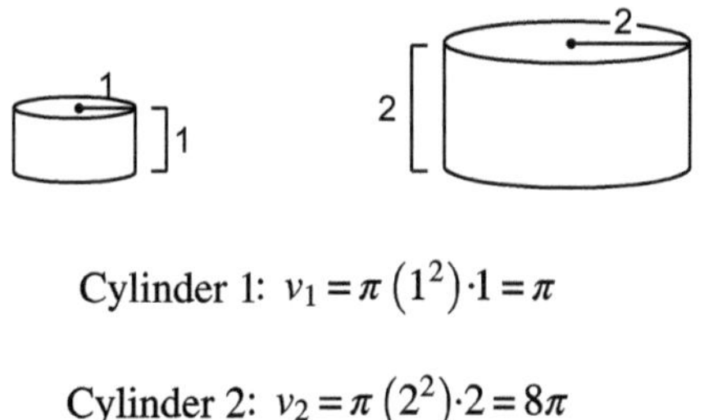

$$\text{Cylinder 1: } v_1 = \pi\left(1^2\right)\cdot 1 = \pi$$

$$\text{Cylinder 2: } v_2 = \pi\left(2^2\right)\cdot 2 = 8\pi$$

While the ratio of height to radius has been held constant, doubling the radius and height produces an eightfold increase in volume.

We will now look at the general case. Consider two more cylinders, with r/h being held constant at a value of k_1. Because of this relation, we will be able to simplify the volume equation by removing one of the variables:

$$r/h = k_1$$

$$\Rightarrow r = k_1 h$$

$$\Rightarrow v = \pi(k_1 h)^2 h$$

$$= \pi k_1^2 h^3$$

Now, if we wish to scale up the dimensions of our first cylinder by a factor of c, still holding $r/h = k_1$, we can determine a general relation for the change in volume:

$$h_2 = ch_1$$

$$\Rightarrow v_2 = \pi(k_1 h_2)^2 h_2$$

$$= \pi k_1^2 h_2^3$$

$$= \pi k_1^2 c^3 h_1^3$$

So, although the heights are scaled by $h_2/h_1 = c$, the volumes are scaled by:

$$v_2/v_1 = \frac{\pi k_1^2 c^3 h_1^3}{\pi k_1^2 h_1^3} - c^3$$

Check this against the original problem, in which the scaling factor was equal to 2. We found that the cylinder volumes $v_1 = \pi$ and $v_2 = 8\pi$, so $v_1/v_2 = 8 =$ (the scaling factor)^3.

Consider the same problem of scale-up for a pair of cylindrical culture vessels, except this time we will deal with adherent cells (cells that are attached to the walls of the culture vessel). For adherent cells, surface area is of key interest. As before, we will impose the restriction that the ratio of radius to height is constant:

$$r/h = k_1$$

$$\Rightarrow r = k_1 h$$

Surface area $\equiv SA = 2\pi rh$ *(Note that we are not considering the possibility of cells growing on the top or bottom of the reactor, for simplicity.)*

$$= 2\pi(k_1 h)h$$

$$= 2\pi k_1 h^2$$

Again, scaling up by a factor of c (still holding $r/h = k_1$), we can determine a general relation for the change in surface area:

$$h_2 = ch_1$$

$$\Rightarrow SA_2 = 2\pi\,(k_1 h_2)\,h_2$$

$$= 2\pi k_1 h_2^2$$

$$= 2\pi k_1 (ch_1)^2$$

$$= 2\pi k_1 c^2 h_1^2$$

$$\Rightarrow SA_2/SA_1 = \frac{2\pi k_1^2 c^2 h_1^2}{2\pi k_1^2 h_1^2} = c^2$$

The point of all of this is that when we hold the ratio of radius to height constant, volume varies as a *third*-order relation with the scaling factor, whereas surface area varies as a *second*-order relation. This may become more intuitive if we recall that volume is a three-dimensional descriptor, whereas surface area can be reduced to two dimensions. These facts take on greater importance when the cells of interest can grow both adherently and nonadherently; certain bacterial cultures are one example. Scale-up of the nonadherent cells is a matter of volume, whereas scale-up for adherent cells is largely a matter of surface area.

Consider a culture in which there are adherent cells growing on the bottom and the sides of the vessel but also nonadherent cells growing in suspension. Scale-up becomes a more interesting problem in this case because the amount of surface area scales with r^2 but the volume scales with r^3. If the adherent cells produce more product per cell than the nonadherent cells, scaling up a cylindrical vessel without a change in geometry will be a losing proposition: the cost of medium and laboratory space requirements will increase faster than the amount of product. A change in geometry should be considered.

Now consider the bioreactor shown in Figure 5.16, in which a stir bar constantly mixes the medium to bring oxygen to the cells. If we were to scale up the volume of this bioreactor while leaving the delivery of oxygen to the cells unaffected, the rate of mixing would have to be increased, perhaps by increasing the speed of the stir bar (rpm). However, this increase in rpm also increase the amount of shear stress placed on each cell. Increasing the volume while keeping the vessel shape, rate of mixing, and shear stress constant cannot be achieved. The biotechnologist should be aware of which parameters are most important and try to control them, realizing that other parameters will not scale at the same rate.

FIGURE 5.16

A spinner flask, commonly used for tissue engineering applications.

A variety of relations must be considered during the scale-up of cell cultures ($\propto$ means "is proportional to," or "scales with"):

- surface area (SA) $\propto r^2$
- volume (V) $\propto r^3$
- pump rate (Q) $\propto nr^3$ (n = rpm)
- energy input (P) $\propto n^3r^5$
- energy input/volume (P/V) $\propto n^3r^2$
- pump rate/volume (Q/V) $\propto n$
- shear rate (laminar flow) (γ) $\propto n$
- shear rate (turbulent flow) (γ) $\propto n^{3/2}$

Mathematically we can see there is a difference between these parameters. Although we can control any single one of them, we cannot simultaneously control them all.

We just saw that rpm comes down to n, and P/V comes down to n, so rpm and P/V can be controlled concomitantly. Because P/V controls oxygen availability, if we increase the pump rate while keeping the volume (and radius) constant, we will have more oxygen available (in a linear fashion). If we have a culture mixed slowly versus a culture mixed quickly, which one will be mixed with the air above it more quickly? The one with the faster mixing; P/V gives you that. Alternatively, you can control the Reynolds number. Consider a bioreactor in which oxygen is being bubbled up through the culture; it has sieve plates that allow the oxygen to get through, but as the oxygen bubbles strike the plates, there is some turbulence, which aids mixing. If we were to scale up this bioreactor, we could do so in a way that would preserve the geometry of the bioreactor, but oxygen availability would be altered. We could try to preserve the oxygen concentration by increasing the flow rate of oxygen, but then we would also be increasing the amount of turbulent force at the sieve plate, which could damage the cells. Once again, you can control some, but not all, of the scale-up parameters.

Box 5.1 Scale-up example

In case you are still not a believer, let us take a sample problem and work it two ways. Make sure you read both methods. Everything should become clear...

Consider a 2-L batch fermentation system in which 75% of the (intracellular) target product was associated with attached cells and 25% was associated with cells in suspension. The total output of this reactor was 2 mg of product per liter per day.

Now suppose we scale up to a 20,000-L reactor with the same height-to-diameter ratio as the original reactor, 2 to 1. Assuming both tanks are cylindrical, that cells will bind to both the walls and the bottom of the tanks, that binding to other internal components of the bioreactor is negligible, and that no cells bind to the tops of the tanks, what will be the yield of the 20,000-L system?

Answer:

Before we get to the scale-up portion of the problem, let's take a look at just what is going on in the system. In the first bioreactor, 2 mg/L/day × 2 L → the total output of the first bioreactor is 4 mg/day.

The problem also states that 25% of the product comes from cells in suspension. We can think of this amount as being directly related to the volume of the bioreactor (*V*). So, the amount of product in terms of reactor volume is 25% of 4 mg = **1 mg/day**.

We also know that 75% of the product comes from cells attached to a surface, such as the walls of the bioreactor. It is common for cells to behave differently when in suspension versus when they are attached. We can think of the amount of product resulting from attached cells as being directly related to the surface area of the bioreactor (*SA*). So, the amount of product in terms of reactor surface area is 75% of 4 mg = 3 mg/day.

Note that the 25/75 split of product coming from volume and surface area is unique to the small bioreactor. When we move to the larger bioreactor, *V* and *SA* will scale differently, so the 25/75 split will no longer apply.

Next, let us utilize the relation between height and diameter. We will use this relation in different ways for solutions a) and b).

Box 5.1 Scale-up example—cont'd

$h : d = 2 = h/d$
$\Rightarrow h = 2d = 2(2r)$ (where r = radius)
$\Rightarrow h = 4r$

Let us now move on to determining the yield of the 20,000-L system:

a) *Solution using a geometric method:*

2-L reactor

We can easily find the basis for scaling the amount of product due to volume:
$V = 1$ mg/day for a 2-Lreactor
$\Rightarrow V = 1$ mg/day/2 L
$\Rightarrow$ **$V = 0.5$ mg/L/day** →

We will use the volume to solve for r:
$V = \pi r^2 h$
$= \pi r^2 (4r)$
$V = \mathbf{4\pi r^3}$ →

$2\,l = 2\ \text{dm}^3 = 4\pi r_1^3$
$\Rightarrow r_1 = 0.5419$ dm

$SA = 2\pi rh + \pi r^2$ (only one area of a circle because no cells attach to the top)
$\Rightarrow SA = 2\pi r(4r) + \pi r^2$
$\Rightarrow SA = 8\pi r^2 + \pi r^2$
$\Rightarrow$ **$SA = 9\pi r^2$** →

Plug in the value calculated for r:
$SA_1 = 9\pi\,(0.5419\ \text{dm})^2$
$= 8.3037\ \text{dm}^2$

In reactor 1, we get 3 mg/day from the surface area:
$\Rightarrow 3\text{mg/day} / 8.3037\ \text{dm}^2$
$\Rightarrow$ **0.3613mg/dm²/day** →

20,000-L reactor

Output from $V_2 = (0.5\ \text{mg/L/day}) \times (20{,}000\ \text{L})$
$= 10{,}000$ mg/day

$20{,}000 = 4\pi r_2^3$
$\Rightarrow r_2 = 11.6754$ dm

$SA_2 = 9\pi\,(11.6754\ \text{dm})^2$
$= 3854.2434\ \text{dm}^2$

Output from SA_2:

$$\frac{\mathbf{0.3613\ mg}}{\mathbf{dm^2}} = \frac{\text{x mg}}{\mathbf{3854.2\ dm^2}} \text{ (per day)}$$

$x = 1392.5$ mg/day

Total output of the 20,000-L bioreactor = (10,000 + 1392.5) = **11,392.5 mg/day**.

b) *Solution using the scale-up factor c:*

The volume scaling factor = 20,000 L/2 L= 10,000 $\Rightarrow c^3 = 10{,}000$
The surface area scaling factor is c^2, which $= (c^3)^{2/3} = (10{,}000)^{2/3} = 464.159$

Now, simply plug in the original outputs due to V and SA from the original reactor and multiply by the appropriate power of c:

1 mg/day (10,000) + 3 mg/day (464.159) = **11,392.5 mg/day**.

This is the same answer as in part *a)* but was a whole lot easier to obtain!

5.7 Summary

We have looked at cells in the context of the changing cell numbers and cell masses that occur during growth and death phases. Cell numbers increase because cells replicate, and replication is described generally in terms of the cell cycle. We discussed cell cycle checkpoints to gain an understanding of why, for example, there is a plateau phase in the growth curve of a cell culture.

We also looked at the growth curve in both a qualitative and a quantitative fashion. The y-axis of a growth curve can represent either cell numbers or total cell mass, so we examined the mathematical difference between these two numbers. We also addressed several different ways of determining cell numbers or cell mass in the laboratory. Finally, we were introduced to the problem of scale-up, and we saw why controlling for a given parameter with larger culture vessels happens at the expense of other parameters. Understanding scale-up is essential for the biotechnologist who wants to take their work from the laboratory to the industrial scale.

This concludes the first unit, The Cell. You should now have a broad enough understanding of cellular structure and function to be able to appreciate the fundamental basis of many biotechnologies. The second unit, Laboratory Applications of Biotechnology, focuses on how we can interact with or manipulate cells to assess how they function in detail. Such detailed understanding will set the stage for expanding biotechnology into large-scale or medical applications. We will continue our journey in the next chapter by looking at microbes and learning different ways to kill them, whether on a surface such as a kitchen table or within our own bodies.

Related reading

Alberts, B., Johnson, A., Lewis, J., Raff, M., Roberts, K., Walter, P., 2008. Molecular Biology of the Cell, fifth ed. Garland Science, New York.

Cooper, G.M., 2000. The Cell: A Molecular Approach, second ed. Sinauer Associates, Sunderland, MA.

Dong, Y., Vanden Beldt, K.J., Meng, X., Khodjakov, A., McEwen, B.F., 2007. The outer plate in vertebrate kinetochores is a flexible network with multiple microtubule interactions. Nat. Cell Biol. 9, 516–522.

Sánchez Pérez, J.A., Rodríguez Porcel, E.M., Casas López, J.L., Fernández Sevilla, J.M., Chisti, Y., 2006. Shear rate in stirred tank and bubble column bioreactors. Chem. Eng. J. 124, 1–5.

Shuler, M.L., Fikret Kargi, F., 2001. Bioprocess Engineering: Basic Concepts. Prentice Hall, Upper Saddle River, NJ.

Zwietering, M.H., Jongenburger, I., Rombouts, F.M., van't Riet, K., 1990. Modeling of the bacterial growth curve. Appl. Environ. Microbiol. 56, 1875–1881.

Questions

1. Name and describe three checkpoints in the cell cycle.
2. Is it possible to tell the difference between G_0 and G_1 using a microscope?
3. What is the difference between absorbance and optical density?
4. Explain what diauxic growth is and why it happens.
5. If a cell suspension is so concentrated that individual cells cannot be resolved for counting in a hemacytometer, how could you still use the hemacytometer to find the concentration of cells?

6. Suppose a particle counter collects medium at 40 μL/min for 30 s and counts 23 cells. What is the cell concentration of the cell suspension in cells/mL?
7. a. Describe why it might be better to pass batch cell cultures in the late log phase rather than in the early log or plateau phases.
   b. Is it always best to pass cells in the late log phase? Explain.
8. Dr. Richtofen puts 100 μL of cell suspension through a particle counter and gets a cell count of 735 cells/mL. He then puts a 100-μL aliquot of the same sample (Dr. Richtofen is obsessive compulsive) onto an agar plate and counts only 55 colonies the next day. Why did the agar plate produce fewer colonies than there were cells counted by the cell counter? (Assume that Dr. Richtofen didn't kill any of the cells between the tests.)
9. When do cryptic growth and diauxic growth occur? What is the difference between the two?
10. Can a cell population grow exponentially for an infinite amount of time? Explain your answer and include a labeled plot of in-population growth kinetics.
11. Is the net change in the number of cells zero, negative, or positive from the beginning of the lag phase to the end of the death phase? Explain your answer.
12. a. If we have a population of cells in early log phase, what phase of the cell cycle are most of them in?
    b. If we have a population of cells in plateau phase, what phase of the cell cycle are most of the cells in?
13. You are asked to quantify the concentration of *E. coli* growing in a culture. You remove 200 μL and streak a plate. After 6 hours, the plate looks like this:

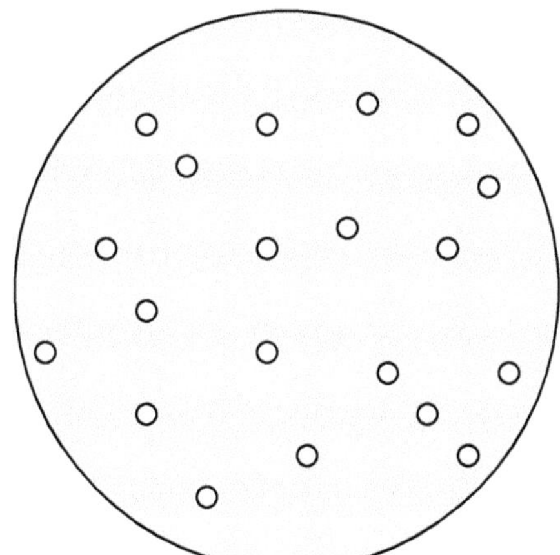

    a. What was the concentration of bacteria, per milliliter, at the time you streaked the plate?
    b. Assuming that the culture has been in (early) log phase since you streaked the plate and that doubling time $\tau_d = 20$ min, what is the concentration of the culture now? (Please leave your answer in the form of an expression; *do not simplify.*)
    c. Assume instead that the culture has been maintained in the plateau phase since yesterday. What is the concentration of the culture now?

d. Forget you ever saw parts b or c of this problem. Suppose your culture was behaving strangely, so you decided to streak a second plate 2 hours after you streaked the plate above. After 5 hours of incubation, your second plate has 160 colonies on it. What is the doubling time for the culture (in minutes)?

e. Assuming that the culture has been in (early) log phase since yesterday and that $\tau_d = 1$ h, what is the concentration of the culture now? What is the value of $\mu_{replication}$?

14. For a transfection experiment, suppose you want to plate 100,000 cells per well, and that you loaded a hemacytometer with 10 μL of a sample of the cells you want to use and obtained the pictured result:

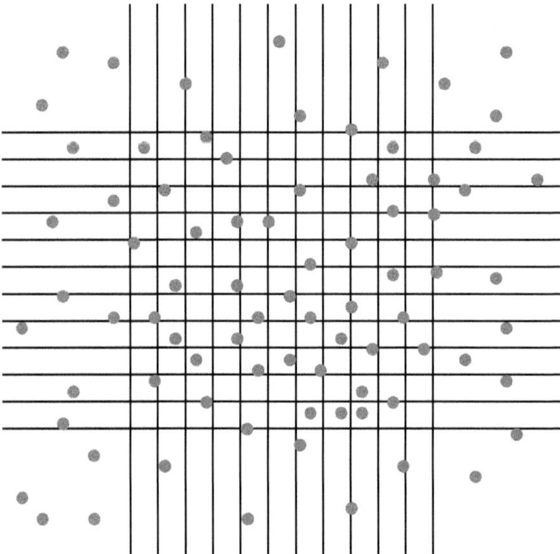

a. How much of the solution will you plate into each well for your transfection experiment? (Show your work.)

b. Given that the area of the (magnified) counting region above is 1 mm × 1 mm, determine the height of the liquid based on the scale-up factor presented in this chapter.

15. Examine the three growth curves. Each curve (A, B, and C) represents a separate strain of microorganism grown in identical culture conditions. (You may consider the lengths of all three lag phases to be equal.)

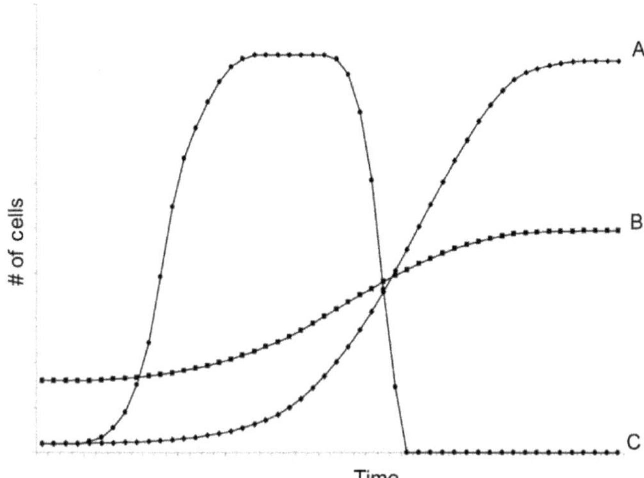

a. If the cells in question break down oil, and you want to use them for this purpose following the explosion of a drilling platform in the Gulf of Mexico, which of the three strains would be best suited for the job? Give two reasons for your answer.

b. Consider the three curves in a different setting. Suppose the cells produce a desired product during plateau phase. (Assume that the same amount of product per cell is produced regardless of strain.) Which strain would you use for your business that makes that product? Give one reason for your answer. State all of your assumptions.

16. If we start two new cultures by taking a set volume out of an existing batch culture at times t_1 and t_2, in which t_1 is in the early exponential phase and t_2 is in the plateau phase, what will the new growth curves look like? Express your answers on a single graph. (Think about how many cells are in the new cultures at t=0 and think about the lengths of the lag phases. Also keep in mind how many cells there are at the two new plateau phases and when the death phase might occur.)

17.

In going from stirred bioreactor 1 to stirred bioreactor 2, we could keep rpm constant and hold oxygen availability steady by bubbling oxygen up through the second tank in addition to stirring. However, another parameter would change because the geometry of the flow patterns changed. Name the parameter.

18.

What is the scale-up factor for volume in going from the first cylinder to the second?

a. $(6-5)^3$
b. $(5/6)^3$
c. $(6/5)^3$
d. $[(6/6)-(5/5)]^3$
e. 5^3
f. 6^3

19.

What is the scale-up factor for volume in going from the first cylinder to the second?
a. $(5-4)^3$
b. $(4/5)^3$
c. $(5/4)^3$
d. $[(7/5)/(6/4)]^3$
e. $(5/4)^2 \times (7/6)$
f. $[(7/6)/(5/4)]^3$

20. In the example in Box 5.1, what percentage of the product comes from volume, and what percentage comes from surface area in the 20,000-L bioreactor? Why is your answer not 75%/ 25%, the percentages of the system before scale-up?

21. Suppose you have two reactors in the shape of cylinders that have the same relative geometry, with the height being 1.5× the diameter. The second reactor has a volume of 100,000 L, and the volume of the first reactor is 12.5 L.
a. What is the scaling factor, *c*, between the two reactors?
b. What is the ratio of the radii of the two reactors (r_2/r_1)?
c. If the surface area of the first reactor is 30.33885 dm^2, what is the surface area of the second reactor?
d. What is the height of each reactor, in inches?

22. Consider a 1-L batch fermentation system whereby 90% of the (intracellular) target product is associated with attached cells and 10% is associated with cells in suspension. The total output of this reactor is 6 mg of product/liter/day.

Now consider a scale-up to a 64,000-L reactor with the same height-to-diameter ratio as the original reactor (2 to 1). Assuming that both tanks are cylindrical, that cells will only bind to the walls of the tanks (not the bottoms) because of the stirring that is taking place, and that binding to other internal components of the bioreactor is negligible, what will be the yield of the 64,000-L system?

UNIT

Biotechnology in the laboratory

CHAPTER

Microbial killing

6

Chapter outline

Microbes ... they can be useful workhorses for the biotechnologist in producing engineered products, but to a biotechnologist performing mammalian cell culture, they can also be contaminants. Biotechnologists spend a lot of time and money killing microbes in the laboratory. It is a very serious issue.

In your everyday life, it might be common for you to wash your hands before you eat or to treat a wound with an antiseptic. You will, I hope, wipe down your kitchen before and after preparing food. A doctor might prescribe an antibiotic for certain illnesses, and surgeons in an operating room insist on sterile instruments. All of these are examples of combating microbes but to different extents. In this chapter, we examine different ways of killing microbes and the different levels of "clean" that are acceptable in different circumstances.

Just as we did in our study of the eukaryotic cell, we will begin our discussion of microbial cells by looking at the outermost portion of the cell exterior. It is important for the same reason here as it was in Chapter 2: it is the first part of the cell with which an agent designed to act on the cell will interact.

Biotechnology and its Applications. https://doi.org/10.1016/B978-0-12-817726-6.00006-X

We will use this information to classify microbes via the Gram stain; this classification will become important when we consider microbial cell death. After the Gram stain, we will address:

- microbial resistance to killing
- sterilization, disinfection, sanitization, and antiseptics
- microbial cell death:
  - death by alcohol
  - antimicrobial drugs (antibiotics), targets

6.1 The Gram stain

Whereas the outermost portion of the mammalian cell is the plasma membrane, prokaryotic cells have one or two additional layers on the exterior of the lipid bilayer that surrounds the cytoplasm. A cell wall surrounds the cytoplasmic (or inner) membrane of the prokaryotic cell, and in some species of prokaryotes, an outer membrane surrounds that cell wall (Figure 6.1). When bacteria are identified, often the first criterion used is whether the cell wall is exposed or hidden by the additional, outer membrane. This determination can be made using the Gram stain.

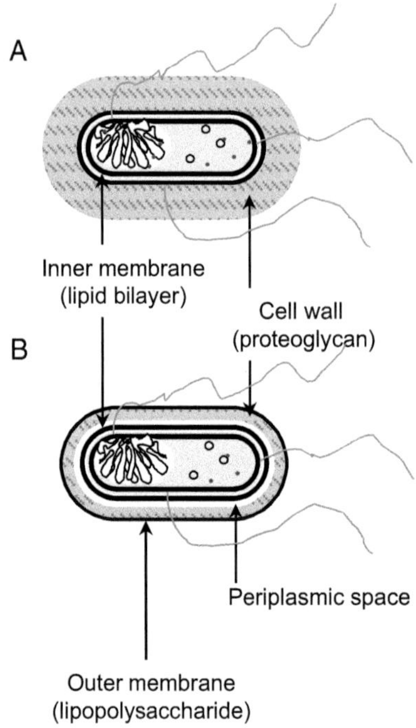

FIGURE 6.1

Proteotypical bacteria. (A) Gram-positive. (B) Gram-negative. Both cells have cytoplasms surrounded by an inner membrane made from a lipid bilayer, and both inner membranes are surrounded by a cell wall made of proteoglycan. Gram-negative calls also have an outer membrane made of lipopolysaccharide. The cell wall in gram-negative bacteria is generally thinner than that of gram-positive bacteria, and there is a periplasmic space between the cell wall and the inner membrane.

The Gram stain was developed by Christian Gram, a researcher in Berlin, in the 1800s. He developed a procedure using a dye, a mordant, and a counterstain to determine whether the bacterial cell wall was exposed. If the cell wall is exposed, the cell is *Gram-positive*, and if not, the cell is termed *Gram-negative*. Interestingly, the Gram stain was first described in published form by Carl Friedlander, not Christian Gram. Friedlander worked with Gram and alluded to the staining procedure used by his colleague in an 1883 paper on *Pneumococci*. The famous paper by Gram was published in 1884. Although Friedlander used the technology and was the first to publish with it, history has since corrected the situation by crediting Gram with the original procedure, now known as the Gram stain.

The Gram staining procedure is simple and only takes about 10 minutes to perform. The first steps are to place a bacterial sample onto a slide and fix the cells in place with heat. The sample might require dilution to help ensure a monolayer of bacteria, allowing for complete penetration of the dyes that will be used. The extracellular fluid (medium and diluent) must be removed without removing the cells, and the cells must be attached to the slide to prevent them from being washed off during the staining and rinsing steps of the procedure. Heat is generally used for such fixation: the slide is gently waved over a Bunsen burner flame, with the heat promoting evaporation of the fluid as well as **denaturation** (unfolding) of proteins. The unwound proteins will bind to the surface of the slide, resulting in adherence of the cells. An alternative method of fixation is via alcohol, which acts to denature proteins, again leading to cellular attachment.

Currently, the dye most commonly used to stain the cells is crystal violet, which has the structure shown in Figure 6.2. Note that the dye is a carbocation. The positive charge will readily interact with anions such as I^-. Gram's iodine, a solution made up of iodine and potassium iodide, will supply the I^-. Once iodide ions have been attached to crystal violet molecules, a precipitate will form. Because it is responsible for forming a precipitate, Gram's iodine is known as a **mordant** (Figure 6.3).

Because the cells have been fixed, crystal violet can penetrate them. The incubation time for staining the slide with crystal violet is only about 1 minute, after which the slide is rinsed with tap water. Next, a solution of Gram's iodine is placed upon the slide for 1 minute, then the slide is rinsed with a **decolorizer** (methanol, ethanol, or acetone). Gram's iodine causes the dye to form a bulky precipitate, and the decolorizer acts as a solvent for the dye. However, certain cells retain the purple dye because

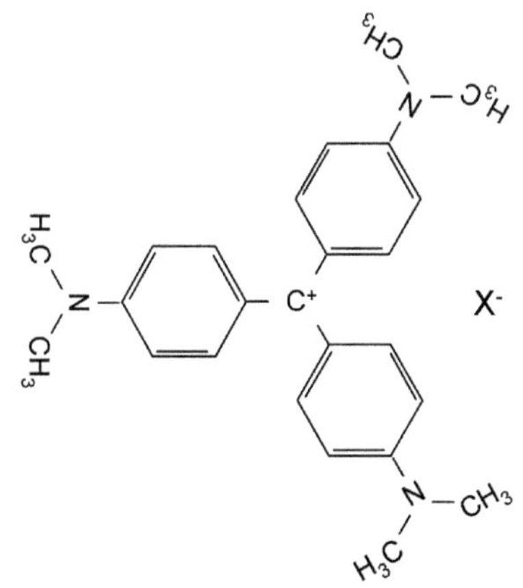

FIGURE 6.2

The structure of crystal violet.

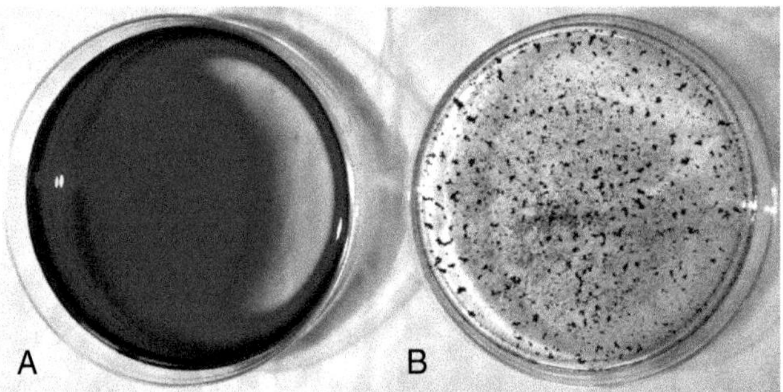

FIGURE 6.3

Crystal violet (A) without and (B) with the mordant Gram's iodine. The mordant causes the dye molecules to precipitate.

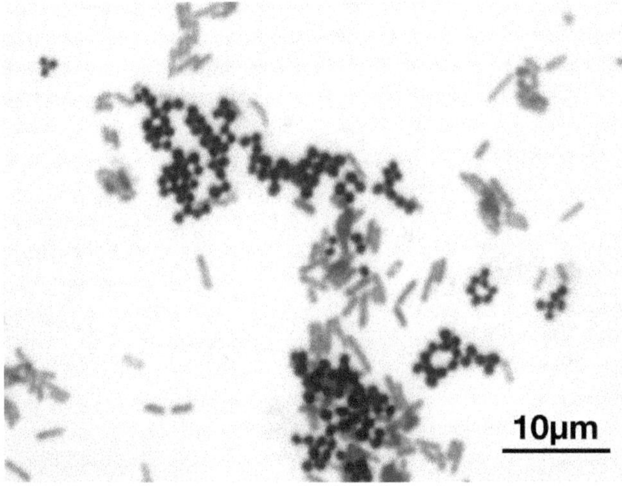

FIGURE 6.4

Gram staining of a mixture of *Escherichia coli* (pink, gram-negative) and *Staphylococcus aureus* (purple, gram-positive).

they have a thicker cell wall; the proteoglycan network of these cell walls traps the dye molecules. However, in other cells, there is an outer membrane and a thinner cell wall that is not sufficient to trap the precipitated dye molecules, so they will be washed away by the decolorizer. The presence or absence of dye molecules after this step is the basis for labeling cells as gram-positive or gram-negative.

The next step is to **counterstain**. The counterstain is performed because, at this point, cells will be either purple or unlabeled, and it is desirable to have every cell labeled for easy identification. The counterstain is a dye such as safranin or basic fuchsin, each of which carries a positive charge. The counterstain is able to get through the cell membrane(s) because the cells have been fixed and are especially permeable after exposure to the membrane-destabilizing decolorizer. The dye enters the cytoplasm and adheres to negatively charged components, coloring the inside of all the cells pink (Figure 6.4).

Box 6.1 General resistance of contaminants to physical and chemical methods of control

Highest resistance

Bacterial endospores

Moderate resistance

Protozoan cysts
Some fungal sexual spores
Naked (nonenveloped) viruses
Some vegetative bacteria

Least resistance

Most bacterial vegetative cells
Fungi (spores and hyphae)
Enveloped viruses
Yeast

At this point, one might ask, "Why do we care about the Gram stain? What good is it, and can you use it on all bacteria?" The reason we might want to perform a Gram stain is because we want to determine whether an infection in a patient or a cell culture is gram-positive or gram-negative, to help us determine which antibiotic to use. But before we can talk about antibiotics, we need some more background on microbial cell killing.

6.2 Microbial resistance to killing

If we want to kill microbes, there are various levels of cleanliness to consider. You've probably heard microbes referred to as "germs" before, but the word "germ" is like the word "stuff"—a catch-all term—and different microbes have different resistances to different cleaning methods (Box 6.1). The hardest microbes to kill or inactivate are the bacterial endospores. You won't be able to destroy bacterial endospores by washing with soap and water, scalding water, or 5.25% bleach (the concentration of regular Clorox); you need to go through some high-end processes to get rid of them. Luckily, bacterial endospores are not huge scourges in our everyday lives. If you work in a science laboratory, especially one that performs cell culture, you need to worry about spores, but they are not so much of a concern in everyday life.

Microbes with moderate resistance to killing methods include protozoan cysts, some fungal sexual spores, naked (nonenveloped) viruses, and some vegetative bacteria. Have you ever had a Staph infection? Such a condition involves contamination with *Staphylococcus aureus* ("Staph") bacteria in the vegetative state. There are Staph microbes on your pencil. They are on your desk. They are on your skin and in your respiratory tract. They are virtually everywhere, and if you get cut, Staph will commonly be present in the cut itself. (Cleaning the wound with soap and water, with the help of a functioning immune system, will typically be enough to protect the body from further invasion by these bacteria.) Other microbes that you may have heard of in the "moderate resistance" classification include *Mycobacterium tuberculosis* and *Pseudomonas*. *Pseudomonas* are notable for their ability to live in soap dishes; one really should clean the soap dish or the inside of the liquid soap dispenser periodically!

Box 6.2 Vegetative versus spore forms

When we refer to "vegetative bacteria" we are pointing out a bacterial state—in this case, the growing and reproducing form. Many bacteria can also transform into a spore form, where they remain dormant until conditions improve. The spore form is a defensive state taken in response to adverse temperatures, pressures, or nutrient availabilities. Thick cell walls and dormant metabolisms make spores resistant to hostile environments, sometimes for millions of years. Spores have been isolated and cultured from the stomachs of extinct bees embedded in amber for up to 40 million years. Spores have also been the underlying cause for the mysterious sicknesses and deaths of those who first entered King Tut's tomb, even though the deaths have been ascribed by some to the "Curse of the Pharaohs" (Figure 6.5).

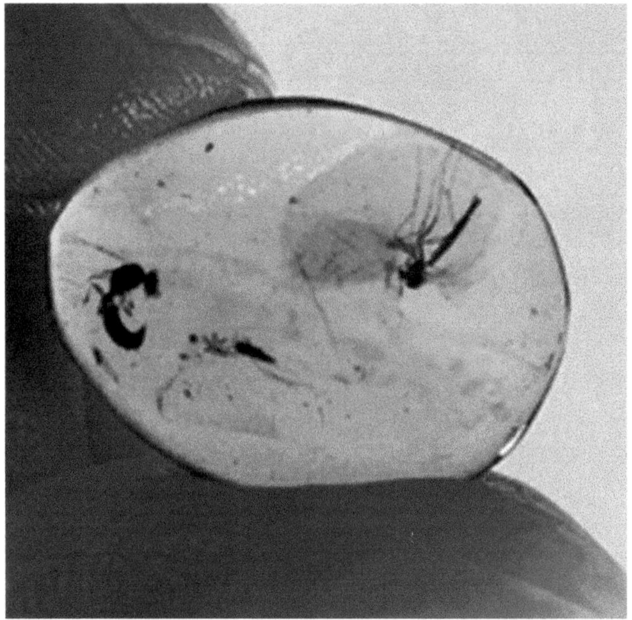

FIGURE 6.5

Insects embedded in amber.

Picture courtesy of Terry Su

Microbes with the lowest resistance to destruction include most bacterial vegetative cells, fungi (spores and hyphae), enveloped viruses, and yeasts. Vegetative bacteria are not in the protected state and are relatively defenseless. Washing your hands, with soap, after you get a cut makes it easier to remove or destroy the types of microbes listed in Box 6.1 as having the least resistance. Of those that might remain, *Staph. aureus* is one of the most common. This partially explains why, if you get a cut and it gets infected, the infection is commonly a Staph infection. Staph are a little more difficult to get rid of, especially compared with fungi and yeasts.

6.3 Sterilization, disinfection, and sanitization

Now that we know what we're up against, let's turn to different levels of "clean" (Figure 6.6).

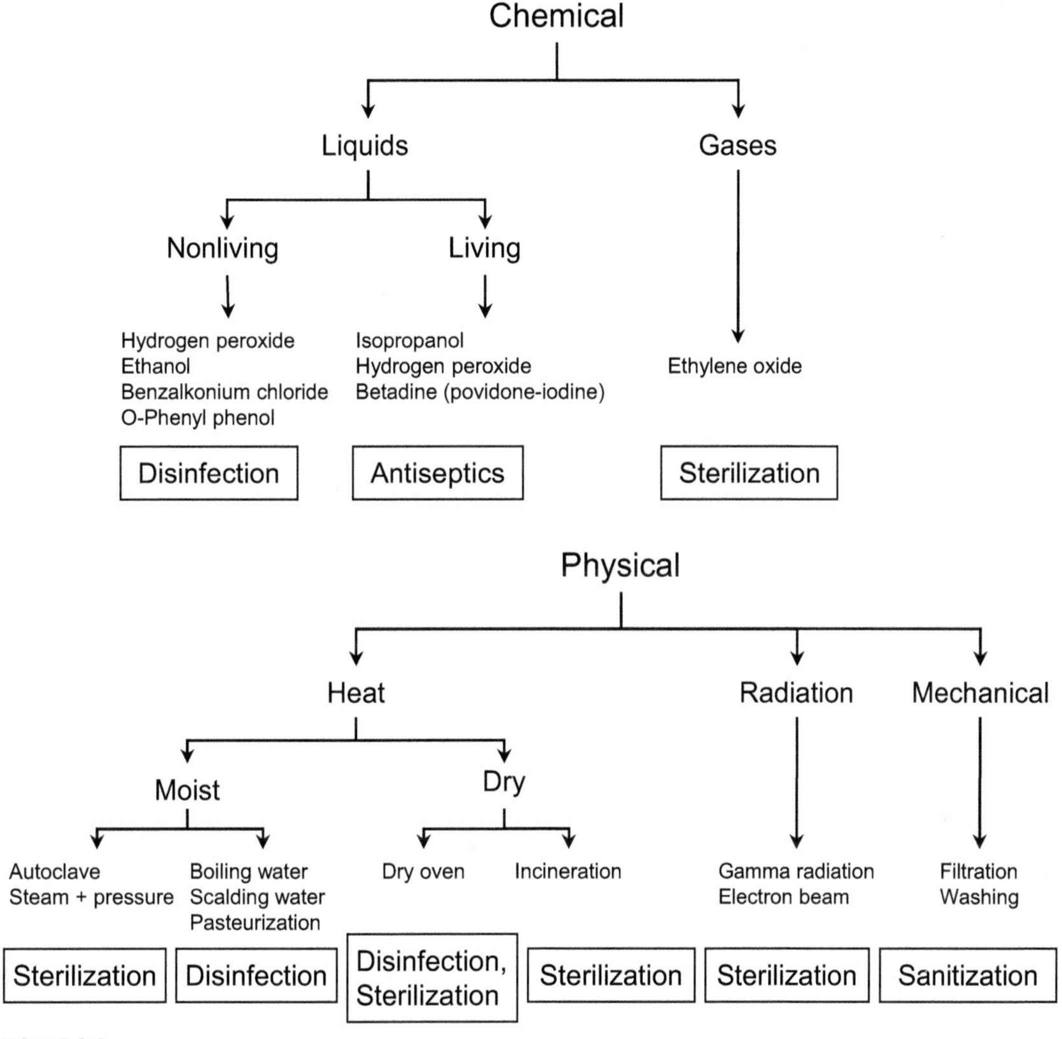

FIGURE 6.6

There are many different means available to remove or destroy microbes. Methods are grouped into chemical and physical types, with the resulting levels of cleanliness shown in boxes.

6.3.1 Sterilization

There are differences between "sterile," "disinfected," and "sanitized." **Sterile** means all viable microorganisms and viruses have been destroyed or removed. Different methods can be used to achieve this level of cleanliness, including autoclaving, treatment with ethylene oxide gas, and irradiation.

6.3.1.1 Autoclaves

An **autoclave** works like a pressure cooker. For sterilization, it's not enough just to boil things. Boiling water will kill most bacteria, but it will not kill bacterial endospores. As a liquid, water cannot get any

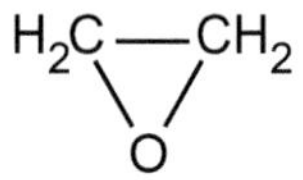

$$H_2C—CH_2 \text{ (ring closed through O)}$$

FIGURE 6.7

The structure of ethylene oxide.

hotter than its boiling point, which is 100°C at standard pressure (1 atm). Steam, on the other hand, can attain temperatures much higher than 100°C, and it can be used to inactivate or destroy bacterial endospores. Stainless steel instruments, glassware, and some surfaces can be sterilized using steam.

To achieve sterilization while keeping water in the liquid phase, however, pressure must be increased. This can be achieved with both pressure cookers and autoclaves. Using Henry's law:

$PV = nRT \Rightarrow P = T(nR/V)$, where

P = pressure,
V = volume,
T = temperature,
n = the number of moles of the substance, and
R = the gas constant.

We see that for a given amount of water held at a fixed volume, raising the temperature will increase the pressure, and vice versa. At a pressure of 2 atm (1 atm above standard pressure), water boils at approximately 119°C. These conditions are commonly used in most autoclaves and are enough to kill all life forms within 15 minutes.

6.3.1.2 Ethylene oxide

Ethylene oxide (EtO) is a flammable, colorless gas at temperatures above 10.7°C (51.3 °F) which smells like ether at toxic levels. Its structure is shown in Figure 6.7. EtO is used in the production of solvents, antifreeze, textiles, detergents, adhesives, polyurethane foam, and pharmaceuticals. Smaller amounts are used in fumigants and in sterilants for spices and cosmetics and during hospital sterilization of surgical equipment. The term **cold gas sterilization** refers to sterilization via ethylene oxide. There are specific reasons why one might want to use ethylene oxide instead of an autoclave. For instance, suppose you wanted to sterilize some plastic tissue culture dishes; the plastic would melt in an autoclave, but ethylene oxide can be used at room temperature, so the dishes could be sterilized without fear of melting.

Although the ease of use and mild temperatures of cold gas sterilization might make it seem like an ideal method, there are drawbacks to the method. Ethylene oxide gas is expensive. Also, because the gas is toxic above certain concentrations, the concentration used for sterilization is dangerous. Special care, equipment, and procedures must be used for ventilation of the gas following the procedure. Finally, ethylene oxide is reactive, so unwanted side reactions can happen during sterilization; for example, ethylene oxide plus water yields ethylene glycol (antifreeze). Cold gas sterilization is therefore not suitable for aqueous solutions.

6.3.1.3 Irradiation

Irradiation is an alternative to cold gas sterilization and is the method of choice for companies producing items like plastic laboratory tubes and culture ware on a large scale. Two main irradiation techniques are used: gamma irradiation and electron beam (E-beam). Gamma irradiation uses radioactive cobalt (^{60}Co) to produce gamma rays that disrupt or destroy microorganisms. Such radiation can easily penetrate a stack

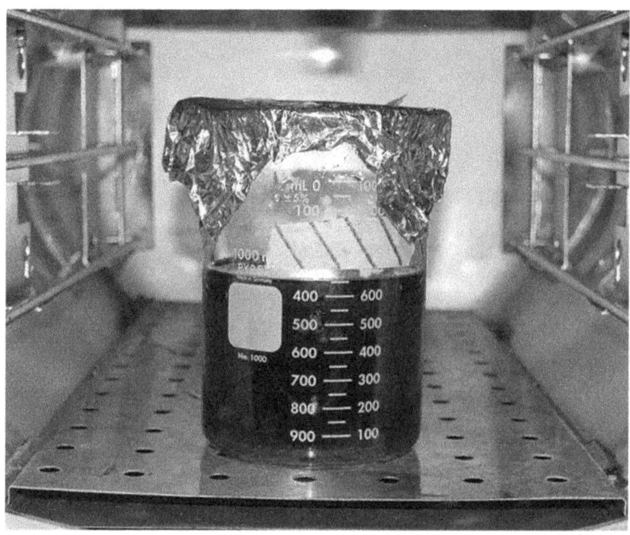

FIGURE 6.8

Mmmmmmm, this delicious-looking green liquid has been autoclaved, so there are no viable microorganisms in it. Is it therefore safe to drink?

of plasticware and its associated packaging. The required doses for sterilization via this type of irradiation are rather high, with a minimum of 25,000 gray (2.5 megarad) needed for items packaged in air.

E-beams can deliver high doses of radiation much more quickly than gamma irradiation, but the treatment volume will be much smaller because the penetrating power of the electron beam is much lower than that of gamma rays or particles. For comparison, gamma irradiation requires minutes to hours to deliver the dosage needed to achieve sterilization, but the particles can penetrate ~50 cm into a sample. E-beam radiation can deliver the same dose in seconds, but the rays can only penetrate ~5 cm - 1/10 of the distance - into the sample.

Keep in mind that "sterile" does not mean "safe to consume." One could run a cup of bleach through the autoclave, but it would still be a cup of bleach and quite toxic to drink after the process. It is a common misconception that "sterile" means "safe" - don't fall into that trap! Just because something is sterilized does not mean it is safe to consume. It only means that all of the viable microorganisms and viruses have been destroyed, removed, or rendered nonviable, not that the substance has been rendered completely safe (Figure 6.8).

6.3.2 Disinfection

Disinfection means the inactivation or destruction of vegetative pathogens, typically by liquid chemical means. This term is typically applied to the cleaning of surfaces. Lysol, 5% bleach, alcohol, and boiling can all be used to disinfect items or surfaces, but bacterial endospores will not necessarily be killed by the disinfection process.

6.3.3 Sanitization

One level lower on the cleanliness scale is **sanitization**. Sanitization refers to any process applied to inanimate objects, including air, liquids, and surfaces such that the number of microorganisms is

FIGURE 6.9

(A) Ahhhh, the hotel toilet, sanitized for your protection. (B) Hotel drinking glasses, also sanitized. If they were sanitized with the same cloth as the toilet, cross-contamination is a distinct possibility. Enjoy your drink.

lowered by mechanical means (Figure 6.9). Examples of sanitization include certain air purification systems, water filtration, dishwashing, and washing clothes with very hot water (with or without soap). The process is not the same as disinfection, which utilizes a chemical agent to *kill* microorganisms (as opposed to removing them), and it is certainly not the same as sterilization because viable microorganisms are left behind after the process.

When you finish washing your dishes at home, there will still be bacteria on them. This may seem disgusting, but in general the residual microbes should not present a problem if you have a functioning immune system. In many restaurants, the cleaning of dishes is taken a step further, with dry heat or steam being used to disinfect them. Keep in mind that sometimes in a restaurant you may get a fork or a knife that still has some old, dry food on it. Although you can see the debris, it has (hopefully) been disinfected because it was exposed to high heat. Do not automatically assume that you will catch a terrible illness from the exposure.

These terms—sanitization, disinfection, and sterilization—refer to methods for lowering the number of viable microorganisms on a surface or in a container or a liquid, but they do not tell us anything about other molecules that might be present, such as toxins. Even though you might have a sterilized cup of motor oil, you still cannot drink it! Similarly, a glass that has just been disinfected with glutaraldehyde might have some residual disinfectant, meaning the glass is currently unsafe for drinking.

On the other hand, you can have a disinfected fork with a piece of dried spinach on it that is still safe to eat. But then-again, you can have a sanitized plate that has had all of the spinach washed off so that it looks clean but has who-knows-what growing on it, so it is not necessarily safe to eat off of it. In this latter case, we have to trust in our immune systems and rely on the person responsible for the plate to have used sufficient cleaning methods to bring the number of microorganisms down to a safe level.

The point is that sanitization and disinfection do not guarantee items are safe for consumption. Having an unexpected item on a disinfected utensil does not mean the utensil is unfit for use during a meal, and the removal of all visible contaminations from a utensil via sanitization does not guarantee the safety of the item.

6.3.4 Antiseptics

To understand antiseptics, one should first understand the root of the word. Consider the word "**sepsis**," a medical term that describes a life-threatening condition brought about by the body's response to microorganisms in the blood and/or the tissues. Sepsis is a serious condition because the infection is often global—meaning it is throughout the body—having been carried by the circulatory system. It must be treated very quickly and aggressively. "Aseptic" simply means "not septic" and is sometimes used as an adjective to imply the prevention of sepsis. Consider a physician about to perform a procedure: he will wash his hands very well and may place a drape over parts of the patient near the area to be worked on. If the procedure is a surgery, the doctor will cover his hair and mouth and wear a gown that has been sterilized. These measures are known as *aseptic technique*. Similarly, when biotechnologists work with cells, we do so in a biological safety cabinet that prevents room air from delivering dust, lint, spores, and so on into the cell medium with which we are working. In addition, a person performing cell culture in a biosafety cabinet will not move his hands or arms over what is being worked on, including any open containers, because particles can fall into the working area or an open container from the lab coat, the skin, or the gloves. The steps taken for the prevention of bacteria growing in cell cultures or in laboratory animals are also known as *aseptic technique*.

Antiseptics are chemicals applied to body surfaces to destroy or inhibit the growth of vegetative pathogens. Their use can be thought of as disinfection for living surfaces. Washing your hands under a faucet using regular hand soap would be sanitizing your skin; you would be lowering the number of bacteria present via mechanical means. Use of an antiseptic—perhaps washing one's hands with a soap containing triclosan or benzalkonium chloride or swabbing the skin with isopropyl alcohol—would be different. Note that disinfectants and antiseptics are different: disinfectants are used on inanimate surfaces, and antiseptics are for body surfaces like your skin. Disinfectants can potentially be harsher than antiseptics because one does not have to worry about the preservation of living tissue. One can always use an antiseptic as a disinfectant, but it is not always safe to use a disinfectant as an antiseptic. Do you really want to pour bleach on that cut?

Hydrogen peroxide (H_2O_2) is a very effective antimicrobial. In fact, the concentration you buy from the store (3%) is very effective, killing a broad spectrum of microbes within 10 to 15 seconds. It is used as both a disinfectant and an antiseptic. However, it is not the best agent to put onto a healing wound because it can damage your own cells. Hydrogen peroxide works by making hydroxyl oxygen radicals, which can oxidize DNA, RNA, proteins, and membrane lipids. It will help to clean a fresh wound by killing microbes. When you first get an open wound, you should clean out any debris, which includes dead cells, tissue, and dirt. You could wash the wound with hydrogen peroxide—but after the healing process begins, hydrogen peroxide will negate some of the body's work in wound healing. At the end of the day, you may have grown fresh granulation tissue to cover the wound; you wouldn't want to strip it away by killing those cells. The healing process is complex, and what might work well on day 0 might not be the best agent on day 2.

6.4 Microbial cell death

We have been discussing how to kill microbes, but just what is microbial cell death? As we saw in Chapter 1, we haven't adequately defined life, so how can we define death? One popular definition is *the permanent termination of an organism's vital processes*. "Vital" means living, so death would therefore be *the permanent cessation of life*. Extending this further, the proposed definition of death is: *not life*. Without a proper definition of life, these definitions for death are not so valuable.

Nevertheless, it's usually easy to tell when a person is dead—their breathing stops, the body is not moving around, the heart is not beating, there are no electrical impulses in the brain—but it can be difficult to tell when a microbe dies. When a microbe dies, it just sits there… but that's quite often what it did when it was alive, so it's not doing anything different as far as motion is concerned. The definition we're going to use for microbial cell death is *the permanent loss of reproductive capability, even under optimal growth conditions*. Such a definition is needed for microbes because of their different states.

Some microorganisms, when they are under assault, armor up in a ball: a spore. Spores have a thick, protective, proteinaceous coat that is hard to penetrate, even with most chemicals. There is no locomotion, and the spores do not respire. Many definitions of "life" state that metabolic function (respiration) is required; if we were merely to say that death is "not life," then "death" would include the cessation of metabolic function. Spores are not performing metabolism, so does that mean they are dead? That's where the second half of the above italicized definition comes in. If you place a bacterial endospore into optimal growth conditions (perhaps a warm, moist environment with ample nutrients), it will desporulate into the vegetative form of that bacterium, which is then actively respiring and replicating. A spore does not respire, and it cannot reproduce—but when you put an intact spore into its optimal conditions, it will demonstrate features that we typically equate with life.

6.4.1 Death by alcohol

Alcohols are germicidal because of their ability to interact with membrane proteins and disrupt lipid bilayers. The number of carbons on the alcohol will help determine how effective it is as an antimicrobial (Figure 6.10). For example, the one-carbon alcohol methanol is not considered to be very microbicidal. Although it can kill some microbes in a population, it is not an effective microbicide because, with one carbon, it is difficult for the methanol to disrupt lipid bilayers such as those found in the plasma membrane.

Ethanol, the two-carbon alcohol, is a far more effective antimicribial and is used in the laboratory as a disinfectant. For example, it is used to disinfect biological safety cabinets by spraying at a concentration between 50% and 90% (typically 70%). Anything less than 50% may be too dilute to kill certain microbes, but pure ethanol isn't ideal because some microorganisms (e.g., the gram-positives *S. aureus* and *S. pyogenes*) are resistant to 100% ethanol. Keep in mind that when you spray down a surface like

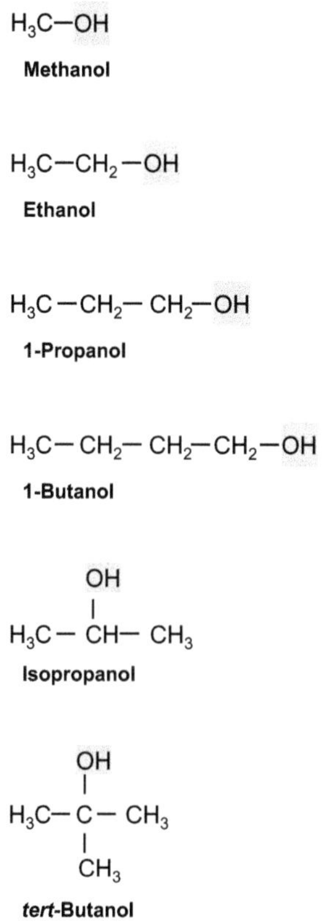

$H_3C{-}OH$

Methanol

$H_3C{-}CH_2{-}OH$

Ethanol

$H_3C{-}CH_2{-}CH_2{-}OH$

1-Propanol

$H_3C{-}CH_2{-}CH_2{-}CH_2{-}OH$

1-Butanol

$H_3C{-}CH(OH){-}CH_3$

Isopropanol

$H_3C{-}C(OH)(CH_3){-}CH_3$

***tert*-Butanol**

FIGURE 6.10

The structures of several common alcohols. Considerations including hydrophobicity, volatility, and cost should be taken into account when deciding on a suitable alcohol for use as a disinfectant or antiseptic.

your desk, the surface isn't instantly "clean." You are only using the alcohol to compromise the integrity of the bacterial cell membranes so that when you apply some shear, you're more likely to lyse the cells. Simply spraying the surface with ethanol will not yield the same level of disinfection.

Using ethanol as a disinfectant is good in the laboratory. However, when you go into the doctor's office, they don't rub your skin with ethanol before an injection; they rub it with rubbing alcohol (isopropanol) because ethanol evaporates too quickly. With two carbons, ethanol is more volatile than alcohols with three carbons or more. During an injection, the needle can carry microbes from the surface of your skin into your body. The skin is rubbed with the less volatile isopropanol because it will have a greater chance to interact with any microbes hiding within the stratum corneum, the upper layer of the skin. In addition, having three carbons renders the molecule somewhat more hydrophobic and allows it to interact with the hydrophobic portions of the plasma membrane (lipid tails and hydrophobic regions of transmembrane proteins) more easily than an alcohol with only two carbons. Thus, the microbicidal activity of isopropanol is greater than that of ethanol.

Cell wall We don't have one
Translation Prokaryotes use different ribosomes
Nucleic acid synthesis We don't manufacture all of our own cofactors
Cell membrane Prokaryotes use different components (e.g., sterols)

FIGURE 6.11

Potential targets for antimicrobial drugs, with reasons why they can be targeted without harming human cells.

Isopropanol also tends to be less expensive than ethanol. One of the reasons is that you can't safely drink isopropanol. "Pure" ethanol, because it can be consumed, can be subjected to taxation. The way that laboratories get around having to pay the expensive ethanol tax is by purchasing ethanol that contains impurities such as methanol. Drinking the typical laboratory ethanol will make one violently ill, but the good news is that it is fairly inexpensive.

You might be thinking that alcohols with even more carbons would be even better choices as microbicidal agents. However, such alcohols have greater hydrophobicity and remove fat from the skin more easily, leading to hardening and cracking of the skin. These alcohols also tend to be more expensive, and as the number of carbons is increased the solubility of the molecules in water goes down. If you need to spray down your biosafety cabinet but have run out of ethanol and isopropyl alcohol, you *could* use butanol, but it is not recommended to swab a patient's skin with butanol before giving an injection.

6.4.2 Antimicrobial drugs

Let us now turn to antimicrobial drugs and discover why we bothered to learn about the Gram stain. Antimicrobial drugs—**antibiotics**—capitalize on differences between microorganisms and humans, which present a variety of potential targets, as outlined in Figure 6.11. For instance, if we had a drug that compromised all cell walls, we could annihilate any pathogenic microbes inside us without harming ourselves, because our cells do not have cell walls. However, sometimes it's easy to get to the cell wall and sometimes it is difficult. Recall that in gram-positive bacteria, there is no outer membrane, so the cell wall is exposed. Such bacteria are more susceptible to drugs that target the cell wall. Penicillin is one example of such a drug.

6.4.2.1 Targeting the cell wall

Consider the prokaryotic cell wall: it is primarily made of the peptidoglycans N-acetyl muramic acid (NAM) and N-acetyl glucosamine (NAG) (Figure 6.12). These molecules are arranged in a very ordered fashion, with alternating layers of NAMs and NAGs connected as shown in Figure 6.13. They are also cross-linked between NAMs via a relatively short protein sequence. This cross-linking is structurally important. Think of a deck of cards; if you were to try to rip it in half, you might not be able to, but it

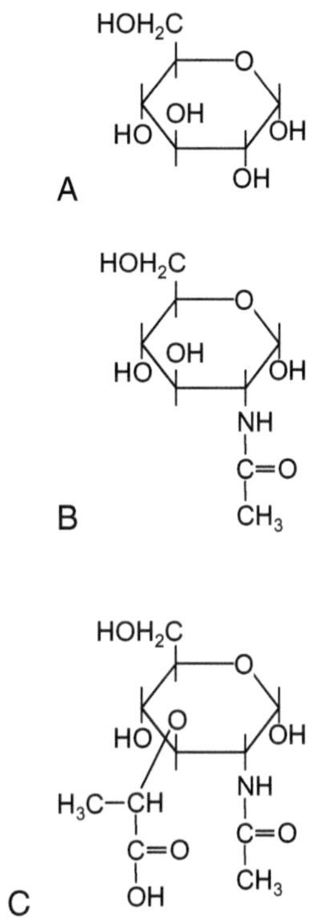

FIGURE 6.12

The structures of (A) glucose (included only as a reference), (B) N-acetyl glucosamine, and (C) N-acetyl muramic acid.

would be easy if you slid the cards off of the deck one by one. However, if there were glue between each of the cards, instead of 52 individual sheets you would have a single thick block, impossible to tear. That is similar to what the cross-linking between NAMs does.

The building of cross-links between NAMs involves polypeptides, commonly pentapeptides. The sequence of the five constituents depends on the species of prokaryote, but there are some commonalities. The third amino acid typically has two amino groups and is often lysine, although diaminobutyric acid and diaminopimelic acid are also common. The fourth member is often an alanine of some sort (either the D or L enantiomer). The fifth member is usually D-alanine. During the cross-linking of two such pentapeptides, at least one of the fifth-position D-alanines will be removed as the fourth member is linked to the diamino member of the other pentapeptide (Figure 6.13B). Some microorganisms use a spacer (e.g., an oligoglycine) to bridge the gap between the two pentapeptides, which affects cell wall porosity and density. Note the presence of D-amino acids. It is thought that bacteria use D-amino acids because it helps them to avoid enzymatic degradation by proteinases that only recognize L-enantiomers, which are more common.

In order for bacteria to grow in size, the cell wall must be broken by autolysins, made larger by the insertion of new NAMs and NAGs (via transglycosylases), then resealed via the formation of new cross-links (using transpeptidases). Penicillins and cephalosporins bind to transpeptidases to prevent

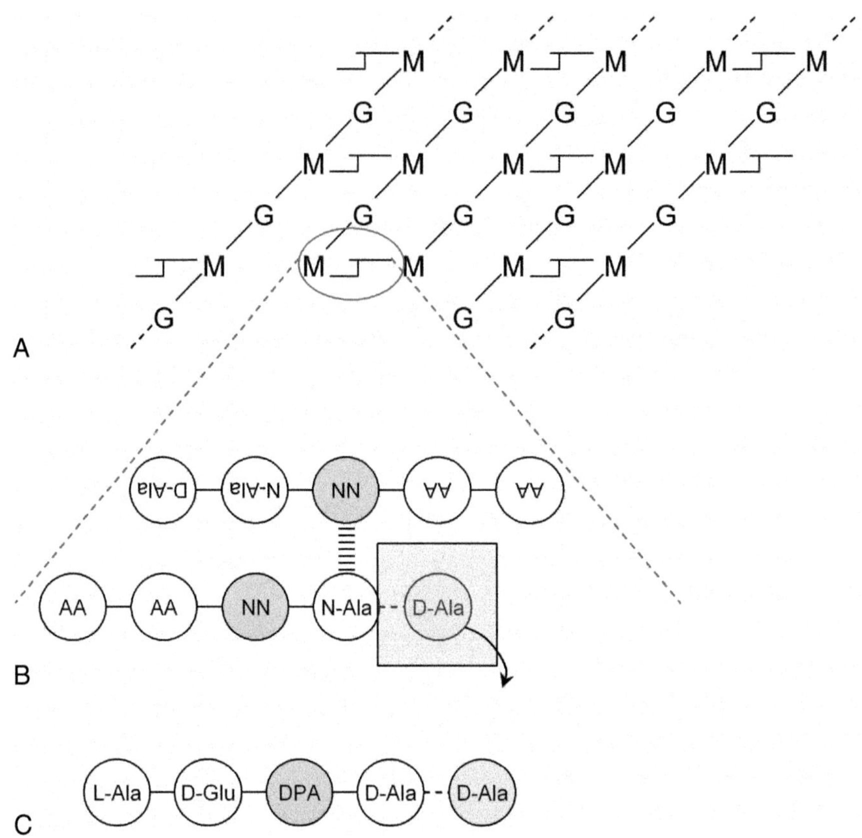

FIGURE 6.13

(A) The bacterial cell wall is composed of alternating layers of N-acetyl glucosamines (NAMs) and N-acetyl glucosamines (NAGs). Interlayer cross-links are formed between NAMs. (B) NAMs contain a species-specific side chain, generally 4–5 amino acids (AAs) in length, that will cross-link with a successive NAM. The third member of the NAM side chain typically has two amino groups (NN); an example is lysine. The fifth member is often a D-alanine, which will be lost during cross-linking of the fourth member to the diamino-containing member of an adjacent NAM side chain. (C) A common pentapeptide used in cross-linking NAMS in the cell walls of *Escherichia coli* is L-alanine–D-glutamic acid–diaminopimelic acid (DPA)–D-alanine–D-alanine. The final D-alanine may be lost during the cross-linking process.

the resealing of the cell wall. With the drug bound, the transpeptidases are no longer active, so there is no further cross-linking, and the cell wall becomes weak. This weak cell wall makes the cell prone to lysis (rupture), commonly due to osmotic forces. The drugs penicillin G, ampicillin, amoxicillin, methicillin, and oxacillin all work in this fashion.

6.4.2.2 Targeting translation

We, as eukaryotes, differ from prokaryotes in how we perform translation. Both prokaryotic and eukaryotic cells use ribosomes, but those ribosomes are fundamentally different from each other.

Box 6.3 When does 60 + 40 = 80?

S stands for the Svedberg unit, which is related to the sedimentation coefficient. The large ribosomal subunit in eukaryotes has a sedimentation coefficient equal to 60×10^{-13}, or 60 Svedbergs, which is where the name "60S subunit" comes from. Keep in mind that sedimentation in this case refers to how far something migrates through a fluid under centrifugal force. It is a function of the relative densities of the particle and the fluid, and the frictional drag, which depends on the geometry of the particle. It should be intuitive that if we drop two rocks having the same shape and size but different densities into a pond, the one with the greater density will settle faster. Consider, however, two pieces of sandstone, each with a mass of 500 g, but one roughly spherical and the other rather flat. If you drop them into a pond at the same time, the round one will sink to the bottom first because it has less frictional drag. Svedberg units are a way to describe migration through a fluid, taking density and geometry into account.

When two subunits come together to make a protein, the final protein may have a geometry that causes less frictional drag than either of its individual subunits. Consider the yin and the yang (in three dimensions); individually, each would migrate through a liquid rather inefficiently, but together they would make a sphere, which has a far lower coefficient of drag. This is similar to the case for ribosomes. In eukaryotes, although the two ribosomal subunits are of 60 and 40 Svedbergs, bringing them together yields a ribosome of only 80 Svedbergs. So 60S + 40S = 80S. Aha—another way to win a bet. Knowing biotechnology pays!

Bacterial ribosomes are made of one 50S and one 30S subunit, which come together to form 70S ribosomes. Eukaryotic ribosomes have a 60S and a 40S subunit, which come together to form an 80S ribosome (refer to Box 6.3 for an explanation of the math). If we target 70S ribosomes with an antibiotic, we can knock out translation in bacteria, which means they can no longer make proteins. Without proteins, the bacteria will die because they are unable to perform metabolic functions; metabolic enzymes are made from proteins. Because we humans do not have 70S ribosomes, our cells will theoretically be unaffected by the drug. We can therefore (in theory) target bacterial ribosomes and destroy them without hurting any of the functions of our own cells. Things just aren't that simple, though. Mitochondria are organelles within eukaryotic cells that behave like symbiotic bacteria. They have their own genomes and carry out translation for some mitochondria-specific proteins. Drugs that target prokaryotic translation can also affect the mitochondria, which can be significantly deleterious because the mitochondria are responsible for most of the energy production in eukaryotic cells.

Examples of drugs that target translation include streptomycin and gentamicin, which bind to the 30S ribosomal subunit of prokaryotic ribosomes to cause misreading during the translation of mRNA. Erythromycin inhibits translocation of the 50S subunit during translation. Tetracycline blocks proper attachment of tRNA to the ribosome.

6.4.2.3 Targeting nucleic acid synthesis

As we discussed in Chapter 4, DNA is made up of four bases, abbreviated A, G, C, and T. Half of the bases (A and G) are purines, and half of the bases (C and T) are pyrimidines, because a purine will pair with a pyrimidine in double-stranded DNA. Note that the synthesis of purines requires N^{10}-formyl tetrahydrofolate, which is generated from tetrahydrofolate. Tetrahydrofolate synthesis, in turn, requires dihydrofolate, which requires para-amino benzoic acid (PABA) for synthesis in microbes (Figure 6.14). In humans, on the other hand, dihydrofolate is derived from folate, which is ingested as vitamin B9. **Vitamins** are molecules that are required for the health of an organism but that must be consumed because that particular organism cannot make them on its own. If we take a drug that impairs the enzyme responsible for making dihydrofolate, we will not hurt ourselves (in theory), because we do not make dihydrofolate in the first place.

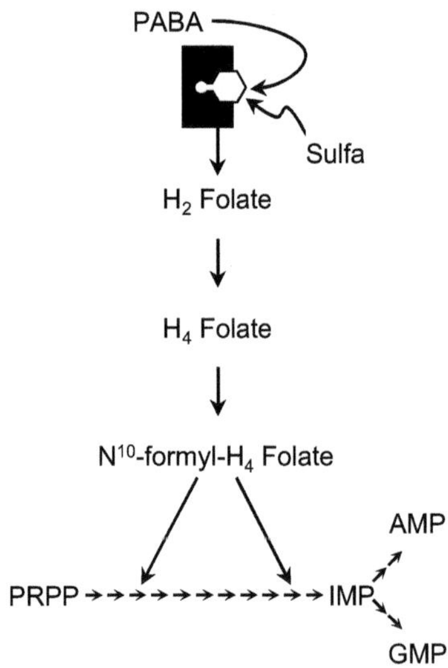

FIGURE 6.14

In microbes, para-amino benzoic acid (PABA) is required for the synthesis of purines (here, AMP and GMP). Sulfa drugs compete for the active site in the enzyme that uses PABA. When the synthesizing enzyme is occupied with a sulfa drug, it is unavailable to incorporate PABA into folate. (Two N10-formyl-H_4 folate molecules are involved in the de novo building of a purine.) *H_4 folate*, Tetrahydrofolate; *IMP*, inosine monophosphate (another purine); *PRPP*, phosphoribosyl pyrophosphate.

In microbes, folate synthesis requires PABA. The sulfonamides (sulfa drugs) are a class of drugs which have structures similar to that of PABA (Figure 6.15). If it binds to one of these sulfonamides, the enzyme that converts PABA to folate will be inactivated, even if just temporarily. By throwing in something else for the enzyme to grab—the sulfa drug—we take some of its attention away from converting PABA to folate. The enzyme therefore catalyzes less folate production, so purine synthesis is ultimately impeded. This is an example of **competitive inhibition**. The enzyme's ability to bind to PABA has been impeded, so less PABA is converted to dihydrofolate. This, in turn, means the bacterial cells have less tetrahydrofolate, which means they become deficient in the number of purines that they can synthesize. The result will be that the bacterium cannot create a copy of its own genome because it cannot produce the needed A and G residues, so cell division will be impossible.

6.4.2.4 Targeting cell membranes

Drugs that target the cell wall are effective on most gram-positive bacteria; however, what can be done to combat an infection of gram-negatives? Apart from broad-spectrum drugs (which will not be covered in this text), one could consider polymyxins. These drugs have a structure vaguely similar to that of phospholipids (Figure 6.16), but the polar head groups are so large that they disrupt membrane integrity upon insertion. They bind to the lipopolysaccharides - the outer membranes - of the gram-negative bacterial cell to cause structural instability, possibly by stripping Ca^{2+} or Mg^{2+} from the phosphate groups of the external membrane. Because they can interact with our own cells and therefore produce unwanted side effects (especially in the kidneys and nerves), polymyxins are generally used as topical treatments.

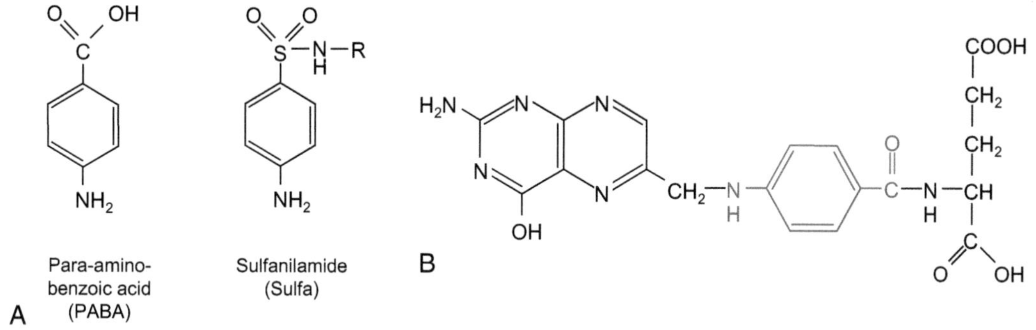

Para-amino-
benzoic acid
A (PABA)

Sulfanilamide
(Sulfa)

B

FIGURE 6.15

(A) The structures of para-amino benzoic acid (PABA) and sulfa drugs. Note the similarity between the structures. (B) The structure of folic acid. The portion derived from PABA is shown in *red*.

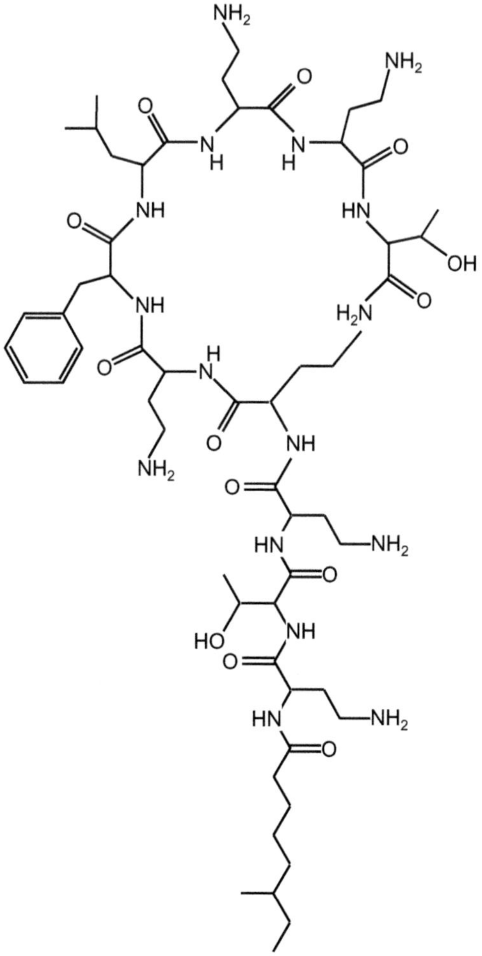

FIGURE 6.16

The structure of polymyxin B. Note the similarity in structure to membrane phospholipids, except that the head group of polymyxin B is much larger.

Polyenes such as amphotericin B and nystatin are used as antifungal agents. Fungi have different sterols in their cell membranes than we do. These drugs bind to ergosterol in the fungal membrane, causing leakage of small ions. Unfortunately, the drugs are not well targeted and can interact with sterols in our own cell membranes. (Can you name one of these sterols?) These interactions can cause destabilization in our own cells, which cause significant side effects. (Check your answer: The main sterol in our own plasma membranes is cholesterol.)

Box 6.4 Penicillin allergy

Penicillin is a drug that, by itself, is too small to spark an immune response. However, the carbonyl carbon of the β-lactam ring (Figure 6.17) can covalently bond with an amino group in a protein to produce a penicilloyl group (breaking the β-lactam ring). Not only will the penicillin lose its antimicrobial activity, but the penicilloyl group on the larger molecule will have also antigenic activity in some people, meaning it will be recognized by the immune system.

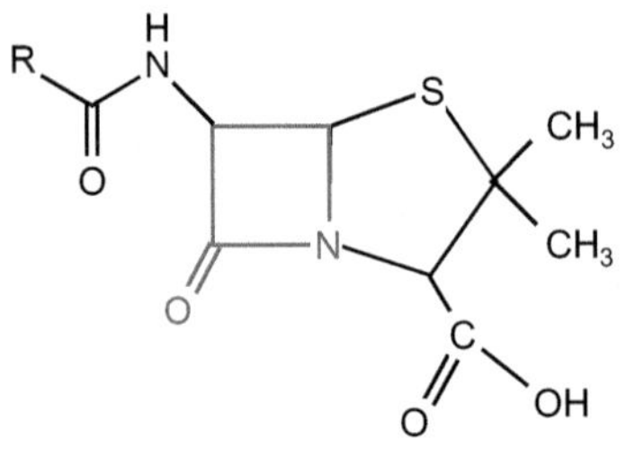

R
H
N
O
S
CH$_3$
CH$_3$
N
O
C
O
OH

FIGURE 6.17

The structure of penicillin. The β lactam ring is shown in red.

6.5 Summary

In this chapter, we examined the differences between bacterial coatings through the Gram stain, which subsequently allowed us to make appropriate choices in antibiotic selection. We also saw that some microbes are easier to kill than others and that not all cleaning applications require (or are able to deliver) the removal of all viable microorganisms.

We went on to discuss different ways of killing microbes, such as via scrubbing with alcohol. Alternatively, for applications of a more medical nature, antibiotics can be used to kill microbes via targeting differences between the microbes and eukaryotic cells. Examples of possible targets include the cell wall, translation, nucleic acid synthesis, and the cell membrane.

In the next chapter, we will move on to eukaryotic cells—mainly mammalian cells. We will learn about cell cultures, primary cells, immortal cells, and cancer cells.

Related reading

Beveridge, T.J., Davies, J.A., 1983. Cellular responses of *Bacillus subtilis* and *Escherichia coli* to the Gram stain. J. Bacteriol. 156, 846–858.

Cano, R.J., Borucki, M.K., 1995. Revival and identification of bacterial spores in 25- to 40-million-year-old Dominican amber. Science 268, 1060–1064.

Chess, B., Talaro, K.P., 2021. Foundations in Microbiology Talaro's Foundations in Microbiology/Barry Chess, Eleventh edition. New York, NY: McGraw-Hill Education, Chapter 12.

da Silva., K.A., 2012. Sterilization by gamma irradiation. In: Adrovic, F. (Ed.), Gamma Radiation. InTech, London, UK. https://doi.org/10.5772/2054. (Chapter 9).

Davies, J.A., Anderson, G.K., Beveridge, T.J., Clark, H.C., 1983. Chemical mechanism of the Gram stain and synthesis of a new electron-opaque marker for electron microscopy which replaces the iodine mordant of the stain. J. Bacteriol. 156, 837–845.

Edwards, R.G., Youlten, L.J., Dewdney, J.M., 1986. Penicillin hypersensitivity: mechanism, diagnosis and management. Indian J. Pediatr. 53, 37–44.

Friedlander, C., 1883. Die Mikrokokken der Pneumonie. Fortschr. Med. 1, 715–733.

Gram, C., 1884. Uber die isolirte Farbung der Schizomyceten in Schnitt-und Trockenpraparaten. Fortschr. Med. 2, 185–189.

Ingólfsson, H.I., Andersen, O.S., 2011. Alcohol's effects on lipid bilayer properties. Biophys. J. 101, 847–855.

Koc, E.C., Koc, H., 2012. Regulation of mammalian mitochondrial translation by post-translational modifications. Biochim. Biophys. Acta 1819, 1055–1066.

Linley, E., Denyer, S.P., McDonnell, G., Simons, C., Maillard, J.Y., 2012. Use of hydrogen peroxide as a biocide: new consideration of its mechanisms of biocidal action. J. Antimicrob. Chemother. 67, 1589–1596.

Morton, H.E., 1950. The relationship of concentration and germicidal efficiency of ethyl alcohol. Ann. N.Y. Acad. Sci. 53, 191–196.

Murray, P.R., Rosenthal, K.S., Kobayashi, G.S., Pfaller, M.A., 2002. Medical Microbiology, fourth ed. Mosby, St Louis, MO.

Popescu, A., Doyle, R.J., 1996. The Gram stain after more than a century. Biotech. Histochem. 71, 145–151.

Rutala, W.A., Weber, D.J., the Healthcare Infection Control Practices Advisory Committee (HICPAC), 2008. Guideline for Disinfection and Sterilization in Healthcare Facilities. http://www.cdc.gov/hicpac/pdf/guidelines/Disinfection_Nov_2008.pdf.

Satlin, M.J., Jenkins, S.G., 2017. 151 – polymyxins. In: Cohen, J., Powderly, W.G., Opal, S.M. (Eds.), Infectious Diseases, fourth ed. Elsevier. 1285–1288.e2.

Zeus, Inc. Sterilization of Plastics. Zeus Technical Whitepaper. pp. 1–7. https://www.thomasnet.com/pdf.php?prid=101488. Accessed December 16, 2020.

Questions

1. What is microbial cell death?
2. Why would one purposely place a vial of bacterial endospores in an ethylene oxide gas sterilizer along with instruments that were going to be used for human surgery?
3. Why would monochloro-*o*-phenylphenol or benzalkonium chloride be good disinfectants but poor antiseptics?
4. Name four ways to sterilize an object.
5. Because of fears of bacteria in the drinking water, a certain town has undergone a "boil water" order. However, boiling will not kill all bacteria.
   a. Why was the town told to boil its water? Give an example in your answer.
   b. What would survive the boiling?
   c. Why don't most people have to worry that the water they drink is not sterile?
   d. How could one go about sterilizing a plastic water bottle?
6. Which would be a better antiseptic for preparing a patient's arm for an injection: ethanol or isopropanol?

7. Name two problems with using methanol as an antiseptic.
8. Why aren't relatively long-chain alcohols (12 carbons or more) used in preparing a patient's arm for an injection?
9. Jim is in the waiting room at the Muddy Hole Hospital, awaiting surgery. The doctor steps out and says the operating room has just been sanitized. Jim is disgusted by the news and insists on having his operation performed at a hospital in the neighboring city. What could have possibly upset Jim?
10. If a bacterium were to stay in the spore state forever, would it be considered dead?
11. Indicate whether the members of each of these sets of three are good antiseptics, disinfectants, or neither:
    a. Bleach, quats, boiling
    b. H_2O_2, soaps, isopropanol
    c. Methanol, triton, DMSO
    d. Dry heat (500°) for 30 minutes, incineration, application of electron beam
12. Consider the sulfa antibiotic class. What is the term used to describe when the enzyme's "attention" is drawn away from PABA and toward a sulfa drug?
13. Consider a soap or detergent: is a longer or shorter hydrocarbon chain the better solvent for the lipid tails found in membrane lipids? Is there a limit to how long or short the chain can be in the soap or detergent? Explain your answer.
14. Why is it not advisable to use soap in your dishwasher?
15. Nystatin targets ergosterol in fungal cells. We mentioned side effects in humans due to nystatin also binding to cholesterol. Could cholesterol, therefore, be considered a competitive inhibitor of nystatin?
16. If I have a drug that doubles the number of cross-links in the gram-positive cell wall and makes them unbreakable, will it be a good or bad choice of antibiotic to treat gram-positive infections?
17. Name four targets of antibiotics and give one specific drug for each target.
18. In Chapter 3, it was stated that L-amino acids are used in our bodies. However, there are instances in nature for which D-amino acids are used. Identify a very important example of this.
19. Explain why the polyene drug amphotericin B has harsh side effects when administered to humans.
20. Joshua has a polypropylene (able to stand temperatures up to 135 °C) water bottle with some kind of black mold growing in it. He autoclaves it, but the black is still present. He washes the bottle a few times, but the mold is still there. Is it safe to drink from the bottle?
21. Chris has discovered some rotten uncooked meat in the back of his refrigerator. Knowing that a pressure cooker works very much like an autoclave, he decides to sterilize the meat by cooking it for 30 minutes. After the meat has been sterilized, is it safe to eat? Why or why not?
22. Shelly has a bottle of strong disinfectant that has been sitting under her sink for a year. If she autoclaves it, will it be safe for her to drink? Would your answer change if she instead used gamma irradiation to sterilize it?

CHAPTER

7 Cell culture and the eukaryotic cells used in biotechnology

Chapter outline

The biotechnologist will often work directly with cells in research or application scenarios. Keeping cells alive and healthy is not always an easy task, and the methods used to keep one cell type alive might not be the same as those needed for other cell types. Even the vessels that cells are kept in will vary in certain instances.

In this chapter, we will look at different types of cells that may be used by the biotechnologist. We will begin by learning that some cells must be attached to a surface before they can grow, whereas other cells must be cultured in suspension. We will then move on to different classes of cells that are grown in culture: primary cells, cancer cells, and cell lines. Finally, we will discuss the main components of cell growth media by looking at what is most commonly used for mammalian cell cultures.

7.1 Adherent cells versus nonadherent cells

Animal cells are typically grown as adherent cultures. Consider a culture dish with medium in it: if you pour the medium away, the cells remain where they are because they are stuck to the bottom of the dish—they are **adherent**. They stay attached to the dish via adhesion proteins such as **integrins**, an entire family of transmembrane proteins used by cells to attach to an extracellular surface. In the body, this surface might be extracellular matrix; in the laboratory, it might be a culture dish or a synthetic matrix covered with collagen. For completeness, we should also mention **cadherin**, which gets its name from "calcium-dependent adherence". The caherins are a similar family of adhesion proteins that are used by cells to attach to other cells.

Specific proteins such as fibronectin, vitronectin, osteopontin, the collagens, thrombospondin, fibrinogen, and von Willebrand factor can be used to promote the attachment of cells to surfaces. These

Biotechnology and its Applications. https://doi.org/10.1016/B978-0-12-817726-6.00007-1

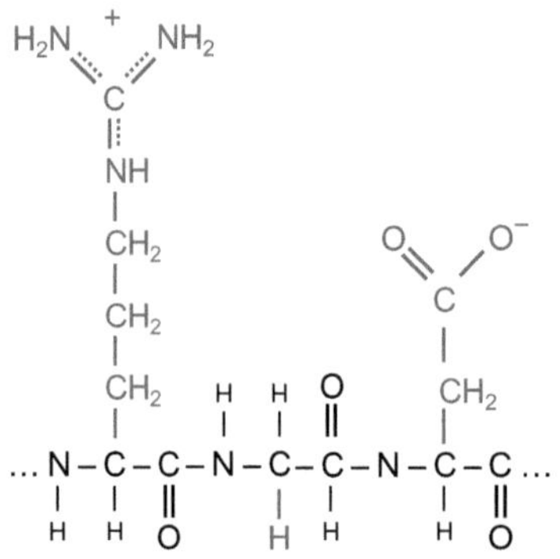

FIGURE 7.1

An RGD sequence: the tripeptide sequence *arginine-glycine-aspartate*. Amino acid identities are defined by their side groups, which are shown in *red*. See Chapter 3 for more information on amino acid identities.

proteins all share a common characteristic: the tripeptide sequence *arginine-glycine-aspartate*, abbreviated **RGD** (from their one-letter amino acid codes) (Figure 7.1). RGD sequences can be used to coat surfaces such as tissue engineering scaffolds to promote cell attachment. Attachment is achieved through the recognition of RGD by integrins, which are transmembrane proteins that act as receptors.

However, not all cells grow as adherent cultures. We have already mentioned the growth of *Escherichia coli* cells for such applications as genetic engineering. When *E. coli* are grown in culture, they can be grown either in a culture dish on top of a Luria-Bertani (LB) agar solid medium or in liquid LB medium as a suspension. Some eukaryotic cells—for example, human hematopoietic cells—are also not anchorage dependent, meaning they can be cultured in the middle of the appropriate cell medium. The culture vessels for cells grown in suspension do not have to be specially treated to allow for cell attachment. However, the user must ensure there is adequate gas exchange (CO_2 and O_2). This can be accomplished by using only a thin layer of liquid (which is impractical for suspension cultures), by stirring the medium and cells inside the culture vessel, or by placing the vessel on a shaker which acts to swirl the cell suspension around inside.

7.2 Primary cells, cancer cells, cell lines

We already know that not all cells are the same; some differences between eukaryotic and prokaryotic cells have been discussed, and for prokaryotic cells, we have seen the difference between Gram-positive and Gram-negative cells. Looking at eukaryotic cells, there are classifications that must be understood if one is to work with live cells, especially in a culture situation. These classifications include primary cells, cancer cells, and cell lines.

7.2.1 Primary cells

Cells that have been taken directly from a body or a tissue are known as **primary cells**, and they can be obtained by biopsy, surgery, or autopsy. Primary cells can also be cultured for a finite amount of time as **primary cell cultures** (cells from primary cell cultures are slightly different from primary cells).

Suppose you have given permission for a sample of your own cells to be collected and grown in the laboratory for research purposes. Your doctor might take a punch biopsy and start a primary cell culture via **explant**. The explant process begins with the breaking down of extracellular matrix in a collected tissue sample; this can be accomplished via mechanical means such as mincing the tissue with a scalpel blade, by chemical means via digestion with an enzyme like papain, or both. Following the mincing/digestion, the processed tissue is placed on a prepared growth surface, covered with growth medium, and allowed to incubate undisturbed for several days. After the incubation, the tissue mounds are examined by light microscopy. If cells are observed on the plate in the area immediately surrounding the tissue mounds, the mounds are gently removed to prevent adverse effects from further tissue degradation or products of cell death. The remaining cells are allowed to proliferate as much as the current conditions will allow in a process known as **expansion**. When cell numbers are great enough, selective media may be employed to prevent the growth of undesired cell types. Also, as the culture is expanded, it may be **passaged** into one or more fresh culture vessels. Passing a culture means the cells have been removed from one culture vessel (by chemical or mechanical means) and placed into a new one. When a primary culture has been passed once, the new culture(s) are called **secondary cultures** (Box 7.1).

Returning to our example, in which your cells were explanted into a cell culture: suppose that several years after you graduate, you return to the laboratory to visit your cells. You are told that there's not much to see because there are no active cultures at the moment. The cells that are still in the laboratory have been frozen, but the active cultures had to be discarded because "they got old." "How can that be?" you exclaim, "I'm in the prime of my life! How could my cells possibly be old?" The answer is related (in part) to something known as the **Hayflick limit**, which helps to explain the mortality of primary cells and cultures.

In 1961 Hayflick and Moorhead published proof of the idea of limited cell division and the eventual **senescence** (old age) of cells in culture. The idea behind the Hayflick limit is that not all of an animal cell's genome is replicated during the cell cycle. As a result, each chromosome is shortened with every cell division (actually, with each S phase; see Chapter 5). After a number of cell divisions,

Box 7.1 Splitting cells, passing cells

Splitting cell cultures was first mentioned in Chapter 5 with regard to the growth curve, and *passing* cells is touched on in this chapter in terms of creating a secondary (or later) culture. The difference between the two terms is one of precision: splitting controls the fraction of cells that is introduced into the new culture vessel. The transfer of cells from one culture vessel to another is a passage, so splitting a culture is also passing a culture. The mechanics are the same for both terms, and commonly involve the following steps for attached cultures:

1. Cells are washed and removed from the original vessel, typically via an enzyme such as trypsin. Trypsin is a serine protease that cleaves polypeptides after the positively charged amino acids lysine and arginine. Adherent cells can be detached from surfaces by cleavage of adhesion proteins. Following detachment, the action of trypsin is slowed by dilution of the cell suspension, often with FBS-containing medium which has proteins that act as competitive inhibitors of the trypsin.
2. The diluted cell suspension undergoes centrifugation to separate the cells from the trypsin. Overexposure to trypsin will cause extensive cleavage of extracellular proteins and lead to cell death. After centrifugation, the trypsin-containing supernatant is poured off, leaving a moist pellet of cells at the base of the centrifuge tube.
3. The cell pellet is resuspended in fresh growth medium. A common volume used for splitting cells is 8mL.
4. An aliquot of the fresh cell suspension is transferred to a new culture flask that already contains warm growth medium. If the aliquot is 1mL, which is 1/8 of the cell suspension created in the previous step, then the cells are said to have undergone a 1:8 split. If a culture is being expanded, up to 8 flasks can be generated from the original flask during a 1:8 split.

chromosomes will have shortened to the point that important parts of the genome are missing, so the cells will not be able to perform as needed to continue to grow and divide.

Consider a human chromosome undergoing replication. When dsDNA is replicated, the parental DNA strands are separated at the replication fork. Each of the daughter DNA strands is polymerized in the $5' \rightarrow 3'$ direction. For one of these strands, the process is straightforward. For the other strand, however, the $5' \rightarrow 3'$ direction of polymerization leads away from the replication fork, so this strand is replicated in segments known as **Okazaki fragments** (Figure 7.2). As the replication fork reaches the end of the chromosome, the final Okazaki fragment will not be formed because its origin would need to be beyond the end of the chromosome. As a result, this daughter DNA strand will be shorter. Shortening will occur with every replication until eventually the unreplicated DNA will contain material that is important to the survival of the cell. At this point, the cell will either cease replicating or will die. This is the concept used to explain the Hayflick limit.

The ends of eukaryotic chromosomes are made up of special DNA sequences that are repeated many times; these specialized ends are known as **telomeres** (Figure 7.3). In humans, the telomere sequence is GGGTTA, and it is repeated approximately 1000 times at the end of each chromosome. It is the shortening of telomeres that is responsible for senescence in some primary cells.

Immortal cells (e.g., cancer cells or cell lines) often utilize the enzyme **telomerase**. Telomerase is an enzyme that adds telomere sequences to the ends of chromosomes each time the cell divides. To make matters more complex, not all immortal cells express telomerase. These cells get around the telomere shortening problem by a telomerase-independent pathway called **alternative lengthening of telomeres (ALT)**. It has even been shown that it is possible to switch between telomerase-positive and telomerase-negative telomere-lengthening processes. A general rule of thumb is that somatic (mortal) cells do not express telomerase, and telomerase expression is seen in many immortal cells.

The subject of telomere shortening was relevant to Dolly the sheep, the first successful clone of a large animal. Dolly was the result of a successful cloning exercise, but she only lived to be 6 years old. Although apparently a newborn lamb at birth, Dolly aged at an accelerated rate. The reason was that the DNA from the nucleus used to clone her already had shortened telomeres, because it was DNA that was taken from an adult somatic cell. When this DNA was transferred into an enucleated oocyte (i.e., an egg cell without a nucleus), the oocyte could be considered a "new" cell, but the transplanted genetic material had already undergone multiple cell divisions. When Dolly started growing and developing and her cells continued to divide, they continued to age from the point at which the original somatic cell had been harvested. In other words, she started her life with adult-age DNA. What seemed like early senescence to the outside observer was actually in line with the real age of her chromosomes.

7.2.2 Cancer cells

Two main factors determine whether a cell can be considered a "normal" somatic cell or a cancer cell: mortality and **contact inhibition**. We have just discussed mortality as being a limit on the number of times a cell can divide in culture. Most somatic cells are mortal, although some types (e.g., stem cells) can be propagated for extended periods and therefore seem to be immortal. This does not make stem cells cancer cells, though. The second factor, contact inhibition, must first be addressed. A typical somatic cell in proper culture conditions will grow and divide (and migrate) until it touches something—another cell, or the edge of the culture plate. Making contact with an object causes the cell to change the direction of its migration. As more and more cells populate the culture plate, there

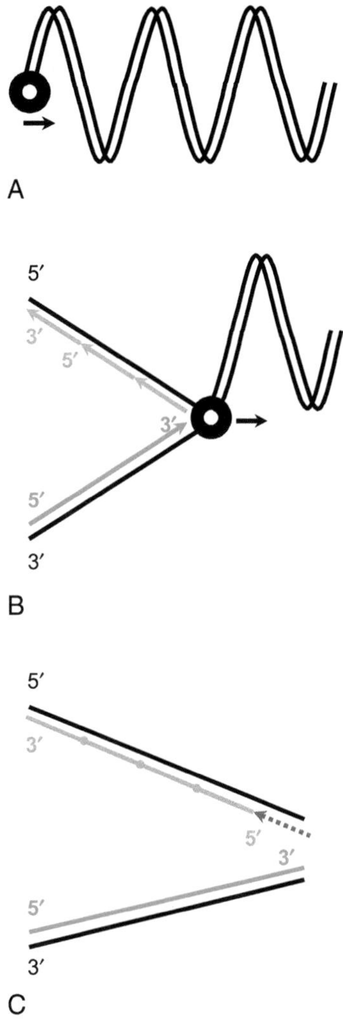

FIGURE 7.2

(A) Replication of dsDNA involves several enzymes, including a helicase, which separates the two strands, and DNA polymerase. The replication machinery is shown as a ring. (B) As the replication fork progresses, the "leading strand" *(blue)* is easily polymerized in the 5′ → 3′ direction. The "lagging strand" is also created in the 5′ → 3′ direction, but this must be done in segments known as Okazaki fragments (orange). (C) DNA polymerase requires a double-stranded primer before polymerization can occur. At the very end of the parent strand, there is not enough room for the primer, so the final Okazaki fragment cannot be created *(red, dashed)*. The end of the lagging strand will not be replicated.

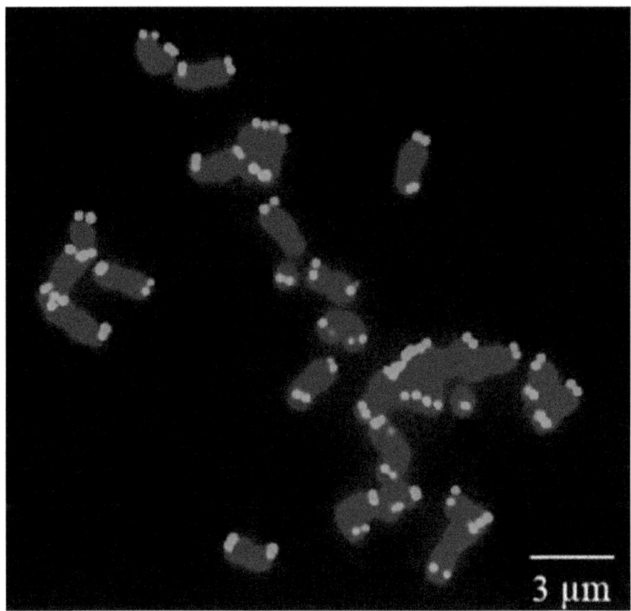

FIGURE 7.3

Chromosomes, shown in *blue*, have been labeled with a probe to detect telomeres, seen here as *bright green dots*. Note that the telomeres appear on the ends of the chromatids.

From: Liu J and Luo X. Hereditas (2019) 156: 13. (Figure 1b).

is an increasing likelihood that they will come into contact with each other. Eventually, each cell will be surrounded on all sides by other cells or the edge of the culture vessel. When the entire culture is in this state, a monolayer has been formed. The cells will no longer migrate and will also stop replicating. In terms of the growth curve, the culture will have reached a plateau phase—not because of a lack of nutrients but because of a lack of space. Similar circumstances occur in three-dimensional cultures and in the body, which helps to explain why we (hopefully) do not continually have large masses of tissue growing out of us. In short, a cancer cell is a cancer cell because it is (1) immortal and (2) not contact inhibited.

In a two-dimensional culture, after forming a monolayer—and sometimes before they reach that stage—cancer cells begin to grow on top of each other. They might form a second layer, or they might begin to grow vertically and then branch out to form a structure that looks something like a mushroom or a ball and chain. It does not take much force to break these structures to release small balls of living cells, which can then relocate to another area and set up a new colony. This is not an uncommon feature in cultures of metastatic cancer cells. Other cultures of metastatic tumor cells may have cells that do not adhere tightly to the culture support, rendering them easily displaceable and allowing them to detach and set up new colonies in areas with more room to grow. Metastasis occurs in similar ways in the body. Tumor cells divide to form a small mass, then metastasis occurs when one or more of the tumor cells lets go and migrates through the bloodstream or the lymphatic system to a distant site, often the liver or a lymph node.

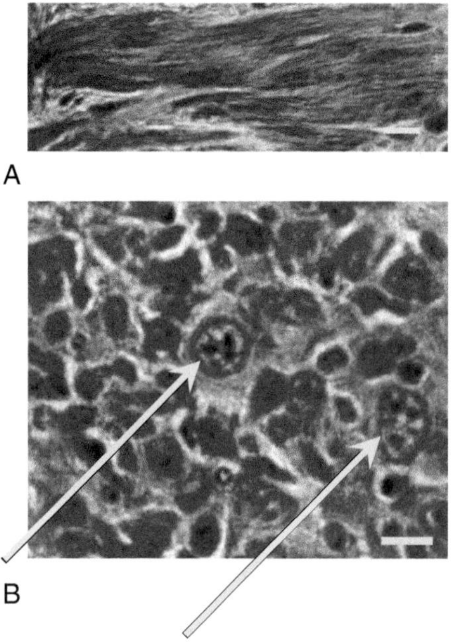

FIGURE 7.4

Stained sections of normal and cancerous cells. Each sample has been stained with hematoxylin and eosin. (A) Smooth muscle cells. Each *(pink)* cell has a single *(purple)* nucleus. (B) Cells from a tumor of transitional cell carcinoma. These cancers stain quite differently from normal tissue. Notice that cancer cells can have multiple nuclei; two such cells are indicated by the *arrows*. Scale bars = 10 μm.

7.2.3 Cell lines

A cell line can be considered a cross between a "normal" cell and a cancer cell, both figuratively and sometimes literally. Cell lines are created in the laboratory to display the key characteristics of a specific cell type while at the same time being immortal. Cell lines are of great value to the research community because they allow for the study of specific cell types without the necessity of returning to the same donor repeatedly as cells reach senescence. They also provide an abundant source of cells that can be used in multiple laboratories around the world with little variation from culture to culture.

Cell lines can be created in a number of ways. A primary culture can be passed into a secondary culture, which gets passed into a tertiary culture, and so on, until the characteristics of the culture change. Eventually, one or more of the cells in the culture will undergo **transformation**: the switch of a cell from a mortal to an immortal state. Transformation of a cell in the body may trigger cancer, but transformation of a cell in culture can be the birth of an immortal cell line. It is common for transformed cells to become **multinucleate** (containing more than one nucleus) (Figure 7.4). Because most primary cells have a limit to the number of times they can replicate, it is easy to select for transformed cells as passage numbers become large—in fact, they will select for themselves. Untransformed cells will either die or be outcompeted as they enter senescence.

FIGURE 7.5

A bottle of Dulbecco's Modified Eagle Medium. The red color is from added phenolphthalein, a pH indicator.

A second way to establish a cell line is by explanting a biopsy of cancer cells. The cells have already been transformed, so letting the cells divide beyond their expected senescence is not necessary. Such cell lines are appropriately referred to as cancer cell lines.

A third way to create a cell line is by fusing a primary cell type with a cancer cell type. Fusion can be accomplished, for instance, by placing the cells in contact with one another in poly(ethylene glycol) and administering an electrical current to cause membrane perturbation. When the current is stopped, the cells' plasma membranes may be permanently joined. Many cells will die as a result, but in theory it only takes one surviving cell to start a new cell line. The goal of this method is to retain characteristics of the primary cell for future study while imparting immortality, thus creating an inexhaustible supply of cells that behave with a standard set of traits. However, let us consider a reality of this method in greater detail. Human somatic cells have 46 chromosomes (23 pairs) in their nuclei. Let's suppose we want to make a fusion cell line from some endothelial cells we obtained from a biopsy. The resulting cell line would not have 46 chromosomes, because we would be fusing two cells that have their own genomes and their own nuclei. In fact, not only will the resulting cell not be restricted to 46 chromosomes, it will also not be restricted to a single nucleus. This is why most attempts at fusing two cells do not produce a viable hybrid cell. When a viable cell is produced, it does not carry with it all the properties of the parent somatic cell. That is a major drawback to using cell lines: data and results obtained from cell lines are not always applicable to primary cells in a living organism. Although cell lines are a great tool for research, data obtained from them should be taken with a grain of salt.

7.3 Care and feeding

Knowing the classes of eukaryotic cells used in biotechnological applications is one thing—but being able to grow and maintain them as cultures is quite another. Cells are living entities. As such, they have some of the same requirements that you have as a multicellular organism. They require food, water, and a safe place to live. The living arrangements are provided by bioreactors and/or incubators, a subject that is covered in Chapter 17. The issues of food and water will be covered here.

The water for eukaryotic cells in culture is typically supplied as part of a liquid medium. Osmolarity of cell media is a primary consideration: isotonic solutions are required to prevent osmotic stress, lysing, or crenation (see Box 11.1 in Chapter 11). This brings us to the second component of most culture media: ions. Again, these are included to reduce or eliminate osmotic stress upon the cell culture. We have discussed the transport of various solutes across the plasma membrane. For instance, the sodium/potassium/ATPase transporter is used to establish an electrochemical gradient across the plasma membrane, and the energy stored via this gradient is responsible for the transport of other ions, such as Ca^{2+}, via the sodium/calcium antiporter (see Chapter 2). Ions are supplied to the culture medium as salts, which dissociate into ions in water. The most common ions included in culture media are:

- cations: K^+, Na^+, Ca^{2+}, Mg^{2+}
- anions: Cl^-
- polyatomic ions (used for adjusting and buffering pH): SO_4^{2-}, PO_4^{3-}, CO_3^{2-}

We also saw in Chapter 3 that proteins are essential components of cells. Although we cannot supply every protein the cells need in the culture medium, it can provide the building blocks and let the cells manufacture their own proteins. All 20 of the naturally occurring amino acids are included in common culture media. The concentration of each amino acid varies according to cell need (e.g., cells use far more glutamine than tryptophan) or the metabolic effect desired by the biotechnologist.

Cells require vitamins, just like multicellular organisms. We saw in Chapter 6 that folate is a vitamin required for DNA (purine) production in humans—hence the basis for the sulfa drugs— so it follows that folate should be included in cell media for human cells. Other vitamins and supplements included in the most common cell medium include:

- B_1 thiamine
- B_2 riboflavin
- B_3 nicotinamide (amide form of niacin, which is vitamin B_3)
- B_5 pantothenic acid
- B_6 pyridoxine
- B_9 folic acid
- Myo-inositol (once called "vitamin B_8," but declassified as a vitamin after it was found to be manufactured in humans)
- Choline (not technically a vitamin; produced in the liver in humans, but the levels produced do not meet the needs of the rest of the body)

Last, we must not forget an energy source. Although some energy could be harvested from the breakdown of amino acids, the point of most cell media is to allow cells to live and perhaps grow with a minimal amount of stress. The preferred energy source for eukaryotic cells is a carbohydrate; for mammalian cells, it is glucose.

The list just described is related to a medium called "Eagle medium." In the 1950s, Dr. Harry Eagle was interested in the idea of growing human cells in culture to permit scientific study at the cellular level outside of the human body. He published seminal papers describing how this could be achieved for the HeLa cell line. In 1959, Renato Dulbecco published work that cultured HeLa cells in a modified version of Eagle's medium. The Dulbecco modification had approximately four times as much of the vitamins and amino acids and two to four times as much glucose. That medium, known as Dulbecco's Modified Eagle Medium (DMEM), remains the basis for a significant number of cell cultures today. Other modifications have been made since then, including the addition of components such as iron and pyruvate, and adjustments to the concentrations of glycine, glutamine, and serine.

Our list—ions, buffers, amino acids, vitamins, supplements, and an energy source—makes up what is known as a **defined medium**. Every component in a defined medium is known, measured, and controlled. If additional factors such as insulin, transferrin, and selenious acid are added, the medium is still a controlled medium because the components are added in known quantities. However, one very common additive is serum, often delivered as fetal bovine serum (FBS). FBS contains growth factors/inhibitors, and molecule classes, including fatty acids, lipids, steroids (e.g., cholesterol), and trace elements that greatly enhance culture conditions for mammalian cells. However, the identities and concentrations of all molecules in FBS are not necessarily known, and they vary from batch to batch. Media that use FBS are therefore classified as **complex media**, because their exact chemical makeup is not known. The biotechnologist loses some control over culture conditions when FBS is used, but its use is so widespread that it is an acceptable component in mammalian cell media. In fact, for many applications, serum is an essential component for maintenance and growth of cell cultures.

7.4 Summary

In this chapter, we saw several aspects that the biotechnologist must consider when eukaryotic cells are to be maintained in a culture environment. We saw that most animal cells are adherent and must be cultured on surfaces, such as collagen-treated flasks or dishes.

We also saw that cells taken from a body and explanted onto a culture dish are known as primary cells and that as the culture expands to cover the surface of the culture vessel, they may be passed into secondary cultures. This process cannot be repeated indefinitely for normal cells, however, because of their own mortality, which was explained in terms of telomere shortening and the Hayflick limit. Immortal cells do exist: cancer cells and cell lines are examples. The three classes of cells were presented to allow the reader to develop a basis for deciding on which cell type to use when planning experiments.

Cells must be grown in a medium that provides for their metabolic needs. Necessary constituents of cell media include the following molecule classes:

- ions
- buffers
- amino acids
- vitamins
- carbohydrates

Several additional supplements were also discussed, such as growth factors, trace minerals, lipids, and cholesterol. Necessary supplements are often added to the media in the form of serum. FBS is a common and accepted component of many cell media, but its inclusion changes the classification of the medium from defined to complex.

Related reading

Dulbecco, R., Freeman, G., 1959. Plaque production by the polyoma virus. Virology 8, 396–397.

Eagle, H., 1955. The minimum vitamin requirements of the L and HeLa cells in tissue culture, the production of specific vitamin deficiencies, and their cure. J. Exp. Med. 102, 595–600.

Eagle, H., 1955. The specific amino acid requirements of a human carcinoma cell (Stain HeLa) in tissue culture. J. Exp. Med. 102, 37–48.

Hayflick, L., Moorhead, P.S., 1961. The serial cultivation of human diploid cell strains. Exp. Cell Res. 25, 585–621.

Kumakura, S., Tsutsui, T.W., Yagisawa, J., Barrett, J.C., Tsutsui, T., 2005. Reversible conversion of immortal human cells from telomerase-positive to telomerase-negative cells. Cancer Res 65, 2778–2786.

Ruoslahti, E., Pierschbacher, M.D., 1987. New perspectives in cell adhesion: RGD and integrins. Science 238, 491–497.

Vogt, M., Dulbecco, R., 1958. Properties of a HeLa cell culture with increased resistance to poliomyelitis virus. Virology 5, 425–434.

Yao, T., Asayama, Y., 2017. Animal-cell culture media: History, characteristics, and current issues. Reprod. Med. Biol. 16, 99–117.

Questions

1. **a.** Suppose that every time you pass cells, you always do so in a 1:8 split (i.e., 1/8 of the plate or flask of cells is passed to a new culture vessel). If your original cells undergo 11 doublings before the first split, you always pass your cells when they reach a certain density, and you pass them into the same size of culture vessel every time, how many passages can you expect to perform before the cells reach senescence? (You may use a Hayflick limit of 60 for this problem.)
   b. Perform part a) again but for a 1:16 split and a 1:4 split.
   c. Write a general equation to predict the number of passages one can expect to get from a primary culture.
2. How many passages can one expect to get out of a culture of cancer cells? Why?
3. Suppose you passed a primary culture of cancer cells, storing a frozen aliquot of them and propagating the rest. After the nonfrozen cells have undergone 50 additional passages:
   a. What is the passage number of the frozen cells?
   b. What is the passage number of the nonfrozen cells?
   c. How many doublings have the nonfrozen cells undergone?
   d. If you thawed the frozen cells and placed them into a new culture, would you expect this culture to behave the same as the culture that had undergone 50 additional passages? Why or why not?
4. What might be a problem with introducing active telomerase into all of your own cells?
5. Name one similarity and one difference between a cancer cell and a stem cell.

6. Give two opposing reasons why a biotechnologist might be concerned with cell adherence.
7. Name three ways in which cell lines are created.
8. What does "DMEM" stand for?
9. Was Eagle medium originally developed for the culture of eagle cells? If not, what cell type was the initial focus for Eagle medium?
10. No fats were mentioned as part of DMEM. How are fats usually delivered in mammalian culture media?
11. Most of the specifics in the discussion of cell culture media were concerned with culturing human or mammalian cells. Are any parts of the discussion also applicable to plant cell cultures? Why or why not?
12. FBS is a very important part of mammalian cell culture. Would there be any benefits to adding it to media used for insect cell cultures? Justify your reasoning.
13. Why is a pH buffer such as sodium bicarbonate important in cell culture media? Once the pH is set, why would it ever change?
14. Name a risk associated with using FBS in a cell culture.
15. Biotechnologists sometimes add penicillin and streptomycin to their cell media. What is the purpose of the addition? If penicillin is added, why would streptomycin also be used?
16. **a.** Name one advantage of using a cell line instead of primary/secondary cells for cell research.
    b. Name one advantage of using primary/secondary cells for cell research as opposed to a cell line.
17. How do cancer cells differ from cell lines?
18. Are stem cells cancer cells? Why or why not?
19. Media additives:
    a. If a medium consists of DMEM with 10% FBS, is it a defined or a complex medium?
    b. If a medium consists of Eagle's Minimal Essential Medium plus a standard amount of ITS (insulin, transferrin, selenious acid), is it a defined or a complex medium?
    c. If a medium consists of DMEM plus a standard amount of chick embryo extract, is it a defined or a complex medium?
20. Name the purpose of each of the following medium components:
    a. ions
    b. buffers
    c. amino acids
    d. vitamins
    e. glucose

CHAPTER

Fluorescence

8

Chapter outline

Apart from being just flat-out cool, fluorescence and the principles behind it are important for the biotechnologist to learn because the use of fluorescence is so prevalent in biotechnological research and applications. Fluorescent molecules (fluorophores) are very often used as reporters or tracers. Fluorescent tags can be attached to other molecules, such as antibodies, to allow for the detection (for quantitation) or tracking (for location) of molecules of interest. Fluorescent reporters like the green fluorescent protein (GFP) can be used to verify that a cell is expressing a plasmid that has been delivered.

The principles covered in this chapter are used in applications including the determination of DNA concentration and purity, quantitative PCR, the demonstration of cell type, and the location of specific molecules inside the cell during specific processes and under certain conditions.

We will start by examining observations made by Sir George G. Stokes that can be replicated to a large extent at home. From these observations, we will be introduced to more formal ways of describing light and the fluorophores that interact with it. We will then move on to fluorescence generation and detection, and laboratory applications of the technology such as fluorescence resonance energy transfer.

8.1 Stokes's experiments

In 1852 Stokes published a 102-page paper that described some interesting experiments involving sunlight and a solution of quinine. You can perform these experiments yourself with a prism, some colored glass, and a tube of quinine (which is a component of tonic water).

In the first experiments, Stokes covered a test tube with black paper. A hole was cut into the paper to allow light to enter through the side of the tube, and the top was left uncovered (Figure 8.1). Sunlight was allowed to enter the tube, which contained a solution of quinine. Examining the solution from the top of the tube, Stokes observed a pale blue arc of light in the quinine. When a smoke-colored piece of glass was

Biotechnology and its Applications. https://doi.org/10.1016/B978-0-12-817726-6.00008-3

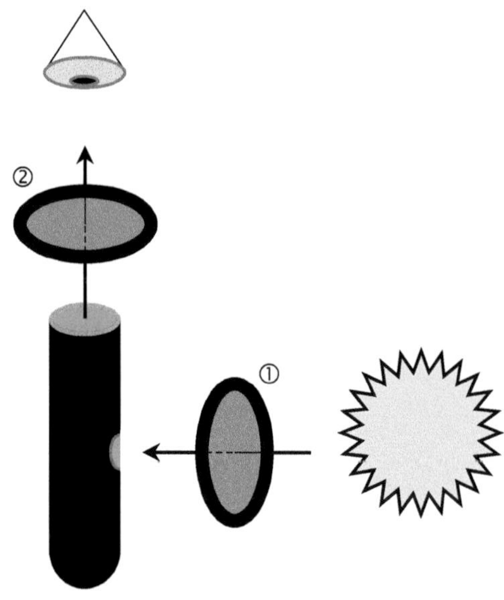

FIGURE 8.1

In this experiment, sunlight enters the side of a tube containing a solution of quinine, and the effect on the quinine is observed from above the tube. Stokes placed smoke-colored glass in either position 1 or 2, as indicated. With the glass in position 1, no fluorescence was detected in the tube. In position 2, a pale blue arc of light was observed in the tube.

placed between the sunlight and the quinine, the arc disappeared. When the glass was moved between the quinine and Stokes's eye, the arc reappeared. The glass was said to have the property of being able to absorb the "invisible rays beyond the extreme violet"—what we would now call ultraviolet (UV) light.

In a later set of experiments, Stokes placed boards in front of his laboratory window so that only a vertical slit of sunlight could enter the room. He then allowed the light to enter a series of prisms so it was dispersed into a spectrum (Figure 8.2). Placing a test tube with quinine in the rainbow of light had very little effect; the solution remained clear like water as it was moved slowly from the red to the blue light. When it reached the extreme end of the violet light, however, a "ghost-like gleam of pale blue light shot right across the tube." When the tube was moved even farther, past the violet region of the visible light spectrum, the quinine solution became opaque with pale blue light of greater intensity. In Stokes's own words, "it was literally *darkness visible*."

So now we know that sunlight can make quinine glow—fluoresce—and that it is the specific UV component that is responsible for inducing the fluorescence. In another set of experiments, Stokes was able to get a better idea of the nature of the light that came out of the fluorescing quinine. First, he had the sunlight go through water before it hit the prism; as expected, the complete visible light spectrum was observed. However, when he replaced the water with quinine, the light hitting the prism was split to reveal only colors in the violet or extreme violet range (Figure 8.3).

It seemed that light in the UV spectrum made the solution of quinine glow, and the light coming out of the glowing quinine was of a different color than the light going in. To verify these observations further, Stokes repeated the experiments described in Figure 8.1, but this time he used cobalt-colored

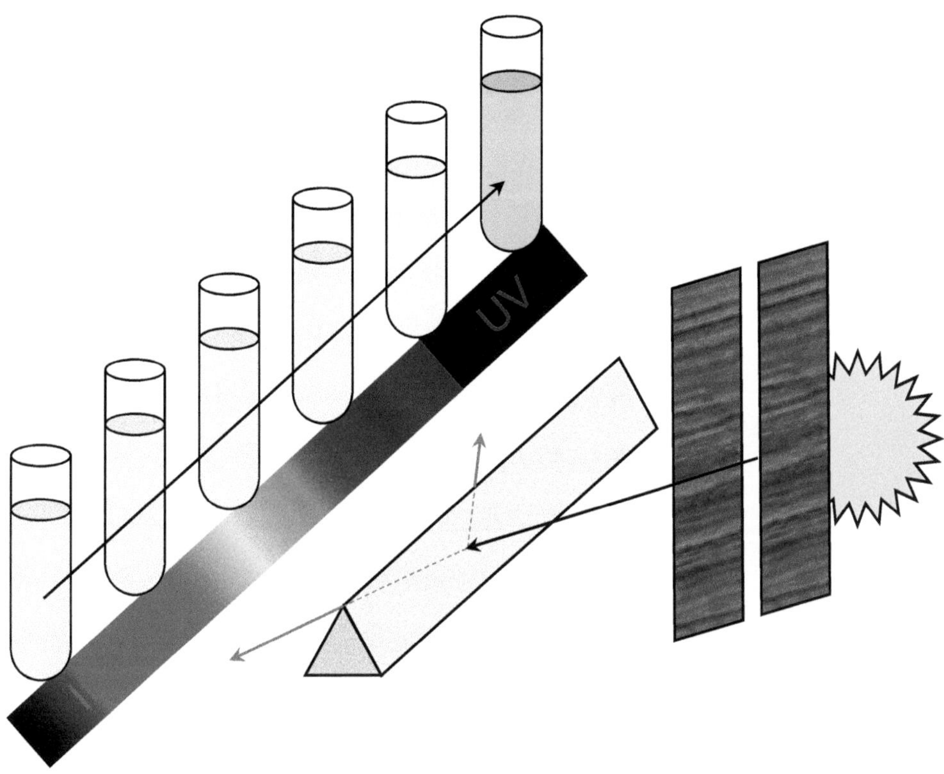

FIGURE 8.2

Stokes's laboratory was darkened, except for a slit of light that was allowed to enter between two boards he put in his window. The light was dispersed by prisms, and a tube containing quinine was run through the resulting spectrum. The contents of the tube remained clear until it entered the extreme violet end of spectrum. When the tube was moved farther still, beyond the visible violet color, the quinine became opaque and glowed even more intensely. *UV*, Ultraviolet.

glass (Figure 8.4). Cobalt glass was selected because it "is highly transparent to the chemical rays." (Once again Stokes was referring to what we would now call UV rays.) The results obtained were in sharp contrast to those obtained using smoke-colored glass: when the cobalt glass was placed between the sunlight and the quinine, the amount of fluorescence remained strong, but when the cobalt glass was placed between the quinine and the eye, virtually all light from the quinine was blocked. Stokes referred to the phenomenon as a change in the refrangibility of light. Today, we describe his observations in terms of excitation and emission wavelengths of a fluorophore.

8.2 Fluorophore properties

It is not uncommon for objects to absorb light; we see it every day. An apple appears red because it absorbs green and blue light while reflecting red. Fluorophores are special, in that they not only absorb light, but they also emit some of the absorbed light at a different wavelength.

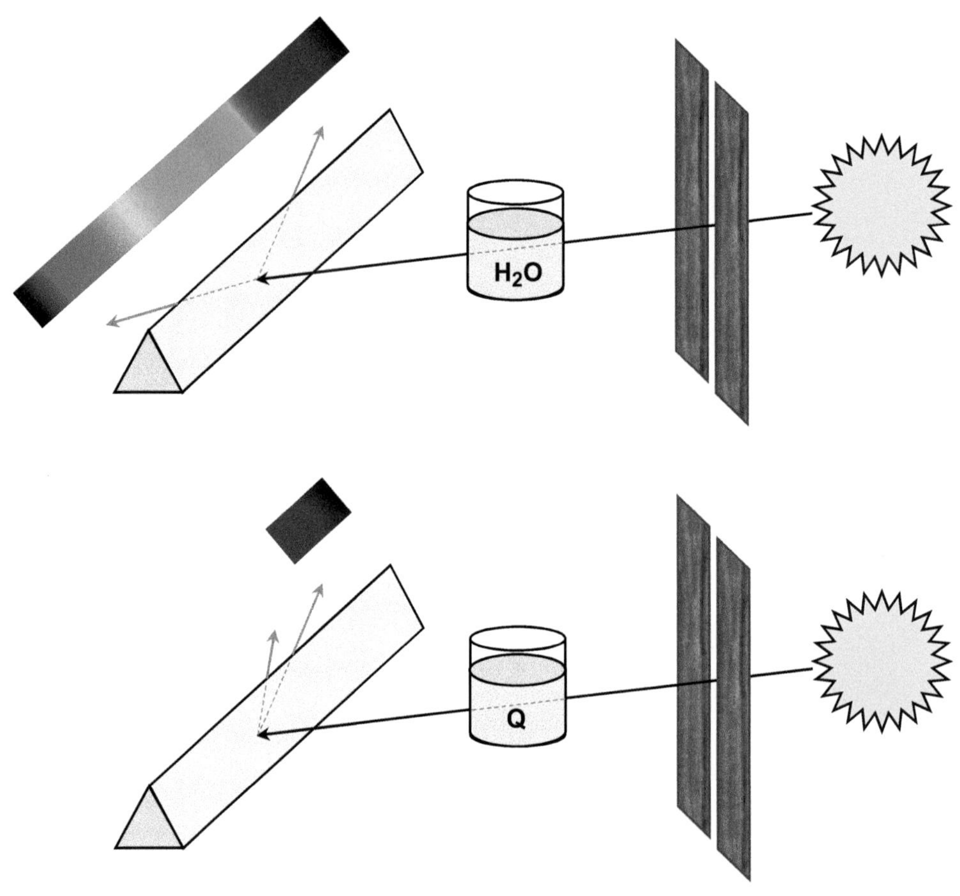

FIGURE 8.3

Sunlight passing through water was dispersed via prisms to reveal the typical visible light spectrum. However, when the light passed through a solution of quinine (Q), only colors in the deep violet portion of the spectrum were observed.

8.2.1 Excitation and emission

If we recall Newton's law on the conservation of energy, it should be intuitive that if a fluorophore absorbs energy from a photon and the energy is then released as a photon, the latter photon will have energy that is less than or equal to the amount of energy held by the first photon prior to absorption. In the visible light spectrum, the photons in the violet region carry more energy (and have shorter wavelengths) than the photons in the red region of the spectrum. This means that if a fluorophore absorbs photons of any given color, the photons emitted will be the same or to the right of that color (as pictured).

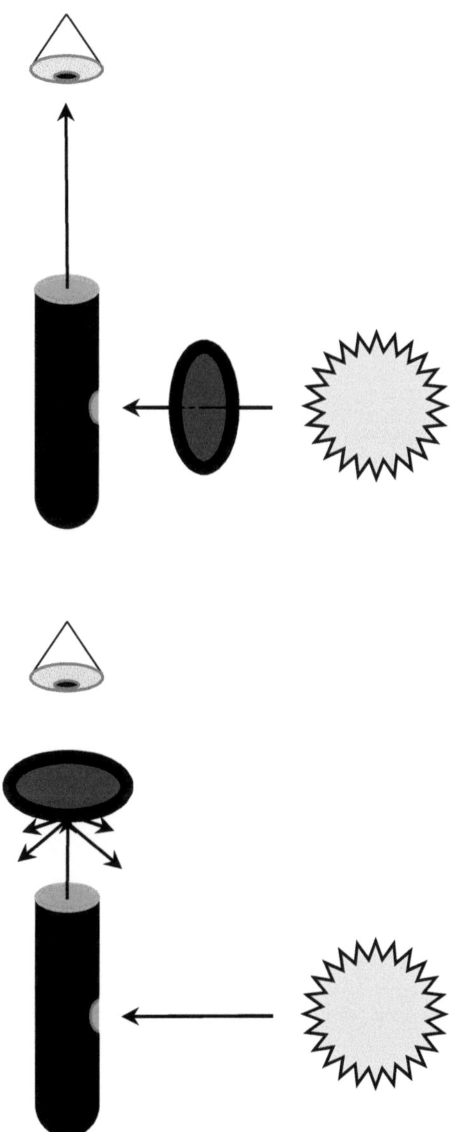

FIGURE 8.4

The experiment depicted in Figure 8.1 was repeated using cobalt glass, which allowed ultraviolet (UV) rays to pass. There was a distinct difference between the light that was needed to make the quinine fluoresce (UV) and the light emitted by the quinine.

We can consider what is going on with fluorescence on the subatomic scale. A fluorophore is a molecule, be it GFP or quinine. Within the molecule are individual atoms, each with nuclei and electron clouds. Let us consider a key electron within the fluorophore. We can illustrate the energy state of that electron with the aid of a Jablonski diagram (Figure 8.5). In the diagram, the lowest energy state of the

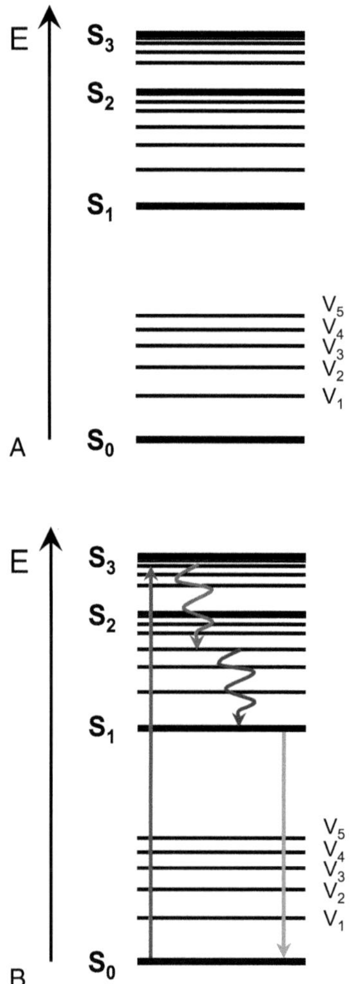

FIGURE 8.5

(A) A typical Jablonski diagram of singlet electron energy states. (B) When an electron is elevated to a higher energy state *(magenta line),* it will return to the ground state by releasing the energy in various ways. *Red squiggle:* internal conversion, which crosses an electronic state (heavy line, S_2). *Blue squiggle:* vibrational relaxation, where the energy descends through vibrational states within the same electronic state. *Green line:* the large energy drop from a higher electronic state to the ground state is achieved via the emission of a photon. This emission is observed as fluorescence.

electron—the ground state—is denoted by S_0. Lower limits of higher electronic energy states are shown by bold lines and the letter "S" with a higher subscript (*e.g.* S_1, S_2 ...). There are also vibrational energy states within the electronic energy states, which are indicated by thinner lines; the vibrational energy states for S_0 are indicated in the diagram by the letter V. When energy is transferred from a photon to an electron, the electron will move to a higher energy state in the diagram.

An electron cannot absorb energy from photons of every single wavelength. Figure 8.5 shows discrete locations for the possible energy states of an electron, which implies that an electron will only absorb photons with energies (wavelengths) that will move it exactly to one of the higher energy states. In other words, an electron will only absorb photons that have energies equal to the difference in energy between the current energy state and a higher one.

Once energy has been absorbed, the electron will eventually return to the ground state. This can be accomplished in several ways, but the most common are **vibrational relaxation**, **internal conversion**, and **fluorescence**. Vibrational relaxation is when an excited electron gives some of its vibrational energy to another electron in the same or a different molecule, in the form of kinetic energy. Because this strictly involves vibrational energy, the energy drop is small: the energy drop will be between vibrational levels but will not go into a new electronic (S) level.

On a strictly mechanical level, internal conversion is the same as vibrational relaxation. The difference is that internal conversion converts the electron to a lower electronic energy state, meaning the energy level can drop below the current S level. Internal conversion happens at higher energy states because the distances between vibrational energy states, as well as in electronic energy states, get smaller and smaller (the lines in the Jablonski diagram get closer and closer).

Vibrational relaxation and internal conversion move the electron down to one of the electronic energy states (S_n, but typically not S_0 if the electron is in S_1 or higher). When the electron moves from an elevated electronic energy state back to S_0, a photon is emitted, and the light energy is observed as fluorescence.

Not every molecule can serve as a fluorophore. Certain geometries are more amenable to absorbing photon energy than others. In addition, just because a molecule absorbs a photon does not mean it will emit the energy in the visible light spectrum; the energy could be released, for example, as heat. In general, ring structures are more able to support the absorbance (and emission) of photons, and the structure of quinine is no different in this respect (Figure 8.6). Proteins that serve as fluorophores also contain at least one aromatic amino acid in the active fluorescent region.

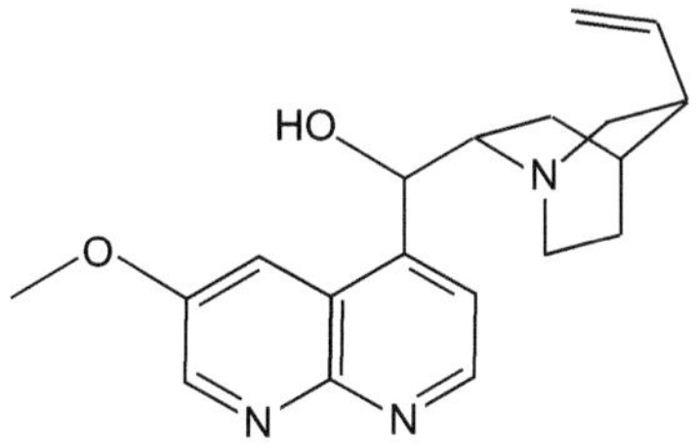

HO
N
O
N
N

FIGURE 8.6

The structure of quinine. Note the rings in the structures, which contribute to the fluorescence potential of the molecule.

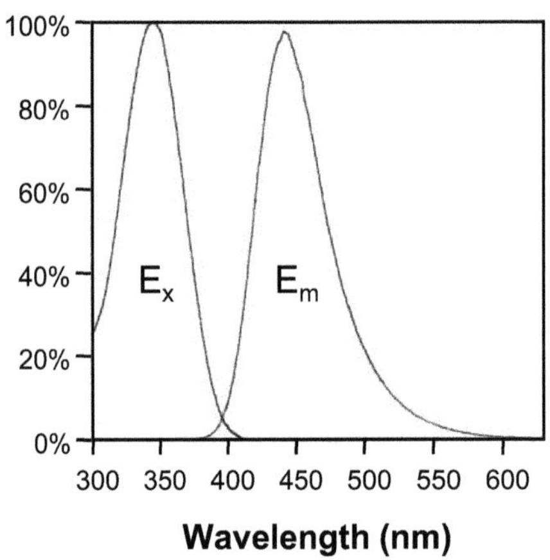

FIGURE 8.7

A hypothetical fluorescence curve. The optimal excitation wavelength for this fluorophore is 346 nm, and the greatest amount of emission is at 442 nm. The Stokes shift is 96 nm. Note that, even though $E_x = 348$ nm, you could use 350-, 360-, or even 370-nm light to excite the fluorophore, albeit less effectively.

Fluorophores are described, in part, by two key wavelengths: the excitation wavelength **(E_x)** and the emission wavelength **(E_m)**. The distance between these two values is, rightly enough, referred to as the **Stokes shift**. As we will see when we discuss light filters, fluorophores with larger Stokes shifts provide greater utility to the researcher.

E_x and E_m are not singular values for a given fluorophore; they indicate the wavelengths of *maximal* absorption and emission. There is a distribution of available wavelengths around E_x and E_m that can be used for fluorescence experiments. From the example given in Figure 8.7, we can see that that E_x, E_m, and the Stokes shift can be easily determined, but also that you have options with regards to the wavelengths you might use for your experiments.

8.2.2 More descriptors of a fluorophore

Simply knowing E_x and E_m of a fluorophore does not provide the entire picture of its fluorescence characteristics. For instance, if you shine light into a 1 M solution of fluorophore, how much of the light has disappeared at a point 1 cm into the solution along the light path? This value is known as the (molar) **extinction coefficient**, ε. It can be determined empirically by noting the absorbance of a solution of the fluorophore and applying Beer's law: $A = \varepsilon lc$, which states that the absorbance of light by a molecule is equal to the extinction coefficient multiplied by how far the light has traveled (the "path length") multiplied by the concentration of the solution. Rearranging, we see that $\varepsilon = A/lc$. Because absorbance is a unitless value, the extinction coefficient ε has the units $M^{-1}cm^{-1}$. To give you an idea of scale, the enhanced green fluorescent protein (EGFP) has $\varepsilon = 55{,}000\,M^{-1}cm^{-1}$.

The amount of light that a fluorophore absorbs is one thing, but will the absorbed light be emitted as a photon? Carrot juice, for example, absorbs light but is not fluorescent. The probability that a fluorophore will emit a photon after a photon has been absorbed is known as the **quantum yield**, Φ. It is a dimensionless number that is expressed as (number of photons emitted)/(number of photons absorbed), similar to a percentage. Φ will always be between 0 and 1. For EGFP, $\Phi = 0.6$ (i.e., 60%).

The extinction coefficient and quantum yield are used together to determine the **brightness** of a fluorophore: brightness $= \varepsilon\Phi$. This relation should make sense; when we determine how brightly a fluorophore will fluoresce, we must consider how many photons it will absorb and what the chances are that it will spit out a photon once one has been absorbed. For EGFP, brightness $= (55{,}000) \times (0.6) = 33{,}000\,M^{-1}cm^{-1}$.

Intensity (I) is not the same as brightness. Although brightness is a descriptor of fluorescence under standard conditions (like the extinction coefficient), the intensity that the investigator observes will depend on the illumination and detection apparatus being used. Intensity is a function of brightness and can be expressed as:

$I = I_0 k\,(\Phi\varepsilon)\,cl$, where

I_0 = lamp intensity,
k = machine constant: how well it gathers light,
$(\Phi\varepsilon)$ = brightness,
c = concentration,
l = path length of light.

Recall from earlier that $\varepsilon = A/lc$. Substituting this into the equation and canceling out the l and c terms gives us:

$I = I_0\,k\,\Phi A$, where A = absorptivity.

8.3 Fluorescence detection

In the laboratory, selecting the "right" fluorophore is not necessarily made by choosing the most intense fluorophore with the largest Stokes shift. Fluorophores will often be detected via fluorescence detection microscopy, fluorescent plate readers, or even qPCR setups. Each of these utilizes a set of filters to sift out unwanted wavelengths and to verify the wavelength of light being detected.

Light filters come in three main categories. **Long pass filters** (LPs) allow light with wavelengths greater than a certain value to pass through. For example, to allow yellow, orange, and red light all to get through a filter, one might select an LP575 filter (Figure 8.8). **Short pass filters** (SPs) follow the same principle, except that all wavelengths shorter than a given value will get through—so an SP530 will allow green, blue, indigo, violet, and ultraviolet light to pass. The third type of filter is the **band pass filter** (BPs). These filters are denoted by two numbers: the wavelength in the middle of the band of frequencies that can pass through the filter and the total width of that band. For instance, BP500/40 describes a filter centered at 500 nm that lets through a band of wavelengths 40 nm wide; specifically, it allows photons in the range of 480 to 520 nm to pass through.

8.4 Fluorescence resonance energy transfer

Consider two fluorophores that are very close to each other in locational proximity: one has an $E_x = 488$ nm and $E_m = 510$ nm, and the second has an $E_x = 510$ nm and $E_m = 590$ nm. If the first

FIGURE 8.8

Visible light spectrum and immediately adjacent wavelengths, illustrating colors of light and their associated wavelengths.

Box 8.1 Excitation and emission example

Suppose you own a microscope setup with a BP480/40 excitation filter and a LP530 emission filter. You are going to tag cells with antibodies that have been labeled with a fluorophore. What would you expect to see if you used fluorescent markers with the following excitation and emission wavelength optima?

	E_x	E_m
1	440 nm	550 nm
2	550 nm	600 nm
3	490 nm	510 nm
4	488 nm	530 nm
5	496 nm	526 nm

Answer:

Fluorophore 1: Nothing would be seen. The band pass filter lets light in the range 460–500 nm through. With an E_x of 440 nm, it is doubtful that the fluorophore will be excited.

Fluorophore 2: Nothing would be seen. Even though 550 nm is in the range of the second filter, that is the emission filter. The excitation filter will not let the 550-nm light pass, so the fluorophore will not fluoresce.

Fluorophore 3: Nothing would be seen. Even though the fluorophore will be made to fluoresce by the 490-nm light, the fluorescent light of 510 nm will not be able to pass through the emission filter, so nothing will be seen.

Fluorophore 4: Cells displaying a beautiful green fluorescence will be seen.

Fluorophore 5: Cells displaying a beautiful green fluorescence will be seen. Although it is true that 526-nm light will not be able to pass through the LP530 emission filter, recall that E_x and E_m are wavelengths corresponding to maximal values on their respective curves (refer back to Figure 8.7). Although the fluorophore is not optimal for the existing filter setup, you should still be able to observe a significant amount of fluorescence.

fluorophore is hit with 488-nm light, it will fluoresce at 510 nm. The emitted 510-nm light can be absorbed by the second fluorophore, which will emit light at 590 nm. This phenomenon is known as **fluorescence resonance energy transfer (FRET)**, and it has been used to show that two molecules are close to one another, perhaps interacting within the cell.

As an example of FRET, imagine that an enzyme has been labeled with the first fluorophore in the previous paragraph, and a certain protein has been labeled with the second. If we then expose a preparation of the enzyme and ligand to 488-nm light and it glows red (more precisely, if we observe light using a red filter), then the two fluorophores are relatively close to each other, which implies that the labeled protein is a substrate for the enzyme. How close is close? That depends on the sensitivity of the detection equipment, but the efficiency of energy transfer depends on the distance between the donor and acceptor fluorophores raised to the power -6. Roughly, this translates to distances in the range 1 to 6 nm.

Conservation of energy applies to FRET as it does to single-fluorophore fluorescence. This means that a FRET pair must have donor excitation energy > donor emission energy > acceptor excitation energy > acceptor emission energy (i.e., traveling in the violet-to-red direction). This may seem straightforward until you try to select a FRET pair for the first time. Figure 8.9 is an illustration of a FRET pair. Note that concessions in efficiency must be made because $E_{m,donor}$ seldom equals $E_{x,acceptor}$.

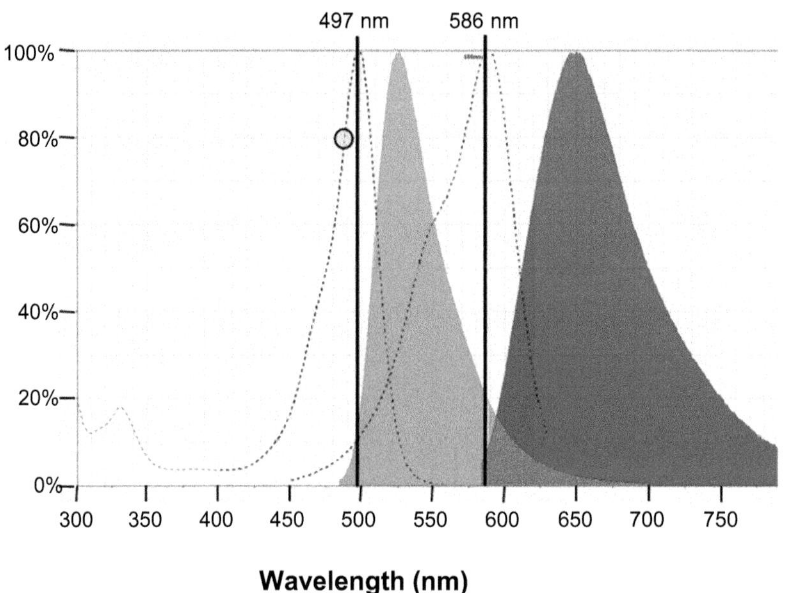

FIGURE 8.9

Fluorescence spectra of two fluorophores that can be used together for fluorescence resonance energy transfer (FRET). The excitation maximum for each fluorophore is given over each excitation curve *(dashed lines).* Emission curves are filled in with the color of each fluorophore, and the height of each is based on excitation with the optimal wavelength for each fluorophore.

The curve amplitudes are not indicative of what would be achieved in a FRET experiment. If we were to excite with a laser of wavelength 488 nm *(yellow circle)*, the *y*-values of the green emission curve would be reduced to ~80% of those shown, which would reduce the amount of excitation of the red fluorophore accordingly. In addition, the excitation energy for the red fluorophore would be dependent on not just a single wavelength but rather the integral of {[the green emission curve (function)] × [the red excitation function]} over the wavelength range spanned by the red excitation curve, plus a similar contribution by the original excitatory wavelength(s) being used. This second term becomes another integral if, instead of a single excitatory wavelength, a range of wavelengths is used—as in the case when a mercury bulb is used with a light filter (such as a BP470/40). The point here is not to teach you how to accurately determine total FRET output but to help you to appreciate the complexity involved in selecting a FRET pair.

Finally, a successful FRET experiment is dependent on filtering the original excitatory signal and the signal from the first fluorophore out of the second fluorophore emission. In this example, the overlap between the green and red curves must be removed. This can be achieved by selecting an emission filter such as an LP650 (or, in theory, LP700) to remove most of the output seen as emission from fluorophore 1.

Image generated, in part, using BD Fluorescence Spectrum Viewer.

8.5 Summary

In this chapter, we were introduced to the exciting (pun intended) world of fluorescence through the experiments of Sir George Stokes, performed in the 1840s and 50s. From Stokes's observations, we were able to examine the physics of fluorescence through the definition of excitation wavelength, emission wavelength, and the introduction of the Jablonski diagram, which is used to explain how photon energy is converted within a fluorophore to electron energy, heat, and the possible emission of lower-energy photons.

We then moved on to discuss how fluorescence is detected in the laboratory through judicious use of lenses, including LP, SP, and BP filters. Finally, we saw how fluorescence can be used to imply interactions between proteins as they become close enough to allow for FRET between fluorophores attached to them.

In the next chapter, we will move to agarose gels and how they can be used to ascertain information about DNA and proteins that interact with genes.

Related reading

Förster, T.H., 1948. [Intermolecular energy migration and fluorescence]. Ann. Physik 2, 55–75 (in German).

McEwen, J. ChemWiki.: The Dynamic Chemistry E-textbook. Maintained by UC Davis. http://chemwiki.ucdavis.edu/Physical_Chemistry/Spectroscopy/Electronic_Spectroscopy/Jablonski_diagram (Accessed March 7, 2014).

McRae, S.R., Brown, C.L., Bushell, G.R., 2005. Rapid purification of EGFP, EYFP, and ECFP with high yield and purity. Protein Expr. Purif. 41, 121–127.

Stokes, G.G., 1852. On the change of refrangibility of light. Phil. Trans. R.S. London 142, 463–562.

Stryer, L., 1978. Fluorescence energy transfer as a spectroscopic ruler. Annu. Rev. Biochem. 47, 819–846.

Questions

1. Suppose you have a microscope with the following filter sets and that you cannot mix and match the filters between sets:

	Excitation	Emission
①	BP 360/40	BP 460/50
②	BP 480/40	BP 535/50
③	BP 535/50	BP 610/75

   a. Which of the following fluorophores could you definitely use with your system?

	E_x	E_m
Pacific Blue	410	455
ParrotGreen 488	488	516
AndiFluor 578	578	530
mmm...Cherry	580	610
mmm...Banana	540	553
mmm...Applesauce	528	577
Red Rain	607	630

   b. At least one of the fluorophores above is obviously made up. Identify it (them) and justify your response.
   c. Which pair of fluorophores from the list in part *a* would work best for FRET analysis? If we could mix and match our filter sets, what pair of the given filters would you use?
2. Why would a small Stokes shift be undesirable if you wanted to use a fluorophore for microscopy?
3. You go out to a nightclub and are very excited to see that some of the drinks are glowing in the darkened room. You buy one of the drinks and immediately drive to your professor's house to show off this seemingly miraculous wonder. Unfortunately, when you present the drink, it is no longer glowing.
   a. Propose a reason why the drink is no longer glowing.
   b. In desperation to save yourself from the embarrassment of having awakened your professor at 2 a.m. for nothing, you shine a flashlight on the drink. Does this make the drink glow? Would a cigarette lighter or match help? How about waiting until the next day and holding the drink up to sunlight? What would you see in each case?
4. If I have an excitation filter that allows 493- to 502-nm light through, can it be used to excite a fluorophore with $E_x = 488$ nm? Why or why not?
5. What could be a use of a molecule that absorbs all wavelengths of visible light but does not emit any? (Think FRET.)
6. Detergent companies used to make the claim that using their product could make your white clothes "whiter than white." There was actually some truth to the claim, and it involved molecules called *phosphors* in the detergent. Come up with a scientific reason why the manufacturers were not lying.
7. Suppose you have a microscope with the following filter set: E_x: BP480/40 and E_m: LP530. Which of the following fluorophores will work with your setup?

Fluorophore	E_x	E_m
1	440	550
2	515	600
3	490	510
4	488	550
5	488	529

8. If a fluorophore absorbs light of wavelength 577 nm, what color(s) might it emit?
9. Explain what fluorescence resonance energy transfer is.
10. If you held a BP500/20 filter up to your eye and looked at a light, what color(s) would you see?
11. Would you be able to see red light through an LP488 filter?
12. What is meant by "BP 480/50"?
13. What is the Stokes shift? Why is it important in terms of a fluorophore?
14. Refer to the Jablonski diagram. When an electron undergoes a large energy drop to the ground state, give two examples of where the energy might go.
15. Fluorescence is most commonly associated with which molecular structure: acids, bases, buffers, long-chain alkyls, aromatics, esters, or peptide bonds?
16. What fundamental principle explains why a molecule that absorbs yellow light will not fluoresce blue?

17. James is having a tough time visualizing his cells that were stained with antibodies tagged with a red fluorescent protein. He looks up the fluorescence spectra for the tag and finds that although his emission filter (LP600) is fine, his microscope filters out the optimal E_x of the fluorophore (510 nm). He decides he can get better stimulation of his fluorophore if he removes the excitation filter altogether. What will he see when he illuminates his sample in this way? (What color(s) and how intense?)
18. James is having a tough time visualizing his cells that were stained with antibodies tagged with a green fluorescent protein. He looks up the fluorescence spectra for the tag and finds that, while his excitation filter (BP 480/40) is fine, his microscope filters out the optimal Em of the fluorophore (510 nm). He decides that he can get better visualization of his fluorophore if he removes the emission filter altogether. What will he see when he illuminates his sample in this way? (What color(s), and how intense?)
19. It is Halloween, and you are visiting a "haunted house," complete with zombies, ghouls, and black lights. There is a genuine-sounding scream from the next room, so you and several other people run in to investigate. You find a man dressed in jeans and a white T-shirt, menacing a group of workers from the haunted house. He has a flask of liquid that is fluorescing yellow under the black lights. "Stay away from me!" he yells. "This flask is filled with concentrated hydrochloric acid, and I am gonna throw it on all of you because I was unjustly fired from my job here last night!"

    You have a choice to make. If he has HCl, you should immediately exit the room to save yourself and the group you are with. However, if he does not have HCl, you should confront him and be a hero. What will you do, and why? (In other words, does the guy have concentrated HCl? Support your answer.)
20. Consider a fluorophore with the E_x and E_m spectra shown below. Propose a set of microscope filters that you could use to observe the fluorophores in a cell culture. (There are many possible answers.)

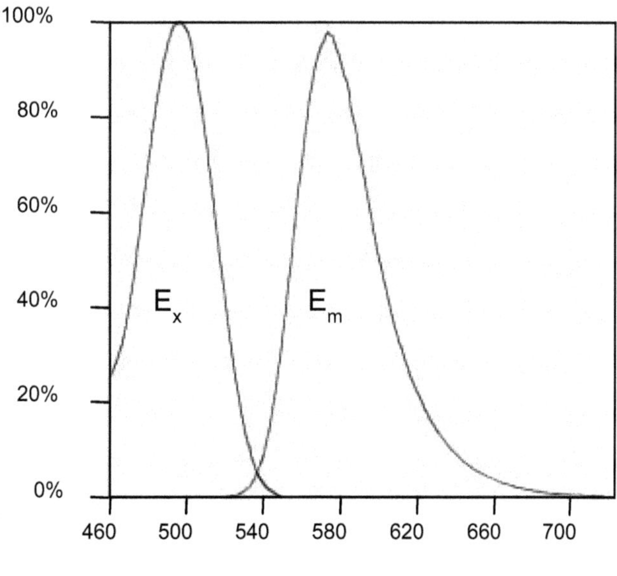

CHAPTER

Agarose gels

9

Chapter outline

The agarose gel is an important tool for biotechnologists working with DNA. Particles of agarose are melted and then cooled, forming a mesh that is loose enough to allow DNA (or other relatively small charged particles) to traverse it after an electrical field is applied. Much like protein movement in poly(acrylamide) gel electrophoresis, discussed in Chapter 3, negatively charged DNA particles travel through the agarose mesh in the same direction as the applied electron current. DNA polymers with lower mass will be able to snake their way through the mesh more quickly, which allows us to separate DNA fragments by their size.

In this chapter, we will examine agarose gels closely to gain an understanding and appreciation of this powerful tool. We will begin with technical considerations, such as what goes into making an agarose gel, how we can visualize the DNA being separated, and the mechanism behind the separation. We will then discuss some of the main applications of agarose gel electrophoresis: the gel shift assay, the DNA footprinting procedure, and perhaps the most widely used application of the technology, restriction analysis.

9.1 Technical considerations

In this section, we will look at technical aspects of agarose gels: how to make them, how to visualize DNA in them, and how to use them for electrophoresis.

Biotechnology and its Applications. https://doi.org/10.1016/B978-0-12-817726-6.00009-5

9.1.1 Mixing and casting

Gels used for DNA analysis are typically made of **agarose** (Figure 9.1), a sugar that can be extracted from red algae. The solvent used to make agarose gels is typically TAE, which is an acronym for tris(hydroxymethyl)aminomethane (Tris), acetic acid, and ethylenediaminetetraacetic acid (EDTA). Tris is a pH buffer, acetic acid is used to help titrate the pH of the solution into the proper range, and EDTA (Figure 9.2) serves as a chelator for divalent cations, such as Ca^{2+}. (Chelation is when one molecule—technically, a Lewis base—binds a central metal atom simultaneously at two or more separate places.) Agarose will be added to the TAE buffer, and the mixture must be heated in order to bring the agarose into the liquid phase. Upon cooling, the agarose will form cross-links to yield a porous gel. The procedure is similar to making Jell-O, which solidifies when it is cooled after cooking. In fact, casting a gel means pouring it into a mold, just like one might do with Jell-O. The mold we use for agarose gel electrophoresis produces a rectangular prism. A "comb" is inserted

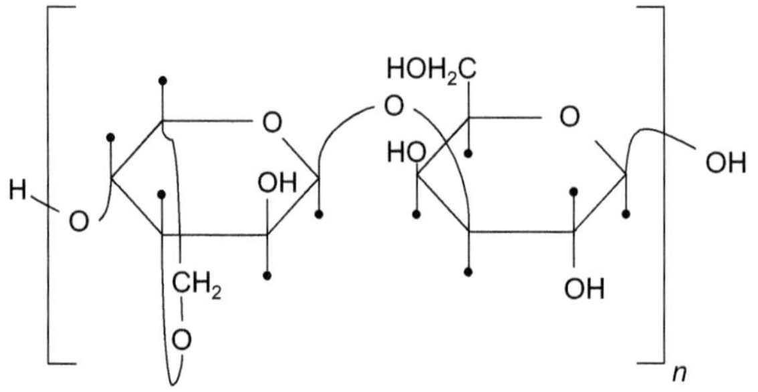

FIGURE 9.1

The structure of agarose.

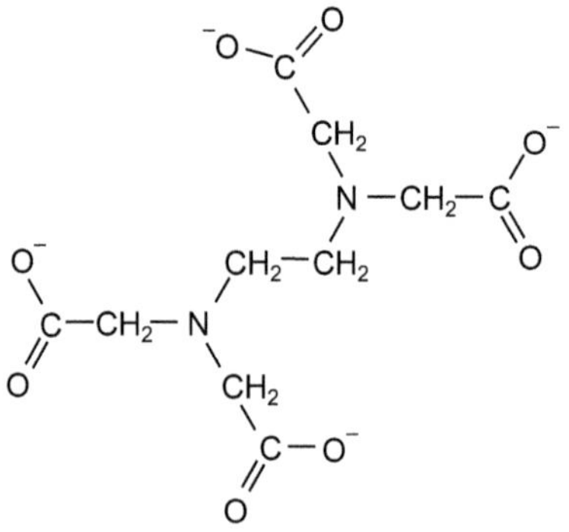

FIGURE 9.2

The structure of ethylenediaminetetraacetic acid (EDTA). Note that in three dimensions, a single divalent cation such as Ca^{2+} can interact with two carboxylate groups at the same time, a phenomenon known as chelation.

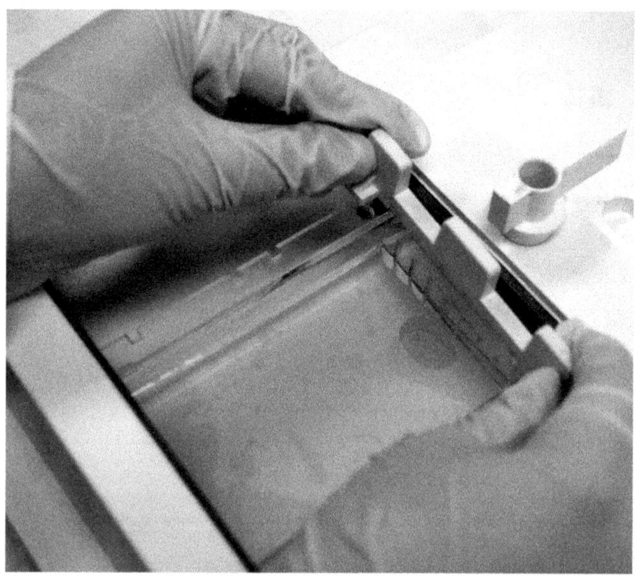

FIGURE 9.3

Removing the comb from a cast agarose gel.

into the gelling liquid to produce cuboid wells for insertion of the DNA samples to be electrophoresed (Figure 9.3).

The percentage of agarose used to make the gel determines its porosity. Higher percentages (1.5%–2%) can be used to separate DNA fragments that are relatively small. Using less agarose (e.g., 0.5%) will yield an agarose network with less cross-linking and therefore larger pores, which will permit easier separation of large DNA fragments. The most common agarose concentration range is 0.8% to 1%; this concentration allows good resolution upon separation of fragments in the range of 500 to 15,000 base pairs (bp) (Table 9.1).

9.1.2 Visualizing DNA

The agarose gel provides a good way to separate DNA based on fragment size, but the DNA must also be visualized. Typically, an intercalating dye is used for this purpose. A very common choice of dye is ethidium bromide (Figure 9.4). Ethidium bromide is a salt, so it separates in solution to yield a positively charged ethidium ion plus a bromide ion. The planar ethidium ion can insert (intercalate) between the stacked bases of helical dsDNA, interacting with them via van der Waal forces. Interactions between an ethidium cation and the negatively charged phosphates on DNA also occur. Although ethidium fluoresces orange when exposed to UV light, it does so much more intensely when intercalated with dsDNA. After it interacts with the DNA and becomes concentrated in the gel where DNA is present, ethidium can be visualized by exposing the gel to ultraviolet (UV) light from a UV light box, as shown in Figure 9.5.

Box 9.1 Beware ethidium bromide

Once intercalated, ethidium can interfere with DNA replication, repair, or transcription, making it carcinogenic. If it gets into your cell nuclei, it will intercalate with your own DNA and remain there perhaps for the life of the cell. It can alter DNA folding and transcription and can even interfere with DNA replication and proofreading. Although the effects of coming into direct contact with ethidium bromide are not as bad as those of, say, phenol, they are cumulative over one's lifetime.

Table 9.1 Size ranges of dsDNA fragments resolvable using different percentages of standard agarose[a]

% Agarose	Resolvable dsDNA fragment sizes (bp)
0.5	700–25,000
0.8	500–15,000
1.0	250–12,000
1.2	150–6000
1.5	80–4000

[a]*Other percentages of agarose can be used.*

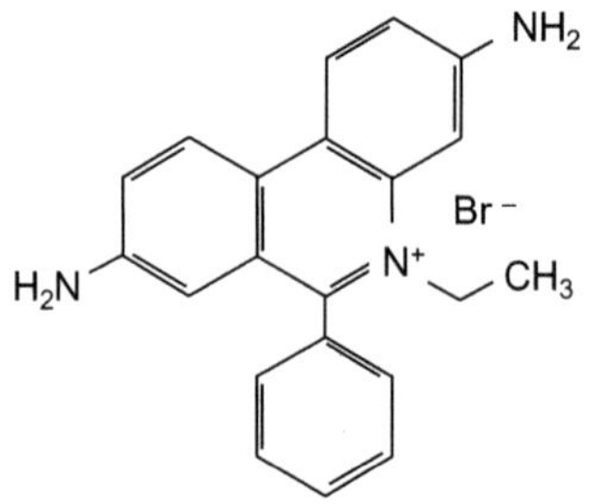

FIGURE 9.4

The structure of ethidium bromide.

Ultraviolet light can induce mutations in DNA, including DNA fragments separated in an agarose gel. If the fragments are being separated for further use, such as the construction of a plasmid, then such mutations can be detrimental to the project. An alternative to ethidium is a class of dyes that fluoresce when exposed to blue light. Members include products such as SmartGlow®, Ez-Vision®, DNAzure®, and SYBR Safe. Although these dyes are more expensive than ethidium bromide and they require the purchase of a blue-light box, many researchers prefer them for DNA visualization because they help to preserve DNA sequences in separated strands.

9.1.3 Loading and running

Once the DNA samples have been loaded into the wells of an agarose gel that has been immersed in TAE buffer, an electrical field (~100 mV) is applied. DNA has the same charge as electrons and will

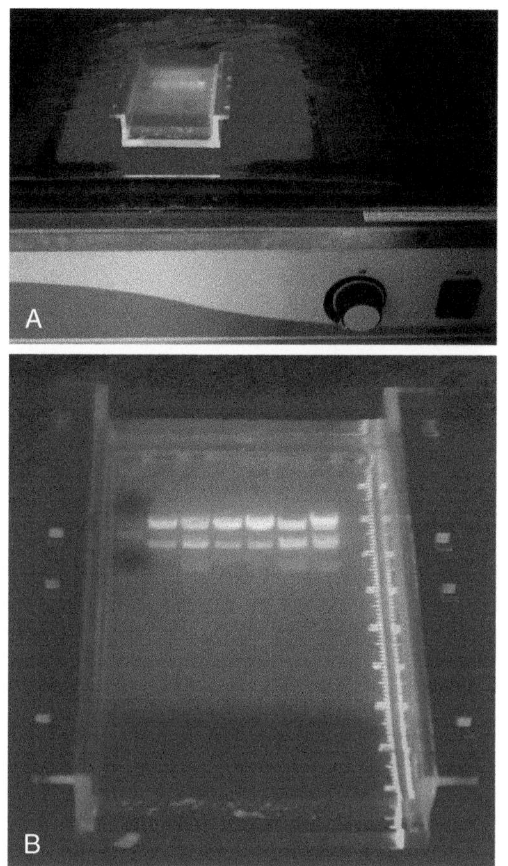

FIGURE 9.5

(A) An ultraviolet light box is used to illuminate an agarose gel containing ethidium bromide. (B) Ethidium that is associated with dsDNA fluoresces brightly. DNA fragments of equal sizes are visible as distinct bands in the gel.

migrate in the same direction: toward the anode. The DNA fragments will be separated on the basis of size rather than the charge of the DNA. Consider two DNA fragments: a 100-bp fragment and a 5000-bp fragment. Although the large fragment is 50 times larger and has 50 times the amount of total charge, it will have the same amount of electrical attraction toward the anode per unit length as the smaller fragment. The charge concentration is the same for both fragments, namely, one negative charge per repeating unit. Separation of the two fragments is achieved because of their differing sizes. The cross-linked agarose polymer is much like an obstacle course that is more easily traversed by the smaller DNA fragment, which can more efficiently snake its way through the twists and turns of the agarose. Smaller fragments travel down the gel more quickly, providing a regular and predictable separation. If one of the lanes is loaded with a **molecular weight marker**—a commercially available standard comprised of a mixture of DNA fragments of known size—then DNA from the sample wells can be compared with the standards, and fragment sizes can be estimated.

9.2 Application of agarose gels: gel shift

Consider two samples consisting of identical DNA fragments, in which the fragments are able to bind a certain transcription factor protein. If we were to mix one of the samples with the protein and then load both samples into separate lanes of a gel, the DNA in the sample containing the protein would migrate more slowly than the naked DNA. This should make sense: running an obstacle course with a monkey on your back will take longer than running the course unfettered. DNA fragments bound to proteins migrate through the gel more slowly, and this phenomenon is referred to as a **gel shift**. The gel shift experiment is used to show that a given DNA sequence will interact with a suspected transcription factor protein or perhaps with another agent—for example, a new polymer designed for gene delivery. However, the gel shift only shows general interactions. Sequence-specific interactions between DNA and a protein can be examined in greater detail via the DNA footprinting experiment, which can yield a more precise location at which a protein binds to a DNA fragment.

9.3 Application of agarose gels: DNA footprinting

DNA footprinting is a way to determine where proteins such as transcription factors will bind on a stretch of DNA. Although the goal of DNA footprinting is similar to that of deletion analysis (see Chapter 4), there are key differences in the procedures that can often trip up the student.

Suppose we have a stretch of ssDNA that is four nucleotides in length. (Such a short fragment was chosen for simplicity of illustration. In practice, the DNA fragment to be analyzed will be much longer.) The stretch of DNA can be labeled on its 5′ terminus via the radioactive phosphorus isotope ^{32}P. This is done because any fragment that contains the ^{32}P can be detected by exposure to x-ray film.

Next, the labeled DNA fragment is exposed to an **endonuclease**. An endonuclease will cut a polynucleotide from within the sequence. This is in contrast to an **exonuclease**, which cuts a polynucleotide from one of the terminal (exterior) bases. The endonuclease of choice is often DNase I, which does not have a specific recognition sequence but cuts randomly.

Consider for a moment the specific case where the DNA fragment is cut only once, between bases 2 and 3. The resulting fragments would be 1-2 and 3-4, in which each digit represents the position of each nucleotide in the original DNA fragment (1-2-3-4). When we run this out on a gel and then expose x-ray film to the gel, the only fragment we will see on the x-ray will be 1-2. The 3-4 fragment will be present in the gel, but because it does not contain the 5′ terminus of the original sequence, it does not have a radioactive label and will not be detectable.

If the original DNA fragment is exposed to DNase I for an extended period of time, the result will include only single bases because all of the phosphodiester bonds will be cleaved. However, if the exposure to endonuclease is limited, the resulting fragments could be of every possible size: 1, 2, 3, 4, 1-2, 2-3, 3-4, 1-2-3, 2-3-4, and 1-2-3-4 (Figure 9.6A). Note, however, that the only fragments detectable via exposure to x-ray film will be those that contain the 5′ label—that is, all fragments that contain base number 1: 1, 1-2, 1-2-3, and 1-2-3-4 (Figure 9.6B). The other fragments will be present on the gel but will not be visible.

Now imagine there is a protein that will bind to the DNA tetramer in such a way as to obscure the phosphodiester bond between bases 2 and 3 (Figure 9.6C). DNase I would still be able to cut between bases 1 and 2 and between 3 and 4, so the set of all possible fragments would include 1, 4, 2-3, 1-2-3, 2-3-4, and 1-2-3-4. Gel electrophoresis of the set would allow us to visualize only the fragments 1, 1-2 -3, and 1-2-3-4 (Figure 9.6D). There would be no visible band that was two nucleotides long. The

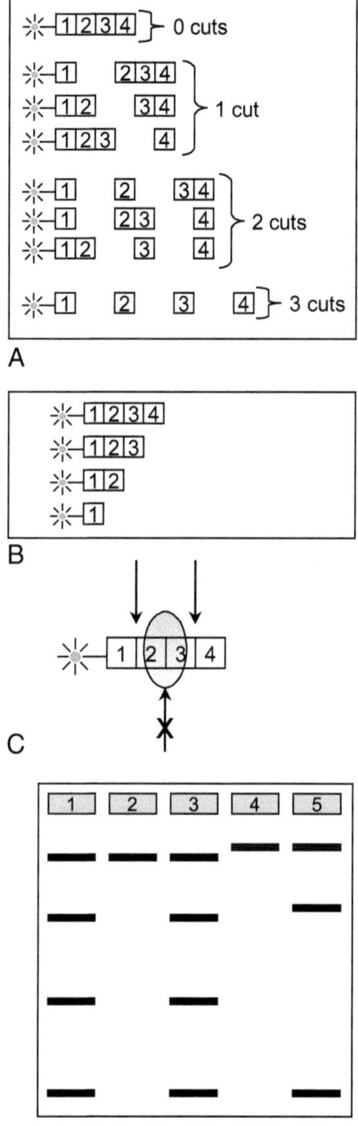

FIGURE 9.6

DNA footprinting. Refer to the text in section 9.3 for a complete explanation of this figure.

absence of the two-nucleotide band (1-2) would indicate that the protein prevented the cut between bases 2 and 3. Using this procedure, one can not only prove that a given protein binds to a specific stretch of DNA, but also estimate where the binding occurs.

9.3.1 A more detailed example of DNA footprinting

In this example, we will again be looking for where a protein binds on a stretch of DNA. The experiment entails the production and isolation of a DNA fragment (likely using the polymerase chain reaction, discussed in Chapter 10), which is then placed into a tube with the protein of interest. The mixture is incubated to give the protein a chance to bind to the DNA, after which DNase I is added to yield the fragments for gel analysis. It is important that the endonuclease is added *after* the protein; if it is added before, the protein will have no chance to protect the DNA from cleavage.

Suppose the protein is an activator, and the DNA sequence contains an enhancer that we are trying to locate and that the protein binds as shown in Figure 9.7. When the protein covers a space between two bases, the endonuclease will be unable to cut at that specific location; the gel shows the pattern of bands that would result from the given coverage. All of the smaller bands will be the same as in the control sample (lane 3) that contains no protein. The protein obscures the bonds between bases 5 and 6, 6 and 7, 7 and 8, and so on, so there will be no corresponding bands in the gel. Also note that there will be a band for the labeled fragment that is 15 base pairs long—because the protein does not extend to the phosphodiester bond between bases 15 and 16—but this band will run slightly slower than the 15-mer in the molecular weight marker lane because the activator protein will be bound; this is an example of a gel shift. Every visible band greater than 15 bp will also be shifted for the same reason. (Also note the shifted bands in Figure 9.6D.) If the protein used binds to the DNA fragment under investigation, the results of the DNA footprinting experiment will show an area on the gel ladder with no bands (lane 5 in Figure 9.7). This area will correspond to the location of protein binding.

A **control** should be run with every experiment. In the control, every parameter will be set to produce a known, predictable, and repeatable result. Controls are run to verify that an experiment was set up properly and that all components are functioning as intended. There are negative controls and positive controls. A negative control is run to show what the result would look like if a certain reaction or response did not take place; a positive control shows what the experimental data would look like if the reaction or response did take place. Experimental runs that test for the presence or absence of an effect under varied but controlled conditions are referred to as test samples. In the DNA footprinting experiment, the molecular weight marker functions as two controls. First, it helps to verify that we made a good gel (e.g., we did not load it before it cooled completely, we added sufficient and functioning ethidium bromide, our TAE buffer was adequate, we did not run the gel upside down), and second, it serves as a standard against which to compare the bands in the lanes that contain the test samples. Other controls to be used for this experiment include a lane for which the DNA sample contained no endonuclease or protein, to verify that the starting DNA material was intact (i.e., was not already cut or sheared), and a lane with endonuclease but no protein (in which every possible DNA fragment size should be visualized), to verify that the endonuclease was functioning and the time of digestion was adequate.

DNA footprinting examples are fairly straightforward when one already knows where the protein binds. In the laboratory, however, we only see bands on x-ray film, and the challenge is to interpret that banding pattern to deduce the location of protein binding. The legend for Figure 9.7 describes an algorithm for determining where a protein binds to a DNA fragment.

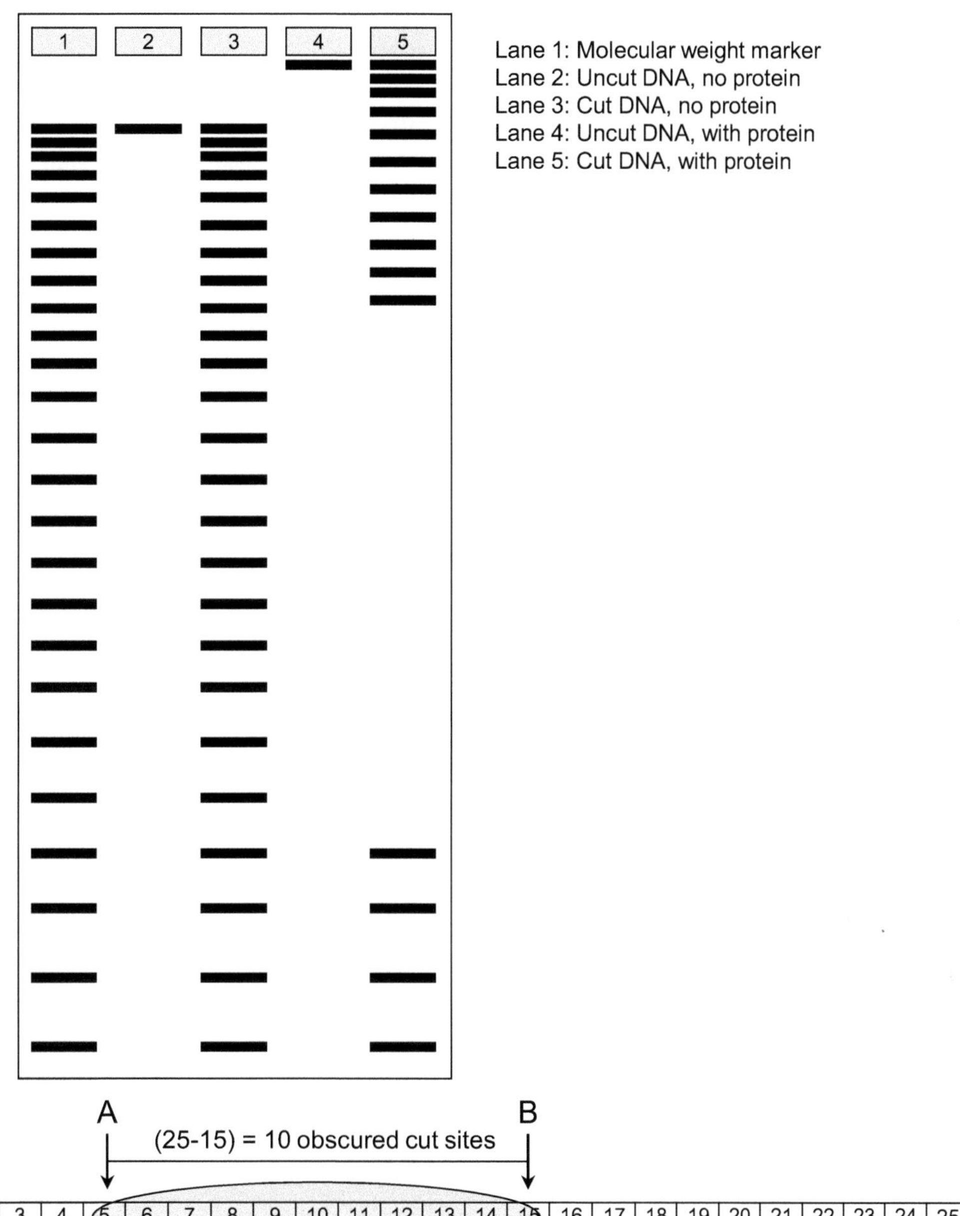

FIGURE 9.7

To determine where a protein binds to a DNA fragment from a DNA footprinting gel:

1. Starting from the bottom, count the number of bands in the lowest group in the sample lane (lane 5). Protein binding is located *after* that many possible cuts (*arrow A* on the DNA fragment). Note that this location is *between* cut sites.
2. Subtract the number of bands in the sample lane from the number of bands in the control cut lane (lane 3). Note that with the protein bound, there is a shift in the apparent weight of the bands with protein bound.
3. The answer in (2) is the number of potential cut sites protected by the protein and determines the 3′ edge of where the protein binds (*arrow B* on the DNA fragment).

9.4 Application of agarose gels: restriction analysis

Agarose gels can also be used to analyze results from experiments involving **restriction endonucleases**—enzymes that cut inside a polynucleotide sequence. They get their name from the fact that they can restrict the proliferation of bacteriophages, which are viruses that attack bacteria. This phenomenon is accomplished by the binding of the restriction endonuclease to a specific sequence in the viral DNA, followed by cleavage. Restriction endonucleases are used by bacteria for self-defense against viral invasions. They recognize specific sequences that otherwise are either modified (Box 9.2) or are not contained at all in the bacterial genome, and they are very effective at preventing the propagation of viruses that contain these sequences. After the viral genome has been cleaved, it cannot be completely replicated or be used to produce more viruses. The sequence that is bound by a restriction enzyme is known as the **recognition sequence** for that enzyme.

We as humans have recognized the mechanism and utility of restriction endonucleases, and specific bacteria that contain a specific endonuclease of interest can be grown and harvested. These bacteria are propagated in the laboratory, after which they are lysed, and the endonuclease proteins are extracted and purified. The first restriction endonucleases were discovered at Cold Spring Harbor Laboratory, but they were first commercialized by the company New England Biolabs in 1975. The initial offering by the company provided scientists with a choice of eight enzymes; today, well over 200 restriction enzymes are available, from many different companies. Many thousands of bacteria and archaea have been screened in the search for new enzymes to sell. During this extensive search, it was found that all free-living bacteria and archaea appear to code for restriction endonucleases, suggesting that they serve as a sort of prokaryotic immune system.

The recognition sequence for a restriction enzyme is typically a **palindrome**—it reads the same forward or backward. Consider the restriction enzyme *Eco*RI, which has the recognition sequence GAATTC. Although the six letters do not spell the same thing when read backward, if we consider the complementary sequence and impose the rule that we only read in the 5′-to-3′ direction, then the recognition sequence becomes:

$$5' - \text{GAATTC} - 3'$$

$$3' - \text{CTTAAG} - 5'$$

which reads the same on both strands in the 5′-to-3′ direction.

There is a systematic naming scheme for restriction enzymes. The first letter is capitalized and denotes the genus of the producing bacteria, and the next two (uncapitalized) letters represent the

Box 9.2 How can a bacterium have restriction endonucleases and not cut its own genome?

The answer is not that the bacterial genome is devoid of recognition sequences; that would be too restrictive (pun intended). In addition to producing restriction endonucleases, bacteria also have enzymes that modify their own genomes. The typical enzyme for such modification is a DNA methyltransferase. The DNA-modifying enzyme will bind to the same recognition sequence used by the associated restriction enzyme and transfer a CH_3 (methyl) group onto one of the bases in the sequence. Because the sequence is a palindrome, one methyl group will be transferred onto each strand. These methyl groups will project into the major groove of the genomic DNA double helix, preventing proper binding or activity of the restriction enzyme. The restriction enzyme and modification enzyme(s) can be separate proteins, or they can be subunits (or even domains) of a larger protein that performs both functions.

species. If there is a fourth letter, it represents the specific strain of bacteria. Following the main name, there may be a number, written in Roman numerals; this indicates that more than one restriction enzyme has been isolated from that particular strain, and the order of discovery. For example, *Eco*RI (pronounced **Ee**-koh ar one) comes from *Escherichia coli*, strain RY13, and it is the first of multiple restriction enzymes to be isolated from this strain; *Eco*RV is the fifth restriction enzyme to be characterized from the same strain, and so on. *Hind*III ("**Hin**-dee three") comes from *Haemophilus influenzae*, strain "d," and is the third characterized restriction enzyme from this strain.

Refer again to the recognition sequence given above for *Eco*RI. This enzyme will cut inside the recognition sequence between the G and the A, indicated by G▾AATTC. Keep in mind that because this is a palindrome, there will be two G▾AATTC cuts at the recognition site, one for each DNA strand. The two fragments will then be separated to yield two fragments that have single-stranded overhangs:

GAATTC → G|AATTC → G + AATTC
CTTAAG → CTTAA|G → CTTAA + G

The fragments above have 5′ overhangs; restriction enzymes also exist that will yield 3′ overhangs. If the DNA fragments have overhangs, they are said to have **sticky ends** because these ends are made up of complementary bases that can form hydrogen bonds. Not all restriction enzymes yield sticky ends. If a restriction endonuclease cuts in the middle of its recognition sequence, as is the case for *Eco*RV (with the recognition sequence and cut site GAT▾ATC), then the product is said to have **blunt ends**.

GATATC → GAT|ATC → GAT + ATC
CTATAG → CTA|TAG → CTA + TAG

The production of sticky ends is more than a random phenomenon included here merely to trip you up on a test; it is a valuable tool that makes the lives of many genetic engineers much easier. It is often the case that we want to remove a section of DNA from one source and transfer it into a plasmid vector. If the source and the vector are cut by the same set of sticky end–producing restriction enzymes, then the desired fragment can be transferred into the plasmid with relative ease because hydrogen bonding will hold it in place. This application of restriction enzymes is presented in greater detail in Chapter 11.

9.5 Summary

In this chapter, we were introduced to agarose gels, which can be used to separate negatively charged particles within a certain molecular weight range. We focused on DNA applications, although RNA and DNA–RNA hybrids have all been analyzed using the technique.

Important technical considerations were presented, including the structure of agarose, the constituents of TAE and why they are used in making agarose-based gels, how intercalating dyes such as ethidium bromide are used for visualizing DNA, and the mechanism behind polynucleotide separation.

After considering the technical aspects of agarose gel electrophoresis, we discussed three technologies that utilize the technology: the gel shift assay, DNA footprinting, and restriction analysis. We were exposed in Chapter 4 to genomic DNA and how cells perform transcription, so the importance of the DNA footprinting technique can now be appreciated. The importance of agarose gels for restriction analysis will become more apparent during our discussion of genetic engineering in Chapter 11.

Before we get to genetic engineering, however, we should first gain an appreciation for another method of DNA manipulation and analysis: the polymerase chain reaction (PCR). Indeed, the classical application of PCR relies upon agarose gel electrophoresis for the visualization of DNA products. A logical step between agarose gels and genetic engineering, PCR is the subject of the next chapter.

Related reading

Chen XG, Fang YL, and Godbey WT, "Molecular Biology Techniques", in Tissue Engineering: Principles and Practices, 1st edition, CRC Press/Taylor & Francis Group, Boca Raton, FL. John P. Fisher [et al.], editors. (2013). Chapter 13.

Green, M.R., Sambrook, J., 2012. Molecular Cloning: A Laboratory Manual, fourth ed. Cold Spring Harbor Laboratory Press, Cold Spring Harbor, New York, NY.

Questions

1. When melting agarose to cast a gel, can water be used as the main solvent? Why or why not?
2. Restriction enzymes are used in the laboratory in conjunction with specific buffers. These buffers contain, among other things, ions (Na^+, Cl^-, Ca^{2+}, and/or Mg^{2+}). Knowing this, suggest why EDTA is a part of TAE, the solution used in preparing agarose gels.
3. How is it that CTCGAG can be considered a palindrome?
4. Given that a restriction enzyme has a hexameric, palindromic recognition sequence beginning with GCA, write the entire sequence. If the enzyme cuts after the G, will it produce blunt ends, a 5′ overhang, or a 3′ overhang?
5. Xinli routinely performs restriction analyses on plasmids in her laboratory. She typically uses plasmids that are less than 6000 bp in length, and she is interested in fragment sizes as small as 500 bp.
   a. Give a range of agarose concentrations that would be appropriate for her experiments.
   b. From the range in part a, Xinli has a definite preference for her routine, daily use. Which concentration of agarose does she probably use? Give two practical reasons why.
6. Why are the effects of ethidium bromide dependent on cumulative exposure over one's lifetime?
7. What is the purpose of using a comb when casting an agarose gel?
8. What is the difference between deletion analysis and DNA footprinting?
9. Consider the gel on the next page carefully. Lane 1 is the result of exposing a set of identical DNA fragments to a 5′ exonuclease for just enough time to cut each individual fragment once. Lane 2 is the result of the same treatment, plus the addition of a protein suspected to bind somewhere on the uncut fragment.
   a. Reproduce the sketch of the uncut DNA fragment shown beside the gel and indicate with an arrow where the protein might bind.
   b. Suppose the protein binds to three consecutive DNA bases. On the sketch you made for part *a*, in lane 3, draw the bands you would expect to see if you were to run a DNA footprinting procedure on the original set of uncut DNA fragments.

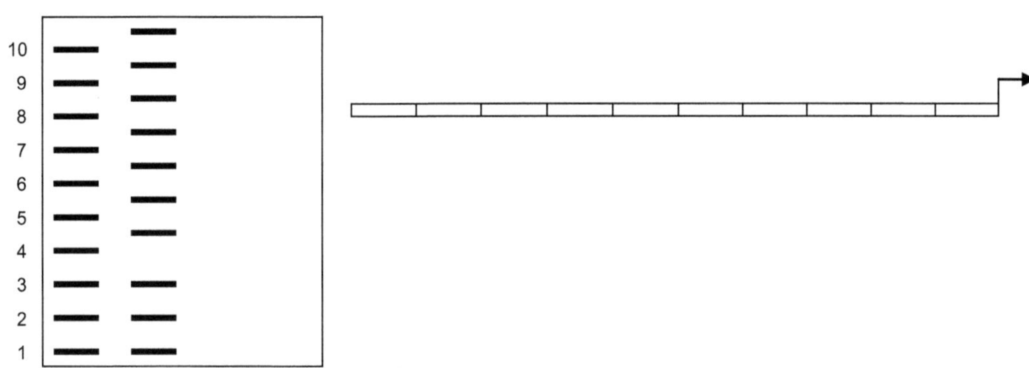

10. Suppose we have isolated a protein that we suspect binds to a specific regulatory region for a certain eukaryotic gene. We perform a DNA footprinting experiment and obtain the following results:

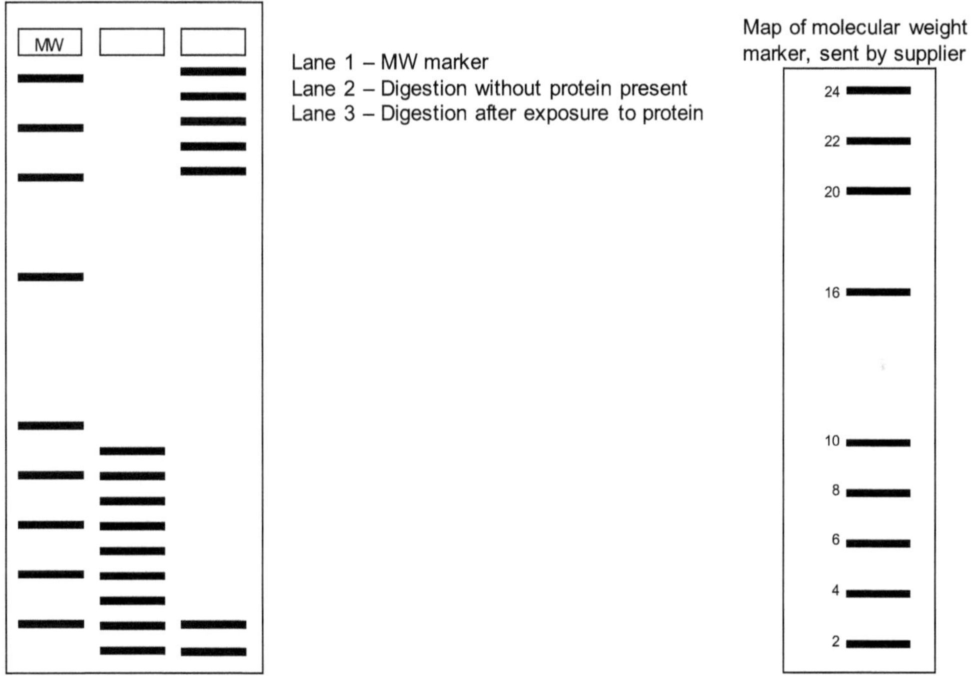

a. Draw a number line like the one below and indicate where you think the protein binds.

919	920	921	922	923	924	925	926	927

b. When the protein binds, it causes an upregulation in transcription of the gene. What is the one-word term that describes the region of DNA to which the transcriptional factor binds? What is the one-word term that describes the transcriptional factor?

c. Is this region of DNA upstream or downstream of transcriptional Start?

11. Suppose we have isolated three proteins that we suspect bind to a specific regulatory region at -100 to -120 for a certain gene. We perform a DNA footprinting experiment and obtain the following results:

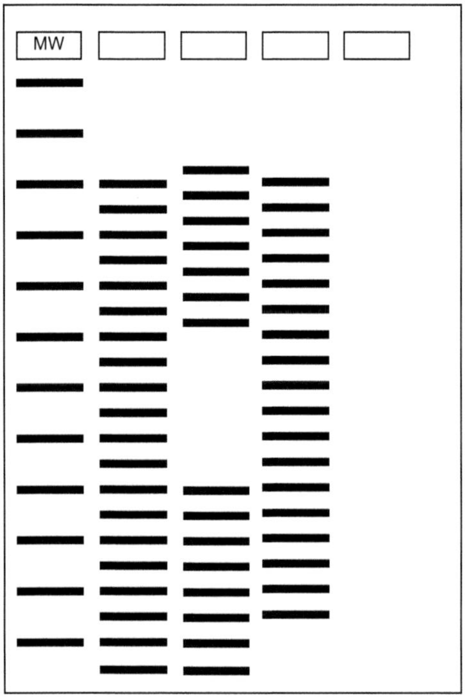

MW

Lane 1: MW marker
Lane 2: Digestion without protein present
Lane 3: Digestion after exposure to protein 1
Lane 4: Digestion after exposure to protein 2
Lane 5: Digestion after exposure to protein 3

Map of molecular weight marker sent by supplier

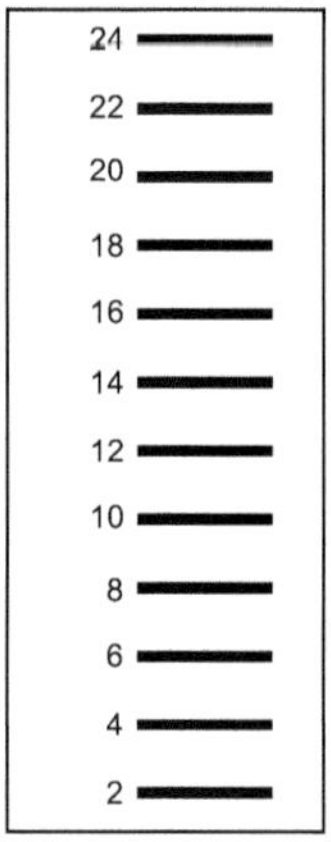

24
22
20
18
16
14
12
10
8
6
4
2

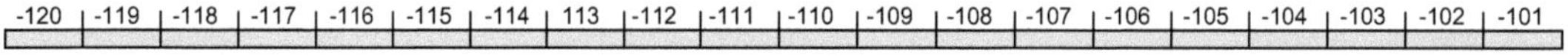

-120 -119 -118 -117 -116 -115 -114 113 -112 -111 -110 -109 -108 -107 -106 -105 -104 -103 -102 -101

a. Draw a number line similar to the one in the figure and indicate where you think proteins 1 and 2 bind.
b. Draw what you think lane 5 would look like if protein 3 bound between (-116 to -112) and (-107 to -104) simultaneously.
c. Name the molecule that permits us to visualize our DNA fragments in this experiment. On a molecular scale, where is it located on the DNA fragments?

12. Agarose gels might also be used as part of the deletion analysis protocol. (This question is intentionally duplicated and expanded from Chapter 4, problem 28.) Suppose we want to locate upstream control regions in a certain gene. We have used an exonuclease to digest the gene from the 5′ end. On the next page is the result of several runs in which we have let the exonuclease chew for different periods of time:

Experiment 1: No exonuclease exposure
Experiment 2: 5-min exonuclease exposure
Experiment 3: 10-min exonuclease exposure
Experiment 4: 15-min exonuclease exposure
Experiment 5: 20-min exonuclease exposure

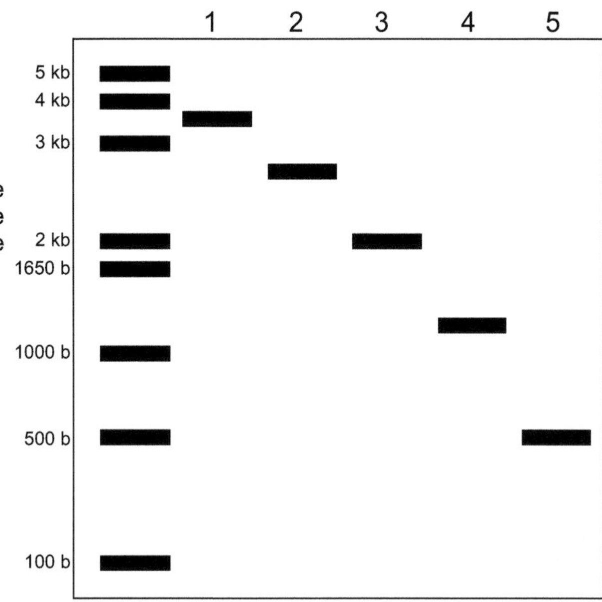

a. This gel has two controls: one for the experiment and one for the gel itself. What are the controls? What does each control show?

b. We then cloned the fragments into a plasmid containing the coding region for a reporter gene, and transfected cells with copies of the plasmid. The following results for reporter expression were obtained:

Relative fluorescence units (RFU)	
Plasmid 1	1.000
Plasmid 2	0.993
Plasmid 3	0.648
Plasmid 4	0.652
Plasmid 5	-0.003

(Plasmid n *was created from fragment* n*, from experiment* n*).*

Estimate the location of each control region.

c. Suppose that RFU = 0 for transfections utilizing the five plasmids above. Give two possible explanations for the result. What control could you have run to prevent the ambiguity?

13. Suppose we want to locate upstream control regions in a certain gene. We have isolated the gene and used an exonuclease to digest it from the 5′ end. Below is the result of several runs, where we have let the exonuclease chew for different periods of time.

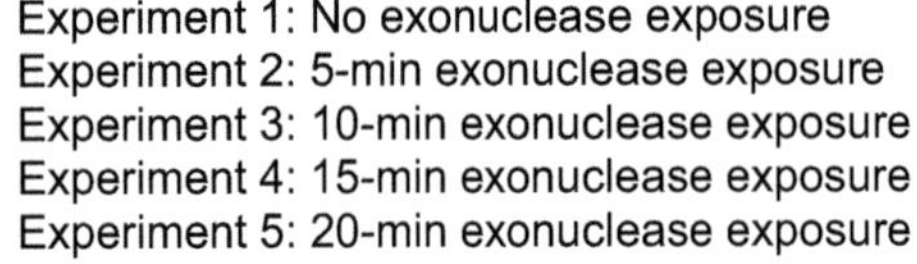

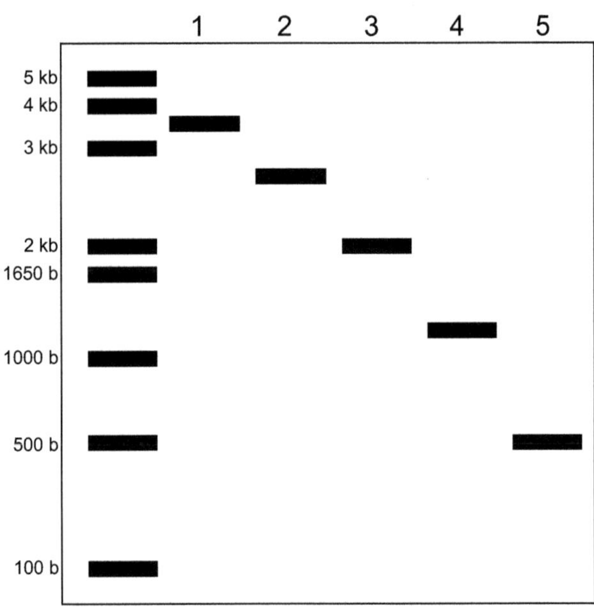

a. This gel has two controls: one for the experiment and one for the gel itself. What are the controls? What does each control show?

b. We then cloned the fragments into a plasmid containing the coding region for a reporter gene, and then transfected cells with copies of the plasmid. The following results for reporter expression were obtained:

Relative fluorescence units (RFU)	
Plasmid 1	1.000
Plasmid 2	0.993
Plasmid 3	0.658
Plasmid 4	0.648
Plasmid 5	-0.003

(Plasmid n *was created from fragment* n*, from experiment* n*).*

Estimate the location of each control region.

c. Suppose that, instead of the values shown above, RFU = 0 for all transfections utilizing the five plasmids. Give two possible explanations for the result. What control could you have run to prevent the ambiguity?

CHAPTER

The polymerase chain reaction (PCR)

10

Chapter outline

In 1993, Kary Mullis was awarded the Nobel Prize in Chemistry "for his invention of the polymerase chain reaction (PCR) method." During his Nobel lecture, Dr. Mullis mentioned his agreement with the statement "if you can understand it, you can explain it." PCR is beautiful in its simplicity. Built from a repeated series of only three steps, PCR is a powerful tool that has been used to assess cellular reactions to stimulation/stress at the transcriptional level.

DNA polymerase, which we have already discussed in relation to cellular replication, can also be used in the laboratory to make multiple copies of a DNA fragment. While the cell uses DNA polymerase to make a single copy of its genome before division, in the laboratory, we can use the polymerase repeatedly, amplifying a region of DNA (the amplicon) at an exponential rate. Suppose we start with a single amplicon in one dsDNA strand. In the first round of amplification, the amplicon will be replicated once. In the second round, the amplicons of both the original DNA and the copy of it made during the first round will be replicated. In the third round, the amplicon and all the previously constructed copies (a total of four dsDNA molecules) will be replicated, leaving us with a total of eight dsDNA fragments at the end of round three. These rounds of replication will continue for a predetermined number of times (n), ultimately yielding (by one theory) 2^n copies of the desired DNA fragment. We will discuss the expected number of copies in more detail later. For now, note that these repeated cycles of replication are known as the PCR.

Biotechnology and its Applications. https://doi.org/10.1016/B978-0-12-817726-6.00010-1

Each core round of a PCR procedure takes place in three steps: melt, anneal, and extend. We will begin this chapter by examining these steps in greater detail. Then we will put the steps together in a recursive fashion to generate multiple copies of the amplicon.

Over the years, PCR technology has evolved to a point where it can be used in near real time to generate quantitative data. The traditional and the real-time techniques will be compared so that we can better appreciate the advantages of the modernized procedure. We will then go into detail about two methods used to measure the amount of dsDNA produced during each round (real-time PCR): one utilizing the fluorescent dye SYBR Green and the other using specially designed primer/probe sets.

Let's jump right into the three steps common to all PCR methods: melt, anneal, and extend.

10.1 Melt

Suppose we have some dsDNA that we want to amplify. (We will discuss the reasons why one might want to amplify this DNA a little later.) Knowing that we're going to use Taq DNA polymerase for the amplification and that this enzyme will require a single-stranded piece of template DNA to dictate the sequence of the new DNA it manufactures, we must begin the PCR process by separating the dsDNA into two ssDNA pieces. The separation is accomplished with heat. Raising the temperature of the solution disrupts the hydrogen bonds between paired bases (A-T and G-C base pairs). We use 95°C to melt the dsDNA because it is a high temperature that is still below the boiling point of water.

It is not advisable to go above 100°C for the melting step because water is a necessary part of B-form DNA structure. B-form is the structure of DNA in an aqueous environment, as opposed to A-form DNA, which is devoid of water (e.g., in a DNA crystal). B-form dsDNA has a double helix structure that includes a major and minor groove. Water molecules intercalate between the two strands of the double helix and pull them together to create a minor groove (Figure 10.1). In pulling these strands together, two other sections of the double helix will be pulled slightly apart, resulting in a major groove in the DNA double helix. Think of the strands as being the same distance apart, but when water gets in there, because of polarity, hydrogen bonding, and electronegativity, two parts of the strands will be pulled closer; this means the two strands will be farther away from one another on adjacent turns, hence the creation of the major and minor grooves. Going above 100°C in the melt step would serve to cook the DNA, just as it alters the structure of the proteins in an egg white when we cook it. At 95°C, the dsDNA is melted into two ssDNA strands without permanently altering the DNA structure.

10.2 Anneal

The next step, annealing, involves the binding of short DNA primers to each of the ssDNA strands generated by the melt step. PCR primers are short DNA sequences that are complementary to the 3′ ends of the amplicon. Keep in mind that there are two 3′ ends in the amplicon (which was, and will be, dsDNA). The primers are designed to bind specifically to sites that flank the region you want to amplify (Figure 10.2).

After the third round of replication, shown in Figure 10.3, five of the six different dsDNA species (see Figure 10.4) involved with the PCR process will be present, including the first appearance of

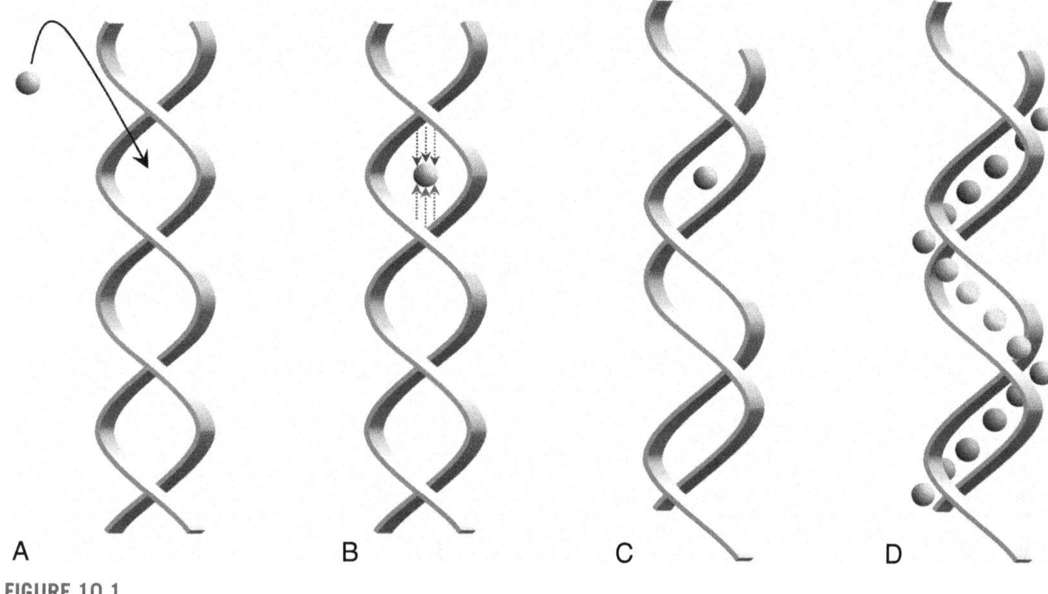

FIGURE 10.1

The existence of a major and a minor groove in B-form dsDNA is due to the presence of water. (A) A single water molecule (not to scale) is shown fitting in between the strands of a DNA double helix. Keep in mind that the water does not lodge on the center axis of the helices—the bases are stacked in there. The bases have been omitted for clarity. (B) Strands of the two helices are drawn toward the water via hydrogen bonding. (C) The shifting of the two helices yields unequal spacing between the strands: a major and a minor groove. (D) Many water molecules are involved in the process, serving to create and line the minor groove.

dsDNA consisting solely of the amplicon. If our goal is to have many copies of only the amplicon, should we be concerned about the presence of the other four extraneous forms of DNA shown in the figure? The answer is no, because they will be present in low numbers. Figure 10.4 illustrates all the different species of dsDNA that occur during PCR. As more and more cycles of PCR are performed, the numbers of each form will be as follows: one of the extraneous forms will disappear, two of them will remain constant at one copy each, and the fourth and fifth extraneous forms will increase by one copy per cycle. However, the number of copies of dsDNA representing only the amplicon will increase exponentially. After 30 cycles of PCR, we will have 60 unwanted molecules of dsDNA that contain extra nucleotides, but we will also have 1,073,741,764 molecules of dsDNA that strictly represent the amplicon.

In Figures 10.2–10.4, the amplicon is represented by darker colors. The original dsDNA is in blue, with the desired amplicon in dark blue and flanking regions in light blue. Primers have been designed to bind at the extreme ends of the amplicon. These primers are designated "forward" and "reverse." The forward primer has the same sequence as the first few (usually 18–24) nucleotides of the sense strand. **Annealing** is the term given to the process whereby the primers attach to the ssDNA. After the dsDNA has been melted and slightly cooled, the forward primer binds to the 3′ end of the antisense strand (the left side as shown in Figure 10.2A). The reverse primer has the same sequence as the first few

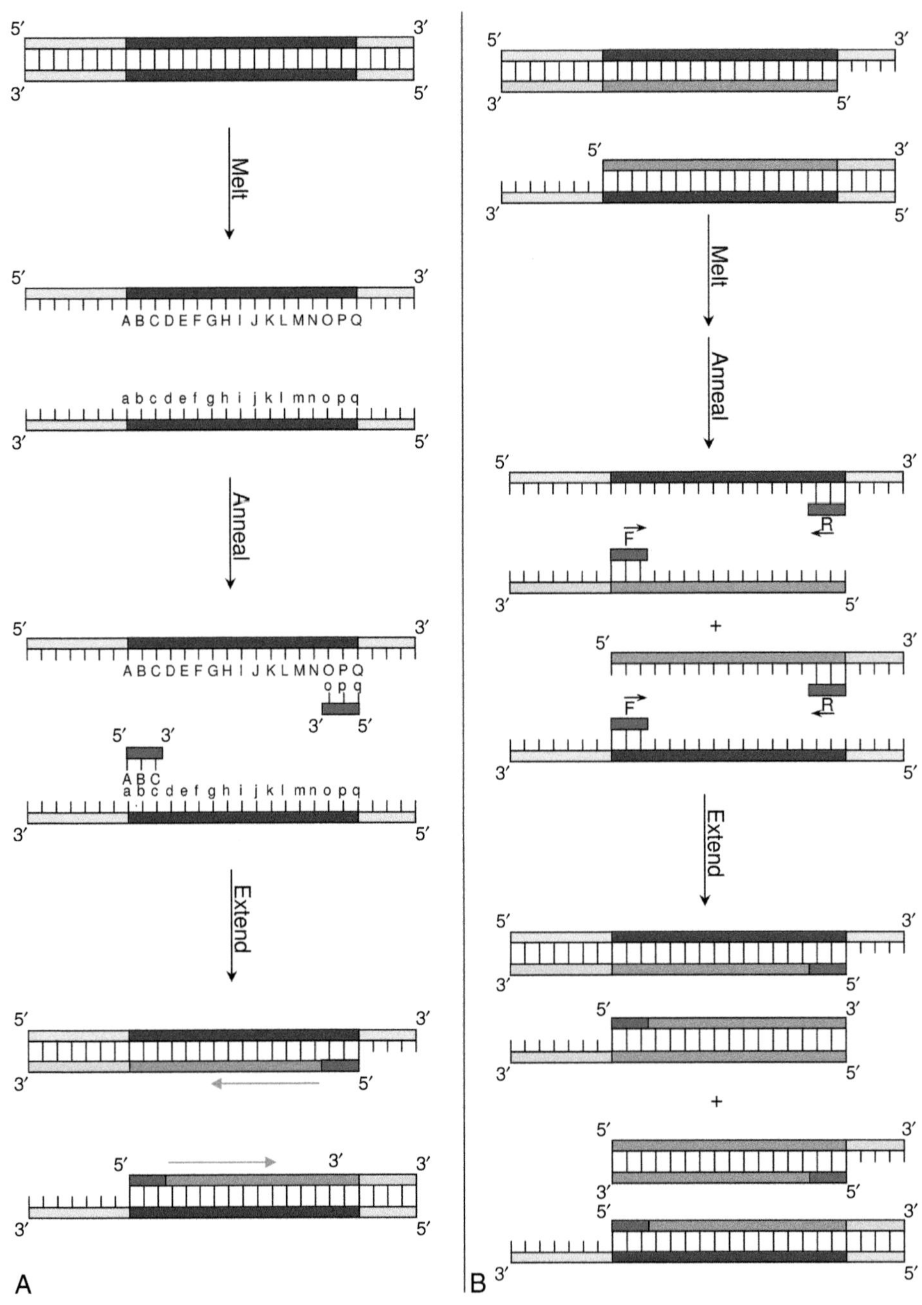

FIGURE 10.2

The first two cycles of a hypothetical PCR set. (A) The first cycle, which begins with the original dsDNA (shown in *blue*). Complementary base pairs are reflected by matching upper- and lower-case letters. Primers are shown in *red*, and newly synthesized DNA is shown in *orange*. Note that the forward primer has the same sequence as part of the coding strand, so it will bind to the complementary strand. (B) The second cycle begins with two pieces of dsDNA, each containing one of the original (*blue*) strands. Sections that will not be replicated are shown in *lighter colors*.

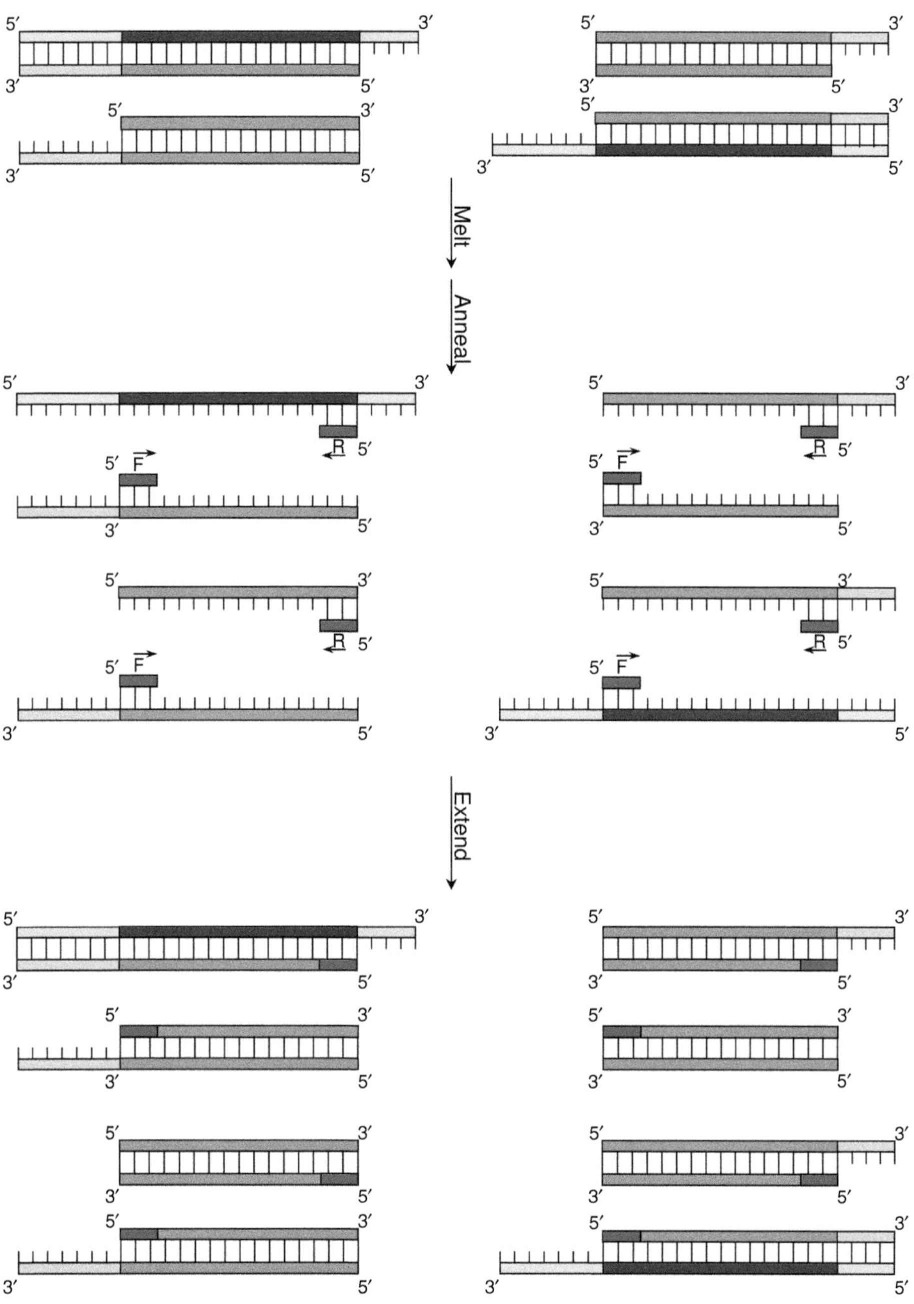

FIGURE 10.3

The third cycle of a hypothetical PCR set (continued from Figure 10.2). Note that although there are several species of dsDNA after this round of replication, dsDNA consisting of only the amplicon first appears after this round. This species will be amplified exponentially in subsequent rounds.

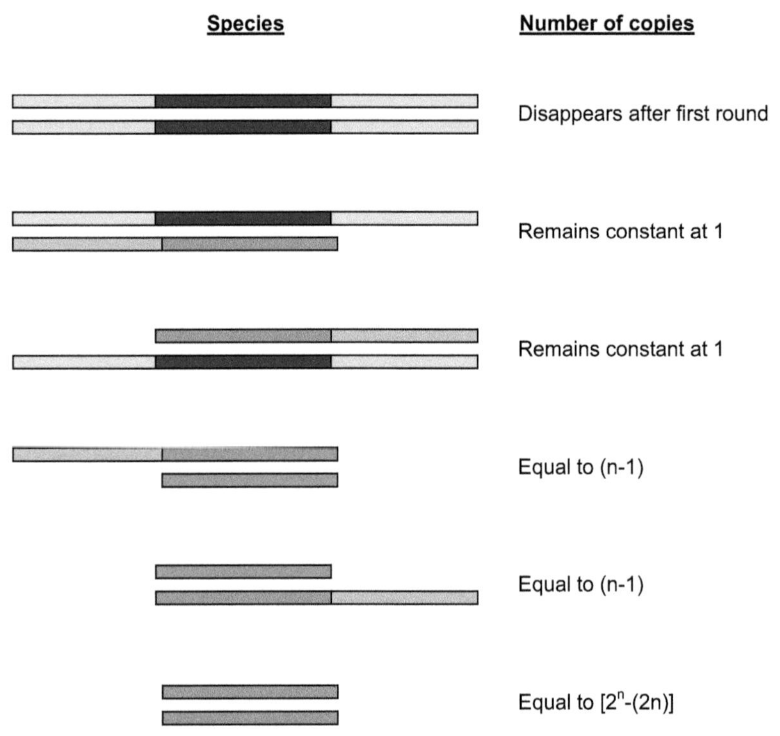

FIGURE 10.4

Different dsDNA species arising during the PCR process. The number of copies of each species after *n* rounds of PCR is shown to the right of each species (n >0).

nucleotides of the antisense strand when reading it from the 5′ end to the 3′ direction (right to left in Figure 2). After the dsDNA has been melted slightly and cooled, the reverse primer binds to the 3′ end of the sense strand (the right side as shown in Figure 10.2). It is crucial to keep track of the strand and direction when referring to DNA sequences. For instance, if you were to order primers from a company, the company will expect the primer sequences to be defined in the 5′→3′ direction. This standard is straightforward and easy to remember, but when you design your own primers, always remember to which strand and which direction you were referring (top versus bottom, sense versus antisense, left to right versus right to left).

We know that a temperature of 95°C allows melting of the dsDNA, but it also prevents the binding of primers to the ssDNA; the temperature must be lowered for annealing to take place. The annealing temperature depends on the specific DNA sequence of both the forward and reverse primers. Consider two 20-nucleotide primers, one made entirely of As and Ts and the other comprised solely of Gs and Cs. Recalling that G-C base pairs involve three hydrogen bonds while A-T base pairs only involve two, it should be apparent that more energy will be required to separate the GC primer after it has bound to its complementary ssDNA. In other words, if we were to slowly heat up a solution of the two primers bound to their targets, the GC primer would separate last. Looking from the other direction, in a hot (95°C) solution of ssDNA plus the two primers, being cooled gradually, the GC-rich primers would bind to their targets first. This principle explains why the annealing temperature for a given primer depends on its specific base content. In designing a primer set (which contains both forward and reverse primers), it is desirable to keep the annealing temperatures of the two primers as close to equivalent as possible; a range of 58° to 62°C is desirable. As the solution is cooled from the 95°C melting step, eventually enough energy will have been removed from the system to allow the binding of the primers with their complementary sequences within the ssDNA.

The optimal temperature for annealing can be determined by the formula:

$T_a = 0.3 \cdot T_m\ (primer) + 0.7 \cdot T_m\ (product) - 14.9$, where

T_a = annealing temperature, and

T_m = melting temperature of the primers or product, defined as the temperature at which 50% of the oligonucleotides and the associated complementary strand are in duplex. T_m can be calculated by:

i. For short sequences (≤ 20 bases), such as primers: $T_m = 2\ (A+T)\ + 4\ (G+C)$, in which A, T, G, and C are the number of adenine, thymine, guanine, and cytosine bases in the sequence, respectively.
ii. For longer sequences, such as amplicons:
$T_m = 81.5 + 16.6\ (\log M) + 0.41\ (\%G + \%C)\ - 0.62\ (\%formamide) - \left(\frac{500}{n}\right)$, where

M = the vconcentration of monovalent cations (e.g., Na+ and K+),
%G and %C are the mole fractions of guanine and cytosine, respectively,
% formamide = the percentage of formamide in the solution, and
n = the number of nucleotides in the sequence.

A relatively high annealing temperature is desirable, but it must be cooler than the temperature to be used for the extension step (discussed in the next section). The reason why a high annealing temperature is desired has to do with preventing **nonspecific binding**. If we have designed a 20-base primer, we expect it to be complementary to its target sequence at all 20 bases. However, if there is another sequence of DNA that is complementary to 17 of the bases in our primer, our primer could bind to this region of DNA despite not being exactly complementary at every single base. Just like the primer made solely of As and Ts versus one made solely of Gs and Cs, it would take more heat energy to remove a primer that was bound to 20 bases as opposed to the same primer bound to only 17 bases. Likewise, in cooling the solution containing primers and ssDNA, the primers will preferentially bind to the DNA region for which 20 nucleotides could interact with the primer, as opposed to a less specific region. The restrictively high temperature (T_m) of annealing prevents nonspecific binding.

10.3 Extend

The extension step of traditional PCR is typically carried out using a DNA polymerase isolated from *Thermus aquaticus* (Taq), a heat-loving (thermophilic) bacterium that was first discovered in a hot spring in Yellowstone National Park. This enzyme, aptly named Taq DNA polymerase, performs optimally at 68° to 73°C and can polymerize DNA at a rate of approximately 1000 nucleotides per minute. So if our amplicon is 300 base pairs long, it will take approximately (300/1000)×(60 seconds)=18 seconds for the Taq DNA polymerase to extend the dsDNA from our primer over the entire length of the single-stranded amplicon. Note that it's fine to increase the length of the extension step to make sure the enzyme has enough time to completely copy the amplicon; extraneous bases will not be a factor in the final solution, as was shown in Figure 10.4.

At this point, we have gone from one dsDNA sequence to two dsDNA sequences, and we can repeat the entire process—melt, anneal, extend—over and over again, each time doubling the amount of dsDNA in our solution. We can continue to repeat this process for a theoretical maximum of 28 cycles. (This theoretical maximum is governed by the concentration of dNTPs that can be present in the starting solution.) In reality, the maximum number of repeated cycles is higher, because the efficiency of every ssDNA binding with a primer is less than 100%, especially at later cycles when primer concentrations are lower and ssDNA concentrations are higher.

Assuming 100% primer efficiency, after one repeat of the cycle, we will have two copies of dsDNA that contain the amplicon (assuming we started with only one piece of dsDNA). After two cycles, we will have four copies (2^2), and after n cycles, we could hope to have 2^n copies.

10.4 PCR loops

The PCR reactions take place in a very small (200-μL), thin-walled tube to allow for better heat transfer, and the total volume of the reaction can be as little as 25 μL. To obtain the changes in temperature, the tube is placed into a thermocycler, which is basically a metal block that can be heated and cooled precisely and relatively quickly. The thermocycler is automated, allowing for the programming of temperatures and times. To get the most efficient transfer of heat into the PCR solution, it is recommended that the walls of the tube be very thin.

The core purpose of the thermocycler is to provide the correct temperatures for each of the three steps of PCR in a repeated fashion. It will heat the tubes to 95°C (melt), cool them off to ~58° to 62°C (anneal), heat them up to 68° to 73° (extend), then back to 95°C, then to 58° to 62°C, then to 68° to 73°C, and so on. A typical run might look something like:

	95°C	10 minutes	
Repeat 34 times →	95°C	15 seconds	(Melt)
Repeat 34 times	59.2°C	30 seconds	(Anneal)
Repeat 34 times	73°C	20 seconds	(Extend)
	73°C	5 minutes	
	4°C	Indefinitely	

The initial temperature of 95°C is held for several minutes to allow the sample to equilibrate. The repeated core process then follows. Recall that the annealing temperature was determined during the primer design process. The time and temperature for the extension step are determined by the specific DNA polymerase and amplicon size being used. The supplier for the DNA polymerase will provide the optimal temperature for the enzyme, and the time for extension is based on the size of the amplicon divided by the enzyme's polymerization rate. The final 4°C step simply preserves the sample until the operator returns to collect it.

When we started out, our sample tube contained some dsDNA, a DNA polymerase, deoxyribonucleic acid triphosphates (dNTPs), a buffer, water, and primers. Pay attention to the dNTPs. The first few times the core steps are performed, they should work very well. However, by the 28th cycle, we will have used up a large number of dNTPs. Trying to run 71 cycles of PCR to obtain 2^{71} copies of a 300-bp amplicon would require that we start with over a mole of dNTP. This is unrealistic. Even running 56 cycles would need over 1 M dNTPs in the reaction tube. The point is that we can only fit so many dNTP molecules in the tube, so we cannot amplify for an infinite number of cycles. The reaction will eventually be limited by the number of dNTPs available. During the later repeats in the PCR loop, the DNA polymerase will operate more slowly because it will take longer for the correct dNTP to diffuse by and bind.

If we were to plot dsDNA concentration versus **cycle number** (how many times the core reactions have been repeated), the data would take on a sigmoid shape (Figure 10.5). During the early cycles, the number of dsDNA strands increases exponentially. However, as the concentration of dNTPs falls, the rate of increase in the number of dsDNA strands will also begin to drop. The rate of dsDNA increase will eventually reach a plateau, meaning a constant amount of dsDNA would be present even if further cycles were performed. The exponentially increasing portion of the curve represents the stage of the reaction at which the number of new dsDNA molecules is approximately doubled during each cycle.

10.5 An application of traditional PCR

Consider an experiment in which we hypothesize that increased glucose levels will cause increased transcription of the endothelin-1 gene in microvascular endothelial cells. If the hypothesis were correct, then endothelin 1 would be implicated in diabetic vascular disease. To test our hypothesis, we run an experiment whereby two cell cultures of microvascular endothelial cells are maintained: one in normal culture conditions and one in conditions that are identical except for an elevated glucose concentration

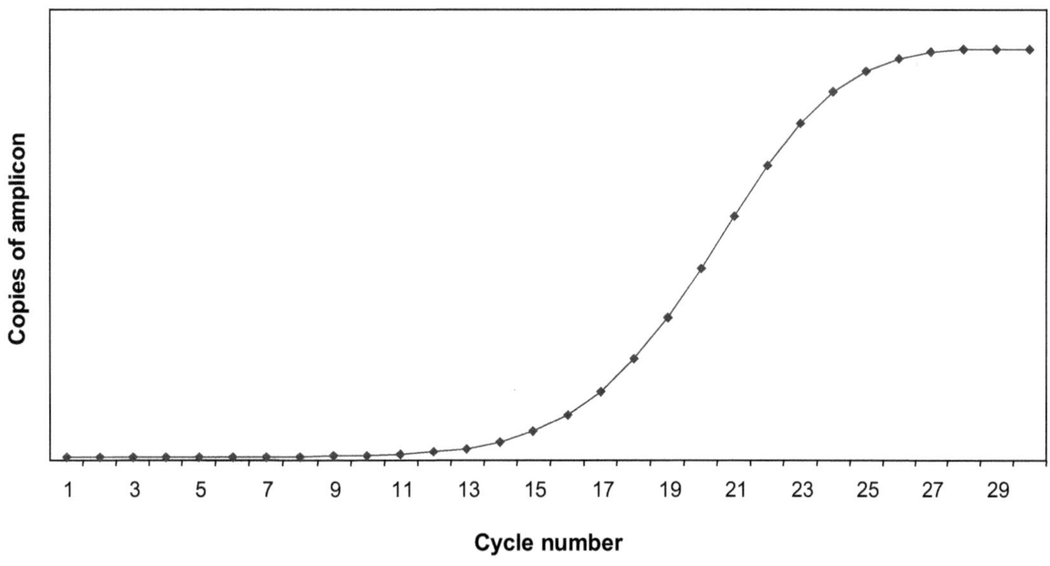

FIGURE 10.5

Although the number of amplicons increases exponentially during the early cycles of PCR, the overall shape of the curve of amplicon copies versus cycle number is sigmoid.

in the medium. We have designed forward and reverse primers to identify *endothelin-1* mRNA and will use PCR to determine the relative amounts of expression between the control cells and the cells cultured with elevated glucose levels.

Recall that PCR requires dsDNA; our hypothesis is concerned with the level of *endothelin-1* transcription, and the product of transcription is not DNA but mRNA. The approach that is commonly taken is to isolate all mRNA from the cells in each culture and generate DNA copies—known as **copy DNA (cDNA)**—using reverse transcriptase. This process, combined with PCR, is referred to as RT-PCR (reverse transcription polymerase chain reaction). Note that if we were to isolate nuclei and perform PCR on *genomic* DNA, the PCR results for the two culture conditions would be the same, because the genomes have been unaltered. What we are interested in is the amount of *transcription* taking place for our gene of interest, so we need to look at the mRNA levels.

For the experiment, cells are grown in cultures under the conditions just mentioned, then they are lysed and the mRNA collected, followed by reverse transcription to yield a cDNA library. The complete set of genes being transcribed in a cell at a particular moment is referred to as the **transcriptome**. Ideally, the cDNA library will represent all the mRNA molecules in the cells from a given culture at the point of cell lysis, and for each transcriptome the cDNA molecules corresponding to individual genes will be in the same proportions as the corresponding mRNAs.

Even though many cDNAs will be generated from the transcriptome, we are only interested in amplifying the cDNA that corresponds to one specific mRNA transcript (*endothelin-1* in this example). We can amplify just this particular cDNA by using primers, one forward and one reverse, that are specific for sequences that are only found in *endothelin-1* cDNA. (Note that a cDNA sequence is not necessarily the same as the corresponding genomic DNA sequence, which may include introns.) The primers must be long enough and specific enough that they will not bind to any of the other cDNAs in the transcriptome library. Verification of specificity is accomplished through database searches of known gene sequences. A popular tool for performing the search is BLAST, which stands for Basic Local Alignment Search Tool. This tool is provided by the National Center for Biotechnology Information in the United States and can be found online at www.ncbi.nlm.nih.gov/BLAST. The need for specificity should explain why we use primers that are 18 to 24 nucleotides long rather than sequences of only 4 to 6 nucleotides.

Again, our tube of cDNAs will contain many sequences in which we have no interest. However, after PCR, we will have amplified our sequence of interest so that it makes up the vast majority of all the dsDNA molecules present (Figure 10.6). We can take a small sample from the tube after PCR and load it into a gel. Following electrophoresis, we should be able to visualize a band that represents the amplification product of PCR using primers directed at the cDNA sequence corresponding to a specific mRNA sequence—which represents the amount of transcription of a given gene at a given time under given circumstances.

Back to our example. We had two cell cultures: one normal and one incubated with increased glucose. Our hypothesis was that the cells exposed to an increased glucose concentration would increase their transcription of the *endothelin-1* gene. PCR will be performed on the transcriptomes of both samples, but if our hypothesis is correct, there will be more copies of the *endothelin-1* cDNA in the tube that corresponds to the cells that were exposed to the increased glucose concentration. Suppose there are seven times as many *endothelin-1* cDNA molecules in the DNA extracted from the treated sample. If we started with 7 versus 1 *endothelin-1* cDNA molecules in the treated versus untreated samples, after one doubling, we would have 14 versus 2 molecules; after two doublings, we would have 28 versus 4 molecules; then 56 versus 8, 112 versus 16, and so on. Even though the number of copies of the *endothelin-1* cDNA is doubled each time, there will always be a sevenfold difference in the number of *endothelin-1* amplicons between the two tubes.

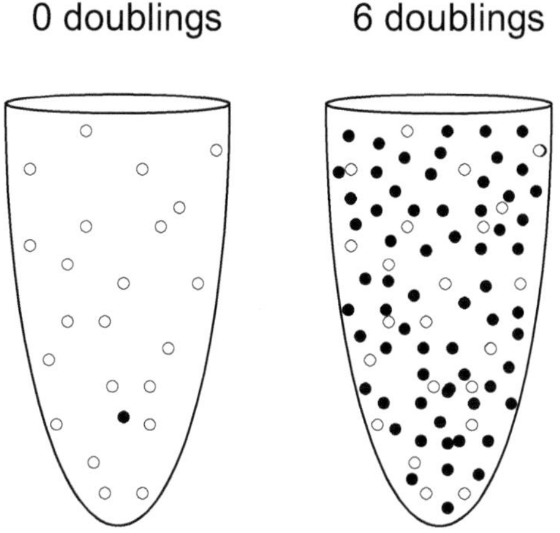

FIGURE 10.6

Even though the cDNA of interest is only a minor constituent of an initial cDNA library, after PCR amplification, it will comprise the majority of the dsDNA in the tube. Notice the prevalence of the *black dots* after only six doublings.

We extracted a set of mRNA for each culture and used each set to produce a cDNA library representing cells from each culture condition. We performed PCR on supposedly equal portions of each of the two libraries using the same conditions and numbers of cycles (the two samples were loaded together, in separate tubes, into the same thermocycler block). Equal portions of the resulting solutions were loaded into a gel and electrophoresed, and the resulting bands were visualized under ultraviolet light. The result might look something like this:

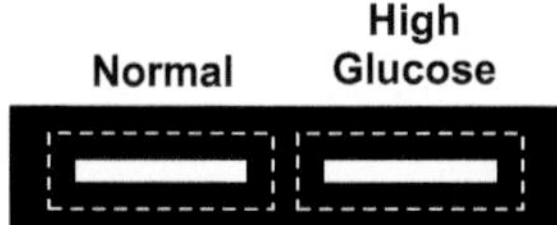

The white bands appear to be of equal thickness and brightness, but computer programs exist that can return a quantitative number to describe the overall brightness in a defined region. These programs measure the intensity, using a gray scale value between 0 and 255, of every pixel in the defined region (shown by dotted lines). Taking a ratio of the two returned values will give an indication of which sample had more *endothelin-1* mRNA in its transcriptome and by how much (fold intensity).

One must always be able to account for mistakes. Perhaps, when loading the gel, slightly unequal amounts of each sample were placed into the respective wells, which is entirely possible considering that we are dealing with microliter amounts of solution. If we were supposed to load exactly 10 μL but each measurement was off by just 1 μL in each direction, then the difference in band intensities could be off by 22% (1 - 11/9). Similarly, if we made a pipetting error while collecting the RNA, either during the generation of cDNA or subsequently while loading the tubes for PCR, there could be significant errors in our final band intensities. Even differences in the number of cells loaded onto each plate at the very beginning of the experiment will make a difference in the final PCR result. What we need is an **internal control** to allow us to account for how much sample was initially loaded. The ideal internal control would capitalize on a gene for which expression would remain constant, regardless of whether the cells were treated or untreated in our experiment. Although there is no perfect internal control, several candidates are commonly used in laboratories. The best candidates are housekeeping genes, because they are needed by the cell at all times for survival. For instance, in cells that use glycolysis as the fundamental pathway for energy production, genes that code for enzymes involved in this vital cascade must always be expressed. One such gene codes for glyceraldehyde 3-phosphate dehydrogenase (GAPdH). Other common genes used as internal controls are those coding for 18S ribosomal RNA—used because this molecule is a part of the functioning eukaryotic ribosome and cells translate the transcriptome at a roughly constant rate—and β-actin, which is always being expressed because it is a part of the cytoskeleton. (Do not confuse this with α-actin, which is part of the muscle contraction machinery.)

(If you want to get way down into the rigorous details of this science, know that there is no perfect internal control. For every system one studies, several internal controls must be evaluated to find the one that remains most constantly expressed during the specific treatment regimen being examined.)

Going back to our experiment, after we isolate the RNA from our two cell cultures and produce two cDNA libraries, we can perform PCR as before, with the exception that we will be amplifying two genes in each sample: *endothelin-1* and, say, *18S rRNA*. The resulting gel might now look something like this:

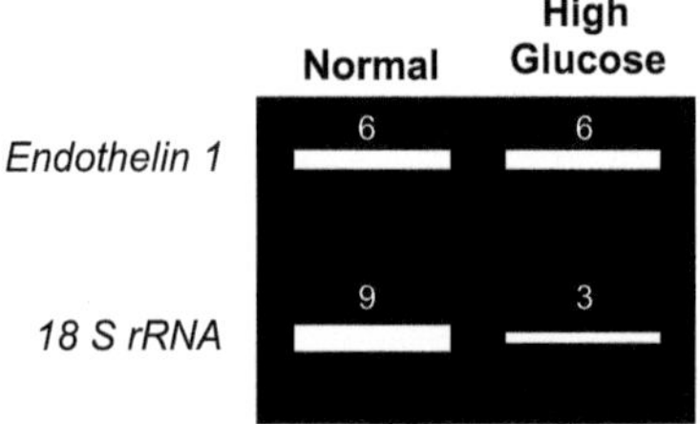

Numbers have been placed over the bands to indicate their relative strengths; notice that the bands for the 18S rRNA are unequal. Based on our assumption that 18S rRNA expression should remain constant regardless of experimental treatment, the above data indicate a pipetting or loading error. This error is easily corrected by normalizing the strength of the bands for the gene of interest to the strength of the bands for the housekeeping gene. In other words, because it appears from the 18S rRNA bands that three times more of the Normal sample was loaded into the gel, the strength of the *EDN1* band in the glucose-treated sample should be normalized up by a factor of three. So, whereas our previous experiment, which lacked an internal control, seemed to indicate no effect of glucose concentration on *endothelin-1* transcription, our properly controlled experiment indicates that increasing the glucose concentration in the culture medium does result in greater *endothelin-1* transcription. The amount of upregulation in this example appears to be (6/3)/(6/9) = 3-fold. A general formula for this relation is:

$$\left(\frac{Gene\ of\ interest}{Housekeeping\ gene}\right)_{Sample} \Big/ \left(\frac{Gene\ of\ interest}{Housekeeping\ gene}\right)_{Control}$$

Keep in mind that the PCR method just described is not recommended for quantitation of gene expression; the above formula is given only to demonstrate the concept behind normalizing to an internal control. For traditional PCR, it is safer to describe findings along the lines of "an upregulation (or downregulation) was observed" rather than "a threefold upregulation was observed."

One *must* normalize to the amount of DNA that was initially loaded into the gel. Although we may *intend* to load the same amount for every sample, the tiny volumes used (~1 μL) lend themselves very easily to measurement errors. Even temperature and pressure differences between tubes and ambient conditions can affect how much fluid is pulled up with a pipette, as will pipette error (for which 4% or less is considered acceptable) and operator error. Even assuming perfect conditions and pipetting precision, the number of cells contributing to each initial RNA sample will affect the ultimate results. The use of an internal reference will help to account for unseen factors such as these.

10.6 Traditional versus real-time PCR

Soon we will discuss real-time PCR, also known as quantitative PCR (qPCR). An advantage of traditional PCR over qPCR is that it is very straightforward, technically simple to perform, and does not require any expensive equipment other than a thermocycler, which can cost under $1000. Real-time PCR involves taking a measurement of fluorescence once per cycle, which means expensive fluorescent tags and an even more expensive fluorescence detector must be used. In addition, the detector must be linked to a computer to store and process the fluorescence data. Real-time PCR is a lot more accurate, as we shall soon learn, but it's a lot more expensive, costing several tens of thousands of dollars for the equipment alone.

One source of error produced by traditional PCR that is not present in real-time PCR stems from the use of postproduction data. In traditional PCR the amplifications are carried out to a predetermined number of cycles before the DNA is examined, and further manipulation is required beyond the PCR process (i.e., a gel must be run). On the other hand, in real-time PCR, a measurement is taken during each cycle. Another problem with traditional PCR is illustrated by the amplification curves shown in Figure 10.7. There is a maximum amount of amplification that can occur before a plateau phase is reached and differences between samples will be obscured. For example, if the optimum number of cycles for visualizing the difference between two samples is 24 but we performed a traditional PCR reaction for 40 cycles, the amplification curves for both samples would be in the plateau phase, meaning a maximum discernible brightness had been produced from the initial cDNA samples and resulting in a gel containing bands of equal strength. The maximum might be due strictly to perception of brightness, or it might be due to depletion of dNTPs as more and more copies of the amplicon were produced. Either way, running the PCR series for too many cycles will lead to saturation of all sample outputs. This might, in turn, lead the investigator to conclude that there is no difference in the expression of a given gene between the samples being analyzed.

10.6.1 Problems specific to traditional PCR

Other problems associated with traditional PCR include odd or unexpected DNA patterns in the gel. Figure 10.8 demonstrates common results that indicate the thermocycler program must be refined. Lane 2 shows a large smear instead of a well-defined band. If the annealing temperature is too low, we have a less restrictive environment for primers to bind to the DNA. In such a case, DNA fragments can be amplified from multiple starting positions in the DNA sequences, in addition to the desired

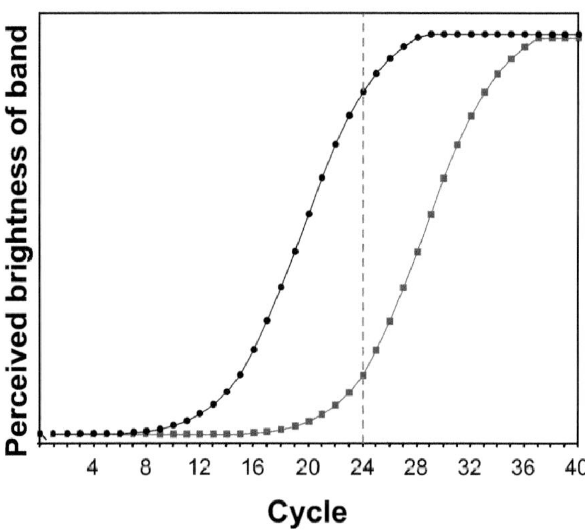

FIGURE 10.7

In traditional PCR, as the number of PCR cycles increases, so does the perceived brightness of the resulting bands—up to a point. Because results are only observed after the total number of PCR cycles have been run, erroneous results can occur if too many amplification cycles have been employed. Compare the differences between the two curves at 24 versus 40 cycles.

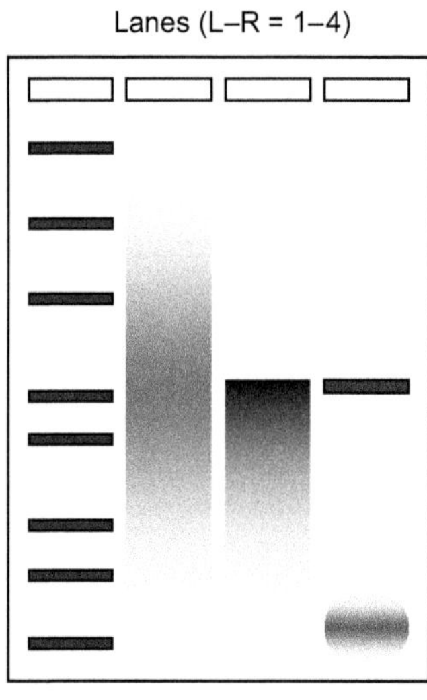

FIGURE 10.8

Common problems with traditional PCR are shown in the above gel.
Lane 1: Molecular weight marker
Lane 2: Annealing temperature too low
Lane 3: Extension time too short
Lane 4: Primer dimers

amplicon. This can result in multiple undesired amplicons that are the wrong size relative to the sequence of interest, producing a smear of fragments on the gel instead of a clean band. The remedy to this situation is to raise the annealing temperature, maybe by only 1°C, which will restrict primer binding to only sites in the DNA that are exactly complementary to the primer sequence (no mismatches).

Lane 3 shows another smear, but this one contains only DNA sequences that are less than or equal to the desired amplicon size. This might indicate that the extension time is too short. Even though calculations might indicate that an extension time of 18 seconds is all that is needed for amplification of the desired fragment, confounding factors such as suboptimal salt concentrations, partial enzyme damage, improperly controlled temperature, or decreasing dNTP availability as the cycle number increases will reduce the rate of DNA polymerization. This situation can often be corrected by increasing the extension time. Recall our discussion of PCR theory and Figure 10.4. Making the extension times longer than needed should not produce an abundance of amplicons that are larger than the DNA segment bounded by the primers, provided there is no nonspecific binding.

Although lane 4 contains the desired amplicon, a second and much smaller band has appeared. This reflects a common problem known as **primer dimers**; if parts of the forward and reverse primer sequences are complementary (especially at their 3′ ends), it provides an additional dsDNA region to which DNA polymerase can bind. As shown in Figure 10.9, the amplicon produced from polymerization using a primer dimer as template can be replicated in future PCR cycles. Although not a significant problem in traditional PCR, other than the fact that dNTP and primer resources will be used to produce meaningless amplicons, primer dimers can be a significant issue in real-time PCR (explained later).

In traditional PCR, even after multiple experiments have been performed to determine the optimal conditions for the reactions—the optimal number of cycles, the optimal annealing temperature, the optimal extension time, and so on—the best result we can hope for is a gel that reflects PCR products that have undergone further processing after the PCR cycles have been concluded. The operator will have to pipette a specific volume of each PCR product into the wells of a gel. Again, there is possible pipette error, plus loss of differing amounts of sample from each well by diffusion or by the turbulence created when the pipette tip is removed after placing the sample into the gel (which is submerged in solution). Every step of processing is associated with some degree of error, so the less manipulation a sample must undergo following a reaction, the closer the resulting data will be to reflecting the true state of the reaction.

10.7 Real-time PCR

Real-time PCR (qPCR) is a method that not only removes the problem of postprocessing but also yields quantitative data. This method utilizes the same thermal cycling steps as traditional PCR, but the amount of dsDNA present is inferred through fluorescence measurements taken directly from the sample during each cycle. There are two strategies for determining the amount of dsDNA with qPCR: the first uses a compound known as SYBR Green, and the second utilizes double-tagged probes.

10.7.1 SYBR Green

If one were to take a known quantity of SYBR Green, load it into a fluorescent plate reader and expose it to blue light, it would fluoresce a little; this is known as background fluorescence. However, when

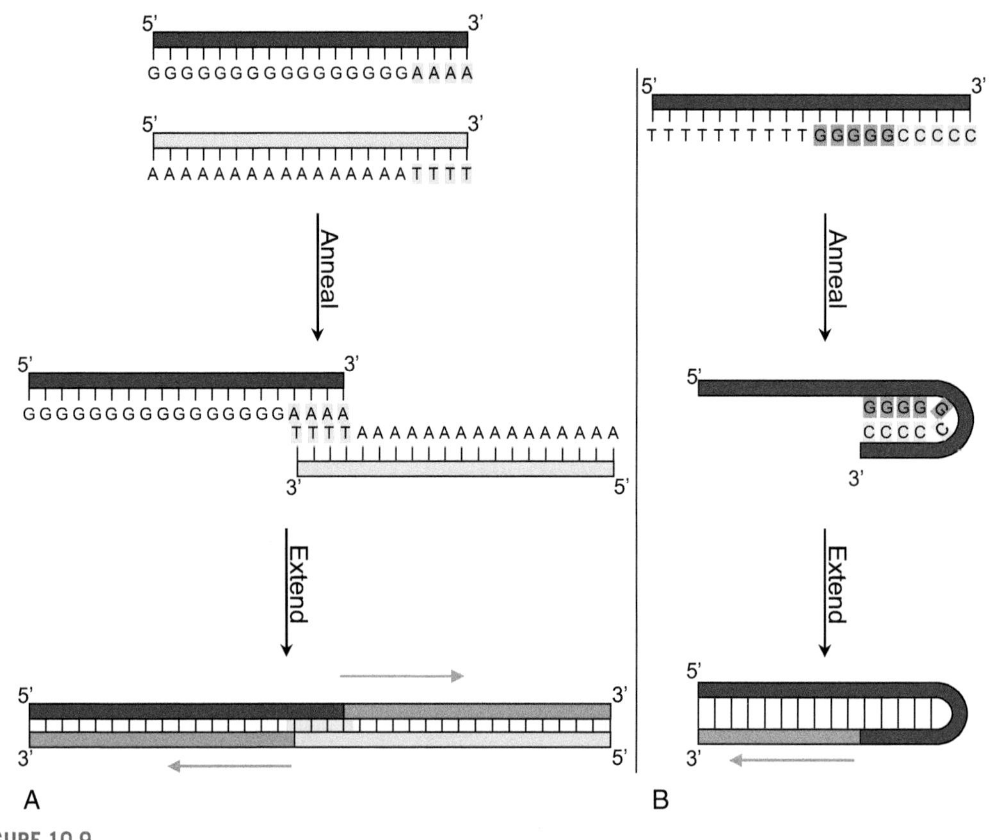

FIGURE 10.9

Two sources of primer dimers. (A) Primers that are complementary at their 3′ ends can pair to allow polymerization in the 5′→3′ direction. (B) Individual primers with 3′ ends complementary to an internal sequence can form hairpins that serve as a double-stranded primer for DNA polymerase. Polymerization is shown in orange.

SYBR Green is intercalated into dsDNA, the dye fluoresces brightly when exposed to blue light. If adding one unit of dsDNA to a sample of SYBR Green yields one unit of fluorescence (after subtracting background fluorescence and assuming an excess amount of SYBR Green), adding two units of DNA will yield twice as much fluorescence. Think of it like this: if there are molecules of SYBR Green floating around in our test tube or sample well, we will observe little fluorescence in the absence of dsDNA. We try to rig the PCR experiments so that there will always be more SYBR Green than dsDNA, so that as the DNA becomes more and more plentiful, there will be more and more sites for the SYBR Green to bind to become strongly fluorescent. The fluorescence will increase proportionally with the amount of dsDNA in the sample. Suppose an experiment commenced with one piece of dsDNA, and a certain amount of SYBR Green fluorescence associated with it (Figure 10.10). When we perform the melt step to obtain two ssDNA molecules, there will only be a background level of fluorescence in the sample. After the anneal and extension steps have been completed, there will be twice as much dsDNA as there was at the beginning of the cycle, providing twice as many binding sites for SYBR Green molecules

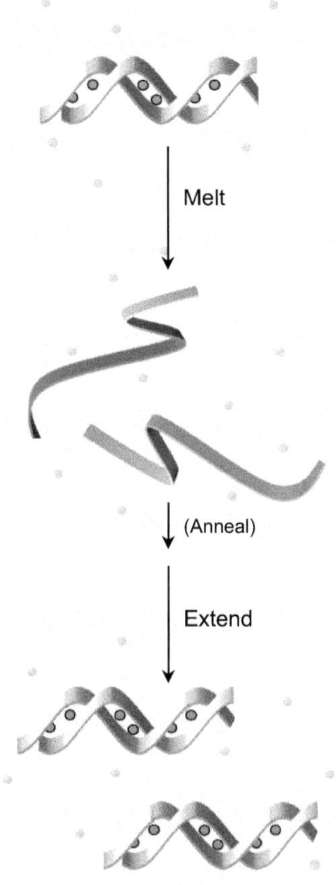

FIGURE 10.10

SYBR Green fluoresces brightly when intercalated in the minor groove of dsDNA. When there is no dsDNA (i.e., after the melt step), there is only a background amount of fluorescence. After the extension step, twice as much dsDNA will yield twice as much fluorescence.

and therefore twice as much fluorescence for detection. Over time, we will be able to track the amount of fluorescence in the sample. Earlier, we saw a graph (Figure 10.5) representing the theoretical amount of dsDNA in a sample as PCR cycles were performed; Figure 10.11 is a graph reflecting the total amount of fluorescence in our sample—fluorescence that is directly related to the amount of dsDNA in the sample.

Returning to our thought experiment—after the first cycle, we will have two units of fluorescence in our sample. Given the vast number of unused SYBR Green molecules in our solution, there will not be much difference between one and two fluorescent units after we subtract background fluorescence. The same is true for the next several cycles, where we will have 4, 8, and 16 units of fluorescence in the sample. Eventually, however, there will be enough fluorescence for the apparatus to reliably detect its presence.

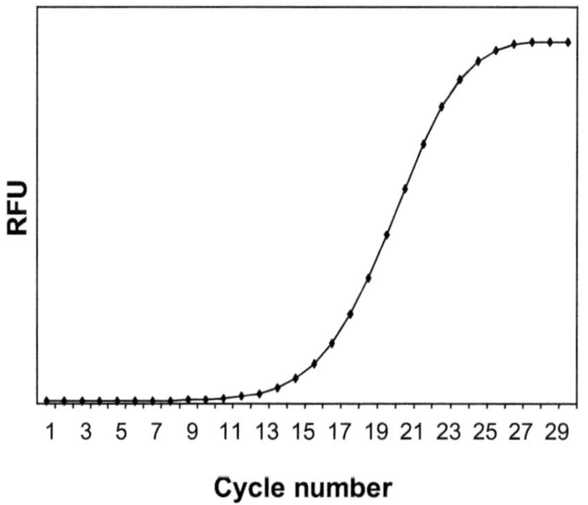

FIGURE 10.11

Note the general shape of the fluorescence curve as the number of cycles is increased in this theoretical real-time PCR experiment. *RFU*, Relative fluorescence units.

The minimum amount of fluorescence that can be reliably detected by the qPCR apparatus is known as its threshold value. The cycle number at which this threshold value is crossed is known as the **cycle threshold** (C_t).

Suppose the threshold limit for our machine is 256 fluorescent units. In the sample we have been discussing, it would take eight PCR cycles (2^8) to yield enough fluorescence to meet or exceed this value. Now consider a second sample of DNA that contains 32 dsDNA molecules at the start of the experiment. It would only take three PCR cycles to generate enough SYBR Green fluorescence for reliable detection ($32 \times 2^3 = 256$). The C_t values for the first and second samples are 8 and 3, respectively: $Ct_1 = 8$; $Ct_2 = 3$.

Now imagine we run qPCR on the same two samples and observe the same C_t values as before ($Ct_1 = 8$; $Ct_2 = 3$), but this time the question is, "By how much does the starting amount of dsDNA differ between the two samples?" The answer can be obtained mathematically from the C_t values: $2^8/2^3 = 256/8 = 32$. In other words, there was a 32-fold difference in the amount of dsDNA in sample 2 versus sample 1. This is consistent with the data presented in the previous paragraph.

The way to use C_t values to determine the fold difference between the amounts of starting dsDNA between two samples can be generalized like this:

$$\text{Fold difference} = \frac{2^{Ct_2}}{2^{Ct_1}} = 2^{Ct_2 - Ct_1}$$

C_t values are returned directly as the output of the qPCR instrument, but they are determined from data such as those in Figure 10.11. The fluorescence curves for all samples will first be generated, and afterward the computer will determine at what point reliable fluorescence increases are detected—the threshold value. (Technically, the computer will take the overall change in RFU from the beginning of the experiment until the curve plateaus, multiply the change by 10%, and add it to the initial RFU value.) The computer notes the cycle number at which each fluorescence curve crosses the threshold value and reports it as C_t. (Note that fluorescence readings are taken in a discrete way, once per cycle, and the threshold value will not always be reached exactly at the end of a given cycle; it may be exceeded during that cycle. Non-integer C_t values are therefore possible and are arrived at via interpolation.)

Consider another example: we hypothesize that drug A acts by causing upregulation in the transcription of a certain gene. To test this, we grow two cultures of cells, administer saline to one culture while adding drug A to the other, isolate RNA from each of the two cell cultures, produce two cDNA libraries, and perform qPCR on the two libraries with primers specific for our gene of interest. Suppose the C_t value for the untreated sample is 24, while the C_t value for the sample treated with drug A is 19. Which sample transcribed more of the gene of interest and by how much?

Answer: By inspection, it should be evident that treatment with the drug causes an upregulation in transcription of the gene. A *lower* C_t value implies a *higher* amount of starting material: it takes fewer doublings to reach the detection threshold of the machine. This is a point of confusion for many students. As long as you keep in mind that C_t values indicate how many doublings were required for detection, the concept of lower C_t indicating higher starting concentrations should present no problem. Using the formula presented earlier, we can roughly ascertain that drug A caused a $2^{(24\text{-}19)} = 2^5 = 32$-fold increase in the mRNA levels associated with the gene of interest.

A few words about semantics: just because the treated cells contained 32 times more mRNA associated with the gene of interest, it doesn't mean they produced 32 times as much of the corresponding protein. The product of this gene may require other polypeptide subunits, from other genes, before a functional protein is produced. It is also possible that the mRNA produced in the presence of the drug will have an altered half-life, meaning it is degraded at a different rate in the presence of the drug. The particular polypeptide may have more than one function, appearing as a subunit in more than one protein or perhaps serving to regulate gene expression in another capacity. It is important to keep in mind that *PCR measures the amount of RNA associated with a particular gene, not the amount of gene expression*. Gene expression is the amount of protein produced from a given gene (by way of the central dogma). Although transcription is a part of this process, simply transcribing a gene is not expressing the gene. PCR only measures RNA levels and therefore can only be used to make inferences regarding gene transcription, not gene (protein) expression.

Returning to the thought example—as with every good experiment, one must not forget the controls. We will not cover every possible control here, such as running the reaction without primers. However, one necessary control is the same as for traditional PCR: the internal control. Although we initially ascertained that the transcription rate for the gene of interest was increased 32-fold in the presence of drug A, we cannot rely on this value because we have not accounted for loading errors. The internal control—a gene for which transcription rates are relatively constant under the experimental conditions—will take care of such errors. The C_t values for the gene of interest must be normalized to the C_t values of the internal control. The example in Figure 10.12 is given to demonstrate that we must normalize each reading for the gene of interest to the corresponding internal control. In other words, what we are interested in is the distance between the two curves—gene of interest and internal control— and we will subtract the distance obtained from a treated sample from the distance obtained for a negative (untreated) control. Simply put, this is a double normalization.

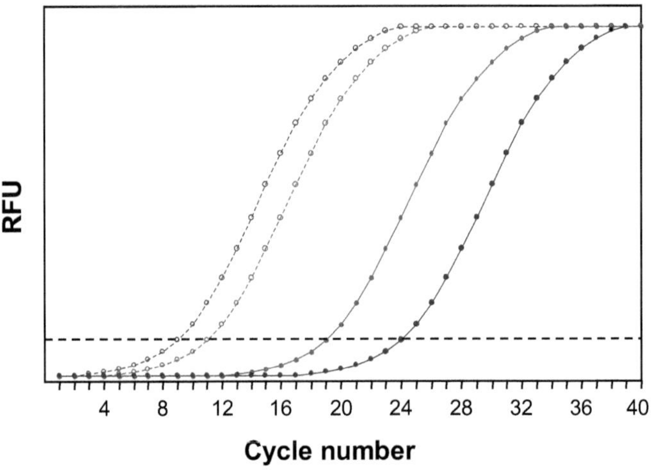

FIGURE 10.12

For qPCR, as with traditional PCR, an internal control must be run to account for inconsistencies in loading or sample concentration. The graph reflects readings obtained for a sample of cells treated with a drug *(red lines)* and an untreated control sample *(blue)*. The gene of interest is shown by the *solid curves* and the internal control by the *dashed curves*. By inspection, we can determine that the distance between the *red curves* is less than that between the *blue curves*. This implies that treatment of the cells with the drug causes an increase in transcription of the gene of interest. Using the formula presented in the text, we can determine that the amount of upregulation is equal to 2^7, or 128-fold.

The formula for determining the effect of the treatment on transcription of the gene of interest now becomes:

Fold difference =

$$\left[\frac{2^{Ct_{GeneOfInterest,NegativeControl}}}{2^{Ct_{ReferenceGene,NegativeControl}}} \Bigg/ \frac{2^{Ct_{GeneOfInterest,Sample}}}{2^{Ct_{ReferenceGene,Sample}}}\right] = \frac{2^{\left(Ct_{GeneOfInterest,NegativeControl}-Ct_{ReferenceGene,NegativeControl}\right)}}{2^{\left(Ct_{GeneOfInterest,Sample}-Ct_{ReferenceGene,Sample}\right)}}$$

$$= \left[2^{\left(Ct_{GeneOfInterest,\ NegativeControl}-Ct_{GeneofInterest,Sample}\right)}\right] \cdot \left[2^{Ct_{ReferenceGene,Sample}-Ct_{ReferenceGene,NegativeControl}}\right]$$

$$= 2^{\left[\left(Ct_{GeneOfInterest,\ NegativeControl}-Ct_{GeneOfInterest,Sample}\right)-\left(Ct_{ReferenceGene,NegativeControl}-Ct_{ReferenceGene,Sample}\right)\right]}$$

This last version of the formula is often called **$2^{-\Delta\Delta}$**, or the "**delta delta method**," in which each delta represents the change in C_t value between the untreated control and the treated sample for each respective gene.

The number for fold difference ($2^{-\Delta\Delta}$) will always be greater than zero. If it's greater than 1, there has been an upregulation in transcription; if it is less than 1, we have downregulation.

If we had no loading error at all, the C_t values for the internal control gene (let's suppose it was GAPdH) would be the same for both the untreated sample and the sample treated with drug A. Suppose this C_t value was 9. Plugging in the numbers from our example into the above formula gives us Fold difference $= 2^{[(24-19)-(9-9)]} = 2^{(5-0)} = 2^5 = 32$, just like before.

Just to check the concept: consider the same example, with the same C_t values, but for some reason only one-fourth as much cDNA from the treated sample was loaded into the PCR tube. Because we only used a quarter of the amount of DNA, we would expect the signal for the reference gene to require two additional doublings to reach the detection threshold, meaning the C_t value would be two higher than that observed for the internal reference gene in the untreated control sample. The C_t values returned from the experiment would be:

	$Ct_{NegativeControl}$ **(untreated)**	Ct_{Sample} **(received drug)**
Gene of interest	24*	19*
GAPdH	9*	11*

**Compare these values to the data presented in Figure 10.12.*

Fold difference $= 2^{[(24-19)-(9-11)]} = 2^{(5+2)} = 2^7 = 128$.

Make sure this example makes sense to you. If we only loaded one-fourth as much treated sample as we intended, whatever fold difference we saw without correcting for loading error would be off by a factor of 4, meaning the values for fold difference we were getting before would only be one-fourth of the actual difference. Likewise, notice that the value of 32 we obtained without utilizing the internal reference was a quarter of the actual value, 128.

10.7.1.1 The fold difference: what it means versus what it implies

The results of PCR, whether traditional or real-time, can be used to infer how much cDNA corresponding to a given RNA sequence, was present. It is often assumed that the cDNA sequences being amplified correspond to mRNA and are unique to a specific gene, neither of which is always true. In the end, if we got to a detectable level of fluorescence in sample 1 sooner than for sample 2, it generally means we started with more cDNA for the gene of interest in sample 1 than we had in sample 2. This implies we had more copies of RNA (assumed to be mRNA) produced from the gene of interest in sample 1 before we ever performed reverse transcription to generate the cDNA. This, in turn, implies that more transcription for the gene of interest took place in the cells used to generate sample 1 (or, far less likely, that for some reason that particular mRNA was being degraded more slowly in sample 1). There are also cases in which some mRNAs are more amenable to reverse transcription than others, but that phenomenon is beyond the scope of this book and will not be covered further.

Using the previous quantitation formula, we may make a conclusion such as, "There is an x-fold upregulation or downregulation of the transcription of our gene in response to the treatment given." The logic tree, progressing backward from output C_t values, is the C_t values indicate relative levels of cDNA in the material after several PCR runs, which indicate the relative levels of cDNA in the starting materials before the first PCR cycle had been run, which imply the relative amounts of specific mRNAs before reverse transcription was performed, which imply the relative amounts of specific mRNAs in the original cell extracts, which imply the relative amounts of transcription of the

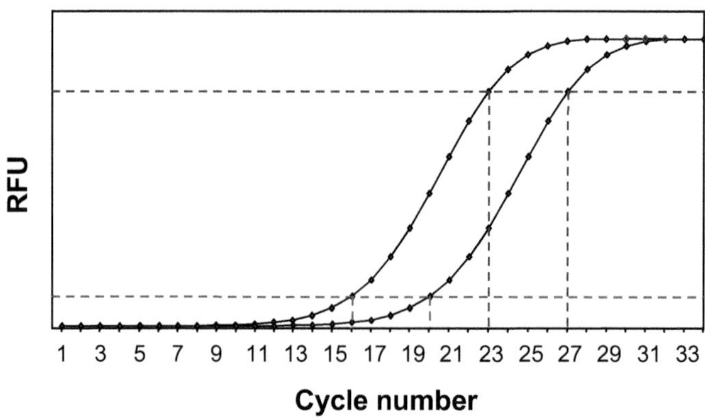

FIGURE 10.13

In analyzing qPCR data, the threshold value chosen is, in theory, arbitrary. In this idealized example, notice it takes four additional cycles for the PCR curve on the right to cross the threshold value *(horizontal, dashed line)* whether the blue or red threshold is used. *RFU*, Relative fluorescence units.

gene of interest in our treated or untreated cells. This tree is a chain where each statement depends on the validity of the previous one.

If the experimental conditions and data from these experiments were absolutely perfect, then the choice of threshold value would be arbitrary. In the idealized example in Figure 10.13, the distance between C_t values is the same between the two curves no matter which of the two threshold values is chosen, meaning the distance between the two curves is the same for a given *y*-value (assuming the values are both taken before the plateau phase). Unfortunately, experiments are never perfect, so we want to choose the lowest threshold value possible. Suppose we had a single sample of cDNA, and we took three aliquots from it for a PCR experiment. The three amplification curves generated would not be identical. Figure 10.14 is an illustration of the three different curves (although in practice, they would not differ quite as much as shown). Notice that, at the higher threshold, the C_t values are quite different, but for the lower threshold, they converge toward the same point. This difference in C_t at higher versus lower thresholds is different from the theoretical values presented in Figure 10.13 because the amount of time taken for individual dNTPs to bind to DNA polymerase differs ever so slightly (because of probabilities involving molecule orientations, diffusion constraints, and so on). The exact rate of polymerization for each DNA polymerase molecule is not a mathematical constant either. As a qPCR experiment proceeds to higher cycle numbers, any such minor differences will be amplified exponentially. To minimize the variance seen in real-world conditions, data for all samples and genes are collected at a low threshold.

The ability to collect reliable data at low cycle numbers is an advantage of real-time PCR over traditional PCR. With traditional PCR, one must allow the reaction to proceed to relatively high numbers of cycles to allow us to see the thickness of bands on an agarose gel. For real-time PCR, data are taken at low cycle numbers to minimize error. Another advantage to taking data earlier in the amplification process is that there are no issues with limited resources, which can skew results. There should be sufficient dNTPs that DNA polymerase can utilize as rapidly as possible. At higher cycle numbers, such

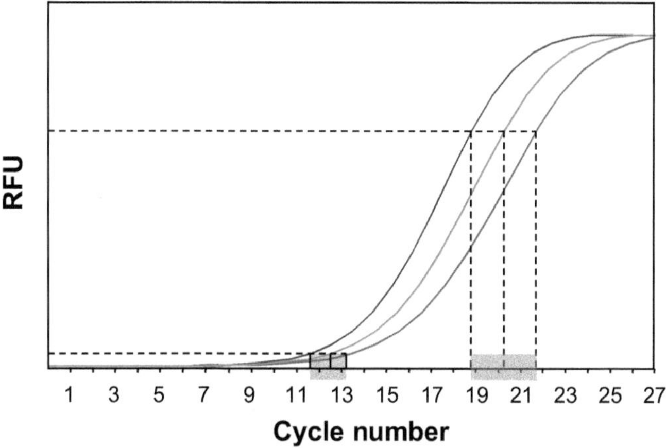

FIGURE 10.14

In practice, there *is* a difference in results obtained using different thresholds. The widths of the *gray boxes* on the *x*-axis illustrate the error in reported C_t value that could result from splitting a single sample into three aliquots and running qPCR simultaneously on each. Having a threshold set at a higher value results in a greater range of C_t values. This principle is different from that illustrated in Figure 10.13. Here, the rate of increase for the three curves is different due to the different amounts of time taken for individual dNTPs to bind to DNA polymerase and the exact efficiency of each individual DNA polymerase molecule in each sample tube. To minimize the discrepancy between theory and practice, the lowest reliable threshold value is used. *RFU*, Relative fluorescence units.

as those used by traditional PCR, dNTP availability can become an issue as these resources are depleted, which may result in data inaccuracies.

10.7.1.2 Primer efficiency

We have assumed so far that for every PCR cycle, we exactly double the amount of dsDNA. This is a fine assumption when learning the theory behind PCR, but in real life, it is seldom the case. One explanation for this relates to the efficiency of the primers that are designed and used. You've got ~20 nucleotides per primer that must find and bind to a sense or antisense strand in the 30 seconds we give them during the annealing step, at the annealing temperature we give them, to allow the DNA polymerase to then find this newly formed 20-mer double-stranded patch for subsequent polymerization. These events do not occur in the given timeframe 100% of the time, especially when primer dimers, secondary structures, or damaged RNA come into play. In these (and other) cases, **primer efficiency** will not be 100%. Different laboratories may consider different ranges of primer efficiencies acceptable, but a range of 90% to 105% is generally used. (Primer efficiencies above 100% often indicate that there was an error in experimental design when determining the efficiencies, commonly because dilutions were too high.) Before, when we used terms such as $2^{Ct_2 - Ct_1}$, the 2 was based on perfect doubling stemming from our assumption of 100% primer efficiency. If a primer set were determined to yield only 95% efficiency, the 2 would be changed to 1.95 (1 + primer efficiency; with a useless primer—0% efficiency—the amount of dsDNA would remain constant, which is why 1 is added to the primer efficiency). A more robust way of expressing $2^{Ct_2 - Ct_1}$ is now$(E+1)^{Ct_2 - Ct_1}$, where E = primer efficiency, because the amount of dsDNA may not be completely doubled in each cycle. For this example, the equation becomes $1.95^{Ct_2 - Ct_1}$. Our formula for fold difference is now:

$$\text{Fold difference} = \frac{(E_{GeneOfInterest}+1)^{(Ct_{GeneOfInterest,\ NegativeControl} - Ct_{GeneOfInterest,Sample})}}{(E_{ReferenceGene}+1)^{(Ct_{ReferenceGene,NegativeControl} - Ct_{ReferenceGene,Sample})}},$$

in which E = the efficiency of the primer set for the gene of interest or for the reference gene, as indicated.

10.7.2 Probes

In the qPCR experiments described up to this point, SYBR Green was used to indicate the amount of dsDNA in each sample. Another method to achieve dsDNA quantitation involves the use of double-labeled probes. For quantitation, the mathematics will be the same as already described, so we can use the same formulae; both the $2^{-\Delta\Delta}$ and the efficiency-corrected methods are valid. The difference between the use of probes and the use of SYBR Green lies in the fundamental use of fluorophores in the test tube.

Imagine we have a single fragment of double-stranded cDNA. Just as before, we will perform a melt step, annealing of forward and reverse primers, and extension via DNA polymerase. However, when primer/probe sets are used, the fluorophore is not an intercalating dye but rather a double-labeled fragment of DNA that is complementary to an interior portion of the amplicon (Figure 10.15). This fragment, labeled with a fluorophore on one end and a quencher on the other, is the **probe**. It is not necessary to use probes that are complementary to each fragment (sense and antisense) in the dsDNA amplicon; only one fragment need be probed. The probe will contain two light-sensitive molecules that serve as labels: a fluorophore and a quencher. From our discussion in Chapter 8, we already know that the fluorophore will absorb light of one wavelength and emit it at a longer wavelength. The quencher works much like the secondary fluorophore of a FRET experiment: its excitation wavelength range should include the emission wavelength of the fluorophore. However, unlike in FRET, the quencher molecule will not re-emit the absorbed photon in the visible light spectrum.

Before a probe molecule has attached to a complementary DNA sequence, the quencher molecule is close enough to the fluorophore to absorb any light it emits. The fluorophore and quencher molecules are attached to the ends of an oligonucleotide that is complementary to a sequence of DNA in the interior of the amplicon, and which will anneal to ssDNA at the same time as the primers. When the DNA polymerase attaches to the primed, double-stranded portion of the DNA fragment, it will polymerize as before in the 5′→3′ direction, but when it encounters the probe (which is now a part of a small double-stranded patch of DNA), its 5′→3′ exonuclease activity will chew up the oligonucleotide probe, freeing the nucleotides that contain the fluorophore and quencher molecules to move about the reaction solution. Because the fluorophore and quencher are no longer tied to the same molecule, they will move apart by diffusion, and the quencher will no longer be able to absorb the photons emitted from the fluorophore, thus allowing for detection of fluorescence.

With SYBR Green, we were able to detect how much more dsDNA was in the reaction vessel by the production of more potential spots for SYBR intercalation. With probes, no fluorescence is detected unless the probe is chopped up by DNA polymerase. Consider a new thought experiment, which starts with only a

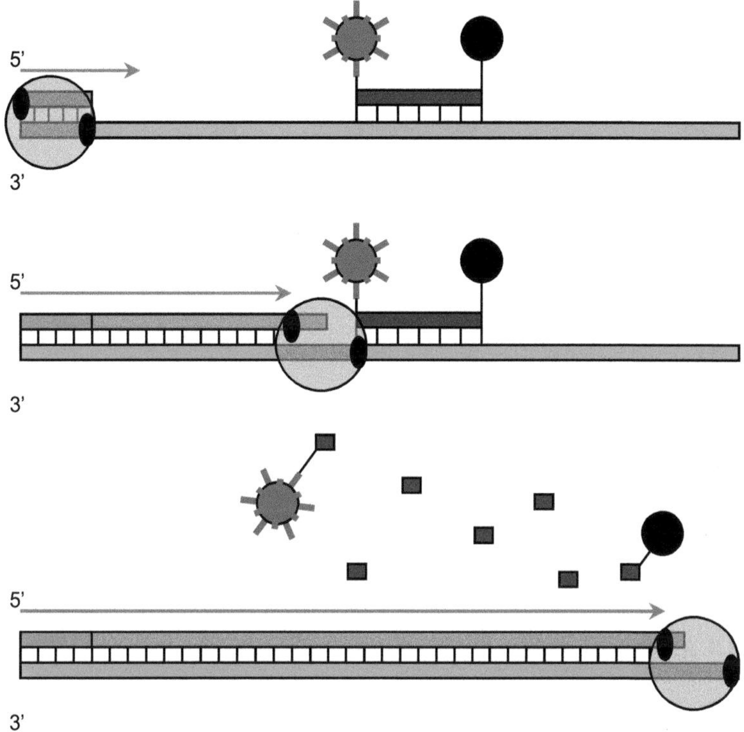

FIGURE 10.15

Primer/probe sets utilize a fragment of DNA that is complementary to an interior portion of one strand of the amplicon, with a fluorophore and a quencher covalently bound to it. The quencher is close enough to the fluorophore to absorb a large percentage of emitted photons, meaning the fluorescence will be quenched down to background levels. When TaqMan® DNA polymerase encounters the double-stranded segment created when the probe binds to the ssDNA segment, the intrinsic 5′→3′ exonuclease activity of the polymerase will chop up the probe, permitting the fluorophore and quencher to diffuse away from each other. The increased distance between them will render the quencher unable to absorb the fluorophore's emitted light, which will therefore become detectable. Increased fluorescence implies that more amplicons have been replicated.

single dsDNA molecule. It can bind our two primers and one probe, so only one probe molecule will be chopped up during the first cycle. (For illustration purposes, assume the probe binds to the antisense strand.) During the second cycle, two more fluorophores will be separated from quenchers—one for each new sense strand produced—and the amount of fluorescence in the reaction tube will be doubled (assuming 100% primer efficiency). Fluorescence will be detected during each cycle and should increase geometrically in a 1:1 relation with the number of new sense DNA molecules produced. The curve showing detected fluorescence versus cycle number should be similar to the curve that would have been generated had SYBR Green been used. In theory, after n PCR cycles, 2^n-1 free fluorophores would have been generated using primer/probe sets, whereas 2^n units of fluorescence would be present had SYBR Green been used.

The advantage of using a primer/probe system is that several primer/probe sets can be used in the same reaction vessel, which allows monitoring of multiple genes at the same time in the same tube, reducing variability. In qPCR instruments that can detect four colors simultaneously, four different primer/probe sets can be designed, each probe with its own unique fluorophore/quencher system, which will allow for the monitoring of amplification of four different genes at the same time under absolutely identical conditions.

10.8 Summary

In this chapter, we learned the three steps common to all PCR procedure: melt, anneal, and extend. We then saw how the steps can be repeated many times to amplify a specific section of DNA. Two methods of PCR were discussed and compared for better understanding, and we saw that each method comes with its own advantages: traditional PCR is relatively inexpensive to perform and yields qualitative data, while the real-time methods yield quantitative data, potentially for more than one gene of interest at the same time.

In the next chapter, we will look at genetic engineering and how we can manufacture complete genes, in usable quantities, from DNA fragments (such as promoters and exons). Using the principles covered in this chapter, you should be able to see how these fragments can be isolated by judicious use of PCR. The ability to manufacture complete genes of interest is one of the most powerful resources for the biotechnologist who wants to harness the power of cells to produce specific products or to be able to modify cellular behavior for medical or industrial reasons.

Related reading

Marmur, J., Doty, P., 1962. Determination of the base composition of deoxyribonucleic acid from its thermal denaturation temperature. J. Mol. Biol. 5, 109–118.

Meinkoth, J., Wahl, G., 1984. Hybridization of nucleic acids immobilized on solid supports. Anal. Biochem. 138, 267–284.

Mullis, K.B., 1993. The Polymerase Chain Reaction (Nobel Lecture). Nobel Media AB. 09.07.20 https://www.nobelprize.org/prizes/chemistry/1993/mullis/lecture/.

Rychlik, W., Spencer, W.J., Rhoads, R.E., 1990. Optimization of the annealing temperature for DNA amplification in vitro. Nucleic Acids Res. 18, 6409–6411.

Questions

1. If the region of DNA you wish to replicate with Taq DNA polymerase is 750 nucleotides long, and Taq works at 1000 nucleotides per minute, how long should you allow for extension during PCR?
2. Is this a proper PCR sequence? Why or why not? Assume there are about 400 bases in the amplicon.

Melt 95°C	45 s
Anneal 42°C	30 s
Extend 73°C	24 s

3. Why is the annealing temperature so precise in PCR? What happens if the temperature is too high or too low? Why does the annealing temperature usually vary from experiment to experiment?

4. Give two reasons why qPCR is more accurate than traditional PCR.
5. Name four advantages of qPCR over traditional PCR.
6. Compare qPCR using SYBR Green versus a primer/probe set.
   a. What is literally being measured with each method?
   b. How does each method indicate the mRNA level?
7. An entire transcriptome is converted into cDNA by reverse transcriptase, and qPCR analysis is performed using SYBR Green and primers specific for a certain transcript. How would you verify that the results of the PCR are not the result of the primers binding to more than one site on the DNA?
8. Why is cDNA not the same as genomic DNA?
9. What does PCR analysis tell us about a gene?
10. What protein is used to turn mRNA into cDNA? For PCR, why is this step performed?
11. Why do we use Taq DNA polymerase and not human DNA polymerase for PCR?
12. Dr. Frank runs a traditional PCR experiment to demonstrate that his new drug increases the activity of a gene that is 10,000 base pairs in length. For his test sample, his gel reflects a band at 300 bp for the gene of interest, and he is happy about it! Is he crazy, did he stay up too late studying, or is there some other reason why he is not upset by the lack of a band at 10,000 bp? If you pick the third option, give the reason. (Hint - he is not crazy, and he went to bed early.)
13. Your instructor will give you a bonus point if you point out this question to him/her. Do not contact the author of this book on this matter; it is between you and your instructor.
14. Name two mathematical models that can be used for real-time PCR quantitation.
15. a. How many cycles must pass in a PCR reaction for the *majority* of your DNA strands to be completely bounded by the target sites for primer annealing?
    b. Will the strands produced in your first round of PCR be longer, shorter, or the same length as your target amplicon?
16. Why is the number of cycles used in a traditional PCR experiment limited? If resources were not a problem, could we use 40 cycles for everything? Why or why not?
17. Suppose you are going to perform PCR in a buffer solution that contains 150 mM of KCl and 100 mM of MgCl (and no formamide). Give the T_m of the following primers:
    5′-CCT TGT CAA GT-3′
    5′-CCA CCT CTG CGA TGC TCT TAC-3′
18. The graph reflects readings obtained for a sample of cells treated with a drug (red lines) and an untreated control sample (blue). The gene of interest is shown by the solid curves, and the internal control is represented by the dashed curves.
    Black lines: untreated controls
    Red lines: cells treated with the drug
    Solid lines: gene of interest
    Dashed lines: GAPdH
    a. What is the fold increase or decrease in transcription of the gene of interest when the cells are exposed to the drug?
    b. What is the fold increase or decrease in transcription of the GAPdH when the cells are exposed to the drug?

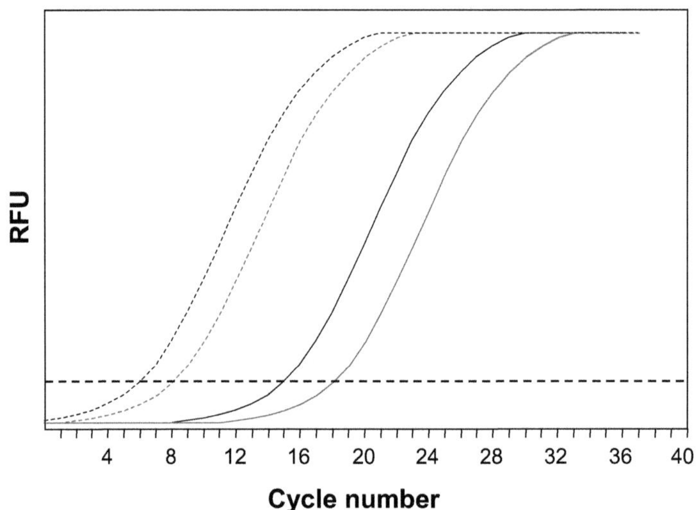

19. Suppose you want to generate a large number of poly(A) polynucleotides (at least 100 A residues per strand) via PCR. You already have a ssDNA template with the sequence of: 5′-$(T)_{120}$ AAA AAA AAA-3′, so you design a primer that has a DraI cut site at the end. (The recognition sequence of DraI is TTT|AAA).
    a. If you use 200 nM of the primer with 1×10^6 copies of the template, how many copies of the template would you have after 30 cycles of PCR? (Assume that this reaction takes place in 25-μL total volume.)
    b. Write the sequence of the primer you will use.
    c. Give two reasons why using a 6-mer as your primer is not a good idea.
    d. Determine the optimal annealing temperature for your PCR.
    e. Write out a modification to the template that would substantially increase the amount of poly(A) you would obtain with the PCR primer that you designed.
20. Suppose you want to use PCR to analyze the transcription of a certain gene. If you use 200 nM of each primer (forward and reverse), sufficient dNTPs and Taq DNA polymerase, appropriate buffer, and 1×10^6 copies of the target cDNA, and the reactions take place in 25-μL total volume, how many copies of the amplicon would you have:
    a. After 10 cycles of PCR?
    b. After 30 cycles of PCR?

What is the theoretical maximal number of cycles you could expect to run under these conditions and still yield meaningful results? (You might want to use a spreadsheet to answer this.)

CHAPTER

Genetic engineering

11

Chapter outline

In this chapter, we will see some basic ways in which custom DNA sequences can be put together or altered, a process known as genetic engineering. The focus will be on circular polymers of DNA. This geometric configuration is routinely used because the biotechnologist can harness bacteria (such as *Escherichia coli*) to manufacture many copies of the DNA. After familiarizing ourselves with the key features that must be included in these circular bits of DNA, we will delve into how the DNA can be cut and pasted to form recombinant DNA.

After the desired DNA sequence has been obtained, we must amplify it to garner enough copies for use in experiments or, potentially, clinical treatments; this is where the bacteria come in. The amplification process consists of transferring copies of our engineered gene into bacterial cells and letting those cells do the work of making many copies of the gene for us. The next step in the process is to separate our engineered DNA from the bacteria, followed by quantitative analysis to determine its concentration and purity. To perform the analysis, we will return to one of our old friends (from Chapter 8): Beer's law.

Biotechnology and its Applications. https://doi.org/10.1016/B978-0-12-817726-6.00011-3

11.1 Plasmid architecture

A **plasmid** is a circular piece of DNA. Much of genetic engineering utilizes plasmids, which can be constructed relatively easily in the laboratory through the judicious use of natural enzymes. A robust plasmid has several elements of great importance. First, there is the gene that the scientist wishes to deliver into target cells. The gene will contain at least a promoter and an exon. The promoter helps to initiate transcription of the encoded gene, and the exon codes for a predetermined polypeptide, which will be expressed using the steps of the central dogma (Chapter 4). Another feature of the plasmid is an **antibiotic resistance marker**, (generally denoted by α^r), which is a bacterial gene that will be useful as a selection marker after many copies of the plasmid are produced in a process called **amplification**. An antibiotic resistance marker codes for a protein that makes any bacterium expressing it resistant to a specific antibiotic (e.g., ampicillin or kanamycin). Another necessary feature of the plasmid is a bacterial **origin of replication**, which will allow the researcher to harness the power of *Escherichia coli* for amplification, using the bacteria as biological factories to churn out many copies of the plasmid. A prokaryote is used for amplification because it lacks a membrane-bound nucleus; this means it is a relatively simple task to get the plasmid into the presence of DNA polymerase, the enzyme responsible for DNA replication. (In eukaryotic cells, DNA polymerase resides in the nucleus.) DNA polymerase does not distinguish between genomic and foreign DNA. The cellular replication machinery will assemble and begin replication upon exposure to an origin of replication (*ori*) in the genome or on a plasmid.

Figure 11.1A is a map of a common reporter plasmid, pEGFP-N1. Plasmid names usually begin with a lower-case p to denote that the DNA is a plasmid. In this example, EGFP is an acronym for Enhanced Green Fluorescent Protein, and N1 indicates that this plasmid can be used to produce a fusion protein, where the protein portion of interest in our final product will be on the N side (N-terminus side) of the complete protein. Later, we will conduct a thought experiment in which we will insert an additional exon into the plasmid. Notice that some of the features in the figure are emphasized with green, and some are emphasized with yellow; this is to illustrate that the plasmid performs different functions in prokaryotic versus eukaryotic cells. The portions of the plasmid denoted in yellow represent sequences relevant to bacteria:

- ^{P}SV40$_e$ is a promoter from simian virus 40 that will dictate the transcriptional Start site for the bacterial gene *Kan*^{r}.
- *Kan*^{r} is a bacterial gene encoding a protein that will confer antibiotic resistance to kanamycin.
- *HSV TK poly A* encodes a signal (obtained from the herpes simplex virus gene for thymidine kinase) that will cause a poly(A) tail to be put onto the 3′ end of the bacterial gene.
- *pUC* is a DNA sequence that serves as a starting point (origin) for bacterial replication.

The portions shown in green represent sequences relevant to eukaryotic (in this case, mammalian) cells:

- ^{P}*CMV IE* is a promoter for "immediate early" genes in the cytomegalovirus genome. This is a **strong promoter** in mammalian cells, meaning it supports a large amount of transcription initiation.
- *EGFP* encodes the gene for an enhanced green fluorescent protein.
- *SV40 poly A* encodes a signal used in the polyadenylation of mRNA transcripts.

(For completeness, the region shown in white encodes the *f1 ori*, which serves as an origin for single-stranded DNA production. Single-stranded DNA is used in some viral applications.)

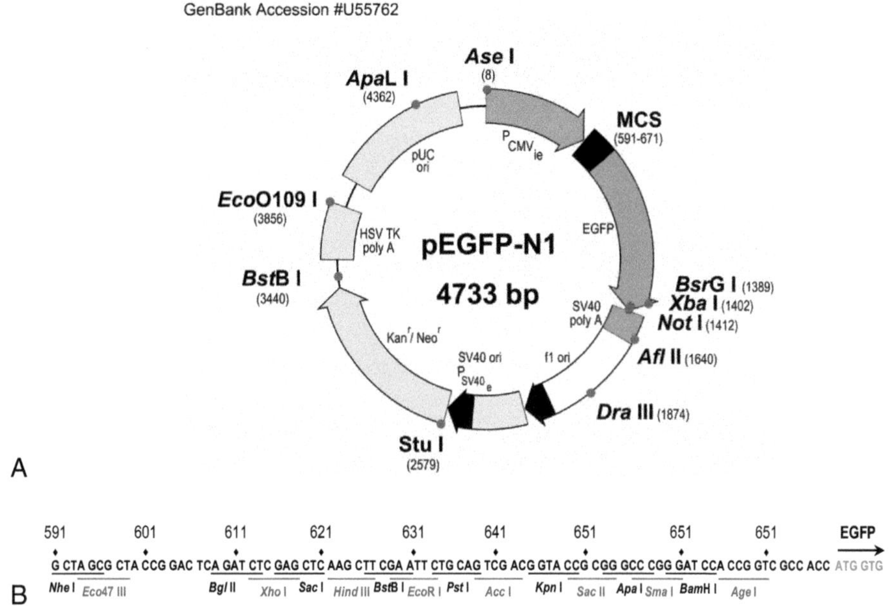

FIGURE 11.1

(A) Map of the plasmid pEGFP-N1. The locations of certain restriction sites are noted by enzyme name, with the precise location of each cut in parentheses. Segments of the plasmid are explained in the text. (B) Specific sequence of the multiple cloning site, with specific enzyme recognition sequences underlined and annotated.

One other feature of commercially available plasmids, including the one shown in Figure 11.1, is the **multiple cloning site** (MCS), a stretch of DNA that is rich in restriction enzyme recognition sequences. It is placed in the DNA sequence deliberately to give researchers the ability to insert DNA strands they have constructed into the plasmid at a useful location with relative ease. The specific bases of the pEGFP-N1 MCS and the names of the enzymes that can be used to cut it are given in Figure 11.1B.

Many plasmids are commercially available to aid the biotechnologist in constructing genes. They may be purchased with or without a promoter, with or without a reporter exon, and with or without an enhancer region. However, without the bacterial features of an origin for replication and an antibiotic resistance gene, amplification of the plasmid inside bacteria in selective media would not be feasible. Commercially available plasmids (vectors) should also have an MCS to allow for insertion of promoters and/or exons with relative ease.

11.2 Molecular cloning

We said the "N1" in the reporter plasmid name meant the vector could be used to produce a fusion protein. This is due, in part, to the location of the MCS. A **fusion protein** is a single protein produced by the expression of two formerly separate genes that have been combined into a single gene. By inserting an exon in the MCS of pEGFP-N1, expression of the resulting gene will produce a protein that has a fluorescent region. The exon could code for almost any protein, but for our example, we will use an exon that codes for bone morphogenetic protein 5 (BMP5). In this example, we are creating a fluorescent protein consisting of H_3^+N–BMP5-EGFP–COO^-.

There are different ways to obtain the DNA sequence for insertion into a vector. One could create a sequence from scratch or use polymerase chain reaction (PCR) to amplify a specific sequence directly from a genomic preparation. In this example, the *BMP5* sequence will reside in a plasmid that was sent to us from a collaborator's laboratory. Our goal is to make a fluorescent version of BMP5 so we can verify its expression in target cells. (As a matter of semantics, note that human genes are written in upper-case italics, and protein names are written in unitalicized upper case.)

When a plasmid is sent from one laboratory to another, a map is typically included to show any regions of interest and known enzyme cut sites. Sometimes removing an exon from the plasmid for insertion into our own vector is easy because the DNA region we are interested in is flanked by two enzymes that are also found in the MCS of our vector, in the same order. Other times the problem is slightly more complex, and a pair of enzymes must be selected from a list of candidates determined by examination of the maps of the donated plasmid and our vector. In this case, it is important to ensure that the enzymes selected are compatible in the same digest buffer and that the cut sites appear in the same order in the donated plasmid and the vector. The order is important because the exon of interest will almost certainly not be a palindrome, meaning that if it were to be inserted backward, the DNA message would be gibberish.

11.2.1 Cutting and ligating sticky ends

Returning to our example, suppose the vector is pEGFP-N1, which has the MCS shown in Figure 11.1B. Also suppose that the plasmid we obtained—call it pBMP5—contains the cut sites shown in Figure 11.2. Which two enzymes would work best for cutting the *BMP5* exon out of pBMP5 for insertion into the vector? *Xho*I and *Nhe*I might initially appear to be the perfect pair, but this selection would be a poor choice because the *BMP5* exon would be inserted in the wrong orientation. The sticky ends we make in the vector and insert are crucial because they will dictate the orientation of the insert in the final plasmid. *Xho*I and *Nhe*I will both produce sticky ends, but the sequence of bases in the overhangs

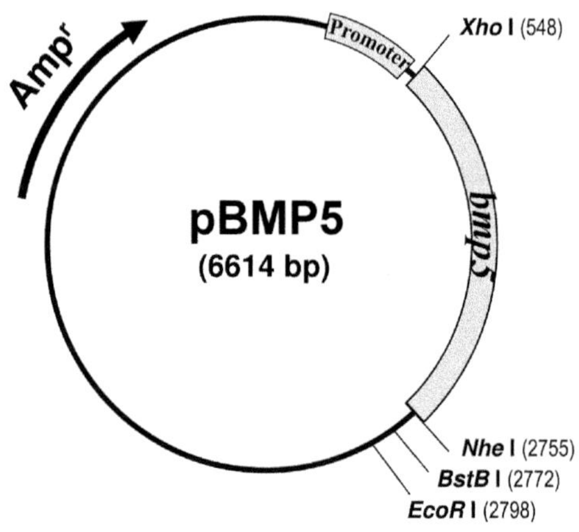

FIGURE 11.2

Partial restriction map of pBMP5, a fictitious plasmid used as an example in the text.

will be different, meaning each sticky end will only pair with the end of a vector produced using the same enzyme. If the exon is inserted backward, the sequence of the insert will be changed to gibberish. For instance, the ATG that would normally produce the Start codon in the mRNA would become CAT (the reverse compliment) and be located perhaps over 1000 bp away from the promoter. Almost every codon would encode a different amino acid, assuming the reading frame was preserved. Even if a functioning mRNA were to be produced, it would encode a useless polypeptide.

Use of the *Bst*BI enzyme with *Xho*I would not be the best choice, either, because it cuts twice in the vector (Figures 11.1A and B)—although there is a low-yield method to get around this. XhoI and BstBI could be used to cut the exon out of the BMP5 plasmid, and this enzyme pair would yield sticky ends in the correct orientation inside the MCS of the vector, but the vector would be cut in three places, essentially rendering it useless.

Next, consider *Xho*I and *Eco*RI; this enzyme pair will effectively remove the exon from the insert, will yield sticky ends in the correct orientation in the vector, and will not generate extraneous cuts in either the vector or insert. The enzymes also happen to be active in the same buffer. This pair is therefore a good choice for the task at hand.

When pBMP5 is cut with *Xho*I and *Eco*RI, two fragments will be produced (Figure 11.3B). The exon for *BMP5* will have sticky ends on either side, and they will match up perfectly with the sticky ends created by cutting the vector plasmid with the same enzymes. Notice that although a large portion of the plasmid containing *BMP5* will be lost, a very small portion of the vector plasmid—the few bases between *Xho*I and *Eco*RI—will also be lost. But how?

The answer is that they will be thrown away. Following exposure of each plasmid to the restriction enzymes, each resulting DNA sample will be run out in separate lanes of an agarose gel (Figure 11.4). Unless the two enzymes cut at sites that are exactly opposite one another in the plasmid—which won't happen because our choice of enzymes will prevent it—the two fragments will be different sizes and will migrate different distances through the gel. Because we have a map of each plasmid, we will be

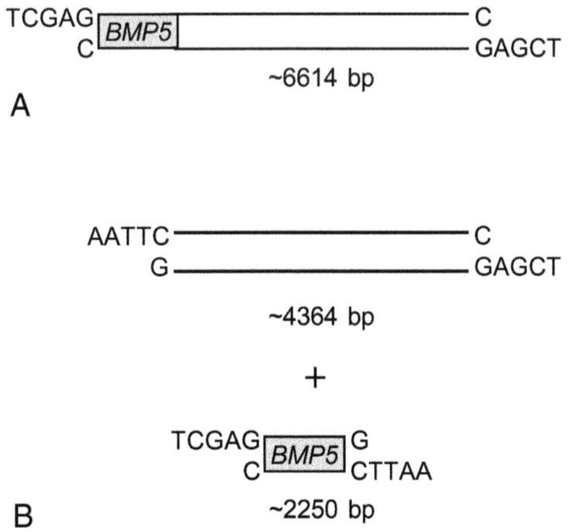

FIGURE 11.3

(A) When the plasmid shown in Figure 11.2 is cut only with *Xho*I, a single, linear, approximately 6614-bp fragment with sticky ends is generated. (B) Cutting the plasmid with *Xho*I and *Eco*RI gives fragments that are approximately 4364 and approximately 2250 bp in length. (Lengths are approximate because not every base is involved in a pairing.)

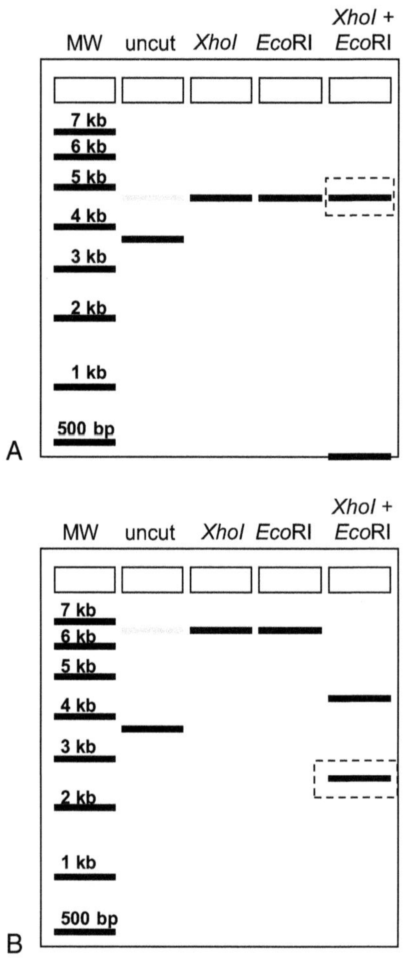

FIGURE 11.4

Results of agarose electrophoresis following *Xho*I and *Eco*RI restriction cuts on (A) the vector and (B) the insert plasmids, as described in the text. Fragments to be excised are denoted with *dotted lines*. Note that the 16-bp fragment in panel A may run off of the gel. *MW*, Molecular weight marker.

able to predict the fragment sizes created from the restriction cuts. We will simply visualize the fragments with ethidium bromide and ultraviolet light (see Chapter 9) and cut out the one(s) we want to keep using a scalpel or razor blade. It is a simple matter to then melt away the agarose to isolate the DNA via an anion exchange resin.

After melting the agarose, the solution will be put into a column containing a matrix that carries a positive charge. The solution will be forced through this matrix via centrifugation, and the DNA will adhere to the matrix by virtue of its negative charges. After washing the matrix to remove residual compounds such as agarose and salts, the DNA will be eluted under slightly alkaline conditions (pH 7–9) and low salt concentrations (<10 mM).

The restriction cuts will be performed in a small tube, which will contain water, various salts, a buffer, possibly an additive such as bovine serum albumin (BSA) or S-adenosyl methionine (SAM), the DNA sample, and the restriction enzymes. These will typically be combined to yield a total volume of 10 µL as follows:

*Water	(10-x) µL
10× buffer	1.0 µL
10× extras	(BSA or SAM. 1.0 µL is used if required by one enzyme)
DNA	y µL (enough for 0.1–1.0 µg)
Enzyme 1	0.5 µL
Enzyme 2	0.5 µL
Total	**10.0 µL**

**Water will be added to bring the total volume up to 10 µL. As written, x = the sum of the volumes of all other additives, including "extras" and* y.

The above formula, although robust, can be confusing at first. If our vector DNA is kept at a concentration of 1 µg/µL, and we are using the enzymes XhoI and EcoRI, then the recipe becomes:

Water	6.0 µL
10× "buffer 4"	1.0 µL
10× BSA	1.0 µL
DNA	1.0 µL
*Xho*I	0.5 µL
*Eco*RI	0.5 µL
Total	**10.0 µL**

"10×" refers to a concentration that is ten times what is needed in the final working concentration. If the final volume of a different reaction was to be 100 mL, we would use 100/10 = 10 mL of a 10× solution in constructing the solution, 1 mL of a 100× solution, or 0.1 mL of a 1000× solution. 1000× concentrations are often used in the laboratory because the math requires only a prefix change; for instance, if the final volume of the solution is to be 2 L, we would need 2 mL of a 1000× solution. If the final volume is to be 58.3 mL, we would use 58.3 µL of a 1000× solution.

Regarding the amount of enzyme, 0.5 µL is often used for practical reasons rather than because of the strict amount of enzyme needed for the given reaction. Restriction enzymes are often measured in **units** (U), in which 1 unit is equal to the amount of enzyme needed to completely cut 1 µg of phage lambda DNA in 1 hour at optimal temperature. The definition includes phage lambda DNA only for standardization purposes; we can expect 1 U of enzyme to cut 1 µg of most plasmids in 1 hour. Most restriction enzymes have optimal activity at 37°C. In terms of our example, suppose that *HindIII* was sold at a concentration of 20,000 U/mL; 1 µL of *Hind*III will contain 20 U. In our example, we are cutting 1 µg of plasmid, so we only need 1 U, which would equate to 1/20 µL of *Hind*III. It's usually impractical, if not infeasible, to measure such a small volume, so this number will be adjusted up to 0.5 µL based on the accuracy of the smallest pipette available in a typical laboratory. Although in theory the enzyme could be diluted, in practice this would impair enzyme activity.

BSA is a common additive to the recipe for restriction digests. This protein has been found to stabilize the folding of some restriction enzymes, and it balances the potential negative effects of enzyme interactions with pipette tips, reaction tubes, and/or the air–liquid interface. The positive effects of BSA have been found to increase the activity of restriction enzymes by as much as twofold. This result is so encouraging that some companies now include BSA in every restriction enzyme sample they sell. Some enzymes (e.g., *Xho*I) are said to require BSA to work properly, whereas others (e.g., *Eco*RI) are not listed as having the requirement. If only one enzyme in a chosen pair requires BSA, then BSA should be included in the reaction mix. Some laboratories use BSA with every enzyme, regardless of whether it is listed as a requirement.

As with any good experiment, the restriction digests must include the appropriate controls. For the current example, we should include a control in which only *Xho*I is used to cut the DNA and another control in which only *Eco*RI is used, adjusting the volume of water for each reaction accordingly. These controls help to ensure that each enzyme is working properly and that it cuts the plasmid as the maps predict. One should also include a control for plasmid that has not been exposed to a restriction enzyme, to demonstrate where the uncut, supercoiled plasmid will run on the gel. More often than not, however, this particular lane shows two bands: one for supercoiled plasmid and one for the linear form that results when a plasmid is cut once. As a plasmid sample undergoes several cycles of freeze–thaw and repeated pipettings, plasmids in solution are occasionally torn open into a linear form. The ratio of brightness between the supercoiled and linear bands serves as a measure of the integrity of the plasmid sample. For this part of the restriction digest experiment, it is also important to visualize where uncut DNA will run on the gel to help verify that each of the enzymes is indeed cutting the DNA in the other control lanes. A third use for the uncut control will be discussed later in the chapter.

After cutting out the appropriate bands and separating them from the agarose, we will have one tube containing vector DNA that is now linear with sticky ends and another that contains the *BMP5* insert, also with sticky ends. In theory, it is best to quantitate the amount of DNA in each tube so that theoretically equimolar amounts of the vector and insert can be combined. (In practice, it is common to use a greater number of insert molecules than vector molecules.) The way to determine the mass of DNA will be discussed later, but this step is often skipped because the DNA yield from gel extractions in this particular procedure is often very low.

Combining aliquots of vector and insert fragments allows them to pair up via their sticky ends. The overhangs from the sticky ends are complementary, so hydrogen bonding will keep the fragments associated. However, the bonding is not covalent, and there will still be gaps at the four ends between the insert and vector, meaning there are two separate phosphoribose backbones for each component. If an enzyme such as DNA polymerase were to attach to this pseudoplasmid, it would not be able to progress beyond the gaps. Many copies of the plasmid will be made in *E. coli* by virtue of DNA polymerase in an upcoming step, so these gaps must be sealed beforehand. The sealing is accomplished by the enzyme **DNA ligase**, which creates phosphodiester bonds between the vector and insert fragments. Following ligation, we will have continuous plasmids that contain both the vector and insert sequences.

11.2.2 Blunt-end ligation

We have now seen a very straightforward example of removing a segment of DNA from one plasmid and inserting it into another. Sometimes, however, the insert will be flanked by enzymes that are not contained in the MCS of the vector. In such a case, the biotechnologist might resort to blunt-end

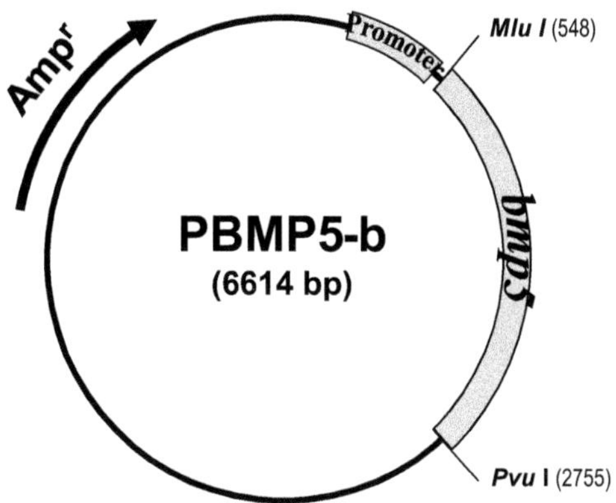

FIGURE 11.5

Partial restriction map of pBMP5-b, a fictitious plasmid used as an example in the text.

ligation. The insert can still be removed as before, but an additional step must be performed to remove the sticky ends. Consider the same example as before, except that the plasmid that contains the insert has the *BMP5* exon flanked by *Mlu*I and *Pvu*I, two enzymes that will not cut the vector (Figure 11.5) Following excision of the insert, the enzymes will be **heat-killed**, which means they will be inactivated by exposure to temperatures in excess of 70°C for 10 minutes. Next, a blunting enzyme will be added to the solution. Two common enzymes used for this purpose are the **Klenow fragment** and **T4 DNA polymerase**. The Klenow fragment is the large subunit of DNA polymerase I, isolated from *E. coli*. In the wild, T4 is a bacteriophage that can infect *E. coli*, and one of its genes codes for a DNA polymerase. Both of these DNA polymerases have 5′→3′ DNA polymerase activity and 3′→5′ exonuclease activity (Figure 11.6). While the two enzymes have very similar activities, T4 DNA polymerase is usually the best choice for blunting for two reasons. First, its 3′→5′ exonuclease activity is roughly 200 times that of the Klenow fragment. Second, unlike the Klenow fragment, T4 DNA polymerase does not displace downstream oligonucleotides as it polymerizes.

The end result from either of these two DNA polymerase enzymes will be a blunt-ended fragment. The process is carried out on both the vector and insert fragments. After blunting, the vector and insert can be combined into a single tube for blunt-end ligation. DNA ligase is still used to carry out this task, as for ligation of sticky ends, but much more time will be needed because the respective DNA fragments will not be held in place by hydrogen bonding.

Besides generating a lower yield because of the lack of hydrogen bonding from sticky ends, another problem with blunt-end ligation is that there is no guarantee that the insert will be introduced in the proper direction. To check for this possibility, one might try cutting the engineered plasmid with a pair of enzymes, one cutting inside the insert and the other cutting outside (Figure 11.7). The sizes of the resulting fragments would be used to verify the direction of the insert.

With blunt-end ligation, it is also quite possible that the vector will seal upon itself without an insert, yielding what is sometimes called an **empty vector**. This phenomenon can be avoided with two different methods. In the first, a molar excess of insert—perhaps five- to eightfold—is used to increase the probability of an insert fragment being lined up with the vector for ligation. The second method involves an **alkaline phosphatase** such as calf intestinal phosphatase, shrimp alkaline phosphatase, or Antarctic

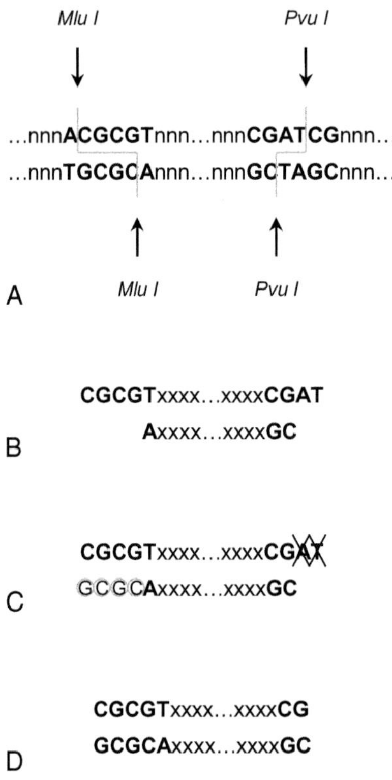

FIGURE 11.6

(A) Recognition sequences for *Mlu*I and *Pvu*I. (B) Sticky ends produced by the two enzymes. Note that *Mlu*I leaves a 5′ overhang, and *Pvu*I leaves a 3′ overhang. (C) The Klenow fragment and T4 DNA polymerase each have 5′ → ′3 DNA polymerase activity and 3′ → ′5 exonuclease activity, which work on the given overhangs as shown. *Circles* represent nucleotides being added. (D) The final result is a linear fragment with blunt ends. Note that both of the original recognition sequences have been lost in this fragment.

phosphatase. These products remove the 5′ phosphates from DNA (or RNA) samples. The alkaline phosphatase is added to the (blunted) vector and later heat-killed or removed by a purification step. Because 5′ phosphates are required for DNA ligase to work (Figure 11.8), self-ligation of the plasmid is prevented. By this reasoning, the only circular plasmids that will be produced will contain inserts. There will still be two nicks in the plasmid, giving the plasmid an architecture known as **open circle** (Figure 11.9). Although DNA deliveries using this architecture are not as efficient as when closed circle plasmids are used, yields are substantially higher using closed circle plasmids versus linear fragments.

11.2.3 Direct extraction of a gene from the genome

Up to now, we have been customizing our plasmids with sequences obtained from other existing plasmids. Although it is common to use commercial vectors that already contain an MCS, a selection

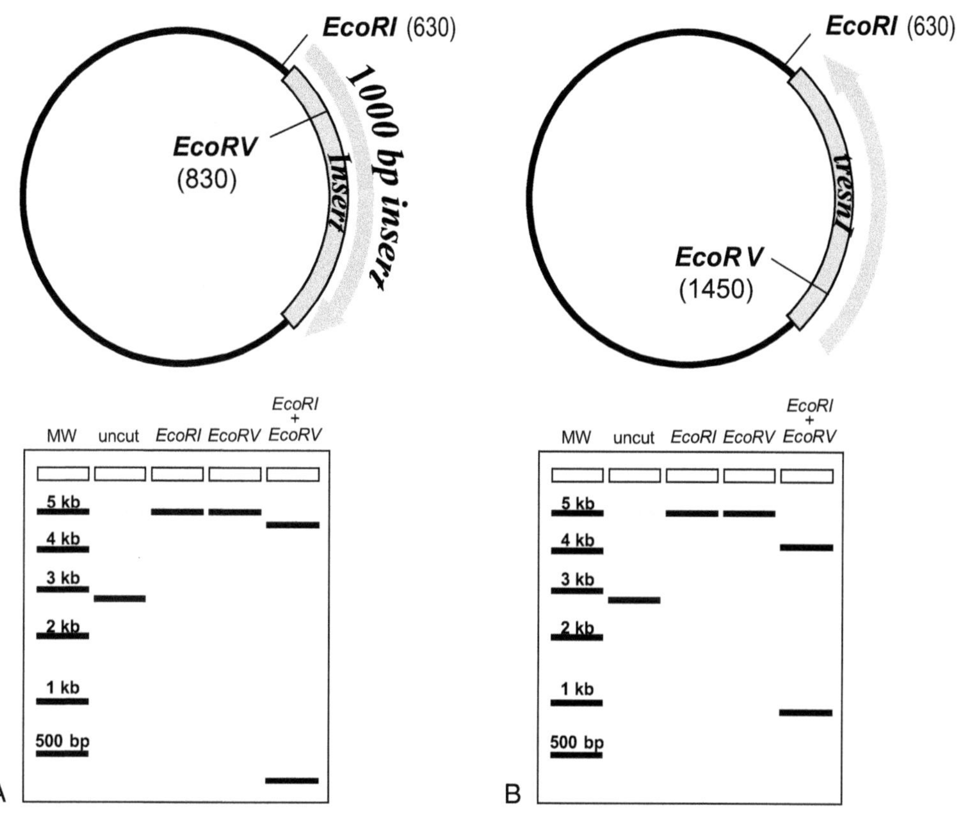

FIGURE 11.7

Detection of insert in reverse orientation. Suppose the entire plasmid, after ligation, is 5000 bp, the insert is 1000 bp, and the insertion occurs 10 bases after the *Eco*RI site. (A) If *Eco*RV cuts 190 bases from the 5′ end of the true coding strand of the insert, then the result of an *Eco*RI/*Eco*RV double digest would be two fragments that are (10 + 190) = 200 and (5000 − 200) = 4800 bp. (B) If the insert were ligated in the reverse orientation, the fragments would be (10 + 1000 − 190) = 820 and (5000 − 820) = 4180 bp. *MW*, Molecular weight marker.

marker (such as a gene for antibiotic resistance), and a prokaryotic origin of replication, scientific research often involves promoters, enhancers, or exons that are not commercially available. Cutting-edge research may use DNA sequences that are not even available from other scientists. So where do these sequences come from? Many times, we can take them from the genome itself, and we have already covered the techniques used for such a feat: genomic DNA can be extracted from a biopsy or a cell culture, and a region of interest can be amplified from the extracted DNA using PCR. By thorough investigation of the literature, including published genome sequences, one can design unique primers for the region of interest and amplify the region to concentrations that will allow for its isolation and insertion into a plasmid. Insertion will occur, as we have already seen, via restriction enzymes and DNA ligase, usually making use of sticky ends; PCR generates blunt-ended

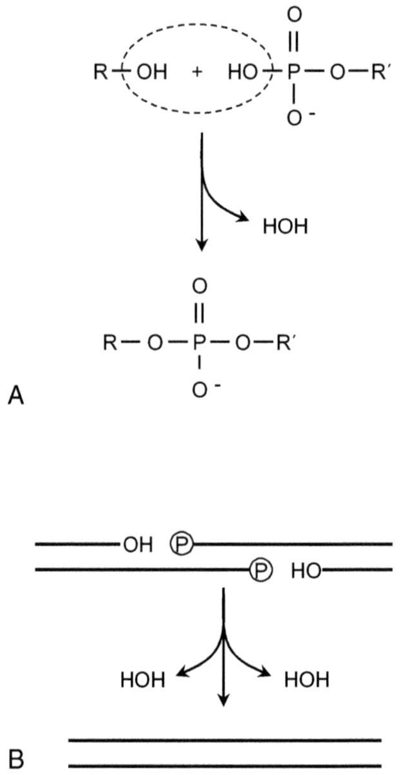

FIGURE 11.8

(A) DNA ligase acts by joining a hydroxyl of a 5′ terminal phosphate with the hydroxyl of a 3′ terminus, releasing a water molecule to create a phosphodiester bond. (B) This mechanism is used to join DNA fragments. Sticky ends are pictured, although the same action is used to join blunt ends. Note the requirement for a phosphate group in the reaction.

amplicons, but there will be restriction sites within these amplicons that allow for the creation of sticky ends.

It is reasonable to wonder how often a genomic region of interest—which must also be flanked by unique sequences to allow for primer binding—will also contain restriction sites for the generation of sticky ends. From a genomic perspective, the answer is "almost never." However, primers can be designed in such a way that additional bases can be inserted into the amplicon (Figure 11.10). These additional bases will include a restriction site (different ones for the forward and the reverse primer), plus a few extra bases (usually around six) to allow the restriction enzyme to bind and operate when needed after PCR. Modification of the amplicon via restriction digest will yield linear, sticky-ended fragments ready for insertion into the plasmid being constructed. This method allows the biotechnologist to engineer plasmids containing genomic DNA sequences, whether the sequence is coding, regulatory, wild-type, or mutated. It is a very powerful tool.

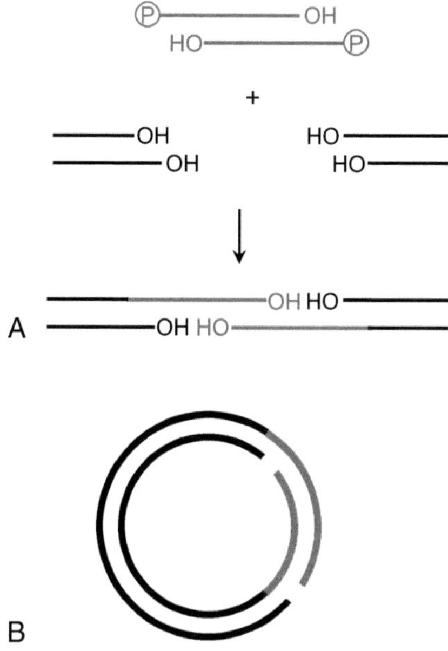

FIGURE 11.9

(A) An alkaline phosphatase has been used to remove the 5′ phosphates from the DNA vector, shown in *black*. Without these phosphates, DNA ligase will be prevented from sealing the ends of the vector together. Only the insert will contain the phosphate groups needed for ligation. (B) Because the insert only contains two phosphates, only two of the four gaps will be sealed. This produces plasmids with an open circle architecture.

11.3 A single plasmid is not enough

After the new plasmid has been constructed, it is not immediately useful. Suppose only one good copy of the plasmid has been created; if it were used for gene delivery, only one cell could receive the plasmid. If an organism were to be treated using gene delivery, transfecting only one cell would have little if any effect. To get an idea of scale, 1 μg of a 5000-bp plasmid contains approximately 1.95×10^{11} copies of the plasmid. These examples are meant to illustrate how relatively useless a single plasmid is. What we need is a way to produce many copies of the engineered plasmid. This can be accomplished via amplification—a straightforward technique whereby a plasmid is inserted into an *E. coli* cell, which then creates more copies of the plasmid using its own DNA polymerase. As the *E. coli* cell grows and divides, the number of plasmids being copied per unit time also increases. The amount of plasmid being produced increases exponentially as long as the *E. coli* cells have ample room to divide and enough food to eat.

Before *E. coli* can begin to amplify a plasmid, the plasmid must first be inserted into one or more bacterial cells. This can be accomplished via **transformation** or **electroporation**. In transformation, *E. coli* cells must first be made **competent**, or able to take up exogenous DNA. This can be accomplished via membrane disruption during exposure to high concentrations of Ca^{2+} or Mg^{2+} and a temperature

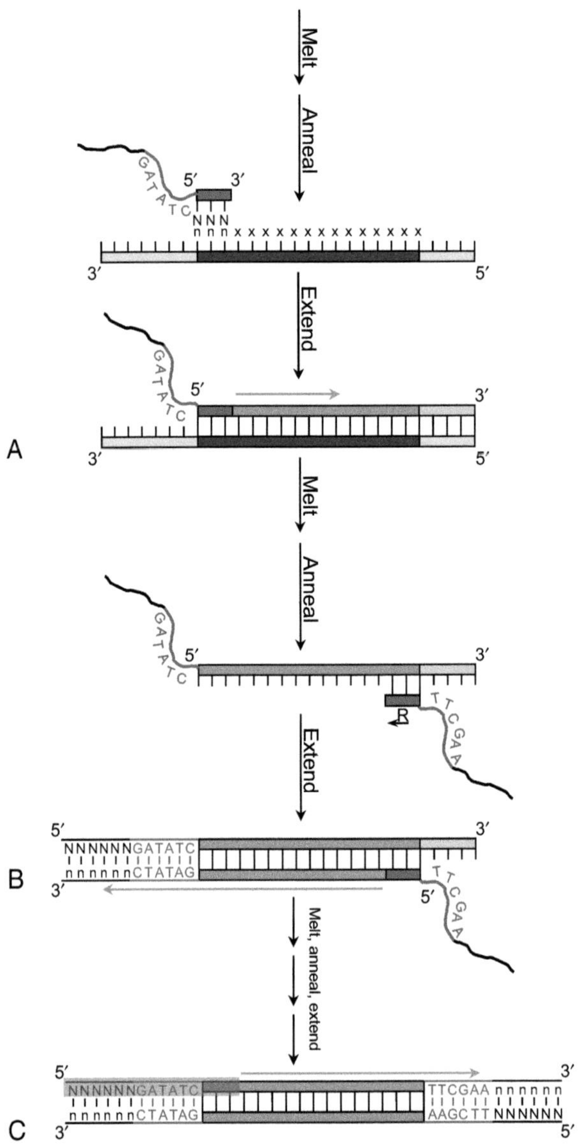

FIGURE 11.10

In adding restriction sites to an amplicon, the PCR steps are the same as those covered in Chapter 10 (see Figure 10.2). (A) Here, extra bases have been added to the 5′ end of each primer. These bases (shown as *red* bases and a nondescript spacer) will not pair with the genomic DNA. "N" stands for any nucleotide, and "n" is the base that is complimentary to "N." The genomic fragment is shown in *blue*. The first cycle of PCR will yield dsDNA fragments with the added cut site and spacer still unpaired. The cycle is shown for only one of the two original ssDNA strands.(B) The second cycle of PCR will yield a completely paired dsDNA fragment on the side where the primer from the previous cycle is located. However, the primer used for this step has an unpaired tail with a second restriction site and spacer.(C) The third cycle of PCR will yield the first completely paired dsDNA fragments. The fragment shown at the bottom of the figure was created by a forward primer, shaded in *red*, binding to the bottom strand of the dsDNA pair shown in the previous step.

shock. The competent *E. coli*, stored frozen but allowed to thaw very slowly on ice in the presence of the transforming plasmids, will be exposed to a temperature of 42°C for 30 to 45 seconds, then returned to the near 0°C environment used for thawing. Ca^{2+} has been shown to be essential for the process because it helps to protect the plasmid DNA from nucleases. Most of the transforming DNA will enter the competent cells during the incubation period following the heat shock.

Electroporation involves the delivery of an electrical current across and through the target cells. Pores will spontaneously open in the cell membranes, and DNA will flow through in the same direction as the electrons. When the current is removed, the pores will spontaneously close, and any DNA in the cytoplasm (or the nucleus, when this technique is used for gene delivery to eukaryotic cells) will be trapped inside the cell.

Once transformed, the *E. coli* are grown in suspension for at least 20 minutes (the time taken for one *E. coli* generation) in a batch reactor. The suspension will consist of cell medium that contains all the nutrients necessary for bacterial growth, typically Luria-Bertani broth (LB; see section 11.3.2), which contains tryptone, yeast extract, and NaCl at isotonic concentrations. Note that no antibiotic is present at this point.

Following this incubation, a sample of the cell suspension is streaked onto plates containing LB agar plus another component that is crucial to permit selection for our transformed cells: an antibiotic. Recall that our plasmids have an antibiotic resistance marker; suppose this marker codes for resistance to kanamycin. Typical *E. coli* cells will die in the presence of kanamycin. However, if one of the cells has been transformed by our engineered plasmid, it will express the antibiotic resistance gene and be able to survive in the kanamycin-laced agar medium. Remember we noted that no antibiotic is present during the first 20 to 40 minutes after transformation; this is to allow the transformed cells enough time to express (transcribe and translate) their new bacterial antibiotic resistance gene. After the kanamycin-laced agar medium has been streaked with the transformed culture, it is incubated overnight to allow each of the transformed *E. coli* to proliferate to form individual colonies. Each colony arises from a single parental bacterium, so all the cells in a colony contain many copies of the same plasmid.

After colonies have grown on the LB-antibiotic agar, small samples from individual colonies can be grown up in a small suspension (~3 mL) of LB plus antibiotic for extraction of plasmid DNA for further analysis. After one or more of the colonies has been verified to contain the correct plasmid, cells from the colony will be permitted to grow in greater volumes of medium (100 mL or more; perhaps liters in the research laboratory, perhaps thousands of liters on the commercial scale) for a mass amplification of the engineered plasmid. Plasmid amplification is the subject of the next section.

11.3.1 Plasmid amplification

When we want to obtain many copies of a plasmid, we harness the power of rapid bacterial replication to perform the work for us. Because the plasmids we engineer will contain a bacterial origin of replication, after we transform the *E. coli* cells with them, the bacteria will make copies of the plasmids just as they make copies of their own genomes (which also have origins of replication). As the bacteria proliferate, more copies of the plasmids will also be created. This is the process of amplification.

How do we extract our plasmids from the *E. coli* cells after amplification? The process will be illustrated here by a classic extraction protocol. Although there are now kits that skip a couple of the following steps and combine a couple of others, a classic protocol is presented here because it has many good chemical principles worthy of consideration. Many people think of the plasmid preparation protocol as a biological technique, but it can also be considered a separation process from a strictly chemical standpoint. We will have to extract our plasmids from the bacteria, and we will do so by lysing the microorganisms. After the *E. coli* cells have been split open, we will first be faced with the problem of separating something hydrophilic from something hydrophobic. We will later be faced with the separation of proteins from nucleotides, as well as the separation of plasmid DNA from RNAs and genomic DNA.

We begin the process with a petri dish loaded up with bacterial food (LB broth, discussed later) mixed with agar and antibiotic. Agar is very similar to Jell-O in consistency. We smear the LB/agar with our bacterial sample, which, after the transformation process, includes both transformed and untransformed *E. coli*. We then place the dishes in a 37°C oven (the same as normal human body temperature) for an overnight incubation, during which the cells will divide approximately every 20 minutes. As the cells continue to divide, they must stay in contact with the food source (the LB/agar), so they will populate an ever-expanding circular region around the initial cell. This cell-containing disk is referred to as a **colony** and will be perhaps 1 to 2 mm in diameter after the overnight incubation (Figure 11.11). As long as the colonies do not overlap on the plate, it is assumed that each originated from a single cell. It is also assumed (although not guaranteed with 100% statistical certainty) that each of these original cells took up only one plasmid during the transformation process. In other words, it is assumed that each colony represents a relatively pure sample of a single plasmid.

Faced with several colonies on a single agar plate, we must determine which ones contain our plasmid in the form that interests us. One way to do this is to take a small sample of bacteria from an individual colony and analyze their DNA on a very small scale. The sample can be taken by "picking" the colony. This entails inserting a sterile toothpick (or pipette tip) into the center of the colony; upon removal, some of the bacterial cells will adhere to the toothpick. By swishing the toothpick around in a tube containing LB medium, we will dislodge some of the bacteria and create a new culture, this time in suspension. After an additional incubation of this new culture for several hours at 37°C, we will end up with a relatively dense suspension of roughly identical bacteria.

So how do we know the bacteria in the culture have the right plasmid? The first method to ensure this is to use a **selective medium**—a medium in which everything except the cells of interest will die.

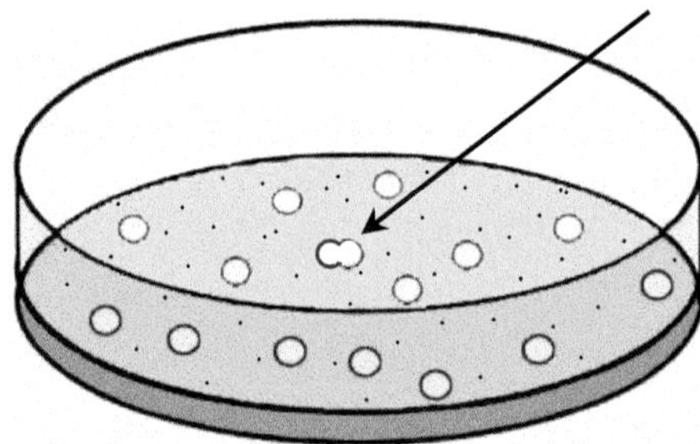

FIGURE 11.11

Luria-Bertani broth agar with bacterial colonies growing. The *black dots* represent untransformed *Escherichia coli*, which are dead because they did not have resistance to the antibiotic that was mixed in with the medium. Note the doublet caused by the overlapping of two colonies *(arrow)*. If this plate is intended for separation of individual bacteria for further study, doublets should be avoided because the origin of each individual bacterium cannot be guaranteed.

In the present experiment, this will entail the use of the antibiotic kanamycin. The LB medium will contain it, as will the LB agar. The only bacterial cells able to survive will be the ones expressing the bacterial antibiotic resistance gene that resides in our engineered plasmid. Cell selection is not restricted to bacterial experiments, nor is it restricted to antibiotics.

Suppose one cell did not take up a plasmid during the transformation process. It will die in the selective medium because it has no resistance to the antibiotic used. Its contents, including its DNA, will still be on the agar plate, but when the time comes to pick colonies, we will not see this bacterium because it will not have proliferated (because it is dead).

The right thing to do is to check many of the colonies growing on the plate because some colonies might contain an unwanted variation of the plasmid we tried to construct. Any colonies that touch each other should not be used because there is no guarantee that, in picking one side of the doublet, we will not take some of the bacteria from the other side; this would mean we had nonidentical microbes in our next culture. After an overnight incubation the colonies will still be rather small, so having colonies that touch should not be a prevalent problem unless we have so many colonies that they must touch each other for spatial reasons. If that ever happens, the researcher may have reason to be quite happy because the transformation and amplification procedure has gone better than expected. In such case, a new plate should be streaked with a lower density of bacteria. A desirable yield for this experiment is 10 to 100 colonies.

Why wouldn't all of the colonies have identical plasmids, given that they were able to grow on the selective medium? Most of them will, but the mutation rate is fairly high in *E. coli*. If a mutation occurred in a single cell soon after plating, it would be propagated into all of the progeny, yielding a colony that contained a very high percentage of mutated plasmids. It is also possible in some situations (especially when blunt-end ligation is used) that the vector reseals on itself without an insert. Such an "empty plasmid" could still survive and multiply in the selective medium because the antibiotic resistance gene typically resides in the vector. It is also possible in blunt-end ligations that the insert is oriented backward. Plasmids created by sticky-end ligation have the possibility of housing multiple inserts. For these reasons, it is necessary to pick several colonies and examine the plasmids from each. One of the common small-scale procedures for doing this—known as a plasmid miniprep—is relatively quick and straightforward to perform. The plasmid prep procedure that will be discussed in detail in the next section is a larger scale of the miniprep procedure.

Each colony picked will be grown in its own tube that contains about 3 mL of LB plus antibiotic (Figure 11.12). Several hours later the bacterial density will be high enough to visualize the microorganisms *en masse*. Each tube is then split into two samples: one that will be frozen (with 40%–50% glycerol to prevent cell lysis from the expansion of water during freezing) and one from which plasmid DNA will be extracted and analyzed. While we keep the first tube in the freezer, let us turn to the contents of the second tube. We are going to perform a restriction digest in the way we have already discussed, to help verify that we have isolated the plasmid we expect. The selection of restriction enzymes will be key in this step. It is important to verify not only the size of the isolated plasmids but also some predicted identity.

Earlier we discussed some of the controls used for the restriction cut experiment. Now that blunt-end ligations have been discussed, we should revisit the subject of controls for the restriction digest experiment. Special problems can occur with blunt-end ligations: the insert can be put in backward, the vector can be sealed with no insert at all, or multiple vectors or vector/inserts can be ligated into a single large plasmid. If we were to simply rely on the control that cut the plasmid once inside the vector, we would not

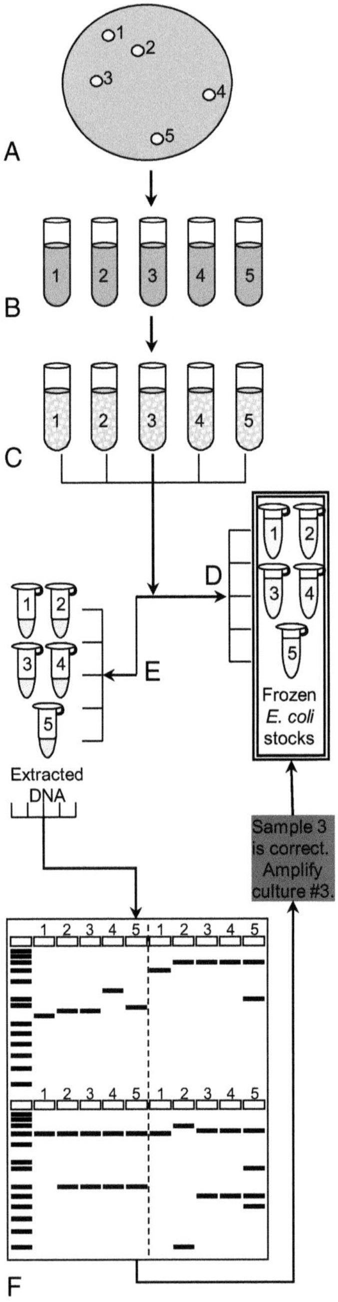

A
B
C
D
E
F
Frozen *E. coli* stocks
Extracted DNA
Sample 3 is correct. Amplify culture #3.

FIGURE 11.12

Logical progression to deciding which colony is appropriate for further amplification. (A) Colonies of transformed bacteria are picked and (B) placed in individual tubes with culture medium, then (C) incubated. (D) After incubation a small amount of each culture is frozen. (E) The rest of the cultures undergo DNA extraction. (F) Extracted DNA is analyzed via restriction digests and gel electrophoresis. If and when a suitable DNA sample is identified, the corresponding frozen culture (in D) will be thawed and amplified for a larger scale extraction.

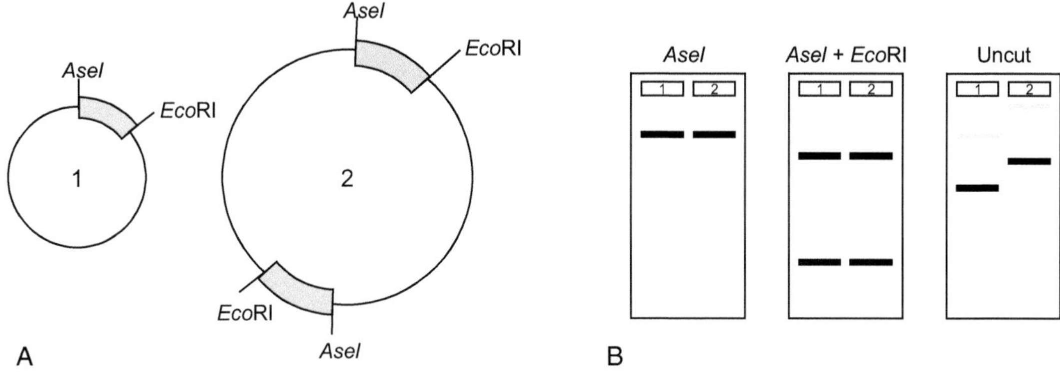

FIGURE 11.13

(A) Plasmid 1 represents a typical vector with a single insert *(blue)*. Plasmid 2 represents a ligation of two (vector + insert) units. The inserts are bounded by recognition sequences for *Ase*I and *Eco*RI, respectively. (B) Routine restriction analysis can be misleading. The first gel shows the results of cutting plasmids 1 and 2 with the enzyme *Ase*I in an attempt to produce a linear version of each vector/insert. However, this technique will not reveal the full picture, because plasmid 2 has two *Ase*I cut sites. Cutting the inserts out of each plasmid (second gel) will likewise obscure the truth because the insert and vector sequences are identical for both plasmids. In this example, only running uncut plasmid samples (third gel) will give a clue as to what is going on. (Note: Sometimes in uncut samples, there will be linear DNA sequences due to plasmids being opened by shear forces. This effect is shown by faint gray bands in the third gel.)

be able to discern instances when two vector/inserts were ligated into a single plasmid (Figure 11.13). However, running our ligation products with the uncut control will reveal these aberrant plasmids, which will be visualized by the supercoiled band running more slowly than usual. We will be picking multiple colonies for DNA analysis and will be able to detect a slow-running supercoiled band because this is a very rare occurrence; most, if not all, of the plasmids from the other colonies will be of the appropriate size. To detect the direction of the insert, we simply run a digest with one enzyme that cuts inside the insert and another that cuts outside it, preferably close to one of the ends (refer back to Figure 11.7).

Sticky-end ligations can yield plasmids with multiple copies of the insert because all sticky ends made by the same enzyme will line up (Figure 11.14A and B). (This statement is made under the assumption that the enzyme is a type II restriction endonuclease with a precisely defined, continuous, symmetric recognition sequence—no wildcard bases or spacers—and that it cuts within its own recognition sequence. These qualifiers are true for the most commonly used restriction endonucleases and for all the sticky end producers named in this text.) Although the insertion of multiple copies is not a highly probable event, it must be guarded against. Notice that a restriction cut to verify proper insertion for the desired plasmid will yield the same results as the same restriction digest performed upon a plasmid with multiple copies of the insert (Figure 11.14C, D [*left*]). This is where the control sample containing uncut plasmid may again come in handy. The migration distance of the supercoiled plasmid will indicate how many vector/insert copies are in the given plasmid. However, if insert sizes are relatively small, cutting the plasmid with only one enzyme would be a better approach for determining the presence of multiple copies of the insert (Figure 11.14D [*right*]).

Our focus on inferring plasmid identity via restriction analysis begs the question, "How do we know that the vector or insert has not gained a mutation?" Although restriction analysis gives us a rough idea

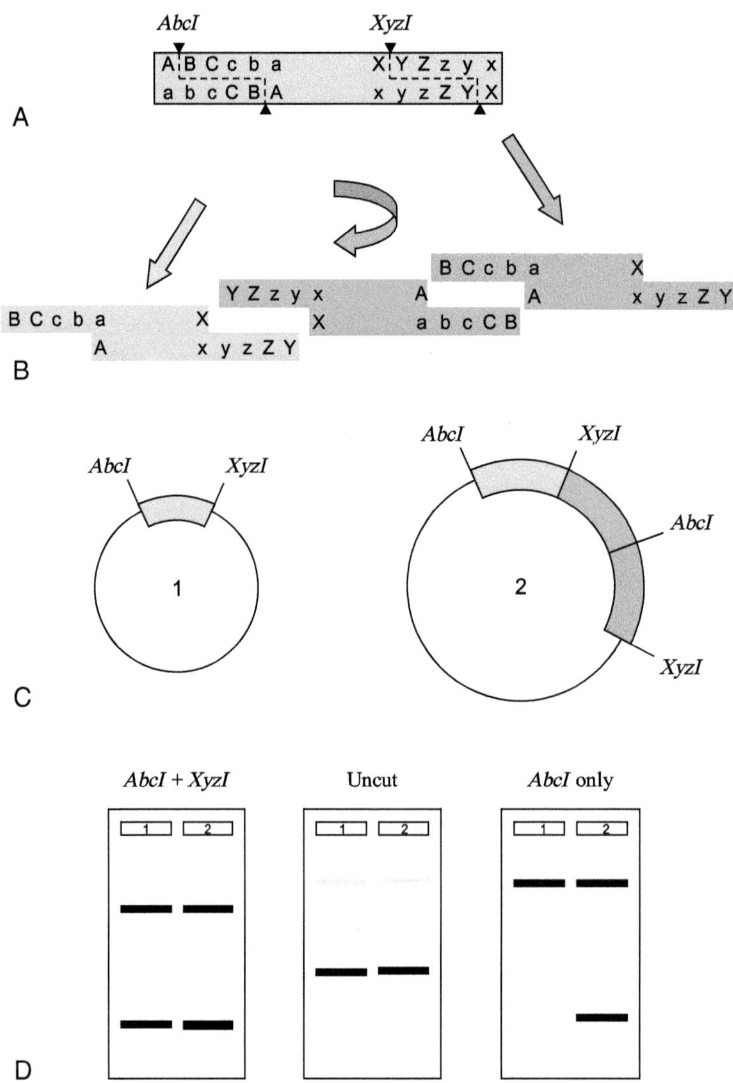

FIGURE 11.14

Detecting multiple sticky inserts. Complementary base pairs are reflected by matching upper- and lower-case letters. (A) Consider a stretch of DNA with recognition sequences for the two hypothetical enzymes shown. (B) Cutting several plasmids will yield several sticky-ended fragments. Note that the middle *(orange)* fragment is the same as the other two but rotated 180 degrees. (C) Although most open vectors will receive a single sticky-ended insert (plasmid 1), there is a possibility that some vectors will receive multiple inserts (as depicted in plasmid 2). (D) It is not always easy to identify a case of multiple inserts. Cutting out the fragment as usual using enzymes that recognize the bounding restriction sites will yield ambiguous results, as shown by the gel on the *left.* Comparing uncut versions of the plasmids will, in theory, show the difference in insert sizes, but if the inserts are relatively small, the difference may go unnoticed *(middle gel).* Cutting at only one of the bounding restriction sites will make the difference between the plasmids clear *(right-hand gel).*

as to the identity of a plasmid, in reality it does little more than verifying the existence of recognition sequences and measuring the relative distance between them. Other methods must therefore be employed if a more rigorous analysis of plasmid identity is desired. DNA sequencing is a fairly reliable way to verify plasmid identity, base by base, and this technique will be employed after the final amplification in our process.

At this point, we know to test for empty vectors, multiple inserts, and insert orientation. We should be able to identify which of our picked colonies are most likely to contain plasmids conforming to our original design. Refer to Figure 11.12, especially part F. The identity of each plasmid represented in the gel will not be revealed here—it's the subject of one of the problems at the end of the chapter—but suffice it to say that sample #2 is the correct one. Each DNA sample is paired with a specific *E. coli* sample that we have stored in glycerol in the freezer. It is now a simple matter to expand the correct culture (#2) by thawing the frozen vial, placing an aliquot of the bacteria into a relatively large volume of LB containing the selection antibiotic, and incubating. What to do with the frozen cultures that contain undesirable forms of our plasmid? We will simply kill the bacteria with one of the disinfection/sterilization methods covered in Chapter 6 (bleach is common) and safely discard them to preserve freezer space.

To expand the once-frozen culture of *E. coli* we just determined to be correct, we will take a sample of it, measured in microliters, and place it into a flask containing LB with a 1000-fold greater volume and the selection antibiotic. For example, a 300-μL aliquot of the *E. coli* could be transferred into 300 mL of LB containing the selection antibiotic. In the laboratory, the volume of LB might be anywhere from 50 to 2000 mL; on the industrial scale, volumes can be much larger. The culture will be incubated for several hours (usually 12–16 h in the laboratory) at 37°C. The cultures will be swirled, stirred, or bubbled to expose the bacterial cells to ample oxygen.

11.3.2 The plasmid prep procedure: the 12-step program for plasmid recovery

The small-scale procedure just described is an application of the plasmid prep procedure. The second, larger-scale procedure will involve the same basic steps. We will now examine a classic version of the procedure in greater detail.

1. Establish a culture This may involve going back to a frozen stock of bacteria (as in Figure 11.12), or repicking a colony that has been shown to contain the correct version of the engineered plasmid. The culture is expanded in a volume of medium appropriate for the amount of plasmid we will need in future experiments. After establishing the expanded *E. coli* culture that contains our engineered plasmid, we must perform several separations to isolate the plasmid. Consider that, at this point, the *E. coli* culture contains cells, cell metabolites, and medium (LB broth).

"LB" is an acronym given to Luria-Bertani broth, a bacterial culture medium whose recipe is attributed to Salvadore Luria and his research associate Giuseppe Bertani. LB broth contains water, sodium chloride, tryptone, and yeast extract. Water is present as a diluent because it is the solvent of life. Sodium chloride is used to create an isotonic solution (Box 11.1). Tryptone, derived from a digest of casein (a protein in cow's milk), provides amino acids to serve as building blocks for protein synthesis and as a carbon source for the bacteria. Yeast extract provides nutrients and growth factors to enable cell growth.

The separation of our engineered plasmid from everything else in the flask is performed stepwise, each step addressing a distinct component of the mixture. The process begins with a modified alkaline lysis, followed by plasmid DNA being separated via an anion exchange column, then elution in a high-salt buffer, and it ends with salt removal via alcohol.

Box 11.1 Osmolarity

Isotonic is a descriptive term that refers to the osmolarity of a solution relative to a cell. It indicates that there is no osmotic gradient between the inside of the cell and the solution in which it resides. A **hypertonic** solution is a solution with a higher osmolarity than the interior of the cell. Cells in a hypertonic solution will lose water and crenate (shrink) as a thermodynamic result of the osmotic gradient being driven toward zero. A **hypotonic** solution has lower osmolarity relative to the interior of the cell. This time, water will rush into the cell, causing it to swell and even lyse (burst) in extreme conditions.

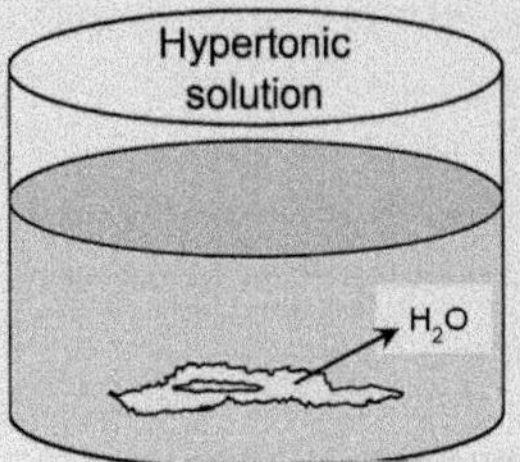

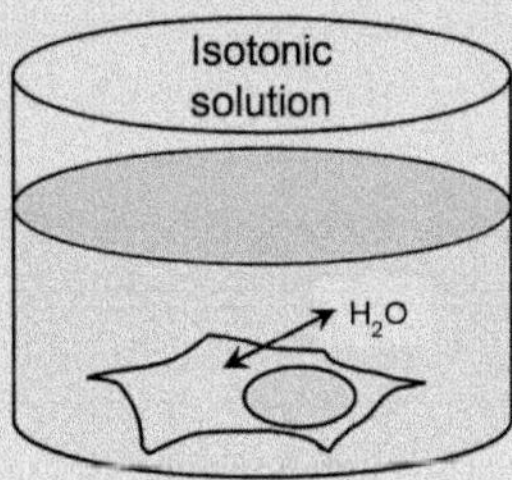

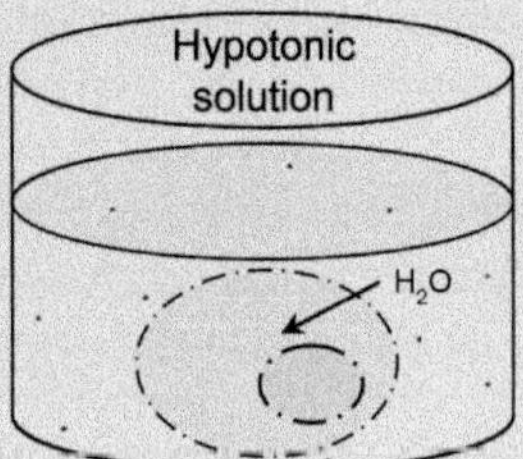

2. *Remove materials surrounding the cells* This step is straightforward: there are no plasmids outside the cells, so separation of the plasmids will include the removal of all extracellular salts, proteins, waste products, and so on. We might consider evaporation for this step, but that would remove only water from outside (and subsequently inside) the cells; salts, proteins, and other molecules from the medium would still remain around the cells. We might consider letting the bacteria grow until the medium became exhausted—but then the cells would be surrounded by waste, and separation would still be required. One could filter the cells: the medium would flow through the pores of the filter while the bacteria would be left behind, awaiting washing and removal from the filter. Although this is a valid possibility, it is relatively expensive.

The method usually employed is centrifugation, which capitalizes upon the difference in density between cells and the surrounding medium. Relative centrifugal force (RCF) is described by:

$$RCF = r \cdot \frac{\omega^2}{G} = r \cdot \frac{(2\pi N)^2}{G} = 1.119 \times 10^{-5} rN^2$$

where ω = angular velocity,
G = gravitational constant of Earth (9.8 m/sec^2 = 3.528 × 10^6 cm/min^2),
r = radius of rotation (distance, in centimeters, from the center of the rotor to the sample), and
N = number of revolutions per minute (RPM).

Practically, what we need to decide upon is the RPM. If a protocol requires a centrifugal force of, say, 6000 × *g*, we need to know how fast to set the centrifuge. This can be determined by:

$$N = \sqrt{\frac{8.937 \times 10^4 RCF}{r}}$$

After centrifugation, we will still have a bacterial soup, but all the chicken and noodles will be in a wad at the bottom of the bowl, with the liquid above them. We can then simply pour off the broth; in

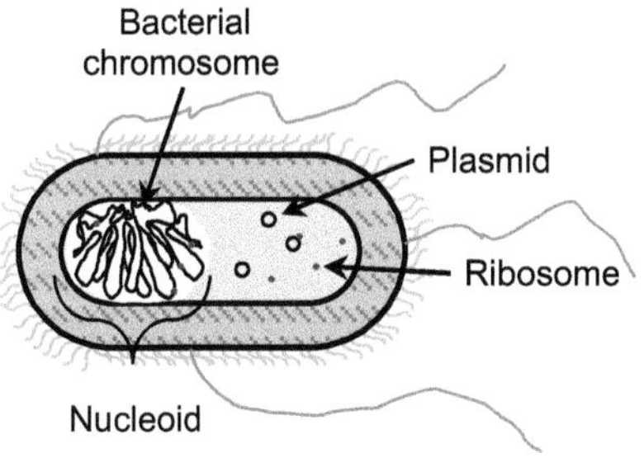

FIGURE 11.15

Location of DNA in the bacterial cell. The chromosome, which is attached to the inner cell membrane, is located in a non–membrane-bound area of the cytoplasm known as the nucleoid.

laboratory terms, we discard the supernatant. With this simple procedure, we have accomplished the removal of the salts, proteins, and wastes from the outside of the cells; we are now left with (virtually) only cells.

3. Resuspend the cells The sample of isolated cells includes our engineered plasmids, plasma membranes, cell walls, and cytoplasm that contains proteins (including enzymes and ribosomes) and mRNA. There is no nucleus to consider because *E. coli* are prokaryotes, which also means their genomic DNA is circular and is attached to the plasma membrane at one or several points (Figure 11.15). Although separating our plasmids from most other cellular components is relatively easy because of differences in size, charge, and hydrophobicity, some conceptual finesse is required to separate plasmid DNA from genomic DNA and RNA. We will need access to the guts of the cells, which means we must lyse them. This is the modified alkaline lysis portion of the protocol, in which cells are split open using a combination of detergent and high pH.

First the cell pellet is resuspended. Prior to the initial centrifugation, we had cells bathed in a soup containing salts, proteins, wastes, and so on, but after centrifugation, the cells are resuspended in a controlled solution of Tris(hydroxymethyl)aminomethane (Tris), EDTA, and RNase A. As we discuss this protocol, the compositions of several buffers or additives will be presented. The point is not to have you memorize the compositions of solutions but to give you an understanding of why certain constituents are used. Tris is a pH buffer. EDTA, introduced in Chapter 9, is a chelator of divalent cations such as Ca^{2+}. RNase A is an enzyme that catalyzes the hydrolysis of RNA. It is important because it will degrade the RNA (including mRNA) present in the cell lysate.

Perhaps the most difficult components to separate from our plasmids in the cell lysate are RNA and genomic DNA because of their chemical similarities as polynucleotides. The first to be removed will be the RNA. This separation can be accomplished because RNA is chemically distinct from both plasmid and genomic DNA in two key aspects: it is single–stranded, and it contains a hydroxyl on carbon 2 of each ribose in the backbone. RNase A acts upon single-stranded RNA in a two-step mechanism that hydrolyzes the molecule at phosphodiester bonds by temporarily having the polymerizing phosphate attached simultaneously to the 3′ and 2′ carbons (using the oxygen atom on the 2′ hydroxyl). This cannot be accomplished by the enzyme on DNA because DNA lacks an oxygen atom on its 2′ carbon (Figure 11.16).

4. Lyse the cells The cells are lysed with 200-mM NaOH and sodium dodecyl sulfate (SDS). The sodium hydroxide raises the pH significantly, resulting in the denaturation of proteins. The SDS, being

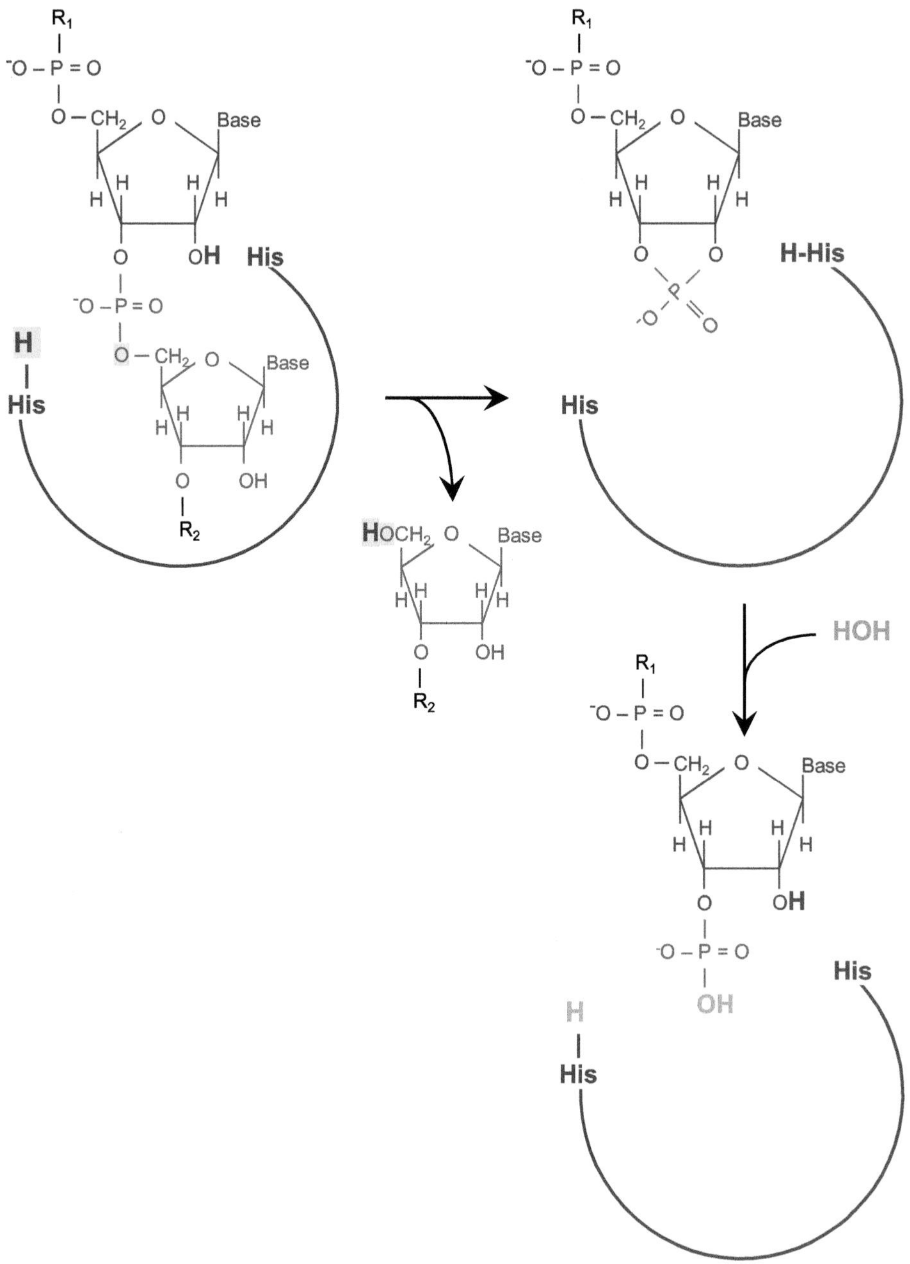

FIGURE 11.16

The mechanism of RNase A. Successive ribonucleotides of the RNA molecule are shown in *blue* and *red* for clarity; the enzyme is shown in *purple*. Two specific histidines (His) in the enzyme act to grab or donate protons, destabilizing the connecting phosphodiester linkage. Notice that the intermediate conformation requires the 2′ oxygen in the *(blue)* ribonucleotide to be possible. In DNA, this oxygen atom is missing, which explains the specificity of RNase A for RNA and protects the integrity of our engineered plasmids from degradation. A water molecule *(green)* restores the enzyme to its original state.

an ionic detergent, both disrupts cell membranes and binds to the denatured proteins (see Chapter 3). The proteins will be coated with SDS to give them a consistent, anionic charge, rendering them soluble in aqueous solutions. Notice that the genomic DNA that is shown attached to the plasma membrane in Figure 11.15 will still be attached to the plasma membrane after cell lysis; thus if we remove the plasma membrane from our solution at this point, we will also remove the genomic DNA. This brings us to the next step in the process: precipitating the plasma membranes and genomic DNA.

When the cells are lysed, the incubation is only allowed to proceed for 5 minutes. Letting it proceed for 5 hours would mean that some of the chromosomal DNA would be hydrolyzed, which would prevent removal of the entire genome during the next step. Some of the plasmid DNA would also be hydrolyzed, which would defeat the purpose of this entire procedure.

5. Precipitate the plasma membrane and genomic DNA The high-pH cell lysate solution is neutralized with 3 M potassium acetate (KAc). KAc dissociates in aqueous solution into K^+ and Ac^-, yielding 3 M of potassium ions.

Recall that in the previous step we used sodium dodecyl sulfate, which dissociates into Na^+ and dodecyl sulfate (DS^-) ions. When the latter are combined with an excess of potassium ions—assured here by the high concentration of KAc—potassium dodecyl sulfate (KDS) will be formed. KDS will precipitate out of the solution, effectively clearing it of detergent.

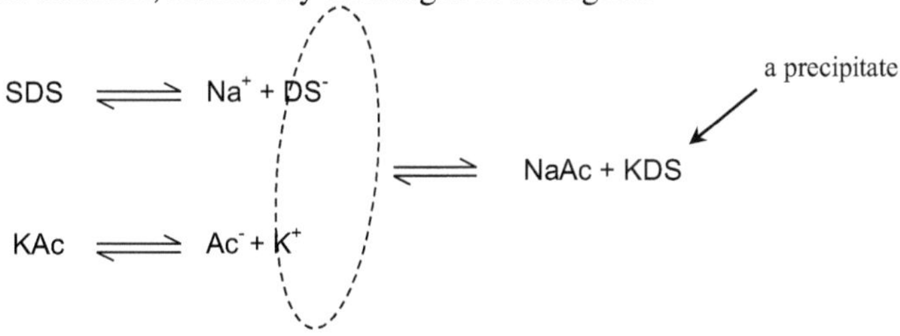

During the precipitation, the above reactions will be drawn to the right, and cellular components that were associated with the hydrophobic portion of SDS will be dragged along as the KDS precipitates. These components will include the plasma membrane—composed mainly of phospholipids—and proteins. Because of the density of the phospholipids, the precipitate will form *upward* out of solution. Also recall that in prokaryotes, organization of the circular genome is maintained through its attachment to the plasma membrane (Figure 11.15); thus, when the plasma membrane is precipitated out of solution, genomic DNA and membrane proteins will also be pulled into the precipitate.

6. Remove membranes, proteins, and genomic DNA At this point, the three types of nucleotides have either been hydrolyzed or separated. We degraded RNAs with RNase A, and we separated out the genomic DNA via KDS precipitation. The bulk solution now contains a mixture of our plasmid, degraded RNA, small proteins, and some salts, in a slightly cloudy solution topped by a fraction that looks like little chunks of Crisco and fried egg whites. The two layers can be separated by pushing the bottom fraction of the solution through a filter into a collection tube. The filtrate will contain our plasmid, degraded RNA, small proteins, and some salts.

7. Remove endotoxins E. coli are gram-negative bacteria. As such, they have a membrane on the exterior side of the cell wall that contains phospholipids, proteins, and lipoproteins but is mainly composed of **lipopolysaccharides** (LPS). Another name for LPS is **endotoxin**. Endotoxins can elicit an inflammatory response in animals, and they can activate the alternate pathway of the complement cascade (not covered here). LPS molecules will almost certainly have been washed through the filter during the previous step of this protocol. If the engineered plasmids will ultimately be used in an animal or clinical application, it is important to remove the endotoxins from the DNA solution to help prevent an adverse response when the DNA is delivered.

There are several different methods for lowering endotoxin levels in pharmaceutical preparations. However, the existence of so many different methods indicates there is still a problem with endotoxin removal. Although endotoxins will most likely not be completely removed from our solution, the quantity can at least be lowered to nonbioactive levels. A method commonly used in DNA purification procedures involves the use of a *nonionic* surfactant such as Triton X-114. Surfactants, as discussed in Chapter 3, are "surface-active" agents, with both a hydrophilic and hydrophobic region in each molecule. As such, they can be used to form micelles around endotoxin molecules to facilitate their removal from our bulk solution.

8. Isolate the plasmid via anion exchange Imagine we have a column packed with beads that carry positive charges. If we surround the beads with a solution that contains water and a salt like NaCl, the salt will split into Na^+ and Cl^- in the aqueous solution. Because of the positive charges on the beads, the Cl^- ions will adhere to them. Now suppose that we pour a solution containing plasmid DNA into the column. The large, polyanionic DNA molecules will displace many of the Cl^- ions from the beads. In other words, the Cl^- *anions* will be *exchanged* for DNA in the *resin*, hence the name "anion exchange resin." (Cationic exchange resins also exist.) In many of the popular plasmid preparation kits, the anion exchange column contains a filter-like material that carries multiple positive charges.

The product from step 7, containing plasmids, salts, small proteins (including RNase A), degraded RNA, and micelles housing endotoxins, is allowed to flow through the column. Most of the proteins, cations from salt dissociation, and endotoxin-containing micelles will flow through completely because they do not contain the correct charge for adhering to the anion exchange surface; it should now be clear why a nonionic detergent was used for the removal of endotoxins. The plasmid DNA will bind to the column preferentially over small molecules like individual ribonucleotides or Cl^- because of its size (high number of negative charges) and charge density.

9. Wash the column The column now contains plasmid DNA plus some small molecules that we do not want to remain in the final isolate. To remove many of these unwanted molecules, we will wash the column twice with column volumes of a solution containing a pH buffer, NaCl, and isopropyl alcohol. The buffer is important because it helps to maintain the negative charge on the plasmid DNA molecules, which in turn maintains their adherence to the positively charged column. The salt, at 1 M in this step, helps to wash out any residual proteins and ribonucleic acids. The isopropanol, at a concentration of 15%, is used to condense the DNA and to prevent nonspecific binding of molecules to the cationic surface of the column membrane. The overall result of this step is that the salt and protein concentrations in the adhered, DNA-containing fraction are greatly reduced.

10. Elute the DNA At this point the column fraction contains our plasmid DNA and little else. We must now collect the DNA by getting it to separate from the beads in a process called **elution**, accomplished using a high-salt solution. Just as we were able to displace Cl^- from the column with plasmid DNA, we can displace the plasmid DNA by washing it with a high concentration of Cl^-. The elution solution contains a buffer (at pH 8.5, up from pH 7.0 in the previous wash step), NaCl at a concentration of 1.25 to 1.60 M (considered a high-salt concentration, and up from 1.0 M in the previous step), and 15% isopropanol.

11. Remove the salts via isopropanol Our working solution now primarily contains plasmid DNA, NaCl, and isopropanol in aqueous buffer. Desalting is relatively easy via the addition of more isopropanol to condense the plasmids further. The condensed DNA is then separated by centrifugation. The

RCF required for separating plasmid DNA from isopropanol is much higher than that used in step 2 of this protocol, at 15,000 × *G* (or more). The pellet will contain primarily plasmid DNA with some residual salt. Most of the salt and isopropanol can simply be decanted from the tube after centrifugation.

12. Remove the residual isopropanol via ethanol The plasmid-containing pellet is now washed with 70% ethanol; the DNA remains condensed in the ethanol and can therefore be pelleted again by centrifugation. The purpose of this step is to remove the remainder of the isopropanol and any salt. After centrifugation, the supernatant can be decanted, leaving the pellet of plasmid DNA, a minute quantity of ethanol and water, and virtually no isopropanol or salt. The advantage of washing with ethanol as opposed to isopropanol in this step is that ethanol is more volatile and therefore more readily removed by evaporation.

The 70% ethanol used in this step contains 30% water. Because ethanol is more volatile than water, the percentage of ethanol around the pellet will decrease during the evaporation step, eventually nearing or reaching 0%, but some residual water will be left behind. It's important never to let the DNA dry out completely because it is relatively difficult to resuspend without inducing physical damage.

With our plasmids now isolated in a relatively pure form, it is a simple task to resuspend the DNA in a suitable buffer. Although one could resuspend the plasmid pellet in pure water, the task will be easier if a buffer such as Tris (pH 8) is used because the slightly alkaline pH will ensure the DNA carries multiple negative charges, making it more soluble in aqueous solution.

11.4 Spectrophotometry

At this point in our journey into genetic engineering, we have created a plasmid, amplified it in *E. coli*, and purified it via the multistep chemical separation procedure just presented; we now have a tube with our engineered plasmid in it. Great! How much DNA, in micrograms, do we have? What volume will we need per experiment? Before we can use the DNA for its intended purpose, such as delivering it into cells, it must first be quantified. This can be accomplished using spectrophotometry.

11.4.1 Beer's law

Double-stranded DNA, like many substances in solution, will absorb light energy if the photons are of the correct wavelength. Different substances have different optima for absorbing light. For dsDNA, the optimal absorption wavelength is 260 nm (Figure 11.17). This implies that if we were to put dsDNA of an unknown concentration into a chamber and then deliver a known amount of 260-nm light, we should be able to determine the concentration of the dsDNA solution in the chamber. This calculation of **absorbance** can be performed using Beer's law:

$$A = alc,$$

$$\text{absorbance} = (\text{molar absorptivity}) \times (\text{path length}) \times (\text{concentration}).$$

Note that molar absorptivity is the same as the extinction coefficient, ε, discussed in Chapter 8.

Suppose we have a known amount of light of a specific wavelength that shines in a defined path toward a transparent cuvette that contains a solution of solvent plus a substance of interest. In the current context, imagine that the solvent is DNA resuspansion buffer (Tris plus EDTA) and the substance of interest is dsDNA (plasmids). The light passes through the solution in the cuvette and comes out the

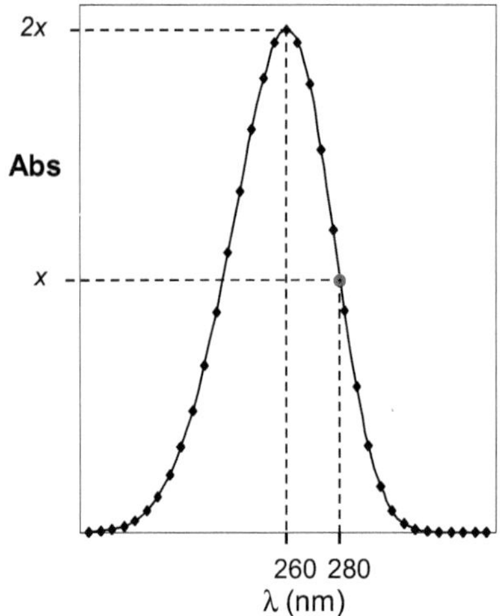

FIGURE 11.17

Absorption spectrum for dsDNA. Note that, for a given concentration of pure dsDNA, the amount of 260-nm light absorbed is approximately twice the amount of 280-nm light that is absorbed.

other side, where it then strikes a photodiode light detector. In the equation of Beer's law, molar absorptivity is a constant for the specific solute at the standard concentration, 1 M. Sometimes referred to as the molar extinction coefficient, it can be found in reference books expressed in either cm^2/mole or $M^{-1}cm^{-1}$, which are both valid because of the unit conversion 1 mL = 1 cm^3. Path length refers to how far the light must travel to get through the sample (in cm) and is represented by the width of the cuvette. Note that values obtained for absorbance are unitless.

Consider that the molar absorptivity of cupric sulfate is around 20, which means a 1 M solution of cupric sulfate, an easily seen blue liquid, will yield an absorbance of 20. For comparison, a 1-M solution of beta carotene (the substance responsible for the orange color in carrots) has an absorbance of ~100,000 (molar absorptivity ~100,000). The reason for the large difference in molar absorptivities between the two is that beta carotene is a huge molecule in comparison to cupric sulfate. One mole of these huge molecules in a liter of water will yield a comparatively opaque solution; a 1-M solution of the much smaller cupric sulfate molecules will allow much more light to pass between the molecules of solute.

We can describe the intensity of light that makes it through the cuvette when the sample molecule is present (I_s) as a function of the intensity of light that makes it through the cuvette when no sample molecule is contained in the solvent (I_0). Using a form of Beer's law:

$$I_s = I_0 \cdot 10^{-alc} \Rightarrow I_s/I_0 = 10^{-alc} \equiv T$$

T is the **transmittance** of the solution at a given solute concentration.

The transmittance is the ratio of light getting through the solution in the presence versus the absence of solute. Because intensity cannot be negative and energy cannot be created, the value for transmittance will always be between 0 and +1. Multiplying T by 100 gives the % transmission, or the percentage of light that gets through the sample in the presence of solute.

Using the above equation, we can solve for c to get:

$$c = \frac{-\log(T)}{al} = \frac{\log\left(\frac{1}{T}\right)}{al}$$

which shows that the concentration of solute has a negative log relationship with the transmittance of the solution.

If we rearrange the Beer's law equation ($A = alc$) to get $c = \frac{A}{al}$ and compare it with the relation $c = \frac{-\log(T)}{al}$ just derived, we see by inspection that $A = -\log(T)$, which means the absorbance of a solution also has a negative log relationship with the transmittance of the solution. Because of the straightforward relationship between absorbance and concentration (no logarithms involved), it should be easy to see why absorbance is typically used to determine solute concentration instead of transmittance. Spectrophotometers use light that is monochromatic, meaning it is of a single color (i.e., has a single wavelength). Values thus obtained for absorbance are written as A_λ, in which λ is the wavelength in nanometers.

11.4.2 Determination of DNA concentration

The above is a general representation of Beer's law. In applications involving dsDNA, certain facts become important. First, the maximal absorbance of dsDNA in solution occurs at a light wavelength of 260 nm. Second, the molar absorptivity of dsDNA (assuming an equal presence of the bases A, G, C, and T) is such that a 50-µg/mL solution has an $A_{260} = 1.0$. Note that this value involves unit conversions to switch from moles to µg/mL. Use of the 50-µg/mL value implies that, assuming a path length of 1 cm:

- A reading of 1.000 for A_{260} indicates a dsDNA concentration of 50 µg/mL.
- A reading of 0.500 for A_{260} indicates a dsDNA concentration of 25 µg/mL.
- A reading of 0.400 for A_{260} indicates a dsDNA concentration of 20 µg/mL.

Note that the above calculations determine the concentration of dsDNA *in the cuvette*. In almost every situation, however, the cuvette will contain a dilution of your DNA stock solution.

Suppose you have just finished purifying some plasmid DNA that was amplified in *E. coli*. Loading the entire volume of product into the spectrophotometer cuvette would make no sense: there would be no DNA left for your intended experiments! So you load a small aliquot (on the order of microliters) into the cuvette. The equation for dsDNA concentration in the cuvette must be multiplied by a dilution factor to yield the concentration of DNA in your initial stock:

$$c = (50 \cdot A_{260}) \cdot \frac{V_{tot}}{V_{sample}},$$ where V_{tot} = total volume in the cuvette (in mL), V_{sample} = volume of sample taken from stock solution (in µL).

(We are assuming the width of the cuvette, *l*, is 1 cm.)

While V_{sample} *can* have units of milliliters, microliter volumes are typically used. The use of µL for V_{sample} in the calculation will yield unit concentrations of µg/µL, which are equivalent to mg/mL. An example is provided in Box 11.2 to help make things clear.

Box 11.2 Determining DNA concentration: example

We have a cuvette that holds 600 μL. We load 598 μL of buffer into the cuvette and take a reading on the spectrophotometer at 260 nm. The reading comes out at 0.000. We then load 2 μL of our DNA stock (stored in the same buffer) into the cuvette, mix, and obtain a value of $A_{260}=0.231$. What is the concentration of our DNA stock?

Answer:

Using $c = (50 \cdot A_{260}) \cdot \dfrac{V_{tot}}{V_{sample}}$, we have

$c = (50\ \mu g/mL * 0.231) \times (0.600\ mL/2.000\ \mu L) = 3.465\ \mu g/\mu L$.

Most spectrophotometer cuvettes are 1 cm wide, which allows for the dropping of l from the Beer's law calculations. Keep in mind that not all cuvettes necessarily hold 1 mL. Smaller volumes allow the user to use a smaller amount of precious stock solution for the analysis. Decreased sample volumes are achieved by using cuvettes with thicker walls and/or shorter heights. These specialized cuvettes will not alter the equation used to determine concentrations.

The above equation is for dsDNA. When RNA is being measured, change the 50 μg/mL to 40 μg/mL (because $A_{260} = 1.0$ for RNA at a concentration of 40 μg/mL). When ssDNA is being measured, the value of the constant is debatable; 34.5 μg/mL is generally acceptable.

11.4.3 Determination of DNA purity

Knowing the concentration of DNA in our stock solution is important, but another important piece of information we can get from absorbance data is the purity of the DNA solution. Proteins represent the major contaminant in DNA preparations. It is important to be aware of how much protein is in a given plasmid solution because proteins can foul downstream procedures when we try to use our DNA. By virtue of the aromatic amino acids (especially tryptophan), we can detect the presence of proteins by assessing the absorbance of 280-nm light. The ratio of plasmid versus protein can be determined using the ratios of maximal absorptions: A_{260}/A_{280}. For dsDNA, ratios between 1.8 and 2.0 are desirable. For RNA, a ratio above 2.0 is wanted. There is an upper limit to the A_{260}/A_{280} ratio because the absorbance curve for pure DNA (or RNA) will have a nonzero value at $\lambda=280$ nm (Figure 11.17). Let's call that value x. (For pure dsDNA, $A_{280}=x\approx A_{260}/2$.) Therefore, as the concentration of protein in the plasmid solution goes to zero, the A_{280} value obtained for that solution will go to x. (For dsDNA, A_{260}/A_{280} will go to 2.) Refer to Figure 11.17 and Figure 11.18.

For plasmid preparations, typical A_{260}/A_{280} values range from 1.6 to 2.0. For quality control reasons, it is best not to use plasmid DNA preparations for which the ratio is below 1.8. In such cases, the DNA solution can be further purified using a technique called phenol extraction or by using commercially available cleanup kits. After purity has been verified and DNA concentration has been determined, it is common to dilute the stock solution to a convenient concentration such as 1 μg/μL.

11.5 Summary

We have discussed the engineering, amplification, and purification of plasmids that are used for biotechnical applications such as gene delivery. The point is to deliver the genes into a cell or organism so that the encoded proteins will be expressed. These techniques are mainstays for the biotechnologist

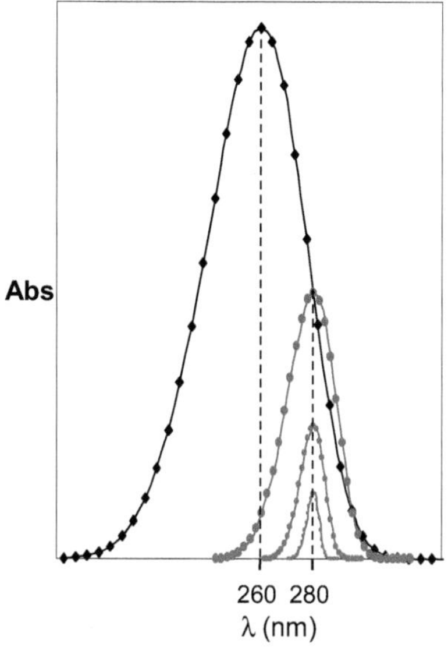

FIGURE 11.18

The absorption spectrum for DNA *(black curve)* is centered at 260 nm, and for protein *(red curves)*, it is centered at 280 nm. To determine DNA purity, A_{260}/A_{280} is calculated. Both DNA and proteins will contribute to the A_{280} value. As protein content goes to zero (pure DNA), A_{260}/A_{280} goes to 2. This is represented by the progressively lower amplitudes of the *red curves*.

who wants to harness cells for the production of specific proteins. The goal of the gene delivery application may be to have the proteins act on the expressing cell itself (e.g., having a stem cell express a growth factor that will aid in differentiation), or it may be the generation of a specific protein product that can be isolated for commercial use (e.g., recombinant insulin). Recombinant insulin is now routinely produced in such large quantities that it's available for home and clinical use. This technology has been great news for patients with type I diabetes, who require insulin injections every day. Although it may seem somewhat tedious that we have covered proteins, cells, restriction enzymes, DNA ligase, bacterial transformations, and chemical separations in such detail, it should be becoming clear that this information is important for the biotechnologist who might want to be able to construct a specific plasmid and put it into a chosen cell or cell type so that a desired product will be expressed. The product could be a drug intended for later distribution, or it could act as part of a treatment for the host cell or organism itself.

This concludes Unit II: Biotechnology in the Laboratory. We have now seen how fundamental physical, chemical, and biologic principles have been utilized to allow biotechnologists to obtain information about cells and cellular processes. In the next unit, we will see how biotechnologists are able to use these principles to manipulate cells for applications of medical, industrial, and societal importance. We will begin our look into real-world applications of biotechnology in the next chapter by examining gene delivery, the process of introducing genes into cells into cells in such a way that the genes will be expressed.

Related reading

Bergmans, H.E., van Die, I.M., Hoekstra, W.P., 1981. Transformation in *Escherichia coli*: stages in the process. J. Bacteriol. 146, 564–570.

Jacob, F., Brenner, S., Cuzin, F., 1963. On the regulation of DNA replication in bacteria. Cold Spring Harbor Symp. Quant. Biol. 28, 329–348.

Lark, K.G., 1966. Regulation of chromosome replication and segregation in bacteria. Bacteriol. Rev. 30, 3–32.

Magalhães, P.O., Lopes, A.M., Mazzola, P.G., Rangel-Yagui, C., Penna, T.C., Pessoa Jr., A., 2007. Methods of endotoxin removal from biological preparations: a review. J. Pharm. Pharmaceut. Sci. 10, 388–404.

Rosenberg, B.H., Cavalieri, L.F., 1968. Shear sensitivity of the *E. coli* genome: multiple membrane attachment points of the *E. coli* DNA. Cold Spring Harb. Symp. Quant. Biol. 33, 65–72.

Sambrook, J., Russel, D.W., 2001. Molecular Cloning, a Laboratory Manual, third ed. Cold Spring Harbor Laboratory Press, Cold Spring Harbor, NY.

Sezonov, G., Joseleau-Petit, D., D'Ari, R., 2007. *Escherichia coli* physiology in Luria-Bertani broth. J. Bacteriol. 189, 8746–8749.

Stec, B., Holtz, K.M., Kantrowitz, E.R., 2000. A revised mechanism for the alkaline phosphatase reaction involving three metal ions. J. Mol. Biol. 299, 1303–1311.

Sueoka, N., Quinn, W.G., 1968. Membrane attachment of the chromosome replication origin in *Bacillus subtilis*. Cold Spring Harb. Symp. Quant. Biol. 33, 695–705.

Takagi, Y., Warashina, M., Stec, W.J., Yoshinari, K., Taira, K., 2001. Recent advances in the elucidation of the mechanisms of action of ribozymes. Nucleic Acids Res. 29, 1815–1834.

Questions

1. How is it that CTCGAG is considered a palindrome?
2. In molecular cloning, what are the consequences of inserting an exon in the wrong orientation?
3. How do we ensure that, following transformation, only bacteria that have received an exogenous plasmid are able to replicate?
4. Commercial plasmid vectors typically contain an MCS. What is an MCS, and why is it useful?
5. What is the Klenow fragment? How is it used in biotechnology?
6. **a.** When performing blunt-end ligation, how does one prevent the vector from sealing upon itself to create an empty vector?
   b. Should the procedure in part *a* be performed upon the insert to prevent plasmids made up of only the insert? Why or why not?

Questions 7–15: the plasmid prep procedure

7. In the neutralization step, why is the molarity of potassium acetate so high?
8. Why is a nonionic detergent used to remove endotoxins?
9. At what points in the plasmid prep procedure are genomic DNA and messenger RNA separated from the plasmid DNA?
10. When is RNase A separated from the plasmid-containing fraction during the plasmid prep procedure?
11. Why is such a high concentration of salt used during the elution step?
12. Why do we resuspend the purified DNA in Tris buffer rather than water?
13. Where do we find lipopolysaccharide? Why should we be concerned with it at all?

14. How is SDS removed from the preparation following cell lysis? (Discuss the "S" as well as the "DS.")
15. Erick misread the procedure in his laboratory notebook and used Triton X-100 instead of SDS when lysing his transformed *E. coli* cells. How will this affect his plasmid purification results? Explain your answer.

Questions 16–20: analysis of plasmid DNA by spectrophotometry

16. When we analyze a DNA sample for the presence of protein contaminants, we look at its absorbance at 280 nm. What are we actually detecting?
17. Suppose you isolated some plasmid DNA and obtained the following spectrophotometric results (using the standard-sized cuvette with path length of 1 cm):

µL sample added	Total volume (mL)	A_{260}	A_{280}
0.0	0.998	0.000	0.000
2.0		0.126	0.062
4.0		0.252	0.120

a. What is the concentration of DNA in the purified sample?
b. Is the sample pure enough for use for gene delivery use? Justify your answer.

18 Consider the following spectrophotometric results obtained from plasmid DNA, purified via the 12-step procedure. In this experiment, a cuvette with a smaller total capacity is used. The cuvette has a pathlength of 0.5 cm.

µL sample added	Total volume (mL)	A_{260}	A_{280}
0.0	0.298	0.000	0.000
2.0		0.096	0.057
4.0		0.192	0.105

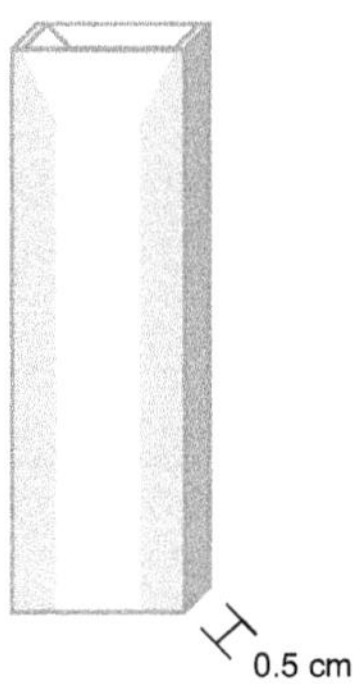

a. What is the concentration of DNA in the purified sample?
b. Is the sample pure enough for use for gene delivery use? Justify your answer.

19. Consider the following spectrophotometric results obtained from plasmid DNA, purified via the 12-step procedure (using a standard-sized cuvette with a path length of 1cm):

μL sample added	Total volume (mL)	A_{260}	A_{280}
0.0	0.598	0.000	0.000
2.0		0.053	0.037
4.0		0.094	0.056
8.0		0.181	0.101

a. What is the concentration of DNA in the purified sample? (Use the correct number of significant digits.)
b. Is the sample pure enough for use for gene delivery use? Why or why not?

20. Consider the following spectrophotometric results obtained from plasmid DNA, purified via the 12-step procedure (using a standard-sized cuvette with a path length of 1cm):

μL sample added	Total volume (mL)	A_{260}
0.0	1.000	0.000
2.0		0.037
4.0		0.086
8.0		0.164

a. What is the concentration of DNA in the purified sample? (Use the correct number of significant digits.)
b. In going from 2-μL sample to 4-μL sample added to the cuvette, we should be doubling the amount of dsDNA, yet the value for A_{260} does not double. Why not?
c. In going from 4-μL sample to 8-μL sample added to the cuvette, the value for A_{260} increased by 90.7%. Predict what value you would expect for A_{260} if an additional 8 μL of sample were added to the cuvette. Justify your prediction.
d. In figuring out the concentration of DNA in the purified sample, you would obtain a different value for each row of data. Which value is the most believable, and why?

Questions 21–29: molecular cloning experiments

21. Suppose you have a plasmid that you obtained from another laboratory, and the only enzymes you can use to excise the mammalian exon are *Asc*I and *Mlu*I. You want to insert the exon into the vector pEGFP-N1, but this vector does not have *Asc*I or *Mlu*I recognition sequences. Using enzymes (no PCR), how might you get the insert out of the donor plasmid and into your vector?
22. During the transformation procedure, why is it important to allow the transformed bacteria to grow in antibiotic-free medium for 20 minutes or so after heat shock?
23. A student in the laboratory performed a transformation experiment and then streaked a plate containing LB, agar/ampicillin with some of the resulting cells. Unfortunately, no colonies were present after sufficient incubation. Give three reasons that could explain why no colonies grew.
24. Sometimes when engineering a plasmid, empty vectors (vectors without inserts) are formed. Can *E. coli* transformed with these empty vectors grow in medium containing antibiotic? Why or why not?

25.

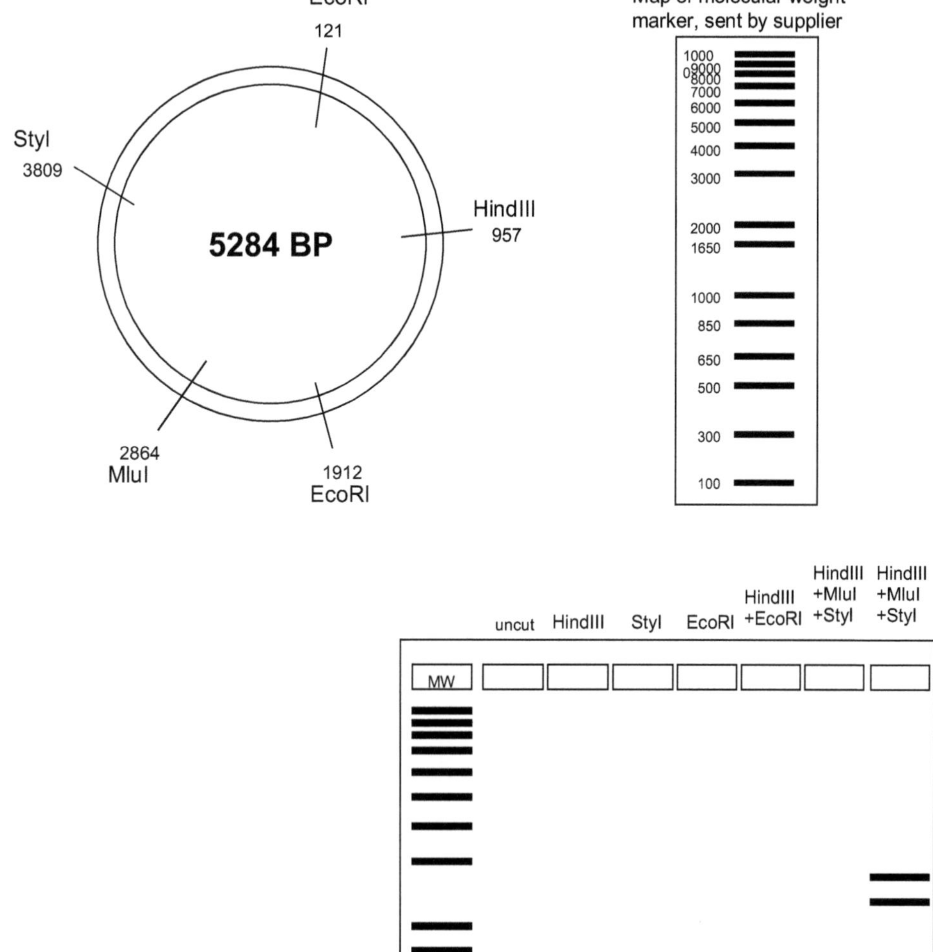

a. Copy the gel and draw in lines to predict the results of the indicated enzyme cuts. Indicate the length of each band you draw.

b. The last lane shows what you actually saw after running the gel, which differs from your prediction. Explain a reason for the difference.

c. What control could you have run to prove your hypothesis from part b?

26. Refer to the figure of a gel below. Suppose your design was to insert a 1000-bp fragment into a 4000-bp vector using *Eco*RI and *Bam*HI sticky ends: 5′–EcoRI–[Insert]–BamHI–3′. Also suppose that there is an *Ase*I cut site 800 bp into the insert. The four quadrants of the gel are upper left, uncut plasmids; upper right, plasmids cut with *Eco*RI; lower left, plasmids cut with *Eco*RI + *Bam*HI; lower right, plasmids cut with *Eco*RI + *Ase*I.

a. Identify the compositions of samples 1–3 in terms of vector and inserts (number and direction), given the banding patterns shown in the figure Sketch the plasmids with inserts, indicating cut sites.

b. Is it possible that an unpicked colony had an insert that was ligated in the reverse orientation? Why or why not?

c. Exactly how many bases make up the bands for all lanes containing sample 3?

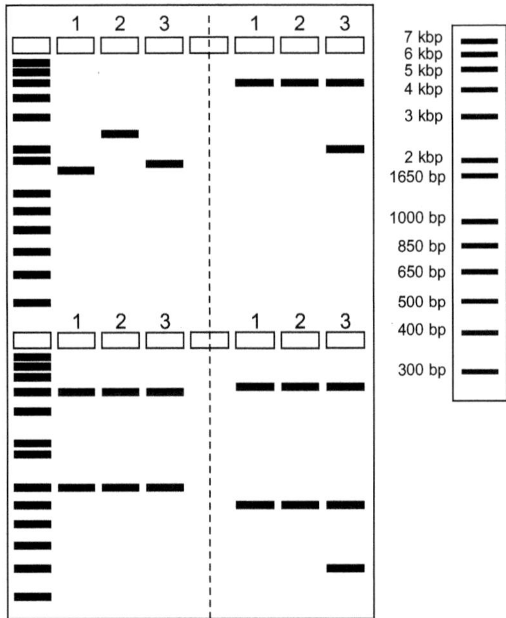

27. Refer to the figure of a gel below. Suppose your design was to insert a 1006-bp fragment into a 3994-bp vector using blunt-end ligation. Also suppose that there is an *Eco*RI cut site six bases into the insert and an *Ase*I cut site 806 bp into the insert and that neither of these enzymes cut the vector. There is also a *Bam*HI cut site in the vector 500 bp after where the insert is to be located. The four quadrants of the gel are: upper left, uncut plasmids; upper right, plasmids cut with *Eco*RI; lower left, plasmids cut with *Eco*RI + *Bam*HI; lower right, plasmids cut with *Eco*RI + *Ase*I.

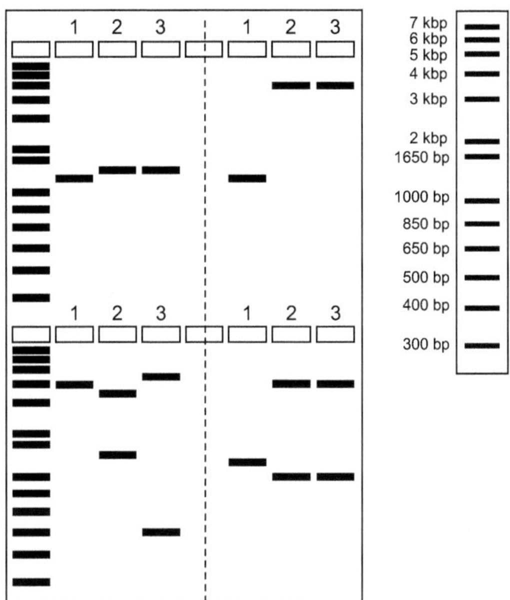

a. Identify the compositions of samples 1 to 3, given the banding patterns shown in the figure.
b. Exactly how many bases make up the bands seen in quadrants 2, 3, and 4?

28. Consider a vector which has unique cut sites in the MCS as shown.

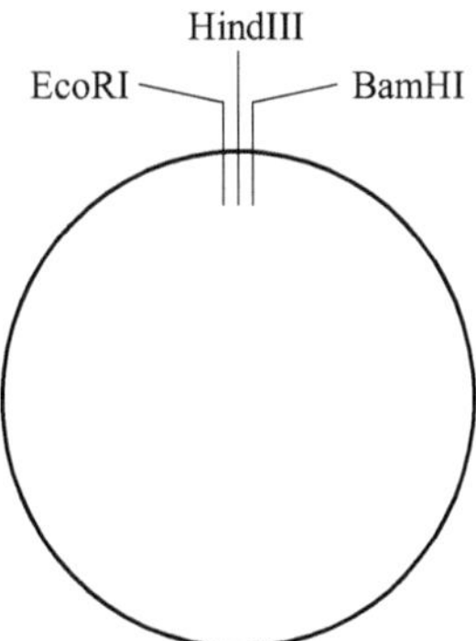

Suppose the vector is prepared for a blunt-end ligation by cutting it with *Hind*III and blunting it with T4 DNA polymerase. The insert is a blunt-ended dsDNA fragment of 800 bp, having a *Sac*II cut site 200 bp from the 5′ end (in the correct orientation).

a. In the style of questions 26 and 27, draw a gel with the following quadrants: upper left, uncut plasmids; upper right, *Eco*RI; lower left, *Eco*RI + *Bam*HI; lower right, *Eco*RI + *Sac*II. Let sample 1 = the plasmid with insert in the correct orientation; sample 2 = the plasmid with insert in the reverse orientation; and sample 3 = an empty vector. Write the number of bases over each sample band in your sketch.

29. Given the following plasmid with a 1000-bp sequence we want to cut out to use for a blunt-end ligation, how large will the insert be after cutting it out of the plasmid using *Cla*I and *Aat*II, followed by blunting with the Klenow fragment? (The recognition sequences of the two enzymes are shown below.)

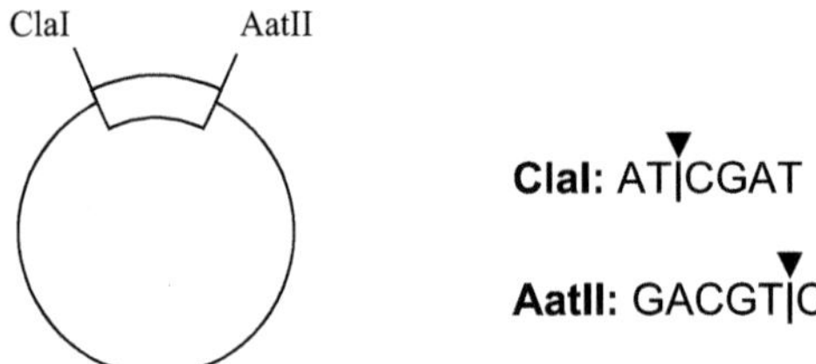

In the vector, the bases of the insert are numbered as follows ("N" = any base):

Base 1		Base 1000
A	TCGATNNNNNNNN ... NNNNNNNNGACGT	C

UNIT

Biotechnology in the real world

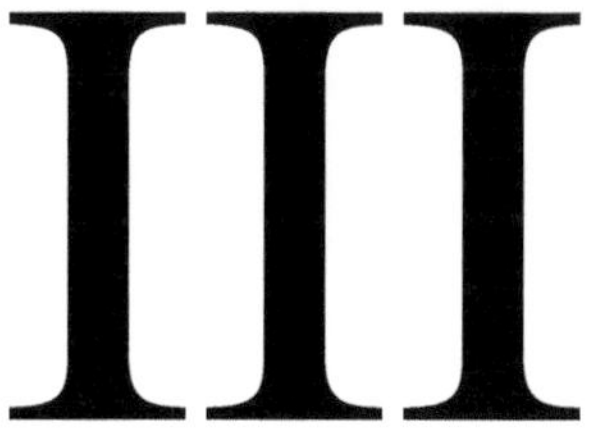

CHAPTER

Gene delivery

12

Chapter outline

Biotechnology and its Applications. https://doi.org/10.1016/B978-0-12-817726-6.00012-5

We have now discussed how to engineer a plasmid and how to amplify it so that there will be enough to use for experiments, processes, or treatments. Next, we will address the topic of gene delivery: *the delivery of genetic material (polynucleotides) into cells to alter the function of the cells*. For example, one might want to deliver DNA constructs to bacterial cells, with the goal of having the bacteria produce a product. A specific example of this is the production of recombinant insulin by *Escherichia coli* or yeast for eventual use in human patients with diabetes. Such insulin would be referred to as a **recombinant protein** because it was produced via the transcription and translation of recombinant DNA. In the laboratory, **recombinant DNA** is produced by the genetic engineering processes we have already discussed that involve *recombining* series of bases from different sources.

A very important application of recombinant DNA is **gene therapy**—the delivery of genetic material into cells for the purpose of altering cellular function. This seemingly straightforward definition encompasses a variety of situations that can sometimes seem unrelated. The delivered genetic material can be DNA or RNA. The alteration in cellular function can be an increase or decrease in the amount of a native protein that is produced or the production of a foreign protein. The delivery of the genetic material can occur directly, as is the case with microinjection, or involve carriers that interact with cell membranes or membrane-bound proteins to facilitate cellular entry. The delivered polynucleotides can be single or double stranded, and they can code for a message or not (as is the case for antisense gene delivery or miRNA, to be covered in Chapter 13). Even the location of cells at the time of gene delivery is not restricted: they can be part of a living organism, they can exist as a culture on a plate, or they can be removed from an organism, transfected, and replaced into the same or a different organism later.

Gene therapy was pioneered after the development of recombinant DNA technology and initially moved forward using viruses to deliver genes to cells in vitro. With the understanding of retroviruses and their possible use as vectors for delivery, gene therapy progressed to applications involving mammalian organisms during the 1980s. Since then, new techniques for gene delivery have been developed that utilize both viral and nonviral carriers. These techniques have worked well enough that proposed disease treatments have evolved to the clinical trial stage, with encouraging success. Although many cellular processing mechanisms remain unclear and the search for the ideal gene delivery vector is ongoing, the tools and methods for gene delivery have provided a strong base from which to build.

Gene delivery is behind many different aspects of biotechnology; many of its applications are not related to gene therapy. Applications that can involve gene delivery include:

- production of proteins via bioreactors for research or pharmaceuticals;
- production of genetically modified plants or animals to help feed the world;
- guidance of stem cell differentiation for tissue engineering or regenerative medicine; and
- modification of microbes to produce biofuels.

In this chapter, we will explore different gene delivery vehicles, the ways in which they can transfer genetic material into cells, the physical properties of each method, and the mechanisms behind successful gene delivery using a given method. Advantages and disadvantages of the different methods will be noted, because no particular gene delivery scheme will be the best for all applications.

12.1 Gene delivery vehicles: an overview

Before considering what cells do with gene delivery complexes, we should consider the makeup of gene delivery vehicles. We will begin by organizing gene delivery vehicles and methods into groups. Figure 12.1 is a basic classification of gene delivery vehicles, with some examples. Although there are examples in the

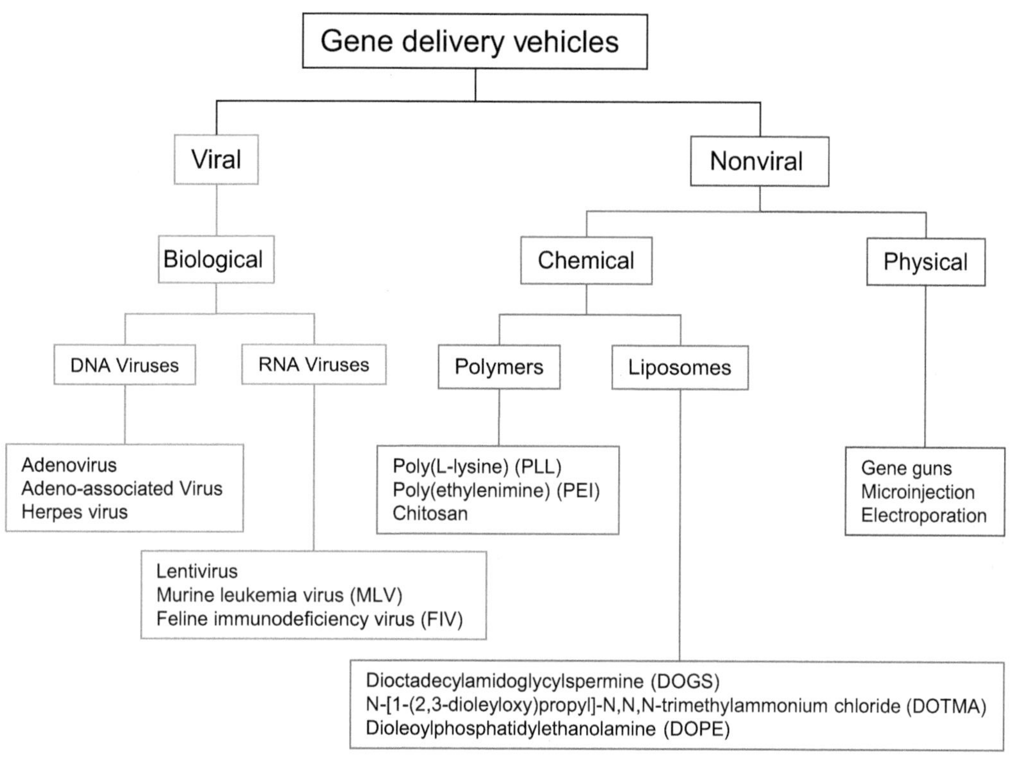

FIGURE 12.1

Organization of some common gene delivery methods and vehicles.

literature of classifications being mixed (e.g., viral proteins being used in conjunction with synthesized polymers), the chart covers the rudimentary ideas for vehicles and approaches that have been used to deliver genes.

If you were to talk to a gene therapist about selecting a gene delivery vehicle for a specific task, the first question that might come up is whether you would like your vehicle to be a virus. Viral or nonviral is the primary delineation for classifying gene delivery vehicles. Methods of gene delivery can be further divided into the three branches of natural science: biology, chemistry, and physics. The viral and biological vectors are nearly synonymous (despite viruses not being alive) because viruses in the wild *are* gene delivery vehicles; they seem to exist in nature only to propagate their own DNA in biological systems.

Moving down one level from the biological classification in Figure 12.1, we see that some viruses deliver DNA into cells, while others deliver RNA. The RNA viruses are known as **retroviruses**; examples include lentivirus, murine leukemia virus, and feline immunodeficiency virus. Examples of DNA viruses used for gene delivery include adenovirus, adeno-associated virus, and herpes virus.

Nonviral gene delivery methods can be divided into chemical and physical approaches. The chemical methods include certain polymers and lipids, whereas the physical methods utilize physical properties and forces to transport genetic material into cells. For instance, you can coat tiny gold particles (similar to the shot in a shotgun) with DNA, load them into a special gene gun, and blast away at cells or tissues. The shot goes into or through the cell, and the genetic material is left behind. As absurd as this description may seem, it represents an actual method of gene delivery that will be discussed later in the chapter. One can think of the physical delivery methods as "brute force" ways of getting genes into cells. Apart from the gene gun, other examples of physical methods are microinjection and electroporation.

The chemical methods can be further subdivided into methods that use polymers and those using lipids. Typical polymers that can be used for gene delivery include, but certainly are not limited to, poly(L-lysine) (PLL), poly(ethylenimine) (PEI), and chitosan. Polymers used for gene delivery can be linear, hyperbranched, or dendrimeric. Whereas PLL is generally a linear polymer, PEI is available in both linear and branched forms. The branched form of PEI is **hyperbranched**, meaning that the branches do not occur in a regular fashion. When the branching is very well controlled and regular, we refer to the branched polymer as a **dendrimer**. Dendrimers can also be used for gene delivery, as is the case for starburst dendrimers. Poly(amidoamine) dendrimers, or PAMAM dendrimers for short, are made in such a way that their sizes are easily controlled and their end groups can be functionalized.

Lipids have also been used widely for gene delivery; many different lipids have been used to create DNA-containing capsules known as liposomes. Some of these lipids are of the same ilk as those found in the plasma membrane, and others have been used for their pH-sensitive, membrane-destabilizing properties. As seen by the examples in Figure 12.1, they can have long, intimidating names that are typically converted to easily remembered acronyms. We will discuss lipids in greater detail in section 12.4, which discusses some of the main chemical delivery methods.

12.2 Viral delivery methods

Why would a biotechnologist want to use a virus for gene delivery? First of all, viruses are known to produce relatively high transduction efficiencies. (Note the use of the word **transduction** here, which denotes gene delivery via a virus. **Transfection** is used to denote gene delivery via one of the nonviral gene delivery methods.) Viruses, in theory, will yield up to 100% transduction efficiency in vitro. In practice, however, a range of 80% to 95% is more realistic. It is often said that viruses naturally introduce genetic material into cells. The use of the word "naturally" may be based on the 4.5 billion years of evolution that produced the efficient gene delivery machines we encounter today, both in the laboratory and in our everyday lives. Some viral applications yield permanent expression of the delivered genes. Retroviruses, which deliver RNA into cells, eventually have DNA copies of the delivered RNA inserted into the host's genome, which yields such permanent transduction.

Some of the DNA viruses used for gene delivery are adenovirus, adeno-associated virus (AAV), and herpesvirus. RNA viruses are the retroviruses, which include lentivirus, murine leukemia virus, feline immunodeficiency virus (FIV), and others. We can use FIV (as opposed to HIV), a very robust virus, to transduce human cells, because humans will not acquire immunodeficiency syndrome from it. Each of these viruses offers slightly different properties, so for any given application, one type might be more suitable than another.

12.2.1 Retrovirus

The distinguishing feature of a retrovirus (Figure 12.2) is that it delivers RNA into cells. Retroviruses are enveloped viruses that can be engineered to deliver up to 8000 bases of single-stranded RNA. They can yield permanent transduction, meaning the delivered genes will be present inside the cell for the

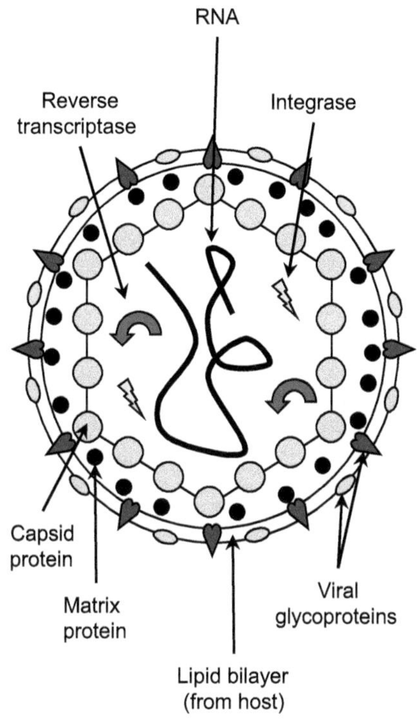

FIGURE 12.2

The basic structure of a retrovirus.

remainder of its life. This is possible because some viral genes and proteins are delivered as a part of retroviral transduction. With a retrovirus, the central dogma can be modified to include a pathway from RNA back to DNA, a process termed **reverse transcription**. When reverse transcription is performed on an RNA sequence, the result is a DNA copy, referred to as **cDNA**. Do not confuse cDNA with genomic DNA: even though the two sequences may code for the exact same polypeptide, cDNA and genomic DNA are different. To understand this point, consider the following discussion of molecular biology.

Introns are eukaryotic DNA sequences that are transcribed but do not code for a polypeptide (or even part of one). (Interestingly, the prokaryotic genome does not contain introns.) These noncoding sequences were at one time called "junk DNA." However, this term is a misnomer because the removal of intron sequences can have deleterious effects upon the cell. Introns appear within a gene as spacers between exon sequences; an exon is a stretch of DNA that codes for all or part of the final protein product. A single gene may have several exons that are separated by introns. Recall from Chapter 4 that the RNA that is first transcribed from such a stretch of DNA is termed the primary transcript, or pre-mRNA. The pre-mRNA undergoes the removal of introns via splicing (see Figure 4.20).

It is the presence of introns that explains why eukaryotic genomic DNA can have a different sequence from cDNA. This distinction should be kept in mind when looking up gene sequences in repository databases. The sequences reported are typically cDNA sequences; this can present problems when trying to isolate genes from genomic DNA preparations using polymerase chain reaction and primers.

When a retrovirus delivers RNA into a cell, reverse transcription of the RNA to produce cDNA takes place via the enzyme **reverse transcriptase**. The cDNA can then be integrated into the host genome via the enzyme **integrase**. After the cDNA has been integrated into the genome, it will be replicated when the cell replicates, ensuring that both daughter cells will have a copy of the virally delivered gene; this is what is meant by "permanent transduction." In theory, the gene will be expressed forever, although in practice expression levels will decrease over time as the inserted sequence becomes a part of the **heterochromatin**.

There are additional problems associated with using retroviruses in an attempt to achieve permanent transduction. These stem from the fact that the location of cDNA integration via integrase is random. First, recall that the proteins encoded by housekeeping genes are always needed by the cell; because of random integration, the cDNA could be inserted into the middle of a housekeeping gene. Such an insertion would render the housekeeping gene useless and would lower the amount of an essential protein, possibly leading to cell death. A second problem with random integration is that the cDNA could be inserted into the middle of a tumor suppressor gene, resulting in the **transformation** of a normal cell into a tumor cell. Third, one should be aware of proto-oncogenes—genes that, if they were to become active, would serve to transform a cell into a cancer cell. If the cDNA were inserted upstream of a proto-oncogene, it could activate or mutate it into an active oncogene, thus transforming the cell. The phenomenon of random integration is a significant reason why retroviral gene delivery has not been used in the clinical setting. However, retrovirally mediated gene delivery is a powerful tool in the laboratory that can be used to produce or enhance cell lines for basic scientific research.

Box 12.1 When fields collide

An inherent problem that crops up when established fields merge is that of overlapping or redundant terminology. We've already discussed the insertion of exogenous DNA into a prokaryotic cell, a process termed "transformation" by molecular biologists and microbiologists. To the oncologist or cell biologist, however, "transformation" refers to the conversion of a mortal cell into an immortal cell. Another such word is "vector." In molecular biology a vector is the main plasmid into which one might insert an exon. To a gene therapist, however, the vector is not the DNA but rather the DNA carrier. An epidemiologist might refer to someone who spreads a disease (e.g., Typhoid Mary) as a vector. Confusing, yes, but reused words can also be celebrated as proof of scientific progress.

The main advantages of retroviral transduction are high delivery efficiency and "permanent" transduction; a key disadvantage is the vector's inability to infect nondividing cells. However, in tissue engineering applications that utilize an ex vivo strategy, this disadvantage could be considered an advantage in that somatic cells are largely protected from accidental transduction following the implantation of a construct containing transduced cells. There are also serious safety concerns with the retrovirus, such as the generation of replication-competent retroviruses and the production of insertional mutations resulting from random integration. Given the substantial level of concern associated with retrovirus use, alternative methods of gene delivery have been developed to reduce or eliminate these specific safety risks.

12.2.2 Adenovirus

Adenoviruses are a group of nonenveloped viruses that carry linear, double-stranded DNA. They occur naturally, infecting mucosal linings in humans and other mammals. There is no treatment for adenoviral infections, but a normal immune system is quite capable of combating them. This point is key to the gene therapist because adenoviral vectors used for gene therapy will be recognized by the immune systems of patients who have combated infections by that specific adenoviral serotype before. (**Serotypes** are the combinations of exposed antigens that make cells or agents serologically distinguishable.) Even if a patient had not been exposed to that specific serotype of virus before, a repeated dose of the same gene delivery regimen would be recognized and attacked by the immune system via a secondary response.

Adenoviruses have icosahedral capsids (shells). In addition to their capsids and DNA cargo, these viruses are also made up of cement and core proteins. Cement proteins stabilize the capsid, and core proteins associate with the double-stranded DNA genome. The DNA is covalently attached to a terminal protein at each 5′ terminus. The structure of an adenovirus is shown in Figure 12.3.

The adenoviral genome is about 36,000 base pairs long. Up to 30,000 bp can be replaced by foreign DNA. When parts of the viral genome are removed, the virus is rendered replication incompetent. It takes several rounds of processing to create such viruses, and propagation requires the use of a "helper" cell line designed to express necessary proteins that are no longer represented in the viral genome. The final generation of viruses have had most of their viral genes removed, leaving a vector that is termed "gutless." These gutless vectors contain only inverted terminal repeats and a packaging sequence surrounding the engineered gene that is to be delivered.

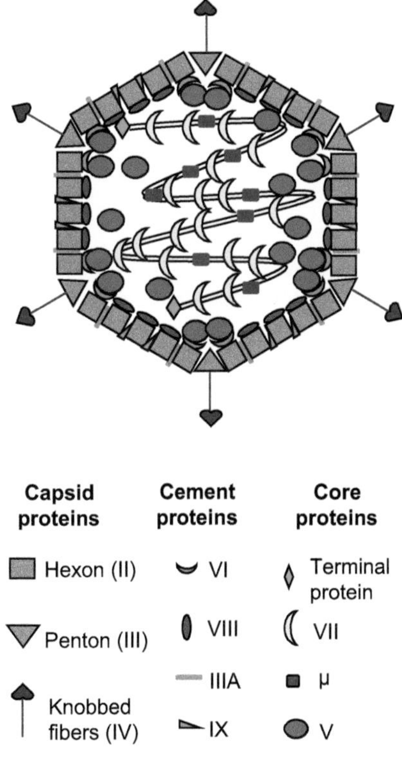

FIGURE 12.3

The basic structure of an adenovirus.

Adenoviral vectors can infect a broad range of human cells, including both dividing and nondividing cells. They are relatively stable and can be concentrated to high concentrations (**titers**). Adenoviruses do not cause the integration of genetic material into host cell genomes, so insertional mutagenesis is not of concern; however, the lack of genomic integration means the resulting gene expression is only transient.

12.2.3 Adeno-associated virus

The human AAV is a small, nonenveloped DNA virus with an icosahedral capsid. It delivers linear, single-stranded DNA that is approximately 4.7 kb long. AAV is naturally replication deficient, with replication typically achieved via a helper virus such as adenovirus or herpesvirus. AAV can mediate long-term transgene expression and transduce a large range of cells. It can also transduce nondividing cells—a feat that retroviruses cannot accomplish—and is replication defective, nonpathogenic, and nonimmunogenic. Limitations of this method lie in the relatively small transgene size that can be delivered, relatively low transduction efficiencies, and the need to use helper viruses for activation.

The infection cycle of AAV begins with the virus binding to the AAV receptor (each serotype appears to bind the same receptor but in different places) and enters the cell via clathrin-mediated endocytosis (see Chapter 2). After endocytosis, acidification of the endosome is followed by release of the AAV particle into the cytoplasm. The mechanism of nuclear entry is not fully understood, but it appears that the viruses interact with nuclear import proteins (e.g., importin-β) to allow them to traverse the nuclear pore complex. AAV is able to integrate its cargo non-randomly into the host genome: in humans, the site is on the q arm (long arm) of chromosome 19.

The serotype AAV2 is the dominant form of AAV, and the AAV vectors for gene therapy were once commonly developed from this serotype. In the laboratory, AAV vectors are constructed by deleting the genes *rep* and *cap* from the wild-type viral genome and using the newly created space for the engineered gene of interest. The recombinant DNA is then cotransfected into host cells with a separate plasmid that contains *rep* and *cap*. Finally, the host cells are infected by an adenovirus, causing them to produce recombinant AAV vectors. After 2 to 3 days of incubation to allow for the production of virus particles, the cells are lysed, and both recombinant AAV and adenoviruses are collected. AAV is then purified from the adenovirus-containing lysate by heat inactivation or gradient centrifugation. (For completeness, this is not the only way to produce AAV; some methods circumvent the use of a helper virus.)

Keep in mind that AAV, like any virus, comes with all of the concerns regarding immunogenicity that we have already mentioned; multiple serotypes of the virus can allow the gene therapist to use AAV more than once to deliver genes if a repeated administration is needed. Several serotypes, including AAV1, 2, 5, and 8, have been used in clinical trials. AAV8 is especially interesting in terms of immunogenicity because research has verified that most (tested) people are devoid of AAV8-neutralizing antibodies in their serum—and even in the presence of such antibodies, there is a lack of AAV8 inactivation on the first administration.

12.2.4 Herpesvirus

The herpesviruses are enveloped viruses with icosahedral capsids. They are able to deliver the largest payloads of any of the viruses used for gene delivery, which allows them to deliver multiple transgenes along with genes encoding specific transcription factors (meaning that expression of the therapeutic gene can be tightly controlled). Among all of the herpesviruses, the human herpes simplex virus type 1 (HSV-1) has been used the most for gene delivery applications and can be used to deliver engineered DNA payloads of up to 130kbp. HSV-1 displays a natural tropism toward neurons, so it is especially suited for tissue engineering applications directed at neuropathy or diseases associated with the central nervous system.

The replication cycle of a herpesvirus begins with the initial binding of the viral envelope to cellular receptors, followed by entry of the capsid into the cytoplasm. The capsid is then able to bind to the nuclear membrane and release its DNA cargo directly into the host nucleus. This is why HSV-1 can be used to transduce nondividing cells.

12.2.5 Baculovirus

Baculoviruses naturally infect insects, so they have been used to produce recombinant proteins based on insect cells. For example, the baculovirus-insect cell expression system has been used to produce EGF-collagen (epidermal growth factor linked to collagen), a biomolecule that can induce cells to grow and divide. Collagen scaffolds are often used for tissue engineering, so a collagen scaffold that also induces supported cells to grow and divide would be very desirable to tissue engineers.

Baculoviruses can carry a large cargo, which makes it possible for the biotechnologist to deliver multiple genes at one time. Because they are insect viruses, they are unable to replicate in mammalian cells. Transduction efficiencies in the range of 75% to 85% are commonly achieved; this is on par with adenovirus in human cardiomyocyte and fibroblast applications and better than adenovirus in human smooth muscle cells.

12.2.6 Summary of viral methods

Each viral vector has its own strengths and shortcomings. Retroviral and adenoviral vectors were initially the most widely studied and applied gene delivery vehicles, but AAV, herpesviruses, and baculoviruses later garnered more interest because of their specific advantages. AAV achieves permanent transduction without the problems of random integration. Herpesvirus can carry a large payload and can transduce neuronal cells, which are difficult to transduce. Baculovirus can also carry a large payload and is considered much safer because it cannot be replicated in mammalian cells.

Adenovirus, HSV-1, baculovirus, and others are used to achieve transient transfection. This means that the delivered gene will only be expressed for a limited period of time, usually on the order of 1 to 3 weeks. An advantage of these viral delivery methods is their high transduction efficiency. Because the transduction is transient, we do not have to worry about random integration into the genome, as we did with retroviruses. The main disadvantage of viral transduction is that the immune system will adapt to the virus's presence and mount a response if the virus is injected into the body a second time. The retroviruses lentivirus, MLV, and FIV are able to achieve permanent transduction, but random integration presents potentially lethal consequences for the cell or host organism. The retroviral methods also require actively dividing cells for integration, which limits their applicability but could also be considered an advantage when we are targeting cancer cells.

12.2.7 Viral methods and the immune response

Suppose for a moment that we use the influenza virus to deliver genes into a human; the person will get sick if we do not first remove some of the viral genes. We remove part of the viral genome for two reasons. First, it would make little sense to inject a pathogenic virus into a patient, and administering an agent that will make a person sick is fraught with ethical problems; the Food and Drug Administration would be very unlikely to approve such an agent. Second, the viral heads of the different types of viruses have evolved to sizes that will just contain a given viral genome. Because the viral head is of finite size, there's only so much DNA that will fit inside. In removing some of the viral genes, we create space that can be filled with our gene(s) of interest.

What genes could we deliver? One example would be nucleotides encoding bone morphogenic protein 4, a growth factor; the associated gene is 7101 bp in length. What could we not deliver? Nucleotides for dystrophin, for which the associated gene is over 2.2 million bp long; that much mass simply would not fit within a viral head. We have already discussed the capacities of several viruses, which vary from type to type. (In case you are interested, the influenza virus is an RNA virus that holds approximately 13.6 kb.)

Even attenuated (weakened) viruses can be delivered into a patient only once; a second dose of the same virus will make the patient sick because their immune system will recognize the viral coat proteins from the first time the virus was administered. These proteins are capsid proteins (or envelope proteins for enveloped viruses). The first time the body is exposed to these viral proteins, the immune system becomes educated and mounts a **primary immune response** to them. This response involves developing antibodies against the specific proteins, and it takes time. The second time the body encounters these foreign proteins, however, genes encoding antibodies against them will already exist as a result of the initial exposure. Because a blueprint for these antibodies is in place, the immune response will be of greater severity and can be mounted in a shorter time. This is a **secondary immune response**. The secondary immune response should be avoided by the gene therapist for two reasons: not only it is very unpleasant for the patient, but the body will quickly be able to detect and destroy the viruses, thus reducing the effectiveness of the gene delivery treatment.

Box 12.2 The flu vaccine

Viruses, such as the influenza virus, are constantly mutating and taking on forms that could possibly go undetected by our immune systems. The flu shot is a way of presenting these new viral proteins to the body so that if we become exposed to the virulent (active) form of the virus, our immune systems have already been educated and can mount a quick response to keep flu symptoms to a minimum. *You do not get the flu from the flu shot!* It's possible that you might experience some mild symptoms if your body mounts a secondary immune response because of prior exposure, but the delivered vaccine does not contain viruses that can replicate, so a full-blown case of the flu is not possible from the vaccination.

An initial transduction with a recombinant virus is similar to getting a flu shot, in that the body will mount a primary immune response to the foreign particles. A second viral transduction (with the same recombinant virus) would be similar to a vaccinated person being exposed to live influenza virus. A secondary immune response, quicker and more intense than the primary immune response, will quickly be mounted to clear the body of the viral particles.

We are encouraged to get a flu shot every year because viruses are constantly changing. They change their capsid proteins so they can fly under the radar; from one year to the next, they can have a completely different set of proteins on display. The flu shot is given every year rather than every week or every month because a complete cycle of viral epidemic takes about a year. As the weather starts to get cold in the fall, people's immune systems become slightly compromised. In addition, we spend more time indoors and in closer contact with other people. It's an opportune time for pathogens to invade. This is one reason why people tend to get sick in the fall or in the winter, although the potential to become ill due to a pathogen exists all year; perhaps you'll make it through the autumn and winter, then get exposed to the virus during the spring or summer. However, by this time the odds are that you will have been either vaccinated or exposed to the virus already, so a secondary immune response will keep you from getting quite so ill. If the virus does not change its displayed

Box 12.2 The flu vaccine—cont'd

proteins, you will have even less chance of getting very ill the following fall or winter because your body has been educated. To combat this, the viruses are always changing. Even though in the spring or summer there might be a new virus that the body has not seen, the weather will be warmer so people will be spending more time outside in conditions where viral concentrations are much lower.

At the beginning of the next flu season, it will be recommended that you get another flu shot to take care of possible exposure to these newly mutated viruses. One might ask how we know what the protein of the year is going to be; the answer is that we don't. However, agencies such as the Centers for Disease Control and Prevention in the United States constantly monitor what illnesses are making the population in general sick. Researchers will isolate and characterize the viruses, taking note of the serotypes that appear repeatedly. "This person is sick with strain 131, and this one has strain 168, and here is a 219, here are two more with 131. … We are seeing a lot of strain 131 lately. We think this strain will be rampant by the fall, so it would be prudent to vaccinate the population against strain 131." That is a broad overview of how scientists pick out which viruses will be selected for vaccination programs each year. Large-scale statistical analyses are performed on samples taken from around the country to determine the most likely candidates for upcoming infections. The analyses are performed in the spring to allow time to prepare enough vaccine for the population by the following autumn. Generally, three or four strains will be selected for the yearly inoculations, but there is an upper limit to the number of proteins that can be used due to time, resource, and titer constraints.

12.3 Physical delivery methods

12.3.1 Gene gun

The actual gene gun setup is very close to our earlier (and seemingly absurd) shotgun analogy, except that the shot used is much smaller than buckshot. Microparticles of around 0.5 to 1 μm are used, typically made of gold. Gold can ionize to carry a positive charge, which will interact well with DNA or RNA thanks to the negative charge on every phosphate group in the polynucleotide.

The biotechnologist should be very aware that if there is a charge on a gene delivery vehicle, it must almost certainly be a positive charge to allow for interaction with the delivered polynucleotides. In this case, we use gold microparticles because not only do they ionize positively, but they are also biologically inert. We will coat the gold microparticles with DNA (for this example) and then load them onto a membrane. After loading the membrane into the gene gun, a sudden application of force is used to move the filter down the barrel of the gun. The force comes from a compressed, inert gas such as helium at a pressure of around 200 to 300 psi. The membrane, loaded with particles, travels down the barrel of the gun until it reaches a position that prevents further movement. However, when the membrane strikes the barrier within the barrel, the inertia of the gold particles causes them to dislodge and travel beyond the barrel of the gun to interact with cells and or tissue. Some of the particles will travel completely through the outer layers of cells, and sometimes an entire particle will remain embedded within a cell. As the particles pass through, some of the DNA can be stripped off the gold by simple shear forces. The result is that DNA is delivered past the plasma membrane into the cytoplasm or nucleus of the cell.

Use of the gene gun is limited to exposed tissues, such as the epidermis, or cells in a plant leaf. Only limited penetration (100–500μm) is attainable, so it would be extremely difficult to transfect cells of an entire organ with this method. Granted, one could set the pressure of the gas to a higher level to impart greater momentum to the gold particles and get them deeper into the tissue, but the cells on the surface will undergo greater damage and the shockwave produced can dislodge cells from the extracellular matrix. There is a trade-off between tissue damage and depth of penetration using a gene gun.

12.3.2 Microinjection

Microinjection is simple to explain: itty bitty needles. The needle gets past the plasma membrane by poking a hole in it. If the procedure is done correctly, the membrane will close back up when the needle is removed, leaving behind minimal cellular damage. In the hands of a skilled operator, microinjection can yield close to 100% transfection efficiency. This technique can be used to transfer oligonucleotides to cell cytoplasms *or* nuclei or to transfer entire nuclei into enucleated eggs (as is the case with cloning). However, the use of microinjection to transfect millions, or even just hundreds, of somatic cells in vivo would be infeasible—both in terms of getting to and visualizing the cells of interest and because of the amount of time and effort required for the procedure itself.

Microinjection is often used with oocytes (egg cells) in suspension; the egg is held in place using a very small tube with an applied vacuum (Figure 12.4). Cells that are already adhering to a support can also be microinjected. Either way, it is important that the cell to be injected is in a position where it can be visualized with a microscope. The microscope has a CCD camera that is linked to a monitor to allow for real-time visualization.

The microinjection needle is positioned just over the cell in the xy plane (at an angle), then lowered slowly in the z direction until a slight depression can be seen on the cell surface. It looks much like lightly poking a water balloon with one's finger. After the correct needle position has been visualized, the needle is then moved slightly in the xz plane to prepare for a stab-and-inject movement that will allow the microinjector to introduce DNA into the cell using predetermined, computer-controlled settings. The injection is made via positive pressure. The injection volume, injection pressure, and postinjection pressure are all tightly controlled; the injection volume should be limited to only the amount of DNA solution that is to be introduced; otherwise, the cell could burst. The postinjection pressure is also important. This pressure is lower than the injection pressure but higher than the baseline pressure used at the beginning of the experiment. Without it, material from the cytoplasm or nucleus would be forced back into the needle as a result of the positive pressure created by the introduction of the injected material. Such a loss of cytoplasmic or nuclear material could spell death for the cell.

12.3.3 Electroporation

A third physical method for gene delivery is electroporation. This technique involves the delivery of a voltage (current) across the cell surface, resulting in the spontaneous creation of pores in the plasma membrane that will close when the voltage is removed. When the voltage is applied and the pores open up, current travels through the cell. Because DNA has a net negative charge, it will migrate in the same direction as the electron flow in the applied current. When the current is removed the pores will spontaneously close, trapping some of the DNA within the cells. Membrane-associated DNA aggregates also form in the presence of the voltage and have been seen to remain at the membrane level for up to 10 minutes after the administration of the current. It is possible that these aggregates enter cells through endocytosis.

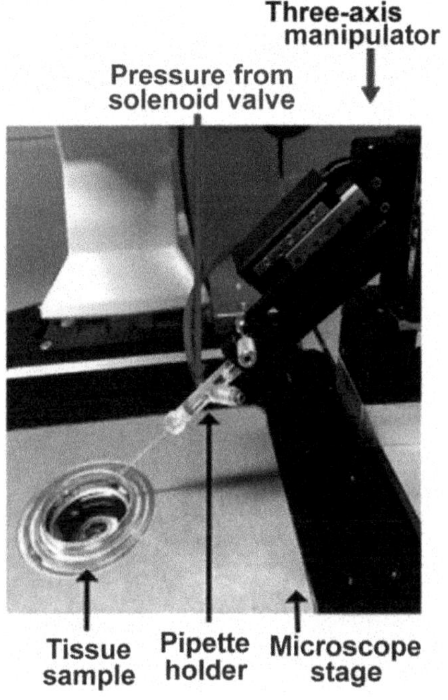

FIGURE 12.4

A typical cell microinjection system.

From: Shull G, Haffner C, Huttner WB, Kodandaramaiah SB, Taverna E. EMBO Rep. *(2019)* ***20:*** *e47880. (Figure 1a)*

Box 12.3 Dolly the sheep

Although not an example of microinjection, the case of Dolly the sheep was the first published result of cloning using the nucleus from a somatic cell harvested from an adult donor. Later examples of animal cloning via nuclear transfer were carried out via microinjection; Dolly is presented here because of historical significance.

For the Dolly experiments, the entire nucleus (including genomic DNA) of a mammary cell taken from the udder of a 6-year-old pregnant ewe was put into an enucleated egg from a different breed of sheep via cell fusion. With a complete genome, the egg cell was then allowed to develop into a morula/blastocyst, which was then implanted into a recipient. The result was the birth of a lamb, Dolly (6LL3), whose genome was unrelated to that of the sheep that donated the enucleated egg cell. The mitochondria and mitochondrial DNA of Dolly presumably came from both the donor and the oocyte, but Dolly's genome (including both sets of chromosomes) came solely from the donor of the somatic cell.

A total of 277 cell couplings were made using adult mammary cell nuclei; of these, 29 progressed to the morula here also blastocyst stage and were implanted into 13 sheep. Of these 29 implantations, only 1 resulted in pregnancy. This pregnancy went to term to produce a live lamb. Other lambs were produced in the study via cloning using fetal and embryo-derived nuclei. The major impact of the study was the discovery that cell differentiation does not involve irreversible DNA modifications that would prevent the cell from being transformed to a totipotent state (see Chapter 17).

Dolly died on February 14, 2003, having been euthanized because of complications due to lung cancer that was caused by a virus. She was 6 years old. Sheep of Dolly's breed (Finn Dorset) typically live for 11 to 12 years, and the sheep that donated the mammary nucleus was 6 years old at the time of cell harvest. It is interesting to note that Dolly's life span was about right for a sheep of her breed, given that she was effectively 6 years old at the time of nuclear transfer. Throughout her life, Dolly had been plagued by health problems, including a weight problem, arthritis, and cancer. However, researchers also noted that her telomeres were prematurely shortened—perhaps the result of shorter telomeres on the chromosomes of the adult DNA donor (see Chapter 7).

For mammalian cells the process uses at least 200 V/cm for plated cells and up to 400 V/cm for cells in suspension, with pulses lasting a minimum of 0.5 milliseconds; multiple pulses are usually delivered. A small amount of cellular material, especially ions and small proteins, may escape during the procedure if the voltage is sufficient and the time of delivery is long enough. Electroporation is very stressful to the cell. Although it can be used on most cell types, it is typically reserved for bacteria because of their high numbers and short cell cycles (meaning they are expendable and have quick recovery times).

12.4 Chemical delivery methods

The chemical gene delivery methods, using polymeric or lipid delivery vehicles, offer several advantages for delivering genes into cells. First, the size of the carried DNA is not limited to what will fit into a finite viral head; artificial chromosomes with millions of base pairs have been delivered by both polymers and lipids. Second, there is potentially no threat of immune response, even with repeated injections of the complexes. Third, polymers and lipids are relatively inexpensive to produce and store. Finally, although nonviral gene delivery usually results in transient gene expression, this can be regarded as an advantage in that there is very low risk of random integration into the genomes of host cells. Of course, there are also disadvantages. The efficiency of chemically mediated gene delivery (25%–40%) is much lower than that achievable with viruses. Some of the delivery vehicles are also somewhat **cytotoxic**, meaning they can cause cell death above certain concentrations.

12.4.1 Polymers

When making **polyplexes** (polymer–DNA complexes) for gene delivery, we need to be aware of the different polymer architectures that can be used. There are linear polymers that consist of basic repeating units strung together (e.g., PLL), and there are branched polymers. The latter group includes hyperbranched polymers such as PEI and dendrimers such as PAMAM starburst dendrimers. We will now take a look at these vectors, and more, in greater detail.

12.4.1.1 PEI

PEI is a polymer that has been used for years in common processes such as paper production, shampoo manufacturing, and water purification. The polymer has the repeating unit shown in Figure 12.5A and can be produced in either linear or hyperbranched forms. In either form, every nitrogen has the potential to be protonated and therefore carry a positive charge; the pH of the environment determines the overall number of positive charges per PEI molecule. Some of the nitrogen atoms are more easily protonated than others, especially in the branched form (Figure 12.5B), which is why branched PEI can serve as a buffer over a wide range of pH values.

Branched PEI comes from the polymerization of aziridine (Figure 12.6A). When aziridine is protonated, it is susceptible to the ring opening at the amine via nucleophilic attack. A convenient nucleophile in this case would be a second aziridine molecule. As the ring from the first aziridine is opened, the positive charge is transferred to the amine of the second aziridine ring (shown in red in the figure). This cycle of ring opening can continue during propagation to yield different architectures—linear or

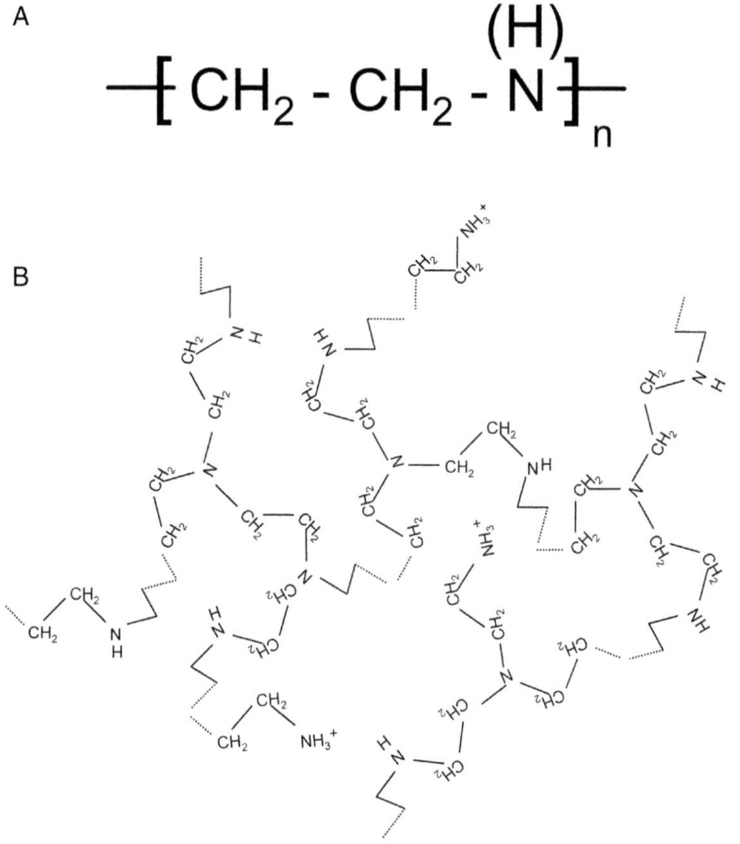

A

$$-\!\![\,CH_2 - CH_2 - \overset{(H)}{N}\,]_n\!\!-$$

B

FIGURE 12.5

(A) Basic repeating unit of linear poly(ethylenimine), PEI. For branched PEI, the hydrogen atom shown in parentheses might instead be another polymer of CH_2-CH_2-N(H). (B) Branched PEI, with 1°, 2°, and 3° amines.

branched—of [CH_2–CH_2–NH] additions (Figure 12.6B). As shown in the figure, it's not guaranteed that any given nitrogen will yield a branch point during polymerization. A polymer formed by this high degree of irregular branching is termed "hyperbranched." At some point, a secondary amino group from within the polymer itself will act as a nucleophile for a terminal aziridine, forming a cyclic structure that can no longer be polymerized. This form of termination is known as **back-biting** (Figure 12.6C).

Again, note that each of the amines within a PEI molecule has the potential to be protonated. At physiological pH (7.2–7.4), PEI is quite cationic. As such, it will readily interact with DNA. In fact, at physiological pH, there will be an excess of positive charges even after the PEI has been complexed with DNA. This is important because positive charges are required for untargeted gene delivery

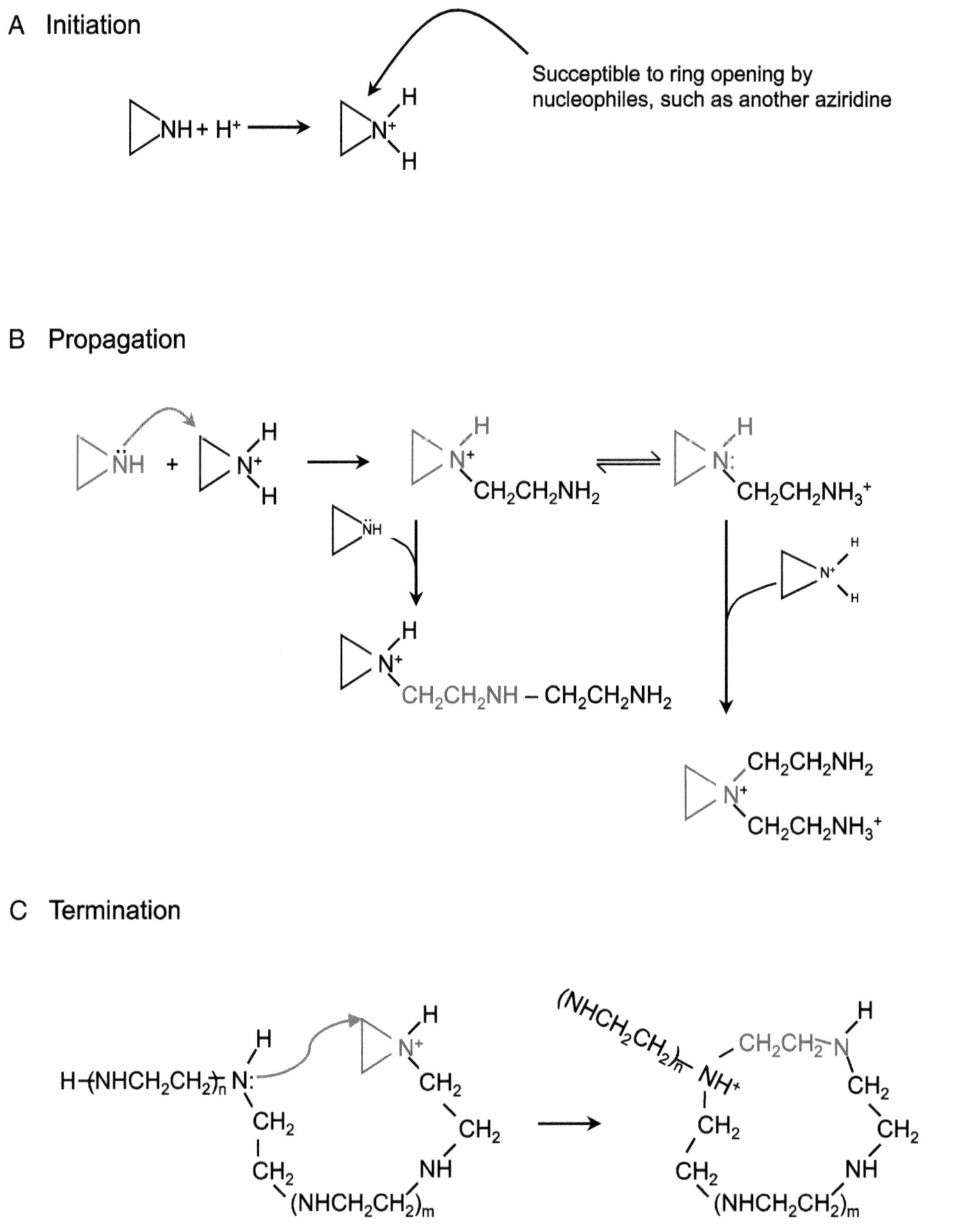

A Initiation

Succeptible to ring opening by nucleophiles, such as another aziridine

B Propagation

C Termination

FIGURE 12.6

Polymerization of branched poly(ethylenimine) (PEI). (A) Initiation occurs after the protonation of aziridine. (B) The ring from the initial aziridine (shown in *black*) is opened by nucleophilic attack by another aziridine (shown in *red*). Chain extension can occur in two different ways, as shown. (C) Termination occurs when a ring is opened by a secondary amine in the same polymer.

complexes to gain entry into cells. Eukaryotic cells typically have a net negative charge on their exteriors. (Don't confuse this with the flipping of phosphatidylserines to the cell exterior during apoptosis. Although these negatively charged phospholipids are primarily found on the cytoplasmic side of the plasma membrane in cells that are not undergoing apoptosis, the cell exterior still carries a negative charge because of transmembrane proteins and other membrane constituents.) This means that charge-charge interactions can occur between PEI–DNA complexes and the cell; the complexes will adhere to a portion of the plasma membrane and be endocytosed, as was described in Chapter 2. (As a quick check of your learning thus far: would the endocytosis be an example of phagocytosis or pinocytosis? The size of the gene delivery complexes will be on the order of 100 nm.)

Following endocytosis, one popular conjecture about what happens to PEI–DNA complexes within the cell is described by the proton sponge hypothesis, which entails the following steps:

1. At some point following endocytosis, endosomes containing the PEI–DNA complexes fuse with lysosomes to create structures known as endolysosomes. (Using our knowledge of endosomal maturation, we will see that this fusion is not a requirement for the hypothesis to still be valid.)
2. V-ATPases pump protons into the vesicular interior.
3. PEI absorbs many of these protons, preventing a change in pH. (Note that, in the case of an endolysosome, the lack of pH drop will prevent the activation of degradative enzymes such as nucleases.) However, protons have still been introduced, so a charge gradient will have been established.
4. Cl^- will flow into the vesicle to neutralize the charge gradient. This, however, will set up an osmotic gradient.
5. Water will enter the vesicle to balance the osmotic gradient.
6. The vesicle will swell due to the entry of ions and water, leading to rupture.

The result of this process is that the PEI–DNA complexes are released into the cytoplasm. However, it is not necessarily the case that all endocytosed PEI–DNA complexes follow this pathway. There is still some debate regarding the cellular processing of such complexes.

12.4.1.2 Dendrimers

Dendrimers are branched polymers that are synthesized in a stepwise fashion to control both monodispersity (homogeneity of molecule sizes) and the exact number of branching layers, or "generations." Dendrimers can be synthesized by either divergent or convergent methods (Figure 12.7). For the divergent method, the dendrimer grows in a stepwise fashion outward from a multifunctional core molecule. Slight structural defects can occur in larger molecules, especially at higher generation numbers. In the convergent method, dendrimer construction begins with the end groups and progresses inward. Because any defects will involve large sections missing from the molecule, defective structures can be more readily separated with this method. Unlike hyperbranched polymers, dendrimers are polymerized in a tightly controlled, stepwise fashion to produce relatively monodisperse sets of macromolecules (for a discussion of monodispersity and polydispersity, see Box 12.4). Dendrimer–DNA complexes are often called **dendriplexes** to preserve terminology analogous with lipoplexes and polyplexes. The low pK_a values of the amines (3.9 and 6.9) afford the dendrimer the potential to buffer pH changes during acidification of the endosome.

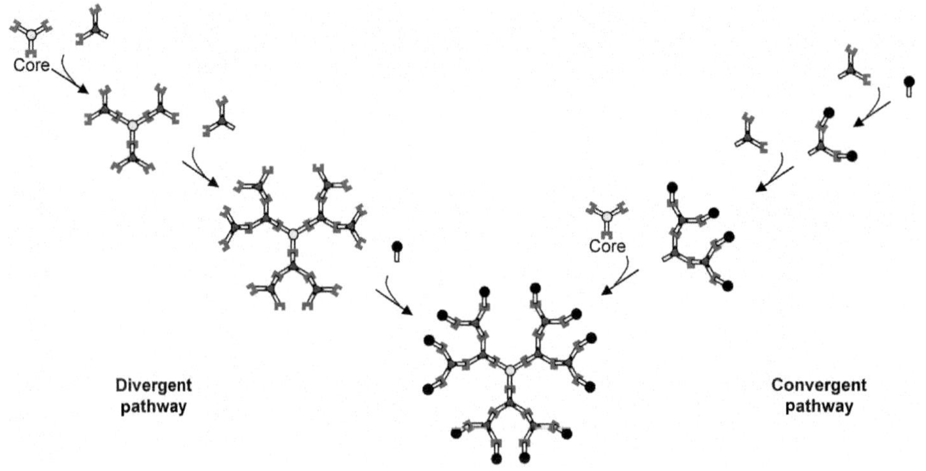

FIGURE 12.7

Dendrimers can be constructed via divergent or convergent pathways. In the divergent pathway *(left)*, the dendrimer is extended outward from a multifunctional core molecule, often ending with a functionalized terminal group. The convergent method *(right)* begins at the outer ends and polymerization extends toward what will be the interior of the dendrimer, ending with the addition of the core molecule.

There is an array of dendrimeric molecules that can be used for gene delivery. A commonly used dendrimer type is the PAMAM dendrimer (Figure 12.8A). Dendrimers having various generation numbers have been used for gene delivery; the optimal number of generations differs by cell type but is typically in the range of 5 to 10 (termed G5–G10). Spherical G5 PAMAM dendrimers have been used successfully in vivo. Other dendrimers, such as poly(propylenimine) dendrimers (Figure 12.8B), have also been used for gene delivery, with transfection efficiencies again being dependent on generation number. Presumably because of the increased density of positive charges at the periphery of the dendrimer, an increase in DNA binding is observed as dendrimer generation number goes up. However, cytotoxicity also increases with generation number. This dichotomy is a common problem with all chemical delivery methods.

12.4.1.3 Chitosan

Chitosan is another polymer that has been well characterized for use in transfection. This natural, nontoxic polysaccharide affords its cargo DNase resistance while condensing the carried DNA to form stronger complexes. The efficiency of chitosan is thought to rely on its ability to swell and burst endolysosomes, which allows the delivered DNA to continue its path to the nucleus.

Chitosan is obtained by the alkaline deacetylation of chitin, the second most abundant polysaccharide in nature. We can obtain chitin in large quantities from crab, shrimp, and lobster shells, considered by the seafood industry to be waste products. Chitosan is a polysaccharide (sugar polymer) composed of randomly distributed β-(1-4)-linked D-glucosamines and N-acetyl-D-glucosamines (Figure 12.9). It is **biodegradable** (meaning it is broken down inside the cell or organism, usually by hydrolysis) and **biocompatible** (meaning the polymer and its degradation products are nontoxic) at "low" molecular weights (10–50 kDa).

Box 12.4 Describing polymer distributions

Polymerization is not an exact process. Not every polymer in a batch will have the same number of repeating units. Because of this, we need a way to describe just how pure a polymer solution is. This is accomplished via the **polydispersity index (PI)**, which takes into account two different molecular weights.

The first molecular weight to consider is the **number average** molecular weight ($\overline{M_n}$). This average is like the averages you are probably already used to. The molecular weight (M) for each polymer in a sample is added, and the sum is divided by the total number of molecules counted.

$$\overline{M_n} = \frac{\sum_i N_i M_i}{\sum_i N_i}$$

The second molecular weight to consider is the **weight average** molecular weight ($\overline{M_w}$). This average is like a weighted average, in which larger polymers are given more influence in the equation for molecular weight determination. The molecular weight for each polymer molecule is squared and added, and the sum is divided by the number of molecules times the sum of the molecular weights.

$$\overline{M_w} = \frac{\sum_i N_i M_i^2}{\sum_i N_i M_i}$$

Because squaring the values of the molecular weights will produce a stronger input from the larger molecules, the resulting weight average molecular weight will always be greater than the number average molecular weight, unless there is only one molecular weight in the polymer sample. As a result, $\overline{M_w} \geq \overline{M_n}$.

The polydispersity index is the ratio of the two molecular weights just described: $PI = \frac{\overline{M_w}}{\overline{M_n}}$.

Note that PI>1, except in the case where all the molecules have the same molecular weight. In that case, PI = 1, and the polymer sample is said to be **monodisperse**.

Example 1: What are $\overline{M_n}$, $\overline{M_w}$, and PI for the following distribution of polymer sizes?

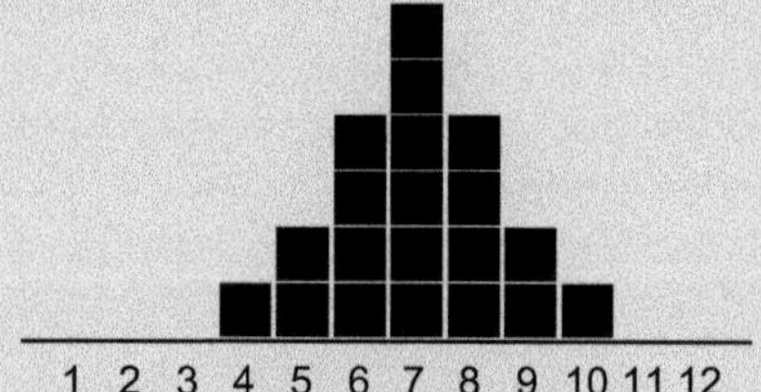

Answer:

$\overline{M_n}$:
(4+5+5+6+6+6+6+7+7+7+7+7+7+8+8+8+8+9+9+10)/20
=[1(4)+2(5)+4(6)+6(7)+4(8)+2(9)+1(10)] / (1+2+4+6+4+2+1)
=140/20 = 7.00

$\overline{M_w}$:
$[1(4)^2+2(5)^2+4(6)^2+6(7)^2+4(8)^2+2(9)^2+1(10^2)] / [1(4)+2(5)+4(6)+6(7)+4(8)+2(9)+1(10)]$
= 1022/140
= 7.30

PI = 7.30 / 7.00 = 1.04

Example 2: Here the distribution is still centered at 7, but it has been stretched out. What are the effects on $\overline{M_n}$, $\overline{M_w}$ and PI?

Continued

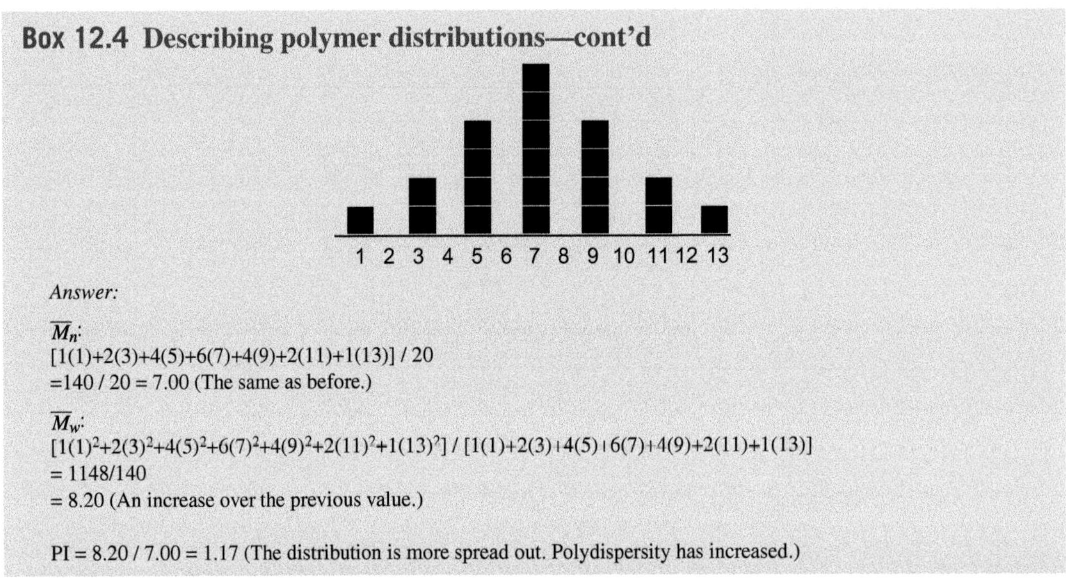

Box 12.4 Describing polymer distributions—cont'd

Answer:

$\overline{M_n}$:
$[1(1)+2(3)+4(5)+6(7)+4(9)+2(11)+1(13)] / 20$
$=140 / 20 = 7.00$ (The same as before.)

$\overline{M_w}$:
$[1(1)^2+2(3)^2+4(5)^2+6(7)^2+4(9)^2+2(11)^2+1(13)^2] / [1(1)+2(3)+4(5)+6(7)+4(9)+2(11)+1(13)]$
$= 1148/140$
$= 8.20$ (An increase over the previous value.)

PI = 8.20 / 7.00 = 1.17 (The distribution is more spread out. Polydispersity has increased.)

Because of chitosan's adhesive and transport properties in the gastrointestinal tract, it has been used for oral gene therapy applications. Chitosan has been used to deliver genes encoding ARAH2 (the dominant anaphylaxis-inducing antigen in mice sensitized to peanuts) as an oral immunization for peanut allergy in mice. Chitosan has also been used to deliver genes encoding Der p 1 (a major triggering factor for mite allergy) as an oral vaccine against allergies to dust mites. Chitosan is not limited to oral administration, however, and has been used for a variety of gene delivery applications, including several in the field of regenerative medicine.

12.4.1.4 PLL

PLL is a well-known polycation that has been widely studied as a nonviral gene delivery vector since the first reported formation of PLL–DNA complexes in 1975. As its name implies, it is a polypeptide of the essential amino acid L-lysine. At physiological pH, each repeating unit of PLL carries a positive charge on the ε-amine of its side chain, a property that has been exploited to allow it to condense plasmid DNA to varying degrees depending upon salt concentration. Although the structure of PLL suggests suitability for gene delivery, unmodified versions are associated with low transfection efficiency and cytotoxicity.

PLL was one of the first cationic peptides used to mediate gene delivery. However, as the length of PLL increases, so does its cytotoxicity. Moreover, the polydispersity of PLL complicates modifications with ligands, making the chemical synthesis of PLL conjugates hard to control.

12.4.2 Lipids

Lipids—particularly cationic lipids—have been used for gene delivery in a process termed **lipofection**. The aqueous environment inside cells causes the hydrophobic tails of the lipids to coalesce to form hollow liposomes, the interiors of which can contain oligonucleotides for cellular delivery. This is different from micelles, which have a single layer of lipids so that the micelle interior is a hydrophobic environment. Liposomes have a bilayer (like a cell) which yields a hydrophilic interior suitable for carrying DNA.

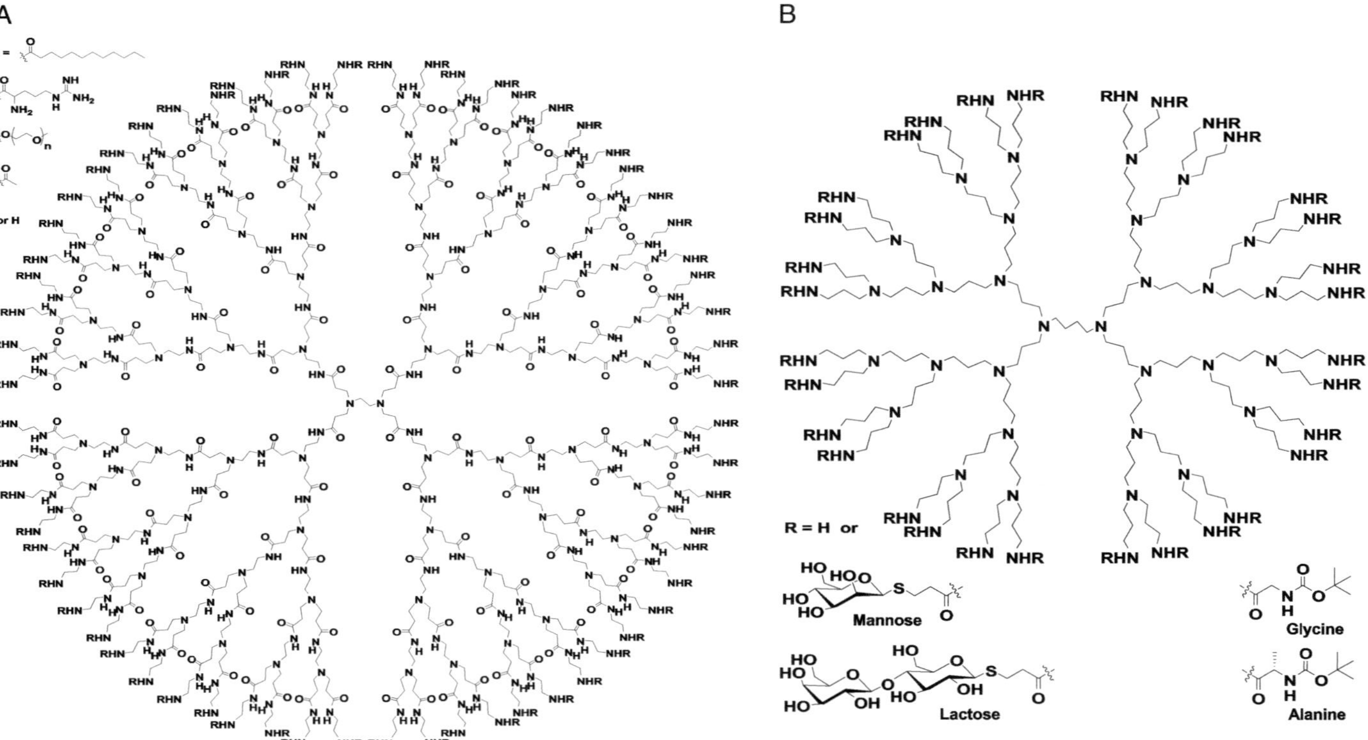

FIGURE 12.8

(A) G4 Poly(amidoamine) (PAMAM) dendrimer with end group functionalization. (B) G3 poly(propylenimine) dendrimer with carbohydrate or amino acid chain end functionalization.

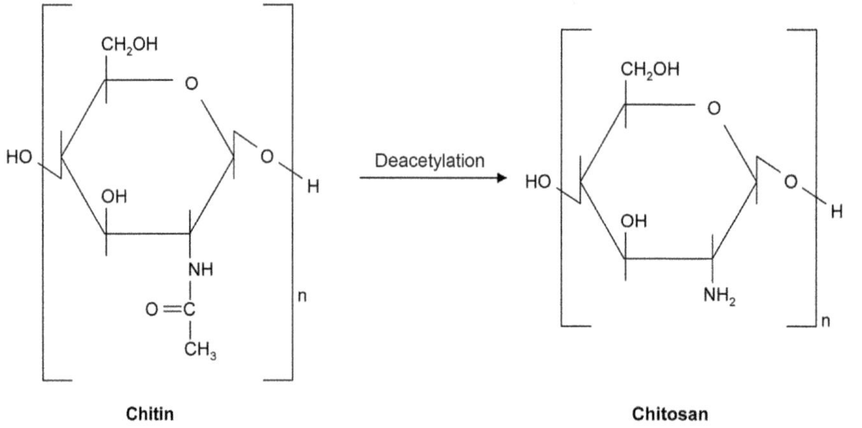

FIGURE 12.9

Chitosan is derived from chitin, which is found in shellfish shells. The figure shows an ideal, complete deacetylation of chitin; however, most chitosan preparations are random copolymers containing both the acetylated *(left)* and deacetylated *(right)* forms of the monomer.

The lipids used for gene delivery typically have a glycerol backbone, one or two fatty acid tails, and a cationic head group. (It might help to think of the phospholipid structure presented in Chapter 2, although most lipids used for gene delivery do not have the phosphate group.) The head group is varied according to the type of target cell and whether the experiment will be performed in vitro or in vivo. Sometimes a molecule like cholesterol is used as part of the hydrophobic portion of the molecule; it is the hydrophobic part of the lipid that allows for self-assembly into liposomes. The morphologies of these structures can vary greatly, as we shall soon see.

12.4.2.1 Liposome geometry

A number of structures can result during the interaction of cationic lipids with polynucleotides to form **lipoplexes** (polynucleotide-containing liposomes). The shape of the lipoplex will be determined by the most thermodynamically favorable conformation, which is described by the **packing parameter** (P), which takes into account the ratio of certain size variables:

$P = v / al_c$ where v= the volume of the hydrocarbon,
a = the effective area of the head group, and
l_c= the length of the lipid tail(s).

We can use the packing parameter to predict the shape of the resulting lipoplex (Figure 12.10):

- $P < \frac{1}{3} \rightarrow$ spherical micelle
- $\frac{1}{3} \leq P < \frac{1}{2} \rightarrow$ cylindrical micelle
- $\frac{1}{2} \leq P < 1 \rightarrow$ flexible bilayers, vesicles
- $P = 1 \rightarrow$ planar bilayers
- $P > 1 \rightarrow$ inverted micelles (hexagonal [H_{II}] phase)

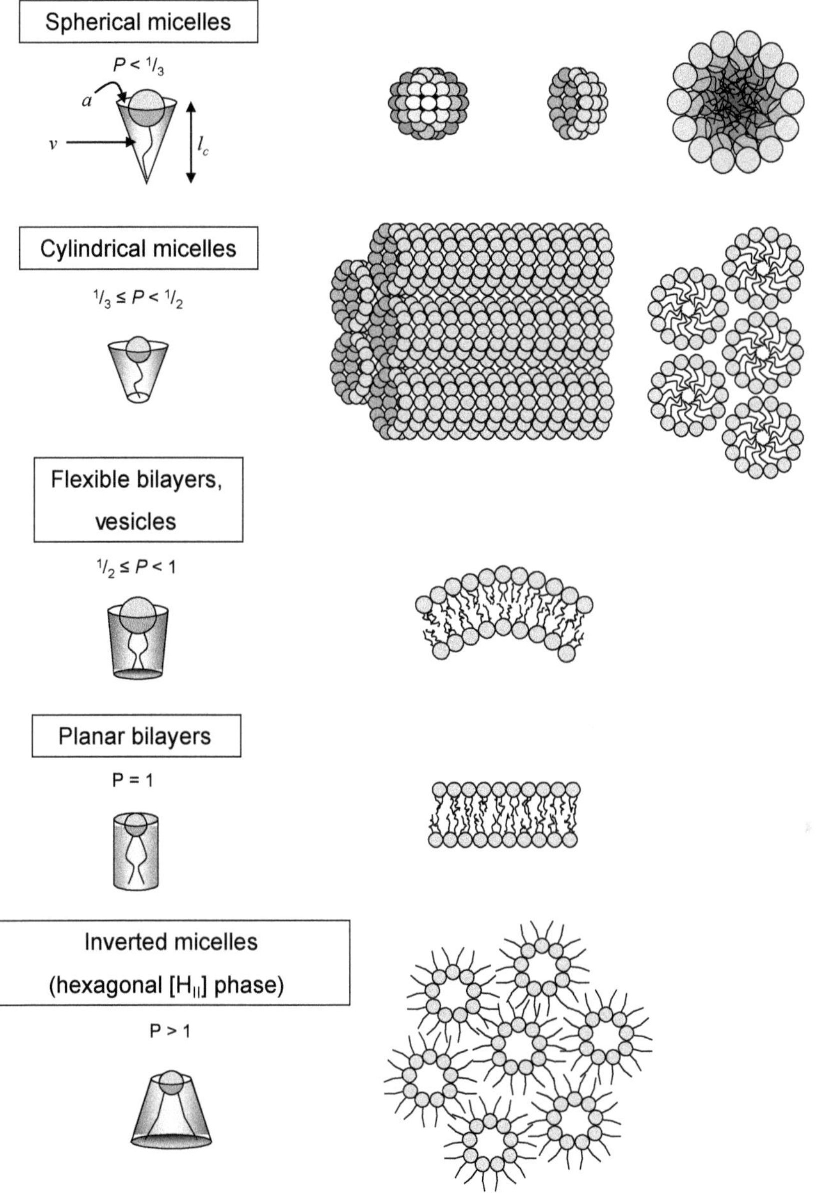

FIGURE 12.10

Structures predicted by the packing parameter *P*.

The lipids used for gene delivery typically have hydrocarbon tails 8 to 18 carbons long. The tails are typically saturated, but sometimes one double bond is present. The combination of hydrocarbon chains attached to glycerol can be asymmetric (as in the plasma membrane phospholipids) or symmetric. Interestingly, asymmetric lipids with a shorter saturated tail and a longer unsaturated tail produce relatively high transfection efficiencies versus lipids with symmetric tails.

The ionizable head group also plays a role in transfection efficiency. The head groups used in gene delivery can carry a single charge or multiple charges, and the net charge is typically cationic. The following discussion presents some of the more common lipids used in gene delivery, separating them by the number of charges in the head group.

12.4.2.2 Monovalent cationic lipids

12.4.2.2.1 DOTMA

N-[1-(2,3-dioleyloxy)propyl]-N,N,N-trimethylammonium chloride, or DOTMA (Figure 12.11), was one of the first synthesized and commercially available cationic lipids used for gene delivery. It contains two unsaturated fatty acid tails that are 18 carbons long, each having a double bond at carbon #9 attached to glycerol via ether bonds. The head group gets its charge from a quaternary amine. The cationic head group provides the ability to entrap DNA or RNA within the liposomal lumen. This feature, first described in the 1980s, significantly influenced and improved the potential of all nonviral gene delivery vehicles. The initial success of in vitro transfection of multiple cell lines with DOTMA sparked a movement to find lipids that were even more effective for gene delivery, and the most successful of these will be discussed later in this section.

DOTMA was eventually paired with the helper lipid DOPE (discussed later) and sold under the trade name Lipofectin. Following the commercial success of Lipofectin, even more improvements in lipoplex formulation were sought, aimed at increasing transfection efficiency and lowering cytotoxicity. The structural changes included different combinations of side chains and attachments to the head group, or head group modification via the replacement of one of the methyl groups (–CH_3) on the head group nitrogen with a hydroxyl (–OH). This latter modification was found to stabilize the resulting vesicles and increase transfection efficiency. Not surprisingly, the length of the hydrophobic tails also affects transfection efficiency: shorter tails appear to be more effective—at least to a point.

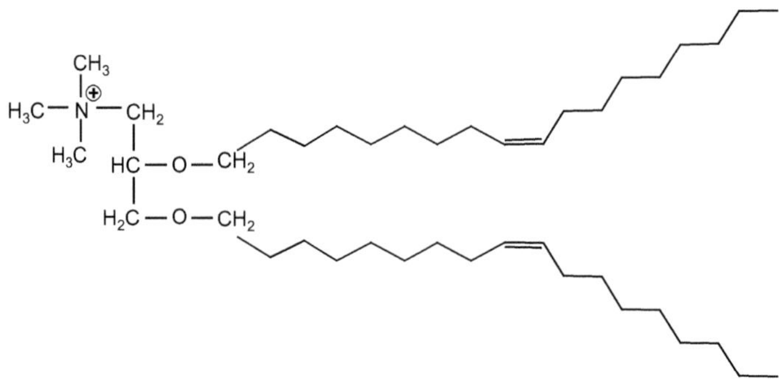

FIGURE 12.11

The structure of DOTMA.

12.4.2.2.2 DOTAP

N-[1-(2,3-dioleoyloxy)propyl]-N,N,N-trimethylammonium chloride, or DOTAP (Figure 12.12), was first synthesized in 1990. The molecule consists of a quaternary amine head group coupled to a glycerol backbone with two oleoyl chains. The only differences between DOTAP and DOTMA are that ester bonds, rather than ether bonds, link the chains to the backbone. It was originally hypothesized that the ester bonds, which are hydrolysable, could render the lipid biodegradable and reduce cytotoxicity.

The use of 100% DOTAP for gene delivery is inefficient, perhaps due to a high surface charge density after lipoplexes have been formed. DOTAP is completely protonated at a pH of 7.4 (which is not the case for other cationic lipids), so it is possible that more energy is required to separate DNA from the lipoplex after delivery into the cell. Thus, for DOTAP to be more effective in gene delivery, it should be combined with a helper lipid such as DOPE, as seems to be the case for most cationic lipid formulations.

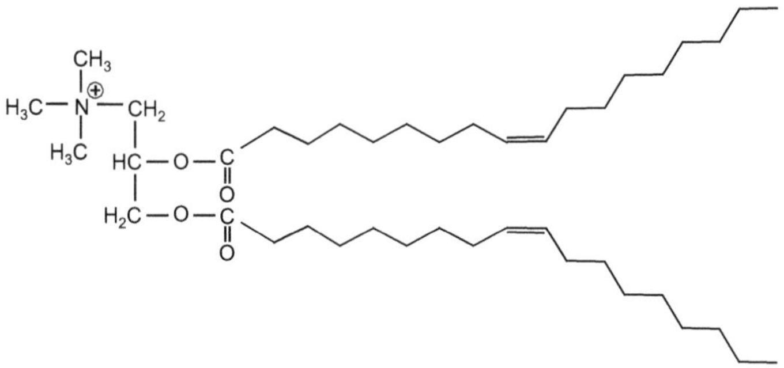

FIGURE 12.12

The structure of DOTAP.

Box 12.5 Naming lipids

Unless you have a strong background in chemistry, the naming of the lipids in this section will be confusing at first. Let us take a quick look at the first two lipids covered—DOTMA and DOTAP—so we can have some idea of where the abbreviations come from. It will be helpful to look at Figures 12.11 and 12.12 as we proceed.

One of the formal names of DOTMA is N-[1-(2,3-dioleyloxy)propyl]-N,N,N-trimethylammonium chloride. Taking the numbers and single Ns out for a moment, we get (dioleyloxy)propyl-trimethylammonium chloride.

To name this molecule, look (in Figure 12.11) for the three carbons of glycerol (to which the two long hydrocarbon tails are attached): these make up the backbone of the molecule. The three-carbon hydrocarbon is propane (referred to as a "propyl" group here). The numbers in the long name refer to the carbon of the propyl group to which everything else is attached.

Next, we describe what is attached to the propane, starting with the long hydrocarbon tails. The tails are 18 carbons long with a double bond between carbons 9 and 10. This specific structure came from the fatty acid oleate (oleic acid), hence the "oleyl" portion of the shortened name. Because there are two such tails, they are described together as "dioleoyl." The "oxy" part of "dioleoyloxy" helps to indicate how they are attached to the backbone (carbons 2 and 3 of the propyl group).

The remaining portion of the molecule is the polar head group, the core of which is a nitrogen atom with four things attached to it; this makes it an "ammonium" group (similar to ammonia, which has four hydrogen atoms attached to it). The only remaining unnamed portions of the molecule are all attached to this ammonium group: three -CH_3 groups. The single carbon makes each of these hydrocarbons "methyl" groups. Because there are three methyls attached to the ammonium

Continued

Box 12.5 Naming lipids—cont'd

group, we refer to this moiety as a trimethylammonium. (The "N, N, N" means the three methyls are attached to the N of the ammonium.) You can now think of the shortened name as **di** **o**leyloxy **t**ri **m**ethyl **a**mmonium propane, or DOTMA.

Turning to the structure of DOTAP (Figure 12.12), we see it is identical to DOTMA except for how the hydrocarbon tails are attached to the backbone—with ester linkages instead of ether.

The formal name for DOTAP is 1,2-dioleoyl-3-trimethylammonium-propane. Taking the numbers out for a moment, we get **dio**leoyl **t**rimethyl**a**mmonium **p**ropane.

Using the same reasoning we applied to DOTMA, it should be a relatively straightforward task to see how we arrive at the abbreviation "DOTAP".

If you are paying *very* close attention, you will notice two differences in the formal names for DOTAP and DOTMA. First, the numbers indicating where the hydrocarbon tails attach are different. This relates to the definition of carbon #1 in each propyl group, which in turn is related to the fact that the attachment sites of the oleate tails have an extra oxygen atom in DOTAP. (The carbon numbering system is outside the scope of this text.) The second difference also has to do with this extra oxygen atom. The formal name of DOTAP has an extra "o" in the oleate descriptor: N-[1-(2,3-dioleoyloxy)propyl]-N,N,N-trimethylammonium chloride (as opposed to "dioleyl" in DOTMA. The DOTMA descriptor is missing an O, and the chemical structure is also missing an O at the junction with the backbone!

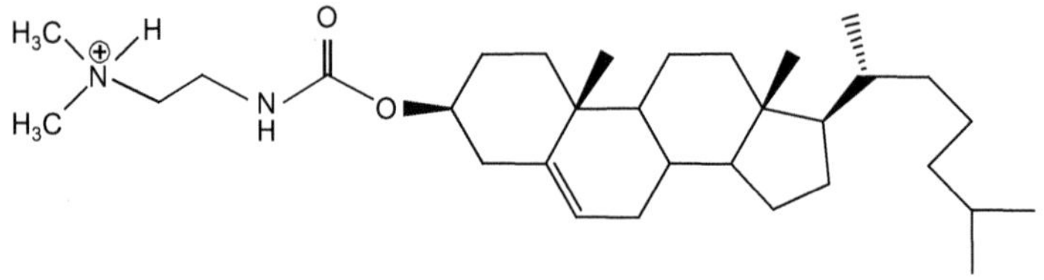

FIGURE 12.13

The structure of DC-Chol.

12.4.2.2.3 DC-Chol

3β[N-(N', N'-dimethylaminoethane)-carbamoyl]cholesterol, or DC-Chol (Figure 12.13), was first described in 1991. DC-Chol contains a cholesterol moiety attached by an ester bond to a hydrolysable dimethylethylenediamine. Cholesterol was reportedly chosen for its biocompatibility and the stability it imparts to lipid membranes (just as in our own cellular membranes), an idea which is supported by DC-Chol's desirable transfection efficiencies and fairly low cytotoxicities.

In contrast to DOTMA and DOTAP, which are cationic because of quaternary amines in their head groups, DC-Chol gets its charge from a tertiary amine. Only half of these nitrogen atoms are protonated at physiological pH, which may result in better DNA dissociation after delivery into the cell. It may also provide lower cytotoxicity and lend the potential to act as a buffer when pH drops (such as in the late endosome or endolysosome).

12.4.2.3 Multivalent cationic lipids

12.4.2.3.1 DOSPA

2,3-dioleyloxy-*N*-[2(sperminecarboxamido)ethyl]-*N*,*N*-dimethyl-l-propanaminium trifluoroacetate, or DOSPA (Figure 12.14), is another cationic lipid synthesized as a derivative of DOTMA. The structure is like that of DOTMA, except for the addition of a spermine molecule ($C_{10}H_{26}N_4$) to the head group via some interesting chemistry (notice the carboxamido ethyl attachment). This cationic lipid, used

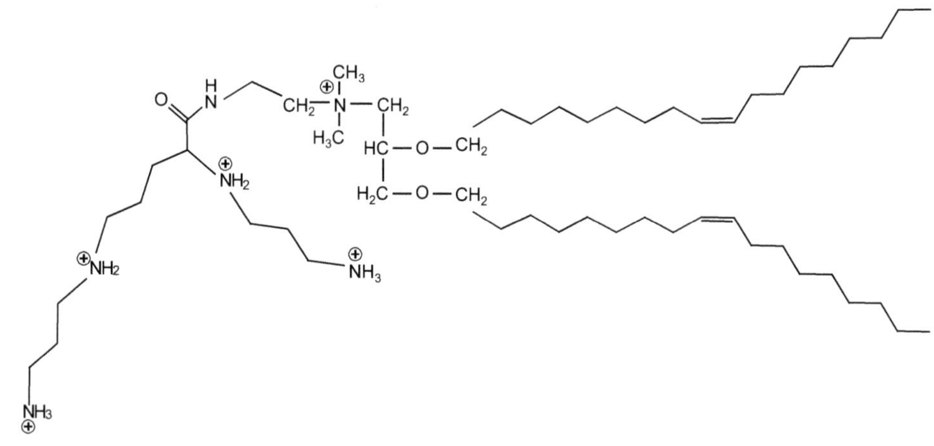

FIGURE 12.14

The structure of DOSPA.

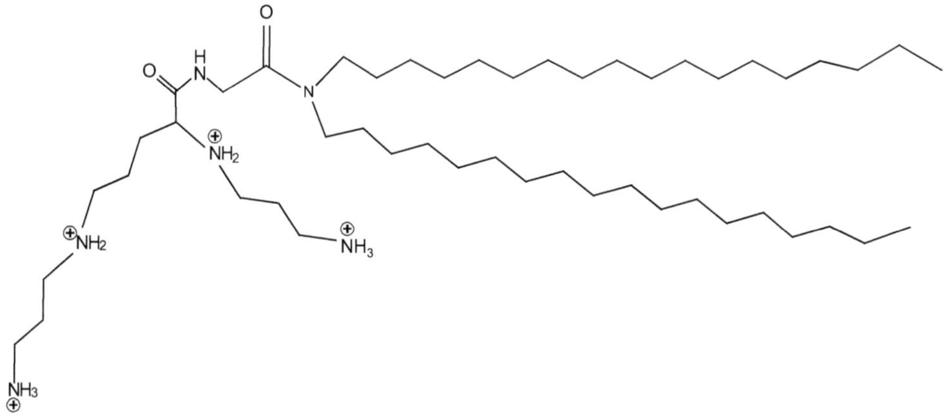

FIGURE 12.15

The structure of DOGS.

with the neutral helper lipid DOPE (we will get to DOPE, I promise!) at a 3:1 ratio, is commercially available as the transfection reagent Lipofectamine.

In general, the addition of the spermine functional group allows for a more efficient packing of DNA in terms of liposome size. The efficient condensation yields a smaller complex, possibly because of the four protonable nitrogens in the spermine. It has been shown that spermine can interact with one strand of dsDNA and wind around the major groove to interact with complementary bases of the opposite strand.

12.4.2.3.2 DOGS

Di-octadecyl-amido-glycyl-spermine, or DOGS (Figure 12.15), has a structure similar to DOSPA; both molecules have a multivalent spermine head group and two 18-carbon alkyl chains. However, the chains in DOGS are saturated, are linked to the head group through a peptide bond, lack a quaternary amine, and do not connect to a glycerol backbone. DOGS is commercially available under the name Transfectam.

Much like the multivalent cationic lipid DOSPA, DOGS is very efficient at binding and packing DNA because of the spermine head group. DOGS is a multifaceted molecule in terms of buffering capacity. At pH values lower than 4.6, all the amino groups in the spermine are protonated, whereas at a pH of 8 only two are purportedly ionized, which promotes arrangement into a lamellar structure.

12.4.2.4 Neutral helper lipids

12.4.2.4.1 DOPE and DOPC

When we are making lipoplexes for gene delivery, lipids with a cationic charge are desirable for their ability to interact with DNA. However, mixing in a class of neutral lipids, known as **helper lipids**, often improves transfection efficiency to a significant degree. Two very common helper lipids are dioleoylphosphatidylethanolamine (DOPE)—the most widely used helper lipid—and dioleoylphosphatidylcholine (DOPC) (Figure 12.16). Look closely and you will see that these molecules are just specific forms of phosphatidylethanolamine and phosphatidylcholine, two of the plasma membrane phospholipids we studied in Chapter 2. The improved transfection efficiencies are due to a conformational change in the liposomes at low pH, from a flexible bilayer to an inverted hexagonal packing structure (refer back to Figure 12.10). It has been shown

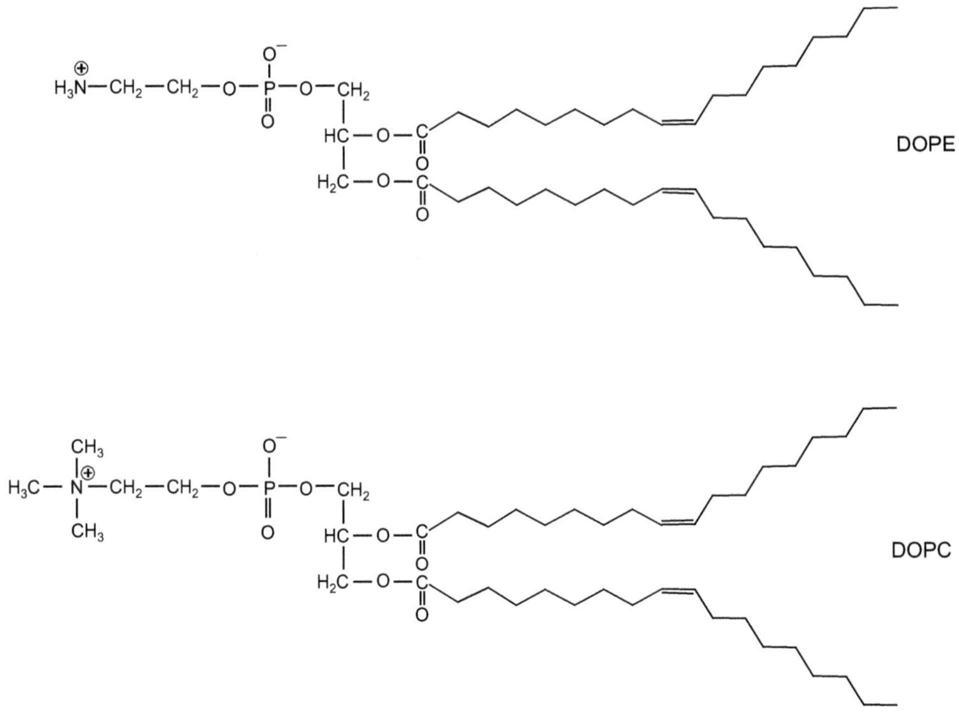

$$H_3\overset{\oplus}{N}-CH_2-CH_2-O-\underset{\underset{O}{\|}}{\overset{\overset{O^-}{|}}{P}}-O-CH_2-CH(-O-C(=O)-R)-CH_2-O-C(=O)-R$$

DOPE

$$(CH_3)_3\overset{\oplus}{N}-CH_2-CH_2-O-\underset{\underset{O}{\|}}{\overset{\overset{O^-}{|}}{P}}-O-CH_2-CH(-O-C(=O)-R)-CH_2-O-C(=O)-R$$

DOPC

FIGURE 12.16

The structure of DOPE *(top)* and DOPC *(bottom)*.

that a hexagonal conformation allows for efficient escape of complexed DNA from endosomal vesicles via destabilization of the vesicle membrane, brought about by exposure of the endosomal membrane to invasive hydrocarbon chains from the helper lipids.

Another advantage of DOPE is that it allows closer contact and packing of DNA helices in liposomal formulations. It is thought that salt bridges form between the positively charged head groups of cationic lipids and the phosphate groups of the DOPE molecules; this association would force DOPE's primary amine to stabilize itself in the plane of the liposome surface, permitting closer interactions with the negatively charged phosphate groups of DNA. Conversely, DOPE could potentially cause the release of a negatively charged counterion from the positively charged head group of a cationic lipid, which would allow easier binding to DNA.

DOPC is used much less often than DOPE. Its choline head group is bulkier than the ethanolamine head group of DOPE, which serves to sterically hinder any interactions that may otherwise have occurred with the primary amine. It would also hinder the alignment of the primary amine in the plane of the liposome surface, thus preventing any liposome-stabilizing effects.

12.4.2.5 Summary of lipids

To help you get a handle on the lipids we have covered, the following table is presented as a summary of the specific lipids mentioned. Note that this is neither an exhaustive list of all the lipids that have been used for gene delivery, nor is it a list of every detail of the lipids presented.

Table 12.1 Lipids covered in this chapter

Lipid	Charge	Structural comment	Why tried
DOTMA	Single cation	Two oleate tails and a trimethylammonium head group	One of first lipids tried
DOTAP	Single cation	The same as DOTMA, but hydrophobic tails are linked to backbone via ester groups	Ester linkages more easily degraded than ethers in DOTMA
DC-Chol	Single cation	A cholesterol, an amino bond, and a choline group (minus one of the three methyls)	Cholesterol for stability Head group: not all tertiary amines are protonated at physiological pH so can act as a buffer in endo(lyso)some
DOSPA	Multiple cations	Like DOTMA, but head group has a spermine attached to one of the terminal methyls	More positive charges, better packing vs DOTMA
DOGS	Multiple cations	Like DOSPA, but hydrocarbon tails are saturated and linked to the head group via a peptide bond Lacks a quaternary amine and does not have a glycerol backbone	Buffers the low pH of endo(lyso) somes, less cytotoxic when released to cytoplasm.
DOPE	Neutral	Simply phosphatidylethanolamine with specific tail groups (two oleoyl groups, like DOTAP)	Helper lipid Aids with endo(lysosomal) escape Improved DNA packing
DOPC	Neutral	Like DOPE but with a choline head group	Not as effective as DOPE, presumably because of the increased bulk of the head group

12.5 Preparation of nonviral gene delivery complexes

At this point in our learning, we should be able to engineer plasmids, amplify and purify them, and dilute them to usable concentrations. If we are to use chemically mediated gene delivery, we now know enough to select the delivery method best suited for our purposes. We must now determine how much of our gene delivery vehicle to combine with our plasmids to make usable gene delivery complexes. For the following example, we will use PEI, although the principles presented apply to any chemically mediated gene delivery method.

In determining the amount of gene delivery vehicle to use for a given amount of DNA, it is important to consider the ratio of charges between the carrier and plasmid because the positive charges of the carrier will be interacting with the negative charges of the DNA. If we consider the charges, it is not so important to discuss the number of carrier molecules versus plasmid molecules in solution, so we can alleviate the necessity of accounting for the number of bases in each plasmid. Although it is true that in virtually all cases, the positive charges on carrier molecules are due to protonated amines, it should not be assumed that every amine will be protonated at physiological pH. Because a precise determination of protonation cannot be determined with a great degree of resolution, we will therefore pay attention instead to the number of amines used to interact with DNA molecules. Specifically, we determine the ratio of carrier amines, which can carry positive charges, to the number of DNA phosphates, which carry all of the negative charges on the polynucleotide. This amine-to-phosphate ratio is abbreviated as the **N:P ratio**.

Nonviral gene delivery complexes are usually delivered to cells in the form of a solution containing two prime constituents (other than salts): the gene delivery vehicle and the DNA or RNA to be delivered. Let us begin by supposing that there are no phosphates in the gene delivery vehicle. There are, of course, plenty of phosphates in the polynucleotides. Although the number of phosphates will vary with the size of the polynucleotide to be delivered, there will only be one phosphate per (deoxy)nucleotide constituent (with the exception of the termini, if we are not delivering plasmids). On the gene delivery vehicle side of things, we must consider the number of nitrogen atoms per **repeating unit**. ("Repeating unit" refers to the recurring portion of a polymer. For nonpolymers such as lipids, this discussion is still valid if we consider the entire molecule to be the repeating unit and the number of repeats as one.) In the case of PEI, whether we are talking about the linear or branched form, the entire molecule can be subdivided into carbon-carbon–nitrogen repeats (the hydrogen atoms have been excluded for clarity). Whether a particular nitrogen atom is a primary, secondary, or tertiary amine, it is a part of a C–C–N unit and has the potential to be protonated. The same is true for the nitrogen atoms on the side chain of PLL or in chitosan; both polymers contain repeating units with one protonable amine each (Figures 12.9 and 12.17). (Note that in PLL, the α amino group in all but the terminal lysine will not be protonated because it will be part of a peptide bond upon polymerization.) Looking at basic repeating units, we can see that controlling the ratio of delivery vehicle nitrogens to polynucleotide phosphates—the N:P ratio—is a relatively straightforward process that can be handled by controlling the total *mass* used for each constituent, alleviating the need to factor in polymer sizes and polydispersities.

If we want to deliver DNA to a known number of cells—say, 100,000—we must determine the appropriate amount of carrier to use, which depends on several parameters. An acceptable N:P ratio must be decided on before the actual transfection procedure. This can often be determined by reading the scientific literature, but we can determine it experimentally for ourselves by trying out several values and selecting the optimum. Suppose we have read several papers stating that an N:P ratio of 5.00:1.00 is optimal for a certain gene delivery vehicle in a given cell type. Keeping in mind that the

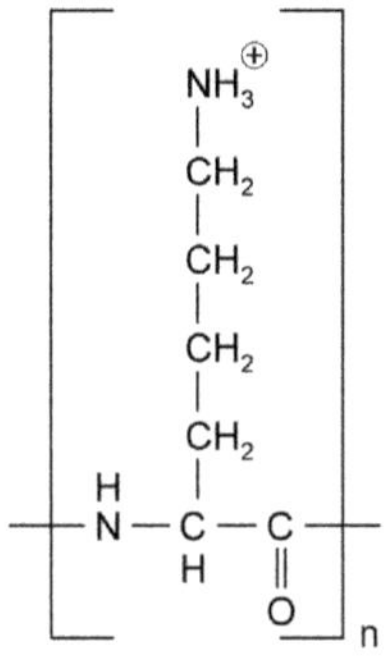

FIGURE 12.17

Basic repeating unit of PLL. Note that only the N in the side chain is protonable.

N:P ratio is based on either a known amount of gene delivery vehicle or a known amount of polynucleotide for a given experiment, using a standard amount of one of these constituents will ensure a set amount of the other. It is common to use a standard amount of DNA per transfection; for this example, let's deliver 2.00 μg. Because the N:P ratio involves a ratio of gene delivery vehicle repeating units to polynucleotide repeating units, we must factor in the molecular weights of these repeating units. The molecular weight for a repeating unit of the polymer can be determined directly from its chemical formula. For PEI, consider the repeating unit to be -[CH_2-CH_2-NH]-, which has a molecular weight of 43.07 g/mol. The average molecular weight of a DNA base is approximately 308.00 g/mol (a more precise determination of this value is addressed by one of the problems at the end of this chapter).

The molecular weight of a repeating unit of PEI is quite different from that of PLL. *Question:* If we were separately to use PLL to deliver the same amount of DNA to cell samples, assuming equal N:P, would we have to use more or less PLL relative to PEI? *Answer:* We would have to use more PLL because the repeating unit is larger than that of PEI. Think of it in terms of delivering a mole of protonable amines: because the molecular weight of a mole of PLL repeating units is larger than that of a mole of PEI repeating units, a mole of protonable amines delivered via PLL will also be associated with greater mass. The formula we are creating is not restricted to polymers—but if we use a lipid, we will consider the size of each lipid molecule rather than the size of a repeating unit.

We must next consider the concentration of gene delivery vehicle in our solution. Pure polymer is a solid, and it would be very impractical to weigh out microgram amounts for each transfection. Sometimes the polymers are sold as very concentrated solutions, but these are often too viscous to pipette with accuracy. It is much more common to have stock solutions from which one can draw during an experiment. The more concentrated the solution, the less one will need to deliver a given number of protonable amines. This inverse relationship accounts for the dilution factor we will use:

$$(\text{concentration of carrier stock solution})^{-1}.$$

To make our formula more robust, we must also take into account the *number* of protonable amines contained in each repeating unit of the gene delivery vehicle. In practice, this number will typically be 1, except for cationic lipids such as DOSPA or DOGS. If there were multiple protonable amines per repeating unit, we would need to use fewer repeating units to achieve the desired N:P ratio. This inverse relationship dictates that the number of positive charges per basic repeating unit should appear in the denominator of our equation.

We now have all of the necessary components to construct the equation we will use to determine how much gene delivery vehicle stock we will use to create one dose of transfection solution:

$$1\ dose\ of\ carrier\ stock =$$

$$(N:P)\left(\frac{DNA}{dose}\right)\left(\frac{MW_{Carrier\ repeating\ unit}}{MW_{DNA\ repeating\ unit}}\right)\left(\frac{1}{Carrier\ stock\ concentration}\right)\left(\frac{1}{Net \oplus Charge/Repeating\ \ Unit}\right)$$

The units for this formula are as follows:

$$\mu L/dose = [\ ]\cdot\left[\frac{\mu g}{dose}\right]\cdot\left[\left(\frac{g/mol}{g/mol}\right)\right]\cdot\left[\frac{\mu L}{\mu g}\right]\cdot[\ ]$$

Using the values mentioned in the previous paragraphs, we can determine the amount of 4.307 mg/mL PEI stock solution to use to deliver 2.00 μg of DNA to 100,000 cells:

$$1\ \text{dose of carrier stock} = 5.00\cdot\ 2.00\cdot(43.07/308.00)\cdot\ (1.00/4.307)\ \cdot\ (1.00)\ = 0.32\ \mu L$$

In practice, this volume is too small to measure accurately. However, experiments typically entail many transfections, so this volume will be multiplied because several transfection doses will be made. Alternatively, stock solutions with greater dilution can be made.

12.6 What is gene therapy, anyway?

Although gene delivery vehicles are typically used to deliver plasmid DNA, they are not restricted to use with plasmids. Recall the definition of gene delivery given at the beginning of this chapter: it is *the delivery of genetic material (polynucleotides) into cells to alter the function of the cells.* Gene therapy is defined as the use of gene delivery to achieve a therapeutic benefit for an organism. The material delivered could be plasmid DNA, linear DNA, or single-stranded or double-stranded RNA, no matter the size.

There is some discussion in the community as to whether delivering small interfering RNAs (siRNAs) would count as gene therapy. Under the definitions just given, there is little question. There is no dispute that using retroviruses to achieve permanent transduction via the delivery of RNA into cells is a form of gene delivery (or gene therapy if it were performed on a patient), but here we assert that delivering siRNA to alter the function of cells also counts. Neither the RNA delivered by a retrovirus nor the double-stranded RNA used as siRNA are genes, but they are both intended to alter the expression levels of a given protein. Adding them to a cell is like delivering a gene, in that the expression of a specific protein in that cell will be altered and a patient could potentially gain therapeutic benefit. One could argue that delivering an inhibitor of some protein could therefore also be considered gene delivery. However, our definition restricts us to the delivery of polynucleotides, so the delivery of proteins (even if they are transcription factors for a specific gene) does not count as gene delivery.

12.7 Summary

In this chapter, we discussed different gene delivery vehicles by grouping them into biologic, physical, and chemical delivery methods. The viruses used for gene delivery today are based on viruses found in the wild. Some are DNA viruses—such as adenovirus, AAV, herpes virus, and baculovirus—whereas others are RNA viruses such as lentivirus, MLV, and FIV. The physical delivery methods are a sort of brute force means of getting genes into cells, and include the application of guns, needles, and high voltage (which sound like they would be a good subject for an AC/DC song). The chemical methods include polymers such as PEI and PAMAM dendrimers, whereas the lipid methods utilize molecules with names like DOGS and DOPE (there's probably a song in there, too). Each of these methods comes with its own set of advantages and disadvantages, which is why there is no universal gene delivery vehicle in use today.

The classical methods for gene delivery focus on the delivery of plasmid DNA to cells. However, this is not the only application of gene therapy. There is a whole world of RNA applications that we have not yet addressed. In the next chapter, we will begin our look at RNA delivery by examining RNA interference: what it is and how it can be harnessed to alter protein expression in a cell. The chapter after that will cover another RNA technology: gene editing via CRISPR-Cas systems. Keep reading—there is more excitement ahead!

Related reading

General gene delivery

Advani, S.J., Weichselbaum, R.R., Kufe, D.W., 2003. Gene delivery systems. In: Kufe, D.W., Pollock, R.E., Weichselbaum, R.R., et al. (Eds.), Holland-frei Cancer Medicine, sixth ed. BC Decker, Hamilton, ON.

Baeshen, N.A., Baeshen, M.N., Sheikh, A., Bora, R.S., Ahmed, M.M., Ramadan, H.A., Saini, K.S., Redwan, E.M., 2014. Cell factories for insulin production. Microb. Cell. Fact. 13, 141.

Munson, J.M., Godbey, W.T., 2006. Gene therapy. In: Bronzino, J. (Ed.), The Biomedical Engineering Handbook, vol. 3. third ed. Tissue Engineering and Artificial Organs. CRC Press/Taylor & Francis Group, Boca Raton, FL.

Verma, I.M., Somia, N., 1997. Gene therapy—promises, problems and prospects. Nature 389, 239–242.

Wolff, J., Lederberg, J., 1994. An early history of gene transfer and therapy. Hum. Gene Ther. 5, 469–480.

Zhang, X., Balazs, D.A., Godbey, W.T., 2011. Nanobiomaterials for nonviral gene delivery. In: Sitharaman, B. (Ed.), Nanobiomaterials Handbook, first ed. CRC Press, Taylor & Francis Group, Boca Raton, FL.

Cloning

Hosaka, K., Ohi, S., Ando, A., Kobayashi, M., Sato, K., 2000. Cloned mice derived from somatic cell nuclei. Hum. Cell 13, 237–242.

Wilmut, I., Schnieke, A.E., McWhir, J., Kind, A.J., Campbell, K.H., 1997. Viable offspring derived from fetal and adult mammalian cells. Nature 385, 810–813.

Viral delivery methods

Buning, H., Braun-Falco, M., Hallek, M., 2004. Progress in the use of adeno-associated viral vectors for gene therapy. Cells Tissues Organs 177, 139–150.

Ding, W., Zhang, L., Yan, Z., Engelhardt, J.F., 2005. Intracellular trafficking of adeno-associated viral vectors. Gene Ther. 12, 873–880.

Gao, G.P., Alvira, M.R., Wang, L.L., Calcedo, R., Johnston, J., Wilson, J.M., 2002. Novel adeno-associated viruses from rhesus monkeys as vectors for human gene therapy. Proc. Natl. Acad. Sci. U.S.A. 99, 11854–11859.

Gonçalves, M.A., 2005. Adeno-associated virus: from defective virus to effective vector. Virol. J. 2, 43–59.

Grassi, G., Köhn, H., Dapas, B., Farra, R., Platz, J., Engel, S., et al., 2006. Comparison between recombinant baculo- and adenoviral-vectors as transfer system in cardiovascular cells. Arch. Virol. 151, 255–271.

Hayashi, M., Tomita, M., Yoshizato, K., 2001. Production of EGF-collagen chimeric protein which shows the mitogenic activity. Biochim. Biophys. Acta 1528, 187–195.

Hu, Y.-C., 2006. Baculovirus vectors for gene therapy. Adv. Virus Res. 68, 287–320.

Jenkins, F.J., Turner, S.L., 1996. Herpes simplex virus: a tool for neuroscientists. Front. Biosci. 1, d241–d247.

Kay, M.A., Glorioso, J.C., Naldini, L., 2001. Viral vectors for gene therapy: the art of turning infectious agents into vehicles of therapeutics. Nat. Med. 7, 33–40.

Kost, T.A., Condreay, J.P., Jarvis, D.L., 2005. Baculovirus as versatile vectors for protein expression in insect and mammalian cells. Nat. Biotechnol. 23, 567–575.

Lachmann, R.H., 2004. Herpes simplex virus-based vectors. Int. J. Exp. Path. 85, 177–190.

Lilley, C.E., Branston, R.H., Coffin, R.S., 2001. Herpes simplex virus vectors for the nervous system. Curr. Gene Ther. 1, 339–358.

Matsushita, T., Elliger, S., Elliger, C., Podsakoff, G., Villarreal, L., Kurtzman, G.J., et al., 1998. Adeno-associated virus vectors can be efficiently produced without helper virus. Gene Ther. 5, 938–945.

Nicolson, S.C., Samulski, R.J., 2014. Recombinant adeno-associated virus utilizes host cell nuclear import machinery to enter the nucleus. J. Virol. 88, 4132–4144.

Russell, D.W., Kay, M.A., 1999. Adeno-associated virus vectors and hematology. Blood 94, 864–874.

Russell, W.C., 2000. Update on adenovirus and its vectors. J. Gen. Virol. 81, 2573–2604.

Santiago-Ortiz, J.L., Schaffer, D.V., 2016. Adeno-associated virus (AAV) vectors in cancer gene therapy. J. Control. Release 240, 287–301.

Smith, A.E., 1995. Viral vectors in gene therapy. Annu. Rev. Microbiol. 49, 807–838.

Zhang, X., Godbey, W.T., 2006. Viral vectors for gene delivery in tissue engineering. Adv. Drug Deliv. Rev. 58, 515–534.

Physical delivery methods

Golzio, M., Teissié, J., Rols, M., 2002. Direct visualization at the single-cell level of electrically mediated gene delivery. Proc. Natl. Acad. Sci. U.S.A. 99, 1292–1297.

King, R., 2004. Gene delivery to mammalian cells by microinjection. In: Heiser, W.C. (Ed.), Gene Delivery to Mammalian Cells, Volume 1: Nonviral Gene Transfer Techniques. Part of Series: Methods in Molecular Biology, vol. 245. Springer, pp. 167–173.

O'Brien, J.A., Holt, M., Whiteside, G., Lummis, S.C., Hastings, M.H., 2001. Modifications to the hand-held Gene Gun: improvements for in vitro biolistic transfection of organotypic neuronal tissue. J. Neurosci. Methods 112, 57–64.

Teissié, J., Rols, M.P., 1993. An experimental evaluation of the critical potential difference inducing cell membrane electropermeabilization. Biophys. J. 65, 409–413.

Chemical delivery methods: PEI

Godbey, W.T., Wu, K.K., Mikos, A.G., 1999. Poly(ethylenimine) and its role in gene delivery. J Control. Release 60, 149–160.

Chemical delivery methods: dendrimers

Grayson, S.M., Godbey, W.T., 2008. The role of macromolecular architecture in passively targeted polymeric carriers for drug and gene delivery. J. Drug Target. 16, 329–356.

Zhong, H., He, Z.G., Zheng, L., Li, G.Y., Shen, S.R., Li, X.L., 2008. Studies on polyamidoamine dendrimers as efficient gene delivery vector. J. Biomater. Appl. 22, 527–544.

Kukowska-Latallo, J.F., Bielinska, A.U., Johnson, J., Spindler, R., Tomalia, D.A., Baker Jr., J.R., 1996. Efficient transfer of genetic material into mammalian cells using Starburst polyamidoamine dendrimers. Proc. Natl. Acad. Sci. U.S.A. 93, 4897–4902.

Zinselmeyer, B.H., Mackay, S.P., Schatzlein, A.G., Uchegbu, I.F., 2002. The lower-generation polypropylenimine dendrimers are effective gene-transfer agents. Pharm. Res. (N. Y.) 19, 960–967.

Chemical delivery methods: chitosan

Chew, J.L., Wolfowicz, C.B., Mao, H.Q., Leong, K.W., Chua, K.Y., 2003. Chitosan nanoparticles containing plasmid DNA encoding house dust mite allergen, Der p 1 for oral vaccination in mice. Vaccine 21, 2720–2729.

Ishii, T., Okahata, Y., Sato, T., 2001. Mechanism of cell transfection with plasmid/chitosan complexes. Biochim. Biophys. Acta 1514, 51–64.

Mansouri, S., Lavigne, P., Corsi, K., Benderdour, M., Beaumont, E., Fernandes, J.C., 2004. Chitosan-DNA nanoparticles as non-viral vectors in gene therapy: strategies to improve transfection efficacy. Eur. J. Pharm. Biopharm. 57, 1–8.

Mao, H.Q., Roy, K., Troung-Le, V.L., Janes, K.A., Lin, K.Y., Wang, Y., et al., 2001. Chitosan-DNA nanoparticles as gene carriers: synthesis, characterization and transfection efficiency. J. Control. Release 70, 399–421.

Roy, K., Mao, H.Q., Huang, S.K., Leong, K.W., 1999. Oral gene delivery with chitosan–DNA nanoparticles generates immunologic protection in a murine model of peanut allergy. Nat. Med. 5, 387–391.

Synowiecki, J., Al-Khateeb, N.A., 2003. Production, properties, and some new applications of chitin and its derivatives. Crit. Rev. Food Sci. Nutr. 43, 145–171.

Chemical delivery methods: PLL

Gonsho, A., Irie, K., Susaki, H., Iwasawa, H., Okuno, S., Sugawara, T., 1994. Tissue-targeting ability of saccharide-poly(L-lysine) conjugates. Biol. Pharm. Bull. 17, 275–282.

Laemmli, U.K., 1975. Characterization of DNA condensates induced by poly(ethylene oxide) and polylysine. Proc. Natl. Acad. Sci. U.S.A. 72, 4288–4292.

Martin, M.E., Rice, K.G., 2007. Peptide-guided gene delivery. AAPS J. 9, E18–E29.

Chemical delivery methods: liposomes

Balazs, D.A., Godbey, W., 2011. Liposomes for use in gene delivery. J. Drug Deliv. 2011, 326497.

Boukhnikachvili, T., Aguerre-Chariol, O., Airiau, M., Lesieur, S., Ollivon, M., Vacus, J., 1997. Structure of in-serum transfecting DNA-cationic lipid complexes. FEBS Lett. 409, 188–194.

Chesnoy, S., Huang, L., 2000. Structure and function of lipid-DNA complexes for gene delivery. Annu. Rev. Biophys. Biomol. Struct. 29, 27–47.

Farhood, H., Gao, X., Son, K., Yang, Y.Y., Lazo, J.S., Huang, L., et al., 1994. Cationic liposomes for direct gene transfer in therapy of cancer and other diseases. Ann. NY Acad. Sci. 716, 23–34.

Farhood, H., Serbina, N., Huang, L., 1995. The role of dioleoyl phosphatidylethanolamine in cationic liposome mediated gene transfer. Biochim. Biophys. Acta 1235, 289–295.

Felgner, P.L., Gadek, T.R., Holm, M., Roman, R., Chan, H.W., Wenz, M., et al., 1987. Lipofection: a highly efficient, lipid-mediated DNA-transfection procedure. Proc. Natl. Acad. Sci. U.S.A. 84, 7413–7417.

Felgner, J.H., Kumar, R., Sridhar, C.N., Wheeler, C.J., Tsai, Y.J., Border, R., et al., 1994. Enhanced gene delivery and mechanism studies with a novel series of cationic lipid formulations. J. Biol. Chem. 269, 2550–2561.

Ferrari, D., Peracchi, A., 2002. A continuous kinetic assay for RNA-cleaving deoxyribozymes, exploiting ethidium bromide as an extrinsic fluorescent probe. Nucleic Acids Res. 30, e112.

Gao, X., Huang, L., 1991. A novel cationic liposome reagent for efficient transfection of mammalian cells. Biochem. Biophys. Res. Commun. 179, 280–285.

Hsu, W.L., Chen, H.L., Liou, W., Lin, H.K., Liu, W.L., 2005. Mesomorphic complexes of DNA with the mixtures of a cationic surfactant and a neutral lipid. Langmuir 21, 9426–9431.

Hui, S.W., Langner, M., Zhao, Y.L., Ross, P., Hurley, E., Chan, K., 1996. The role of helper lipids in cationic liposome-mediated gene transfer. Biophys. J. 71, 590–599.

Israelachvili, J.N., 1991. Intermolecular and Surface Forces, second ed. Academic Press, London, UK.

Jain, S., Zon, G., Sundaralingam, M., 1989. Base only binding of spermine in the deep groove of the A-DNA octamer d(GTGTACAC). Biochemistry 28, 2360–2364.

Kolašinac, R., Kleusch, C., Braun, T., Merkel, R., Csiszár, A., 2018. Deciphering the functional composition of fusogenic liposomes. Int. J. Mol. Sci. 19 (2), 346.

Leventis, R., Silvius, J.R., 1990. Interactions of mammalian cells with lipid dispersions containing novel metabolizable cationic amphiphiles. Biochim. Biophys. Acta 1023, 124–132.

Malone, R.W., Felgner, P.L., Verma, I.M., 1989. Cationic liposome-mediated RNA transfection. Proc. Natl. Acad. Sci. U.S.A. 86, 6077–6081.
Wasungu, L., Hoekstra, D., 2006. Cationic lipids, lipoplexes and intracellular delivery of genes. J. Control. Release 116, 255–264.
Zabner, J., Fasbender, A.J., Moninger, T., Poellinger, K.A., Welsh, M.J., 1995. Cellular and molecular barriers to gene transfer by a cationic lipid. J. Biol. Chem. 270, 18997–19007.
Zuhorn, I.S., Bakowsky, U., Polushkin, E., Visser, W.H., Stuart, M.C., Engberts, J.B., Hoekstra, D., 2005. Nonbilayer phase of lipoplex-membrane mixture determines endosomal escape of genetic cargo and transfection efficiency. Mol. Ther. 11, 801–810.
Zuidam, N.J., Barenholz, Y., 1997. Electrostatic parameters of cationic liposomes commonly used for gene delivery as determined by 4-heptadecyl-7-hydroxycoumarin. Biochim. Biophys. Acta 1329, 211–222.
Zuidam, N.J., Barenholz, Y., 1998. Electrostatic and structural properties of complexes involving plasmid DNA and cationic lipids commonly used for gene delivery. Biochim. Biophys. Acta 1368, 115–128.

Naming

The formal name for DOTMA came from PubChem, CID: 6438350.
The formal name for DOTAP came from PubChem, CID: 107898.

Questions

1. Chloroquine is a lysosomotropic agent, meaning it shuts off the proton/ATPases in lysosomes to prevent their acidification. It has been used to enhance the transfection efficiency of certain nonviral delivery agents, presumably by preventing the activation of pH-sensitive degradative enzymes (acid hydrolases). However, chloroquine reduces the transfection efficiency of lipoplexes that use the helper lipid DOPE. Explain this phenomenon.
2. Name two advantages of using DOPE in liposomal formulations.
3 What are $\overline{M_n}$, $\overline{M_w}$, and PI for the following distribution of polymer sizes?

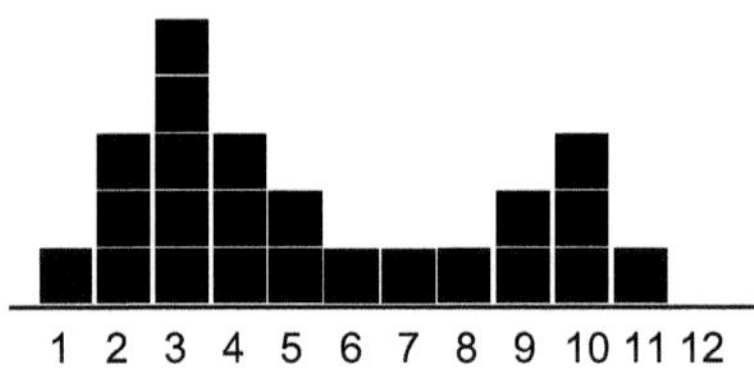

4. Refer to Chapter 4 for the structures of the nucleotide bases. Suppose the percentages of A, T, C, and G bases in the genome of a certain cell type were all the same. Determine the average molecular weight of a DNA base in that genome.
5. Suppose we have two tubes. The first contains a solution of plasmid DNA in a buffer. When we put 2 μL of this solution into a cuvette containing 600 μL of the same buffer (no DNA), we find it has an A_{260} of 0.343. The second tube contains a 10 mg/mL solution of PLL.
   a. What volume of the DNA solution in tube 1 will contain 4 μg of DNA?
   b. How much of the PLL solution in tube 2 will be needed to create one dose of gene delivery complexes containing 4 μg of DNA at a 3:1 N:P ratio?

6. Suppose that we have an 8.614 mg/mL PEI solution. What is the molarity of the solution? How many amines are in the solution?
7. Which of the listed options would be the best gene delivery vehicle or method for each application (a–f)?

a. Repeated deliveries to a patient with diabetes	Adenovirus	Retrovirus	Liposomes
b. Repeated deliveries to skin cells in culture	Adenovirus	Retrovirus	Poly(L-lysine)
c. Permanent delivery to muscle cells in culture	Adenovirus	Lentivirus	Liposomes
d. Delivery of RNA to a cell	Herpesvirus	Adenovirus	FIV
e. Transfection of a frog oocyte	Microinjection	Electroporation	Gene gun
f. Delivery to treat liver disease	Microinjection	PEI	Gene gun

8. Give three reasons (related to one principle) why permanent transduction via a retrovirus can harm a patient.
9 Suppose you want to use the polymer in the figure to deliver 4 μg of plasmid DNA to 50,000 cells at 37°C and a pH of 7.4. You will use N:P = 6.0 to deliver the 7000-bp plasmid. The polymer, for which $n = 1000$, is stored at 2.0 mg/mL in 90 mL of solution. You may use 300 for the molecular weight of a single DNA monomer. If we are to let the transfection proceed for 200 minutes, how much of your polymer solution will you use for the transfection procedure? (Round atomic masses to the nearest integer.)

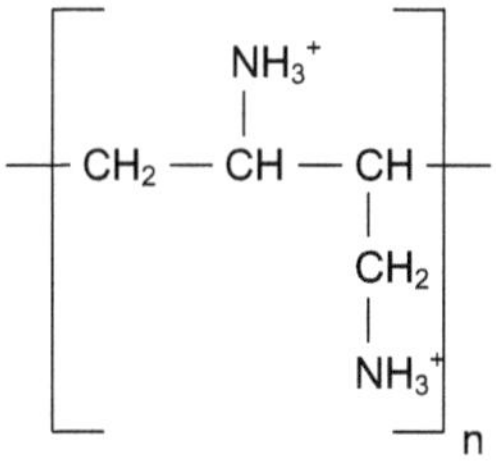

$$\left[\, CH_2 - CH(NH_3^+) - CH(CH_2NH_3^+) \,\right]_n$$

10. What is a main difference between micelles and liposomes?
11. In terms of gene delivery, what is the difference between transformation, transfection, and transduction?
12. Some viral gene delivery regimens can yield permanent transduction. Will cells undergoing such transduction express the delivered genes permanently? Why or why not?
13. A researcher delivers an engineered gene to some furry black laboratory mice via transduction. Unfortunately, the mice do not express the gene in sufficient quantities. Can the researcher go ahead and repeat the procedure on the same mice? Why or why not?
14. **a.** Why is DOPE called a "helper lipid"?
    b. How does DOPE improve transfection efficiency?
15. Dr. Burgess wants to deliver some genes into his laboratory rats and insists on using PEI as his gene delivery vehicle. He has been warned of the low transfection efficiency versus viral methods, but he says he will make up for it by using a little extra PEI in his PEI–DNA complexes. Do you approve of this approach? Why or why not?

16. Name an advantage of each of the following viral delivery methods:
   a. AAV
   b. HSV-1
   c. MLV
   d. Baculovirus
17. Jon tries to deliver genes with only DOTMA and gets acceptable results. A friend tells him to try adding a little DOPC to her DOTMA, so he changes his lipid formulation to 10% DOPC 90% DOTMA, which gives better transfection efficiency. Encouraged by the improvement, he uses even more DOPC in the next trial (50% DOPC and 50% DOTMA) and achieves transfection efficiencies better than he has ever seen with his lipids. Jon is so encouraged by this new wonder lipid that her next transfections use a lipid solution containing only DOPC.
   a. Predict the transfection efficiency of this next transfection. Justify your answer.
   b. Predict the degree of toxicity that will accompany this next transfection. Justify your answer.
18. Consider the molecule shown, which is used for gene delivery.

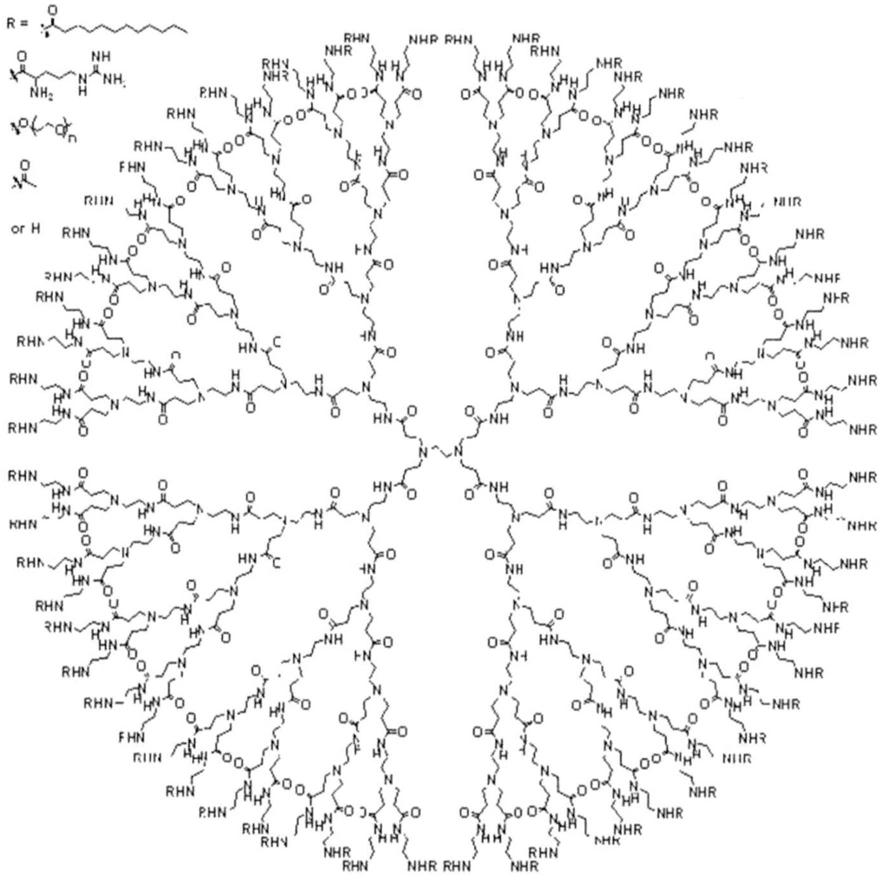

a. To what general class of molecules does it belong?
b. What generation is it?
c. How do you know?
d. Name one advantage and one disadvantage of using this molecule instead of an adenovirus as a gene delivery vehicle.

19. When making chitosan, what is the purpose of the deacetylation step?
20. Suppose we have a virus that only targets dividing cells. Name three targeted applications of this virus for human gene therapy.
21. Consider the structure of DOTMA and that we will modify it by attaching a molecule to one of the methyl groups at the terminus of the polar head group.
    a. Predict the properties of the resulting lipid if the added molecule were a spermine.
    b. Predict the properties of the resulting lipid if the added molecule were a cholesterol.
    c. Predict the properties of the resulting lipid if the added molecule were a 10-repeat linear PEI.
22. DOTMA contains 18-carbon tails. Predict the effectiveness of a similar gene delivery molecule that has 24-carbon tails. Frame your answer in terms of the packing parameter. (Do not consider miscibility issues.)
23. a. How much of a 4.307-mg/mL stock solution of PEI would we use to deliver 3.6 μg of DNA at a 7.5:1 charge ratio?
    b. Your answer from part *a* is too small to pipette accurately. How might you get around the problem?
24. Suppose you have calculated the correct amount of PEI to use for delivering a 4000-bp reporter plasmid to a set number of cells. If you wanted to change your experiments to deliver an 8000-bp plasmid instead, will the amount of PEI needed be doubled, halved, or stay the same? Justify your answer.

CHAPTER

RNAi

13

Chapter outline

So far, we have considered DNA as the molecule central to gene therapy applications. However, we should be aware that RNA is processed in special ways inside the cell not only to yield protein production but also to limit or halt protein production. Using the techniques of gene delivery (Chapter 12), we can deliver RNA into cells to effect changes in protein production and cellular behavior. The interference of protein production via RNA is the focus of this chapter, in which we will examine:

- the odd observation that led to the field of RNA interference technologies;
- how the cell processes dsRNA to protect itself from certain types of viral infection;
- the use of double-stranded RNA to reduce the amount of a specific protein produced by a cell;
- the role of antisense RNA in gene silencing, and how Slicer, Dicer, and RISC are involved in the process; and
- a more permanent form of RNA interference, originating at the genomic level to produce molecules called micro RNAs.

13.1 Co-suppression

Back in the late 1980s (the famous paper came out in 1990) a fellow named Richard Jorgensen was interested in agricultural biotechnology, and he wanted to show that genetic engineering could be used in plants. He wanted to attract investors to his company, DNA Plant Technology Corporation, because—as with any company—money was needed for the company to thrive. In biotechnology, an important way to get money is through investors. As biotechnologists, this could mean we might go to a venture capital group to pitch an idea, and maybe the venture capital group will invest cash into the company in exchange for **equity** (a percentage ownership of the company).

As an entrepreneur, Dr. Jorgensen wanted to demonstrate genetic engineering in a way that investors (who quite often are not scientists) would understand and *be excited about*; after all, if you can't excite people about your project, they most likely won't give you any money for it. By altering the appearance

Biotechnology and its Applications. https://doi.org/10.1016/B978-0-12-817726-6.00013-7

of some flowering plants, Jorgensen would establish "**proof-of-concept**" demonstrating that protein expression patterns could be altered in plants via the delivery of RNA. Proof-of-concept experiments do not necessarily produce the exact end result a company or scientist is interested in, but they do show that the scientific idea is valid and can be built upon.

With the help of his research group, which included Carolyn Napoli and Christine Lemieux, Dr. Jorgensen set about showing that the expression of a gene could be reduced or silenced through the delivery of RNA. He selected the chalcone synthase protein (CHS), which plays a part in the production of a purple pigment in some plants, and he chose pink petunias as his model organism because changes in expression of the pigment would be easily detectable. The group delivered plasmids designed to produce a complementary (antisense) copy of *Chs* mRNA on transcription. The plasmids contained a strong promoter (from the cauliflower mosaic virus), the antisense exon, and a gene for kanamycin resistance. (Yes, a promoter, an exon, and an antibiotic resistance gene—just like in Chapter 11.) As a control, they delivered similar plasmids that contained a sense version of the *Chs* exon instead of the antisense version.

For the plants in which the antisense plasmids had been delivered, the group was successfully able to knock down the expression of the purple pigment. However, some of the plants receiving the sense version of *Chs* produced beautiful *white* flowers rather than more colorful pink ones! The plasmids had somehow knocked out the expression of the purple pigment, and they had done so more effectively than the plasmids coding for antisense mRNA. The experiment was repeated in dark purple petunias to investigate the strength of the gene suppression, and some white flowers were again produced. Flowers with very interesting combinations of purple and white patterns also appeared (Figure 13.1). Jorgensen termed this phenomenon "co-suppression," and it served to confound a significant portion of the scientific community for several years.

Dr. Jorgensen recounts the research in his own words:

> To give you the full picture, we were doing two experiments—one that introduced an antisense *Chs* transgene, intended to silence endogenous *Chs* transcripts, and a second that introduced a sense *Chs* transgene that intentionally had been engineered to produce high levels of CHS protein expression, in order to overexpress CHS protein and possibly produce more pigment, assuming the endogenous *Chs* gene was rate-limiting to pigment synthesis (which we did not know for petunia, but was known for corn kernels and snapdragon flowers). The original purpose of these experiments was to try to produce plants with a visual alteration of a normal phenotype, merely as an illustration of plant genetic engineering in flower crops that we could use in fund-raising efforts when talking to potential investors. We chose petunia as the target flower crop species because:
>
> (1) it was easily transformable with transgenes, at a time (1987) when very few plant species could be transformed
>
> (2) a clone of the *Chs* gene was already available, and
>
> (3) it had flowers with anthocyanin-pigmented petals.
>
> We chose a specific variety of petunia that had light pink flowers because we wanted to see if the sense *Chs* transgene could produce more deeply pigmented flowers (and whether the antisense transgene could reduce pigmentation). To our surprise, both sense and antisense transgenes reduced pigmentation, and the sense transgene was more effective at doing this than the antisense transgene! We did subsequent experiments with a deeply pigmented purple petunia line, and were still able to block all pigmentation in some of the transformed plants.

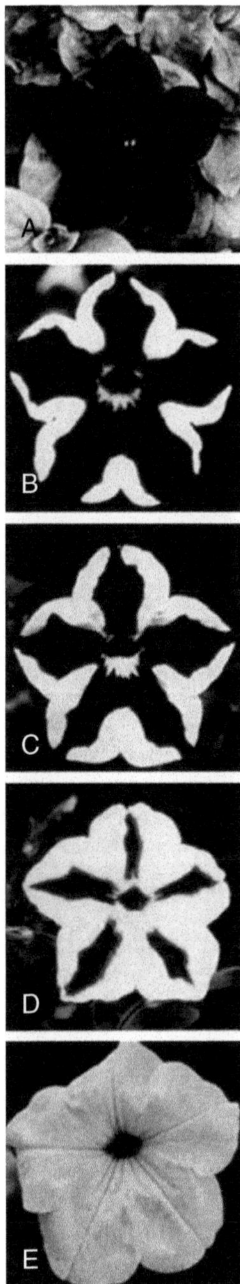

FIGURE 13.1

When Napoli, Lemieux, and Jorgensen transfected purple petunias like the parent shown in (A) with a gene they intended would amplify the production of a purple pigment, some very surprising results were obtained (B–E).

It was significant that our sense transgene was engineered for overexpression of protein expression; otherwise we probably would not have discovered that a sense-oriented transgene could silence a homologous, endogenous gene. The frequency and degree of cosuppression by chalcone synthase transgenes are dependent on transgene promoter strength and are reduced by premature nonsense codons in the transgene coding sequence.

13.2 RNA interference

A few years later, Andrew Fire and Craig Mello performed a famous set of experiments to further investigate this type of suppression of gene expression, which by this point was referred to as "interference." What they found was this: they injected sense single-stranded RNA into *Caenorhabditis elegans*, and nothing happened. They injected antisense single-stranded RNA into these worms, and not much happened. As a control, they injected double-stranded RNA, and the worms no longer expressed the protein associated with that gene: the worms were curled and they twitched. The gene in question was involved with structural elements inside the body; when the worms lacked this structural element, they were no longer straight. The important result was that an effect was seen when double-stranded RNA was injected into these worms. Note that these results were different from what Jorgensen achieved.

Let's take a moment to examine a more straightforward example of antisense technology. Single-stranded RNA can be injected into a fertilized frog oocyte. The egg will divide, eventually forming a blastocyst in which the genomic gene in question will be expressed to the same degree as in the wild-type animal, unless the injected ssRNA was antisense to the mRNA for the corresponding gene. When antisense RNA is injected, there will be reduced expression of the protein related to the gene. In terms of the central dogma, this should make intuitive sense: the cell takes a genomic message in the form of dsDNA and transcribes it into mRNA, which is then transported into the cytoplasm for translation. If one were to inject a sequence of ssRNA complementary to this particular mRNA, would it not follow that the two complementary strands would bind together? The ribosome needs ssRNA to perform translation, so some would predict that this particular mRNA will not be translated and the cell will experience a reduction in the amount of the protein encoded by the gene.

If we deliver dsRNA into the cytoplasm, as was done by Fire and Mello, there will likewise be reduced expression of the gene. This indicates that there is something more going on than simply the prevention of ribosomal binding.

The Jorgensen experiments were not the same as the Fire and Mello experiments. Jorgensen called the effect he observed "co-suppression." He delivered sense ssRNA encoding CHS; not only was the delivered RNA suppressed, but also the endogenous gene was suppressed. Fire and Mello, after determining the important role of dsRNA, termed the suppression they observed "**RNA interference**" (which we now abbreviate **RNAi**). For their discovery and characterization of gene silencing via double-stranded RNA, Fire and Mello were awarded the Nobel Prize in Physiology or Medicine in 2006. It is this dsRNA-mediated suppression that is the focus of this chapter.

Why would dsRNA reduce the amount of protein produced from the associated gene? The answer has to do with three proteins, known as **Dicer**, **Slicer**, and **RISC**. If dsRNA is delivered into a cell, the cell will destroy it. This may be the result of a primordial immune response, given that many naturally occurring viruses infect cells via the delivery of double-stranded RNA. The cell can recognize the viral RNA because it is double-stranded; the only RNA produced by the cell itself is single stranded, so dsRNA is seen as an oddity that must have come from a foreign source. The cell will not only chop up the dsRNA, but also destroy any mRNA associated with the sequence of bases found in the dsRNA (even if it was produced by the cell itself), ostensibly because it may also be linked with a foreign source.

At the heart of RNA interference is the RNA-induced silencing complex (RISC). There are many different types of RISCs, but at least three features are common to all of them:

1. a subunit that serves as an RNA helicase;
2. a subunit that binds to small ssRNA; and
3. a subunit that acts as an endonuclease (the Slicer).

When dsRNA is introduced into the cytoplasm, Dicer binds to it and cleaves it into smaller pieces. RISC associates with the smaller dsRNA fragments (Figure 13.2), and its helicase serves to separate them. The sense strand of this RNA is discarded, leaving RISC primed with a short, single-stranded, antisense RNA molecule. The primed RISC can now interact with other (single-stranded) RNAs, such as mRNA, and use Slicer to degrade the ones that pair with the sequence being held.

If we designed the original dsRNA based on a specific gene we wish to silence, the antisense piece of RNA held by RISC will allow it to bind with the mRNA transcribed from the target gene. Slicer will degrade the bound mRNA strand. By reducing the number of mRNA sequences from the gene in question, the number of proteins expressed from that gene will be reduced.

(It can be difficult to remember the functions of Dicer and Slicer. It may help to remember that Dicer degrades double-stranded RNA, and Slicer severs single-stranded RNA.)

Dicer degrades relatively long dsRNAs into much smaller fragments before RISC binds to and discards the sense strand. Depending on what is delivered into the cytoplasm, the endonuclease step involving Dicer might not be necessary. If we deliver a small (19–24 nucleotide) ssRNA sequence, the phenomenon of interference can still be observed; RISC can bind these small RNAs and continue with the process as already described. This RNA is called small interfering RNA, or **siRNA**.

13.2.1 Determination of "antisense"

How does RISC "know" which strand to keep and which to discard? The first part of the answer appears to have something to do with 3′ overhangs: any siRNA produced via cleavage by Dicer will have 3′ overhangs. In the laboratory, it has been shown that siRNAs with short (2–4 base) 3′ overhangs are more effective at gene **knockdown**, so synthetic siRNAs are now typically made as 21-mers that overlap by 19 bases, meaning they have a 2-nucleotide overhang at each 3′ terminus. Whether the siRNA is produced by Dicer cleavage or is chemically synthesized, Dicer will introduce the siRNA to RISC for further processing. The identity of one of the overhangs influences how easily the transfer takes place. However, there is no agreed-upon sequence for the 3′ overhangs regarding strand selection.

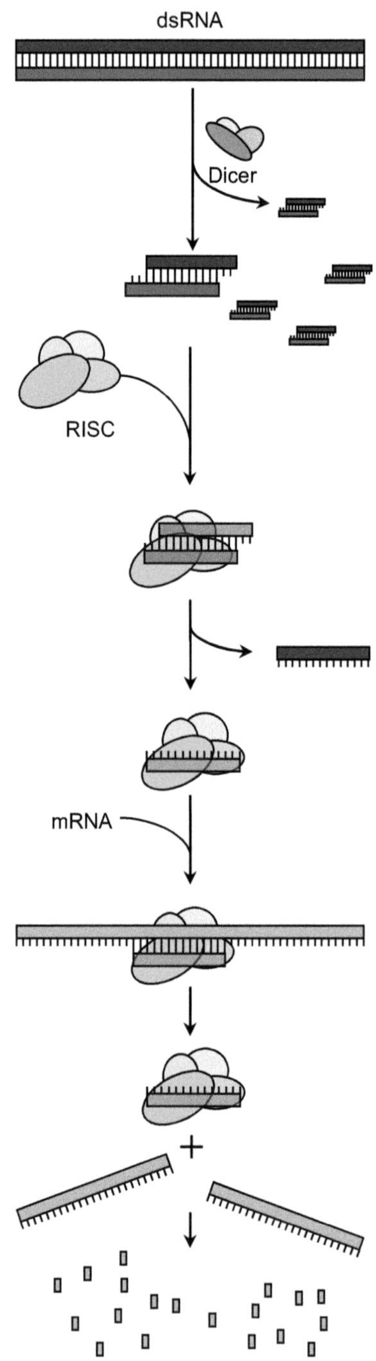

FIGURE 13.2

RNA interference hinges on the recognition of double-stranded RNA. If the dsRNA strand is sufficiently long, Dicer will cleave it into smaller fragments of small interfering RNA (siRNA). One of the smaller dsRNA fragments will be picked up by the RNA-induced silencing complex (RISC), and the sense strand (shown in *blue*) will be discarded. Any mRNA containing a sequence that matches up with the retained antisense strand will be degraded by Slicer, a member of the RISC. Sliced mRNA fragments will be further degraded in the cytosol. *dsRNA*, Double-stranded RNA.

Looking more deeply into the matter, we see that the RNA helicase of RISC also plays a role in determining which strand of siRNA will be retained within the complex: it will be the strand having lower internal stability at the 5′ end. Consider this 21-bp siRNA (a blunt fragment is shown to make the point clearer):

5′–GnnnnnnnnnnnnnnnnnnnT
CnnnnnnnnnnnnnnnnnnnA –5′

There are two 5′ termini, one for the top strand and one for the bottom strand. Notice that the 5′ terminus of the top strand is participating in a G-C base pair, while the 5′ terminus of the bottom strand is participating in an A-T base pair. The A-T base pair is less stable because it has one less hydrogen bond, so the bases are more easily separated. This fraying provides the starting point for the RNA helicase. The helicase always moves in a 5′-to-3′ direction, and the strand that is oriented in the same 5′-to-3′ direction (right to left in this case) will be retained by RISC. To be precise, all of the first four bases are important in this process, but the 5′ terminus is the most important.

RISC does not know which strand is the antisense strand; it could retain either strand if conditions were just right, which could lead to cleavage of the wrong mRNAs. However, the 3′ overhangs created by Dicer do play a role. The end-fraying detected by the RNA helicase determines which 5′ terminus will be retained (along with the rest of the bases in that strand), and it is Dicer (or a biotechnologist) that is responsible for creating the 5′ termini.

13.3 miRNA

We have already discussed several examples of gene delivery. We can use it to deliver plasmids to express a protein of our choosing or to switch on a molecular cascade once the delivered plasmid is expressed. We now have a new application to consider: the delivery of siRNA to knock down the expression of a gene.

Rather than delivering siRNA directly, could we deliver a gene encoding an RNA sequence that is antisense to a given mRNA to not only knock down gene expression? Could we also integrate the gene into the genome to achieve permanent knockdown? Such a scheme has been tried, and it works to some degree. However, a more effective route mimics something that already occurs in the cell—a process that involves **micro RNA (miRNA)**. Micro RNAs have corresponding DNA coding regions in the genome. They are transcribed just like other genes, but contrary to the principles of the central dogma, they are not translated. The primary transcript of an miRNA, referred to as **pri-miRNA** (short for "primary-miRNA"), contains one or several stem-loop structures (Figure 13.3). One (or more) of the stem loops will be cleaved by an enzyme complex called the Microprocessor, which contains the protein Drasha, while the pri-miRNA is still in the nucleus. The result is a free-floating hairpin called a **pre-miRNA**. The pre-miRNA is then exported out of the nucleus and is acted upon by Dicer, which cleaves the loop from the pre-miRNA to yield an miRNA duplex. Because of their similar structures at this point, the rest of the processing story is the same as for siRNA: binding to RISC, separation of the dsRNA, retention of one of the strands, and inactivation of specific mRNA molecules via Slicer-mediated cleavage (compare with Figure 13.2). The cell uses this process to regulate protein expression, for example, to suppress the expression of oncogenes. Biotechnologists have used miRNA technology in an attempt to achieve permanent knockdown of genes (**gene silencing**) by inserting DNA sequences into the genomes of viral vectors for eventual integration into host genomes. The intended result is to produce engineered miRNAs to a selected gene.

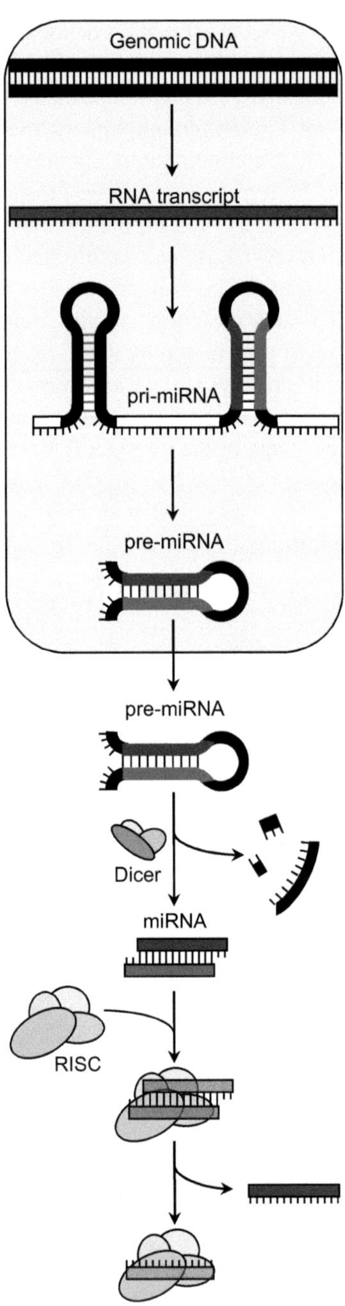

FIGURE 13.3

The origin of miRNA in the cell begins at the genome level. The primary RNA transcript contains inverted repeats that form stem-loop structures (pri-miRNA), which are cleaved into independent structures (pre-miRNA). Pre-miRNA is transported from the nucleus (shaded in *green*) to the cytoplasm, where it is acted upon by Dicer to produce miRNA. Note that the structures of miRNA and siRNA cannot be distinguished at this point.

13.4 Summary

We have seen the mechanism behind RNAi, which utilizes Dicer, Slicer, and RISC to process dsRNA. We have also learned about a more permanent form of RNA interference that originates at the genomic level and yields an RNA transcript that is processed through pri-miRNA and pre-miRNA forms to eventually yield miRNA.

Several types of RNA interference were covered, including delivery of dsRNA, which can directly interact with Dicer (or RISC, if small enough, as is the case with siRNA); antisense RNA, which can pair with mRNA to create dsRNA; and miRNA, a form of dsRNA that arises from inverted repeats in the gene and forms stem-loop structures in the transcript that are processed further. At the start of the chapter, we also saw RNAi arising from the overproduction (or addition) of sense RNA, as was the case in the Jorgensen petunia experiments. This type of RNA interference requires a special RNA-dependent RNA polymerase that recognizes and copies overexpressed transcripts to produce dsRNA molecules. This type of interference is beyond the scope of this chapter, although one could argue that it sparked the field of RNA interference.

In the next chapter, we will take a look at another RNA technology: CRISPR-Cas systems. These are more commonly referred to as gene editing, and they are at the center of many hopes and fears regarding direct, targeted alteration of the human genome.

Related reading

Béclin, C., Boutet, S., Waterhouse, P., Vaucheret, H., 2002. A branched pathway for transgene-induced RNA silencing in plants. Curr. Biol. 12, 684–688.

Cook, H.A., Koppetsch, B.S., Wu, J., Theurkauf, W.E., 2004. The *Drosophila* SDE3 homolog *armitage* is required for *oskar* mRNA silencing and embryonic axis specification. Cell 116, 817–829.

DiNitto, J.P., Wang, L., Wu, J.C., 2010. Continuous fluorescence-based method for assessing dicer cleavage efficiency reveals 3′ overhang nucleotide preference. Biotechniques 48, 303–311.

Denli, A.M., Tops, B.B., Plasterk, R.H., Ketting, R.F., Hannon, G.J., 2004. Processing of primary microRNAs by the Microprocessor complex. Nature 432, 231–235.

Felekkis, K., Deltas, C., 2006. RNA interference: a powerful laboratory tool and its therapeutic implications. Hippokratia 10, 112–115.

Fire, A., Xu, S., Montgomery, M.K., Kostas, S.A., Driver, S.E., Mello, C.C., 1998. Potent and specific genetic interference by double-stranded RNA in *Caenorhabditis elegans*. Nature 391, 806–811.

Gregory, R.I., Yan, K.P., Amuthan, G., Chendrimada, T., Doratotaj, B., Cooch, N., et al., 2004. The Microprocessor complex mediates the genesis of microRNAs. Nature 432, 235–240.

Hammond, S.M., Boettcher, S., Caudy, A.A., Kabayashi, R., Hannon, G.J., 2001. Argonaute 2, a link between genetic and chemical analyses of RNAi. Science 293, 1146–1150.

Heasman, J., Torpey, N., Wylie, C., 1992. The role of intermediate filaments in early *Xenopus* development studied by antisense depletion of maternal mRNA. Dev. Suppl. 119–125 1992.

Jorgensen, R.A., 2003. Sense cosuppression in plants: past, present and future. In: Hannon, G.J. (Ed.), RNAi: A Guide to Gene Silencing. Cold Spring Harbor Laboratory Press, Cold Spring Harbor, NY, pp. 5–21.

Jorgensen, R.A., Doetsch, N., Müller, A., Que, Q., Gendler, K., Napoli, C.A., 2006. A paragenetic perspective on integration of RNA silencing into the epigenome and its role in the biology of higher plants. Cold Spring Harb. Symp. Quant. Biol. 71, 481–485.

Kennedy, S., Wang, D., Ruvkun, G., 2004. A conserved siRNA-degrading RNase negatively regulates RNA interference in *C.* elegans. Nature 427, 645–649.

Khvorova, A., Reynolds, A., Jayasena, S.D., 2003. Functional siRNAs and miRNAs exhibit strand bias. Cell 115, 209–216.

Liu, Y.P., Berkhout, B., 2011. miRNA cassettes in viral vectors: problems and solutions. Biochim. Biophys. Acta 1809, 732–745.

Napoli, C., Lemieux, C., Jorgensen, R., 1990. Introduction of a chimeric chalcone synthase gene into petunia results in reversible co-suppression of homologous genes in trans. Plant Cell 2, 279–289.

Peters, L., Meister, G., 2007. Argonaute proteins: mediators of RNA silencing. Mol Cell 26, 611–623.

Que, Q., Wang, H.Y., English, J.J., Jorgensen, R.A., 1997. The frequency and degree of cosuppression by sense chalcone synthase transgenes are dependent on transgene promoter strength and are reduced by premature nonsense codons in the transgene coding sequence. Plant Cell 9, 1357–1368.

Que, Q., Wang, H.Y., Jorgensen, R.A., 1998. Distinct patterns of pigment suppression are produced by allelic sense and antisense chalcone synthase transgenes in petunia flowers. Plant J. 13 (3), 401–409.

Rose, S.D., Kim, D.H., Amarzguioui, M., Heidel, J.D., Collingwood, M.A., Davis, M.E., et al., 2005. Functional polarity is introduced by Dicer processing of short substrate RNAs. Nucleic Acids Res. 33, 4140–4156.

Schwarz, D.S., Hutvágner, G., Du, T., Xu, Z., Aronin, N., Zamore, P.D., 2003. Asymmetry in the assembly of the RNAi enzyme complex. Cell 115, 199–208.

Torpey, N., Wylie, C.C., Heasman, J., 1992. Function of maternal cytokeratin in *Xenopus* development. Nature 357, 413–415.

Winter, J., Jung, S., Keller, S., Gregory, R.I., Diederichs, S., 2009. Many roads to maturity: microRNA biogenesis pathways and their regulation. Nat. Cell Biol. 11, 228–234.

Zhou, J., Song, M.S., Jacobi, A.M., Behlke, M.A., Wu, X., Rossi, J.J., 2012. Deep sequencing analyses of dsiRNAs reveal the influence of 3′ terminal overhangs on dicing polarity, strand selectivity, and RNA editing of siRNAs. Mol. Ther. Nucleic Acids 1 (4), e17.

Questions

1. Explain the difference between siRNA and miRNA. Are Dicer and Slicer used for both siRNA and miRNA?
2. Considering that it only acts on double-stranded RNA, how is Dicer involved in the production of miRNA that originates from single-stranded RNA?
3. What main functions of RISC are common across species boundaries?
4. What are the differences and similarities between siRNA and miRNA?
5. How did RNA interference contribute to the production of white petunias?
6. Why might a mutated Slicer gene make a cell unhealthy, prone to viral infection, and potentially cancerous?
7. How does RISC determine which strand is sense and which is antisense?
8. Is it possible for miRNA to code for a protein?
9. A 9- to 24-bp strand of dsRNA was found in the cytoplasm, being held by an RNA-induced silencing complex. Is the strand siRNA or miRNA? Explain.
10. What is the difference between pri-miRNA and pre-miRNA?
11. Researchers want to fight a single-stranded RNA viral infection in a patient. The viral RNA is single stranded and has the sequence $(C)_{15}$GGUGCA. The cells that the viruses "attack" produce an essential protein which has an mRNA sequence that contains ACGUGG$(C)_{15}$.
    a. The researchers are considering delivering the siRNA:

    5′-UGCACC$(G)_{15}$-3′
    3′-ACGUGG$(C)_{15}$-5′

Should they use the siRNA or not? Explain.

- **b.** Now consider that the cells instead produce an essential protein which has an mRNA sequence that contains UGCACC$(G)_{15}$. Should the researchers use the siRNA or not? Explain.

12. Why is "knocking out" a gene the wrong term for RNAi? What would be a better term?

13. RNAi has been described as a primordial immune system. Explain the reasoning behind this statement.

14. After successfully using RNAi to lower gene expression in a certain batch of cells, Dr. Harmanis performs a nuclear runoff assay, which isolates RNA from cell nuclei, to help determine transcription amounts and rates. Predict what she will find. Explain your answer.

15. Predict which strand (top or bottom) is more likely to be retained by the RISC complex:

ATTA ... CGGC	TA ... GCAT	CGCG ... GCGCTT
TAAT ... GCCG	GCAT ... CG	GGGCGC ... CGCG

16. Suppose a cell is infected with a virus that releases dsRNA into the cytoplasm, and this RNA contains a subsequence that is complementary to a section of *Drasha* mRNA. Assuming this complimentary subsequence is a 21-mer that will be held by RISC:

- **a.** What would be the effects on the host cell?
- **b.** Would the cell lose the ability to fight infections via a different RNA virus? Why or why not?

CHAPTER

Genome editing

14

Chapter outline

Genome editing is the modification of the genome within a living cell through the insertion, deletion, or replacement of one or more segments of DNA. The most common way of doing this is via nucleases that cut the DNA in site-specific locations. Examples of these targetable nucleases include CRISPR-Cas systems, TALENs, and ZFNs. (These acronyms will be decoded in the text.)

In addition to Cas proteins, TALENs, and ZFNs, we will discuss genome editing techniques that seek to either swap out or insert strands of DNA. Three additional types of enzymes will be introduced: transposase, recombinase, and integrase.

The fundamental steps in these methods are first, the location and cutting of a specific section of DNA, and second, repair of the cut DNA, often using the cell's own DNA repair mechanisms. It is during this repair step that new or altered DNA sequences may be introduced into the genome.

We will end the chapter by briefly discussing what the biotechnologist will deliver into the target cells. Following from the central dogma, we can deliver DNA, RNA, or proteins. It is good to have choices, but to make a good decision, we must be aware of the ramifications of each alternative.

Biotechnology and its Applications. https://doi.org/10.1016/B978-0-12-817726-6.00014-9

14.1 Targetable nucleases

All genome editing approaches that use targeted nucleases share a similar mechanism. First, a double-stranded break is introduced in the genomic DNA at a specific location, and then the cell responds via its own repair mechanisms to fix the break. As we shall soon see, this repair step is when the magic happens.

There are three major entities that currently can be used for targeted genome editing:

- **CRISPR**-Cas— CRISPR stands for **c**lustered **r**egularly **i**nterspaced **s**hort **p**alindromic **r**epeats. Note that these are features found in DNA; the entities used with CRISPR for gene editing are known as CRISPR-associated systems, hence the proper term "CRISPR-Cas."
- **TALENs**—**t**ranscription **a**ctivator-**l**ike **e**ffector **n**ucleases
- **ZFNs**—**z**inc **f**inger **n**ucleases

Of these three systems, CRISPR-Cas has attracted the most attention, partly because of its simplicity. However, we should not discount TALENs and ZFNs; these are also good methods for directed gene editing, with a considerable amount of research behind them. Interestingly, all three systems act on the cell's own DNA repair mechanisms, so a therapeutic application that uses one particular system could be used with either of the other two systems, provided certain conditions are met.

14.1.1 CRISPR

CRISPR elements were first noted (but not named or immediately pursued) in *Escherichia coli* in 1987. CRISPR elements involve sequences of DNA that are repeated, although the repeats do not appear one after the other. As time went on, more repeat sequences were identified in different organisms such as archaea, bacteria, and mitochondria, which led to increased interest in these types of repeats. The term "CRISPR" was not coined until 2002.

Like RNAi, CRISPR-Cas systems are thought to be parts of a primordial immune system, whereby bacteria acquire immunity to certain viruses. The bacterial immune process basically consists of chopping up the genome of a viral invader into small fragments of around 20 bases. A single fragment is placed into a special location within the bacterial genome—a CRISPR locus. The locus, now modified with the viral gene fragment, can then be transcribed and processed into CRISPR-RNA (crRNA). When we use the CRISPR-Cas system in the laboratory for genome editing, this special RNA is referred to as **guide RNA (gRNA)**.

Cas proteins are restriction enzymes (endonucleases) that have a *tunable* recognition sequence. The specificity of the enzyme is determined by the gRNA (or crRNA) sequence. The Cas protein binds to a region on the 3′ side of the gRNA known as the scaffold sequence, leaving a section at the 5′ end of the crRNA (gRNA)—termed the spacer sequence—free to act as a probe for locating homologous regions within DNA (Figure 14.1). In wild-type bacterial systems, crRNA is used to identify DNA as foreign; in engineered systems, gRNA is used to locate a region in the genome for editing in a site-specific way. After binding of the Cas-bound gRNA spacer sequence to a homologous region of DNA, the Cas protein cleaves the DNA in one of several ways. There are three different ways in which gene editing with CRISPR can be carried out: via double-stranded breaks, nickases, or homology-directed repair.

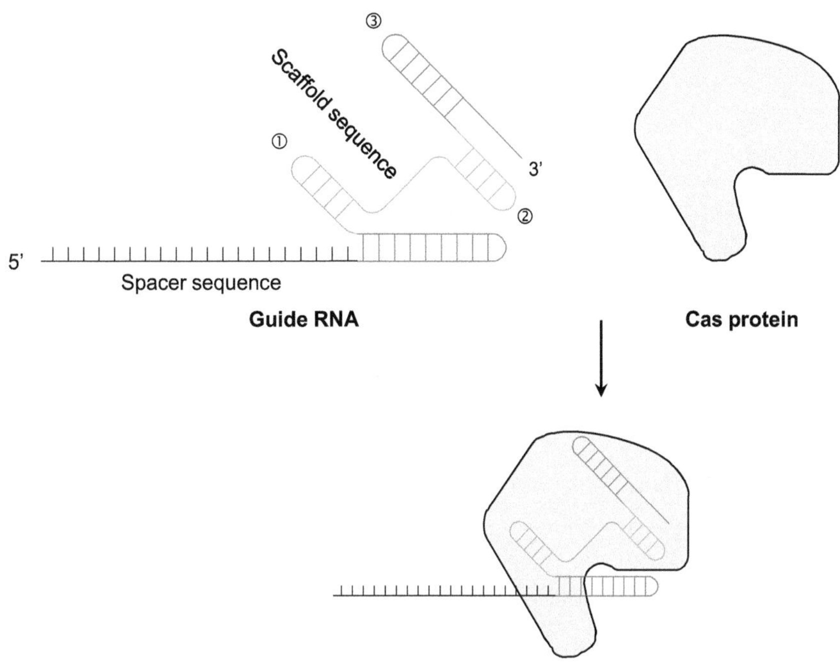

FIGURE 14.1

The DNA recognition and cleavage functions carried out in the CRISPR gene editing systems come from a complex consisting of a Cas protein interacting with a gRNA (or crRNA). The gRNA consists of a scaffold sequence, which binds to the Cas protein, and a spacer sequence, which extends from the complex. (The structure of the gRNA includes three stem-loops, denoted by *numbers in circles.*) The spacer sequence acts as a probe by interacting with chromosomal DNA via complementary base pairing. The wild-type Cas protein has domains that are responsible for inducing double-stranded breaks in DNA after homology with the spacer sequence has been established.

14.1.1.1 CRISPR-Cas and double-strand breaks

When using CRISPR-Cas, there are three requirements for the transcribed gRNA:

1. It must have a scaffold sequence that will bind to a given Cas. Cas will not cut DNA if it is not bound to a gRNA.
2. The target of the spacer sequence must be unique; otherwise, several genes may be bound and cut. This would produce off-target effects.
3. The target must be just upstream of a **protospacer-adjacent motif**, also known as a **PAM sequence.**

For discussion purposes, we will focus on a specific Cas, Cas9, which comes from *Streptococcus pyogenes* and is abbreviated as SpCas9. The CRISPR-SpCas9 system is currently one of the most well-characterized and widely used gene editing systems. The PAM sequence for SpCas9 is 5′-NGG-3′ and is an *absolute requirement* for this system; if the SpCas9/gRNA binds to a target sequence that is not next to a PAM, the Cas will not cut the DNA. However, with the PAM present, cleavage will take place

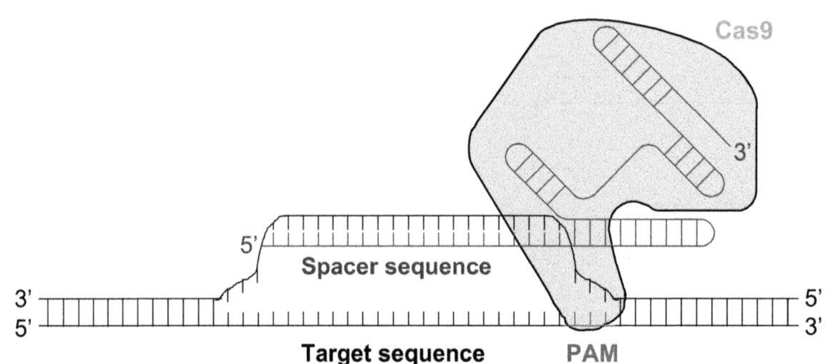

FIGURE 14.2

Interaction of Cas9/gRNA with dsDNA. Although a homologous DNA sequence may be recognized by a Cas9/gRNA complex, DNA cleavage will only take place if a protospacer-adjacent motif (PAM) sequence is present in the DNA. It is the PAM sequences, which appear in numerous, near-evenly spaced copies in the DNA, that are referred to by the CRISPR acronym (clustered regularly interspaced short palindromic repeats).

three to four bases upstream of the PAM (Figure 14.2). SpCas9/gRNA can bind to any portion of the DNA with a PAM, but it is the matching with the spacer sequence that determines whether DNA cleavage will take place.

The SpCas9 protein has six domains, including two endonuclease domains, HNH and RuvC. If there is a good match between the gRNA spacer sequence and the target DNA sequence, a conformational change in the Cas9 occurs to activate the endonuclease domains. HNH cleaves the target DNA strand, and RuvC cleaves the nontarget strand. In this way a double-strand break is introduced into the DNA.

Cells repair double-strand breaks via one of two pathways—**nonhomologous end joining (NHEJ)** or **homology-directed repair (HDR)**—of which NHEJ is the more common in the cell. It is also more error prone; insertions and deletions can easily occur, and these can cause frameshift mutations. Frameshifts can cause codons to be translated differently (i.e., different amino acids are assembled onto a growing polypeptide), the elimination of a Stop codon, or the introduction of a premature Stop codon. In each case a nonfunctional protein is the very likely result. Altering one copy of a gene in this manner would result in a knockdown. Altering both copies, which is very likely because the gRNA spacer sequence should target both alleles in the genome, would knock out the gene in that cell.

HDR has a much higher fidelity, but it is not used as much in the cell. We will return to HDR in a moment.

14.1.1.2 CRISPR-Cas and nickases

One problem with generating knockouts via double-strand breaks is that, despite the method's use of a 20-base recognition (spacer) segment on the gRNA, sequences with near-homology may also be cut. One logical solution to this problem would be to use gRNAs with longer spacer sequences; however, the CRISPR-Cas system is set up for ~20-nucleotide spacers, so using a gRNA that has a 40-nucleotide spacer sequence would not work well. An alternative approach is to modify the Cas protein so that it only cuts one DNA strand. By creating a Cas9 mutant with an inactivated HNH (or RuvC) domain, and

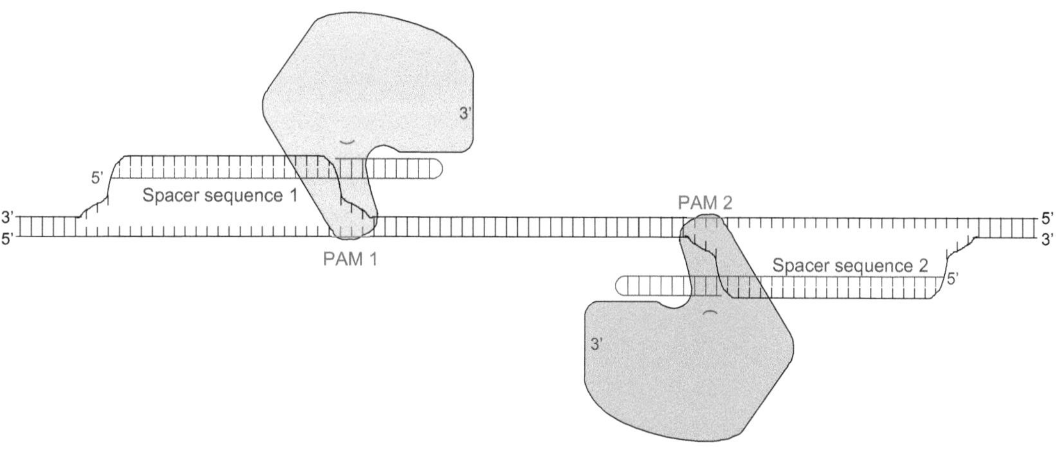

FIGURE 14.3

Nickases working simultaneously to generate single-stranded breaks in a targeted region of DNA. The requirement of homology matching in two distinct gRNA spacer sequences improves specificity over wild-type Cas9 systems, resulting in reduced off-target cleavage. *PAM*, Protospacer-adjacent motif

by loading two mutated Cas9 proteins, each with one of two different gRNAs that are specific to separate but close areas on the two strands of DNA, one can essentially require homology of 40 nucleotides (20 for each of the two gRNAs) before the formation of a double-strand break (Figure 14.3).

The single-stranded breaks introduced by the mutated Cas proteins are known as "nicks," hence the term "nickases" used to describe the proteins. Nicks are usually repaired quickly and with high fidelity by the HDR system. With the Cas nickase system, a double nick, where both strands are cut, will only occur if both spacer sequences line up to their respective targets with good homology.

14.1.1.3 CRISPR-Cas and homology-directed repair

Whereas gene editing using CRISPR with NHEJ is used for knocking out gene expression, HDR is the means by which CRISPR is used for specific gene edits, such as altering the identity of a single base. Three components are requisite to this system: a Cas protein such as SpCas9 or Cas9-nickases, gRNA, and a repair template with homology arms.

The mechanism used for this precise form of gene editing is similar to what was described in section 14.1.1.1 (double-stranded breaks), except that DNA (in either linear or plasmid form) containing the desired insertion/repair is codelivered with the Cas9/gRNA. Consider a linear repair template designed to correct a point mutation in the genome. Even though its sequence will differ from the genomic sequence by only one base pair, the length of the repair template must be longer than one base pair. The bases to the left and right of the repair base pair, called the "homology arms," will have the same sequence as the genomic DNA. The length of each arm depends on the size of the insert (Figure 14.4).

It is critical that the repair template does *not* contain a PAM sequence; otherwise, it would be cut by the Cas proteins. If the section to be inserted does contain a PAM sequence, then it should be re-engineered with a silent mutation to remove the PAM. For example, an AGG sequence, which codes for arginine and is also is a Cas9 PAM sequence, could be modified to AGA, which still codes for an arginine but is no longer an "NGG" PAM sequence.

Table 14.1 Cas9 variants/orthologues.[a]

PAM sequence	Specific Cas protein
NGG	Cas9 from *Streptococcus pyogenes* (SpCas9)
NGCG	SpCas9 VRER variant
NGAG	SpCas9 EQR variant
NGAN	SpCas9 VQR variant
NNAGAAW	Cas9 from *Streptococcus thermophilus* (St1Cas9)
NNGRRT	Cas9 from *Staphylococcus aureus* (SaCas9)

[a]*When designing a CRISPR-Cas experiment, we are not limited to only one PAM sequence. Variants and orthologues of Cas9 offer more flexibility for gene editing.*
N = any nucleotide; *PAM*, protospacer-adjacent motif; *R* = purine *(A or G); W* = "weak" (A or T).

Recall that HDR, although having high fidelity, is not used by the cell as much as NHEJ, which means NHEJ will still be at work. Because mammalian somatic cells contain two alleles of every gene, the result of this type of gene editing for a given cell will be a combination of

- wild-type,
- NHEJ-mutated, and/or
- HDR-inserted (repaired) alleles.

It is therefore imperative for the researcher to perform a screen to isolate only cells that have undergone HDR for both alleles. Such cells are termed HDR/HDR clones.

14.1.1.4 SpCas9 variants and orthologues

Although the SpCas9 system has been widely accepted for laboratory applications worldwide, some limitations still must be overcome. For example, SpCas9 depends on the specific (*S. pyogenes*) PAM sequence 5′–NGG–3′, and the cleavage site must be close to this PAM. Although NGG is abundant throughout the mammalian genome, the appearance of the sequence might not be close enough to the desired target to allow for the desired modification. To address this issue, multiple novel Cas9 variants and orthologues that utilize different PAM sequences have been engineered or discovered (Table 14.1). These proteins provide great convenience to scientists by expanding the editable areas within the genome.

Off-target effects occur when the nuclease induces a double-strand break at an unwanted site, leading to unwanted cellular changes such as transformation. High-fidelity mutants of SpCas9 have been created to reduce the possibility of these effects. One example of such a mutant is *S. pyogenes* Cas9 High-Fidelity variant 1 (SpCas9-HF1) which operates by reducing potential interactions between the Cas and the phosphate backbone of DNA. SpCas9-HF1 has editing activity like that of wild-type SpCas9, but off-target events have been nearly eliminated. A separate mutant was designed to reduce the affinity between a specific area in Cas (located between the HNH, RuvC, and PAM-interacting domains) and genomic DNA. The mutated enzyme was named "enhanced specificity" SpCas9 (eSpCas9) and, like SpCas9-HF1, has activity similar to the wild-type enzyme but with a clear reduction in off-target events.

Another interesting system employs CRISPR-C2c2, which comes from the bacterium *Leptotrichia shahii*. Instead of working on DNA, CRISPR-C2c2 cleaves certain single-stranded RNAs through anchoring via a 28-nucleotide sequence in the crRNA. This system could be used to downregulate gene expression via posttranscriptional knockdown, similar to RISC-mediated RNA interference. Moreover, fluorescent labeling of C2c2 makes it possible to tag specific mRNA sequences, and linking C2c2 with a splicing factor makes it possible to affect mRNA splicing. So far, the system has been used in bacteria and plants.

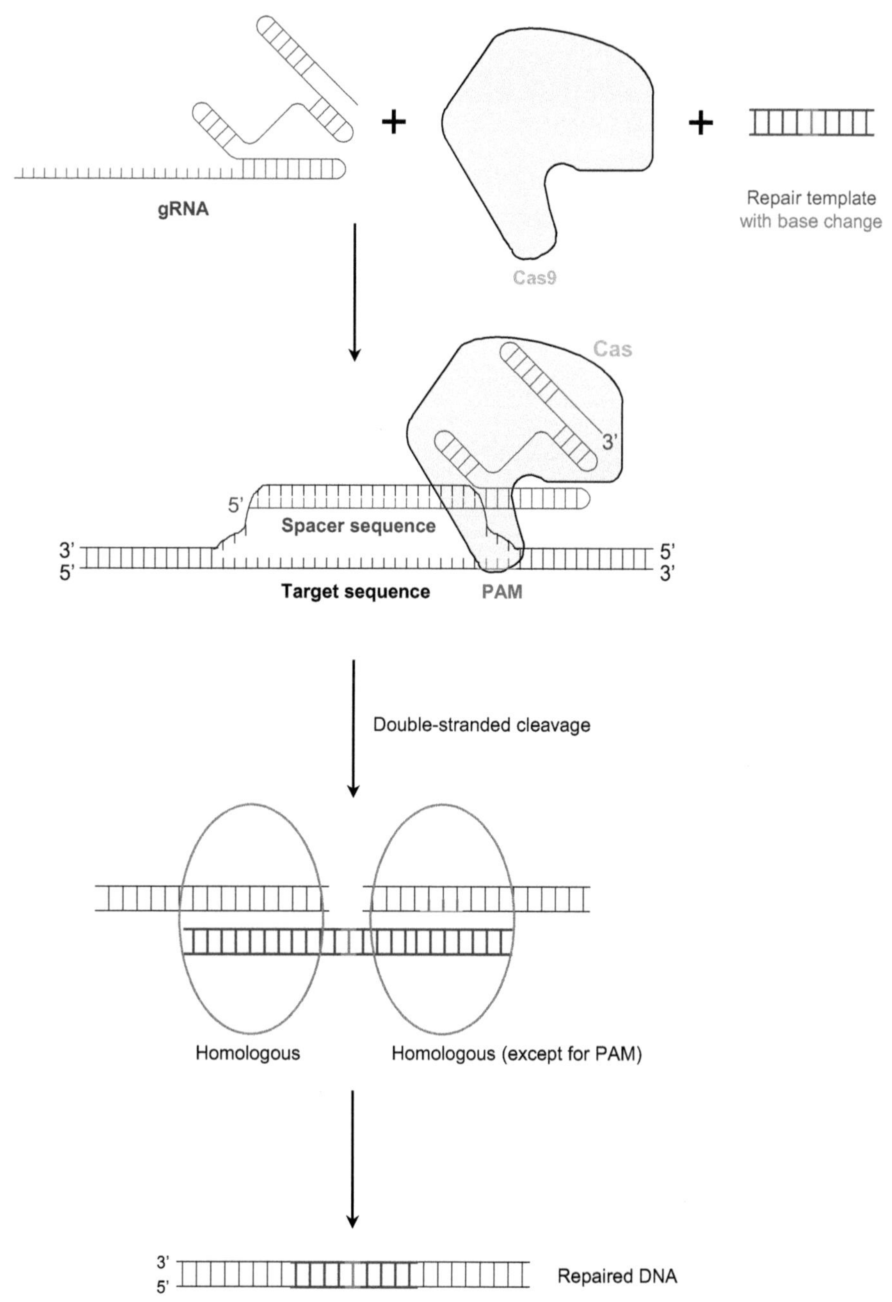

FIGURE 14.4

Homology-directed repair. In this method of genome editing, Cas/gRNA complexes (Cas9 shown here) are codelivered with a repair template. The template is introduced into the genome via homologous recombination following the introduction of a double-stranded break. *PAM*, Protospacer-adjacent motif

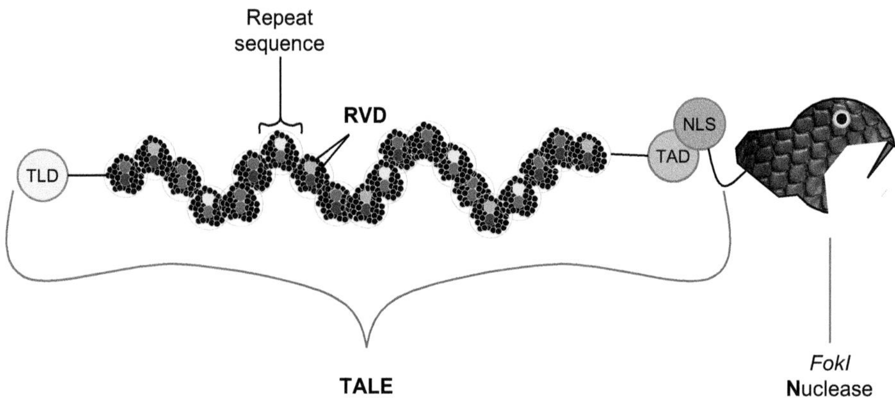

FIGURE 14.5

Structure of a TALEN, which is made up of a transcription activator-like effector (TALE) and a nuclease (a portion of the FokI restriction enzyme). The TALE has a translocation domain (TLD) at its N-terminus and a transcription activation domain (TAD) and a nuclear localization signal (NLS) near its C-terminus. In the middle of the TALE are several repeats, each consisting of 33 or 34 amino acids *(black circles)*. Within each repeat are two amino acids that can be varied to give the repeat a specificity for a particular DNA nucleotide *(colored circles)*—the repeat variable di-residue (RVD). Keep in mind that this is a schematic of a single TALEN; two TALENs are required for nuclease activity because *Fok*I must dimerize to be active.

14.1.2 Transcription activator-like effector nucleases (TALENs)

TALENs are chimeric proteins that contain two functional domains: a DNA-recognition **t**ranscription **a**ctivator-**l**ike **e**ffector (TALE) and a **n**uclease domain. They work for gene editing by recognizing a specific sequence, which the user can design, and introducing a double-stranded break with an overhang. TALENs therefore serve as a form of customizable restriction enzyme.

The basic TALEs used in TALEN technology are secreted by the genus *Xanthomonas*, a type of bacterium that is pathogenic to plants. A TALE consists multiple domains, including a translocation domain at the N-terminus, a transcription activation domain and a nuclear localization signal hat are close to the C-terminus, and multiple repeats in the center that recognize DNA molecules (Figure 14.5). The repeats are a highly conserved sequence of 33 or 34 amino acids, with positions 12 and 13 being variable; this variable portion is termed the "**repeat variable di-residue" (RVD)**. By changing the RVD sequence of a particular repeat, that repeat can be made to bind a specific nucleotide. As examples, a monomer with an RVD consisting of asparagine-asparagine will bind with a purine, asparagine-isoleucine will bind with adenine, asparagine-glycine will bind with thymine, and histidine-aspartate will bind with cytosine (Table 14.2). By manipulating the RVDs of its repeated sequences, a TALE can be made to serve as a probe for a specific site within the genome.

The nuclease portion of a TALEN is the catalytically active domain of the restriction enzyme *Fok*I, minus the DNA recognition domain. *Fok*I can be used in mammalian cells to cut genomic DNA, but it must be dimerized to be functional.

One challenge in using TALENs used to be that the enzyme construct was hard to build due to the numerous repeat sequences that had to be customized for targeting. A solution came with the advent of

Table 14.2 The identity of the variable sequences (RVDs) in a TALE determine the nucleotide targets

RVD sequence	Nucleotide bound
Asparagine-asparagine	Purines (mainly guanine)
Asparagine-isoleucine	Adenine
Asparagine-glycine	Thymine
Histidine-aspartate	Cytosine

a method called Golden Gate assembly—a way to assemble multiple DNA fragments into a single construct. The method depends on the use of a type IIS restriction endonuclease, but one that yields relatively long, 4-base overhangs, such as *Fok*I, *Bsa*I, or *Esp*3I (which is why *Fok*I was selected for TALENs). Type IIS restriction endonucleases cleave *outside* of their recognition sequences, which allows for the enzyme's recognition site to be removed from the fragment of interest, thus allowing for digestion and ligation within the same reaction mixture. Similarly, because the site of DNA cleavage is not inside the recognition sequence, the resulting overhangs are not dictated by the enzyme, so multiple, specific overhangs can be created simultaneously, allowing for the creation of multiple assemblies at the same time.

Compared with the CRISPR system, which requires only the design of a single guide RNA, the TALEN system requires much more effort in sequence design and cloning. As a result, CRISPR is the more widely accepted method for genome editing. However, TALENs do have certain advantages over CRISPR in specific situations. For example, TALENS can theoretically target any site within the genome, whereas CRISPR requires the presence of a PAM site. Therefore, when there is no PAM site available for a given application, one can still turn to a TALEN system. TALENs are also associated with lower off-target effects versus wild-type SpCas9. For perspective, a comparison of CRISPR-Cas9 (using SpCas9) versus TALENs showed TALENs producing no off-target cleavage, while half of the applications (three of six) yielded off-target events for SpCas9 nucleases. However, do not forget about high-fidelity Cas enzymes, which continue to be developed.

14.1.3 Zinc finger nucleases (ZFNs)

ZFNs are another type of DNA recognition/cleavage construct that acts like a restriction enzyme. The recognition domain of the ZFN is the Cys2-His2 zinc finger motif, and the domain that carries out DNA cleavage again uses *Fok*I (without the *Fok*I recognition domain). The Cys2-His2 zinc finger motif consists of a single zinc atom with ~30 amino acids in a beta sheet-beta sheet-alpha helix conformation (see Chapter 3). Each zinc finger is able to come into contact with three DNA bases by way of amino acids on the alpha helix part of the motif. As with TALENS, we can tune the zinc fingers to recognize specific triplets of DNA bases by adjusting the primary sequence of amino acids in the alpha helix. A library of zinc finger domains that can target nearly all of the possible triple nucleotide combinations has been developed. Stock amino acid sequences are used as linkers to connect 3–6 zinc finger motifs, thereby creating monomers that are specific for DNA targets 9–18 nucleotides in length. Recall from our discussion of TALENs that *Fok*I is only active as a dimer. This means two ZFN monomers must be used to achieve cleavage, which also means the total number of nucleotides being targeted is $2\times(9$ to $18) = 18$ to 36 nucleotides.

The primary hurdle when using ZFNs is that of enzyme design and construction. Not only is the cloning process used to produce the correct amino acid sequence labor intensive, but the correct amino acid sequence is also imperative for achieving high targeting specificity. As a result, the number of studies applying this technique has dwindled since TALENs and CRISPR were introduced to the field. However, unlike the TALENs and CRISPR systems, ZFNs originated from mammalian cells, so there has been less concern over immunogenicity. As of early 2020, there had been 14 clinical trials involving ZFNs in the United States (and one in China), 9 of which were aimed at combating HIV (information from *ClinicalTrials.gov*).

14.2 Other genome manipulation tools

14.2.1 Transposons and transposase

Transposons, sometimes called "jumping genes," are stretches of DNA that can change their location within the genome. They are classified according to the way in which they achieve translocation. Class I transposons (also known as retrotransposons) are transcribed into RNA, with the products being reverse-transcribed back into DNA for insertion. They are basically genes that encode themselves. For Class I transposons, the original transposon DNA remains in the genome during the process, so the number of transposons increases with every cycle. Class II transposons do not require transcription/reverse transcription but rather use transposases to cut the transposable element out of the genome and to cut the DNA at the new insertion site. The transposase is commonly encoded within the transposon itself. Class II transposons are basically genes that encode transposase, which acts to move the original transposon. Note that Class I transposons amplify themselves, whereas Class II transposons move themselves. The transposons used for genomic engineering are in Class II.

Combined with the enzyme transposase, a type II transposon can be moved efficiently between plasmid and genome in a cut-and-paste fashion. Transposase acts by recognizing a sequence of inverted terminal repeats (ITRs) that appear at each end of the transposon and introducing a pair of double-stranded breaks into the DNA. An engineered gene, flanked by ITRs, can be cut out of a plasmid and transferred to a site within the genome that is flanked by the same ITRs. This method can also be used to delete genes that appear between two ITR sites in the genome. There is no size limit for transposons. An example of biotechnological use of this system is the reprogramming of primary cells into induced pluripotent stem cells (iPSCs).

14.2.2 Recombinase

The major function of recombinase is to exchange strands between two DNA segments with partial sequence homology. The most widely used recombinase systems for genome manipulation are Cre-loxP and Flp-FRT. The Cre and Flp enzymes work in very similar fashions, so only the Cre-loxP system is described here.

Cre is an enzyme that catalyzes recombination, hence the proper name "Cre recombinase." This enzyme acts in a site-specific manner to introduce two cleavage sites in one or two dsDNA molecules and then to rejoin them with the two cut sites swapped (Figure 14.6). The recognition sequence for Cre (as well as Flp) is a 34-bp sequence. For Cre, this sequence is known as a loxP site. The first and last 13 bases of the 34-bp loxP site make up an inverted repeat that surrounds an 8-base spacer sequence,

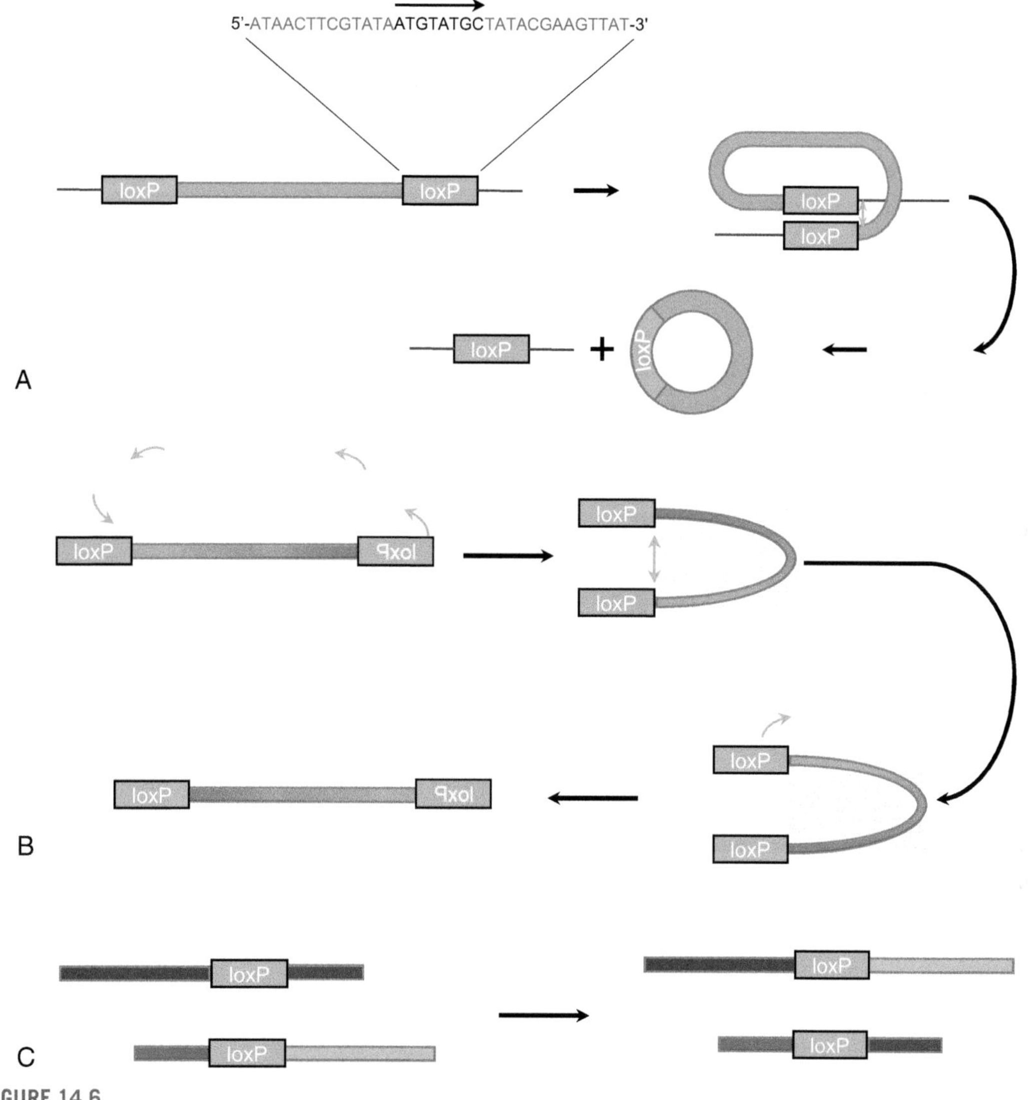

FIGURE 14.6

Cre recombinase (not shown) is active at loxP sites in DNA to catalyze recombination. (A) The loxP site consists of the 34 bases shown. Note that the outer 13 bases on each side of the loxP site *(red letters)* are inverted repeats of one another and that they flank an 8-base spacer sequence (in *black*) that determines directionality. (Several spacer sequences now exist. Only one is given for clarity.) Two loxP sites in the same orientation in the same DNA strand will allow for self-excision of the surrounded sequence to produce a plasmid. (B) Two loxP sites in the opposite orientation will allow for a reversal of direction in the DNA between the two sites. Note the green and orange portions of the DNA in the figure. (C) Two loxP sites on different DNA strands will allow a recombination event between the two strands.

used by Cre to determine directionality. Two loxP sites in the same orientation in the same DNA strand will allow for self-excision of the DNA between them, thus producing a plasmid because of the way the DNA is looped to bring the two loxP sites together (Figure 14.6A). If, on the other hand, two loxP sites in the same dsDNA fragment happen to be in opposite orientations (and pointing toward each other), recombination will reverse the direction of the DNA sequence between the two loxP sites, again because of how the DNA will be looped to bring the two sites together in the Cre enzyme (Figure 14.6B). A third situation is seen when two separate dsDNA strands each have a single loxP site, allowing a recombination event to occur between the two strands (Figure 14.6C).

One application of the Cre-loxP system is to insert a gene that is flanked by loxP sites into the genome (perhaps via CRISPR-Cas). The gene should be expressed until Cre recombinase is present, at which time the gene will be excised from the genome. Alternatively, a Stop codon flanked by loxP sites could be inserted upstream of a gene of interest. In this case the gene would not be expressed until Cre recombinase was present, at which time the Stop codon would be excised, and gene expression could commence. This latter example is called "lox-stop-lox." (Note that Cre must be introduced to the cell for Cre-loxP applications because Cre is a viral gene. Delivering a gene encoding Cre recombinase is a common way to introduce Cre into the cell.)

14.2.3 Integrase

In the past, gene insertion was typically achieved by either random plasmid integration (plus selection via antibiotics) or via infection with retroviral vectors. Insertions occurred arbitrarily, so the copy number, location, and direction of insertion were difficult to control. Homologous recombination (with or without a targetable nuclease) addressed the problem of directionality, but there are restrictions on the size of a gene that can be inserted in this manner. The efficiency of homologous recombination is also low, especially in primary cells, meaning the number of modified cells that can be produced by this method is low.

A recent technology, "dual integrase cassette exchange" (DICE), provides an alternative route whereby copy number, location, and the direction of gene integration are controlled, and with higher efficiency. DICE adopts the phage integrases phiC31 and Bxb1, which have the capability to insert a gene into their own recognition sites by unidirectional recombination. However, these recognition sites do not exist in the mammalian cell genome. To get around this problem, a "landing pad" was introduced into a safe region of the mammalian genome by integrating the recognition sites using TALENs. Desired genes could be efficiently implanted into the landing pad by codelivery of the two integrases and a donor template.

14.3 Delivery cargo

The genome editing tools just described can be delivered into cells via multiple routes based on whether DNA, RNA, or protein is the tool being used. We could deliver DNA in the form of transposons or genes that encode the desired nucleases, we could deliver RNA that can be translated to produce the nucleases, or we could deliver the nuclease proteins, recombinases, or integrases themselves.

14.3.1 DNA

DNA is often delivered as a plasmid, which is typically constructed via standard molecular cloning techniques (see Chapter 11). When properly preserved, DNA has a long shelf life; the major limiting factor for the associated methods is the efficiency of DNA delivery into cells. Instead of using plasmids, we can insert DNA encoding genome-editing proteins into a viral genome (although plasmids can be adapted for use with viruses for delivery into cells). Although viruses typically generate higher gene delivery efficiencies than do nonviral gene delivery agents, immunogenicity is a concern for in vivo applications, especially if repeated delivery events must be performed.

Another concern regarding the delivery of DNA into cells is that the foreign DNA could integrate into the host genome through homologous recombination. If this occurs in an untargeted fashion, vital host genes may be disrupted or inactivated, tumor suppressor genes may be knocked out, or oncogenes might be activated. Whether targeted or untargeted, genomic integration could cause the encoded nuclease to be expressed in a sustained manner, potentially causing continuous formation of double-strand breaks and serious off-target effects.

14.3.2 RNA

The editing methods that utilize delivered RNA avoid the possibility of genomic integration. RNA can be produced by cells transfected in vitro, with the resulting RNA being harvested for delivery to cells of interest in subsequent experiments. With RNA delivery, the efficiency of enzyme production may be enhanced because genetic materials do not have to reach the nucleus, and the transcription step is eliminated. However, RNA is more difficult to handle and store than DNA because RNases are abundant in the cell, on the skin, and throughout most research environments. The immunogenicity of RNA produced in vitro should also be considered.

14.3.3 Proteins

Instead of delivering genetic materials, delivering proteins into cells for gene editing is a commercially available option. There are multiple CRISPR-Cas9 protein delivery kits on the market. Like Cas proteins, TALENs and ZFNs can also be delivered into cells.

For the CRISPR-Cas system, the ribonucleoprotein complexes are formed in vitro by combining a Cas nuclease with the appropriate gRNA, with complete constructs being delivered into cells without risk of genomic integration. Another advantage to delivering proteins instead of (deoxy)ribonucleotides is that the amount of nuclease enzyme that will reside within the cell is more precisely controllable. We do not have to speculate on the amount of gene that will be expressed in the cell, because we know how many protein/RNA complexes were administered.

14.4 Summary

Gene editing involves the cleavage of double-stranded DNA in a targeted fashion. Although restriction enzymes have been used in molecular biology for approximately 50 years, the field of genome editing has arisen relatively recently, with the discovery of endonucleases that have tunable targeting and can be used in living cells. Three major forms of such systems were presented in this chapter: CRISPR-Cas, TALENs, and ZFNs.

For the researcher, the choice of editing system should consider both the efficiency of a given method and the number of off-target effects generated. For example, a system that produced the desired genetic change in 99% of treated cells would not be desirable for clinical use if an off-target insertion, deletion, or base change caused 0.1% of the cells to be transformed into cancer cells. In this chapter, we addressed recent advances in Cas research that have led to enzymes that operate with greater specificity with the goal of eliminating off-target effects.

In addition to Cas proteins, TALENs, and ZFNs, we discussed genome editing techniques that seek to either swap out or insert strands of DNA. Three additional types of enzymes were introduced: transposase (with transposons), recombinase, and integrase. The fundamental tenets of the chapter also apply to these alternative methods: DNA must be cut in a highly targeted fashion, and the cleaved DNA must be repaired. It is during this repair step that new or altered DNA sequences may be introduced into the genome.

Gene editing requires proteins, in the form of enzymes. The biotechnologist has choices about how to introduce these proteins into target cells. We could deliver them directly by encapsulating them in liposomes or PEI, we could deliver them indirectly in the form of the genes (DNA) that encode them, or we could deliver an intermediate product (RNA) that will be translated into the proteins without the problems of nuclear entry and transcription associated with DNA delivery. Regarding CRISPR-Cas systems, we also have choices regarding how to deliver gRNA into cells: we could deliver it directly, complexed with a Cas protein, or we could deliver a gene that will be transcribed into a gRNA molecule.

In the next chapter, we will leave the realm of RNA technologies to look at particular features found in our DNA that allow the identification of individuals via their genomes. The technologies, known collectively as "DNA fingerprinting," have been used by archaeologists, police, attorneys, and parents to establish identities of peoples or individuals. The techniques to be discussed all build upon things we have already learned in this book and further illustrate the broad impact of biotechnology.

Related reading

CRISPR-Cas

Abudayyeh, O.O., Gootenberg, J.S., Konermann, S., Joung, J., Slaymaker, I.M., Cox, D.B., et al., 2016. C2c2 is a single-component programmable RNA-guided RNA-targeting CRISPR effector. Science 353, aaf5573.

Fu, Y., Foden, J.A., Khayter, C., Maeder, M.L., Reyon, D., Joung, J.K., et al., 2013. High-frequency off-target mutagenesis induced by CRISPR-Cas nucleases in human cells. Nat. Biotechnol. 31, 822–826.

Ishino, Y., Shinagawa, H., Makino, K., Amemura, M., Nakata, A., 1987. Nucleotide sequence of the iap gene, responsible for alkaline phosphatase isozyme conversion in *Escherichia coli*, and identification of the gene product. J. Bacteriol. 169, 5429–5433.

Jansen, R., Embden, J.D., Gaastra, W., Schouls, L.M., 2002. Identification of genes that are associated with DNA repeats in prokaryotes. Mol. Microbiol. 43, 1565–1575.

Kleinstiver, B.P., Pattanayak, V., Prew, M.S., Tsai, S.Q., Nguyen, N.T., Zheng, Z., et al., 2016. High-fidelity CRISPR-Cas9 nucleases with no detectable genome-wide off-target effects. Nature 529, 490–495.

Kleinstiver, B.P., Prew, M.S., Tsai, S.Q., Topkar, V.V., Nguyen, N.T., Zheng, Z., et al., 2015. Engineered CRISPR-Cas9 nucleases with altered PAM specificities. Nature 523, 481–485.

Mojica, F.J., Diez-Villasenor, C., Soria, E., Juez, G., 2000. Biological significance of a family of regularly spaced repeats in the genomes of Archaea, Bacteria and mitochondria. Mol. Microbiol. 36, 244–246.

Slaymaker, I.M., Gao, L., Zetsche, B., Scott, D.A., Yan, W.X., Zhang, F., 2016. Rationally engineered Cas9 nucleases with improved specificity. Science 351, 84–88.

Zhang, D., Li, Z., Yan, B., Li, J.F., 2016. A novel RNA-guided RNA-targeting CRISPR tool. Sci. China Life Sci. 59, 854–856.

Zhang, K., Raboanatahiry, N., Zhu, B., Li, M., 2017. Progress in genome editing technology and its application in plants. Front. Plant Sci. 8, 177.

TALENs

Boch, J., Scholze, H., Schornack, S., Landgraf, A., Hahn, S., Kay, S., et al., 2009. Breaking the code of DNA binding specificity of TAL-type III effectors. Science 326, 1509–1512.

Cermak, T., Doyle, E.L., Christian, M., Wang, L., Zhang, Y., Schmidt, C., et al., 2011. Efficient design and assembly of custom TALEN and other TAL effector-based constructs for DNA targeting. Nucleic Acids Res. 39, e82.

Joung, J.K., Sander, J.D., 2013. TALENs: a widely applicable technology for targeted genome editing. Nat. Rev. Mol. Cell Biol. 14, 49–55.

Mahfouz, M.M., Li, L., Shamimuzzaman, M., Wibowo, A., Fang, X., Zhu, J.K., 2011. De novo-engineered transcription activator-like effector (TALE) hybrid nuclease with novel DNA binding specificity creates double-strand breaks. Proc. Natl. Acad. Sci. U.S.A. 108, 2623–2628.

Wang, X., Wang, Y., Wu, X., Wang, J., Wang, Y., Qiu, Z., et al., 2015. Unbiased detection of off-target cleavage by CRISPR-Cas9 and TALENs using integrase-defective lentiviral vectors. Nat. Biotechnol. 33, 175–178.

ZFNs

Durai, S., Mani, M., Kandavelou, K., Wu, J., Porteus, M.H., Chandrasegaran, S., 2005. Zinc finger nucleases: custom-designed molecular scissors for genome engineering of plant and mammalian cells. Nucleic Acids Res. 33, 5978–5990.

Gaj, T., Gersbach, C.A., Barbas 3rd., C.F., 2013. ZFN, TALEN, and CRISPR/Cas-based methods for genome engineering. Trends Biotechnol. 31, 397–405.

Gonzalez, B., Schwimmer, L.J., Fuller, R.P., Ye, Y., Asawapornmongkol, L., Barbas 3rd., C.F., 2010. Modular system for the construction of zinc finger libraries and proteins. Nat. Protoc. 5, 791–810.

Kim, H., Kim, J.S., 2014. A guide to genome engineering with programmable nucleases. Nat. Rev. Genet. 15, 321–334.

Pavletich, N.P., Pabo, C.O., 1991. Zinc finger-DNA recognition: crystal structure of a Zif268-DNA complex at 2.1 A. Science 252, 809–817.

Other genome manipulation tools

Branda, C.S., Dymecki, S.M., 2004. Talking about a revolution: the impact of site-specific recombinases on genetic analyses in mice. Dev. Cell 6, 7–28.

Coluccio, A., Miselli, F., Lombardo, A., Marconi, A., Tagliazucchi, G.M., Gonçalves, M.A., et al., 2013. Targeted gene addition in human epithelial stem cells by zinc-finger nuclease-mediated homologous recombination. Mol. Ther. 21, 1695–1704.

Maizels, N., 2013. Genome engineering with Cre-loxP. J. Immunol. 191, 5–6.

Metzger, D., Feil, R., 1999. Engineering the mouse genome by site-specific recombination. Curr. Opin. Biotechnol. 10, 470–476.

Skarnes, W.C., Rosen, B., West, A.P., Koutsourakis, M., Bushell, W., Iyer, V., et al., 2011. A conditional knockout resource for the genome-wide study of mouse gene function. Nature 474, 337–342.

Woltjen, K., Michael, I.P., Mohseni, P., Desai, R., Mileikovsky, M., Hamalainen, R., et al., 2009. PiggyBac transposition reprograms fibroblasts to induced pluripotent stem cells. Nature 458, 766–770.

Zhu, F.F., Gamboa, M., Farruggio, A.P., Hippenmeyer, S., Tasic, B., Schule, B., et al., 2014. DICE, an efficient system for iterative genomic editing in human pluripotent stem cells. Nucleic Acids Res. 42, e34.

RNA delivery

Sahin, U., Kariko, K., Tureci, O., 2014. mRNA-based therapeutics-developing a new class of drugs. Nat. Rev. Drug Discov. 13, 759–780.

Questions

1. A common misnomer is that we use CRISPR to edit the genome. Why is this a misnomer? Explain what does the actual editing.
2. Hans has designed a gRNA that matches perfectly with a gene he is trying to knock out in his model cells.
   a. When we say "matches perfectly," do we mean all of the bases in the gRNA match up with the gene of interest? Explain your answer.
   b. If the genomic target sequence contains 5′–AAAAGAGAGACCCC–3′, which of the following would be a part of the gRNA sequence?
      i. 5′–AAAAGAGAGACCCC–3′
      ii. 5′–TTTTCTCTCTGGGG–3′
      iii. 5′–UUUUCUCUCUGGGG–3′
      iv. 5′–CCCCAGAGAGAAAA–3′
      v. 5′–GGGGTCTCTCTTTT–3′
      vi. 5′–GGGGUCUCUCUUUU–3′
3. a. Hans pairs up his perfectly matched gRNA (question 2) with SpCas9 and delivers the complexes in a way that gets them into the cell nucleus. Unfortunately, the SpCas9 does not cut the genomic DNA. Why not?
   b. Propose a solution to the problem in part *a*.
4 Consider the following genomic target sequence: 5′–CACACACACACCGGCACACACA–3′. Xi has designed a gRNA with a sequence that matches up perfectly with the gene of interest, but he gets very poor results when he uses the gRNA with SpCas9.
   a. Give a reason for Xi's poor results. (Hint: The answer lies within the sequence that was given and it is not due to self-pairing.) Explain your answer.
   b. Would Xi be able to use the same targeting sequence with a ZFN system? Why or why not?
5. The *Fok*I enzyme used in TALENs and ZFNs is similar to the other restriction enzymes we have covered in this book, in that it uses a specific recognition sequence (GGATG), and it creates an overhang (5′ in this case) by cleaving dsDNA. However, it is special in that the identity of the overhang is not the same for every cut. How can this be?
6. Refer to Figure 14.5. The tail of the snake is the TALE portion of a TALEN.
   a. What is the function of the TALE?
   b. How does the biotechnologist create specificity for a given TALEN?
   c. What TALEN sequence would bind to the DNA sequence ATCG? (Make sure you indicate all of the members of the repeat sequence, even the entities you do not know specifically.)
7. Even though we can have a single TALEN or ZFN, in practice we always design TALENs or ZFNs for use. Why are the plural forms of the acronyms being used?
8. How can we use CRISPR-Cas to knock out the expression of a gene without using a repair template?
9. Suppose you have some cells in culture and you want to modify a single nucleotide in a specific gene of interest (that codes for a cell surface protein) to create a culture for further testing. You have decided to modify the base using a CRISPR-Cas system with a repair template. The wild-type sequence in the cell is GGGGGG, but you want to change the sequence to GGGTGG. (Shorter and nonrealistic sequences were chosen for clarity.)

a. After the editing procedure, you notice six types of cells with respect to the gene of interest. What are they?
b. The results in part *a* are not the gene-edited culture you are seeking! What further processing can you perform to obtain a culture containing only cells expressing the modified version of the edited gene?

10. The zinc finger motif is present not only in ZFNs but also in some transcription factor proteins. Why is this not surprising?
11. a. Consider gene editing that uses nicks instead of double-strand breaks. How do we modify the associated endonuclease to only cut one DNA strand?
    b. Which method of gene editing–double-strand breaks or nicks–is considered to be more specific? Support your answer.
12. Make a sketch to explain how a plasmid could be used with the Flp-FRT system to insert a gene into the genome.
13. What would be the result of exposing the following DNA sequence (only the sense strand is shown) to Cre recombinase?

GGGGGGGGATAACTTCGTATAATGTATGCTATACGAAGTTATCCCCCCTTTTTTATAA CTTCGTATACGTATGTATATACGAAGTTATA AAAAAA

Sketch the product(s), or provide the resulting DNA sequence(s).

14. What would be the result of exposing the following DNA sequence (only the sense strand is shown) to Cre recombinase?

GGGGGGGGATAACTTCGTATAATGTATGCTATACGAAGTTATCCCCCCTTTTTTATA ACTTCGTATAATGTATGCTATACG AAGTTATAAAAAAA

Sketch the product(s), or provide the resulting DNA sequence(s).

15. a. In the chapter on gene delivery, we discussed the disadvantages of inserting (c)DNA into the genome via a retroviral method. One such disadvantage is associated with the retroviral integrase and three specific outcomes that could arise from the insertion. What were the three outcomes?
    b. Biotechnologists might want to use integrase to insert genes into the genome. To do so, they must avoid the problems associated with part *a*. How is this accomplished? Explain how the technology works.
16. Name three ways to get Cre recombinase into the cell when a Cre-loxP system is to be used.
17. Why would reducing potential interactions between a Cas protein and the phosphate backbone of DNA yield a system with higher fidelity?
18. The cell can use two different methods for repairing DNA strand breaks—one with high fidelity and one with low fidelity that repairs the break without regard to whether the two ends are homologous.
    a. What are the names of the two different repair systems? Which one has lower fidelity?
    b. Why would one ever want to use CRISPR-Cas with the lower-fidelity repair system?
19. In SpCas9, what is the name of the nuclease domain that cleaves the genomic DNA that is hydrogen bonded to the gRNA spacer sequence?
20. Suppose that a patient has sickle cell disease (Sβ thalassemia disease), which is the result of a mutation in the genes associated with the β subunit of hemoglobin, the oxygen-carrying protein in red blood cells. Propose a gene editing scheme to treat the patient. Which editing system will you use? How will you get the appropriate molecules to the appropriate cells? What are the appropriate cells?

CHAPTER

DNA fingerprinting 15

Chapter outline

DNA fingerprinting is a technology that is used to confirm identity. It is used by police crime labs to match evidence with suspects to help establish guilt or innocence, it is used to establish the identity of human remains, and it can also be used to determine paternity. In this chapter, two different methods of DNA fingerprinting will be discussed to show the evolution of the technology. We will begin with a method first described in the 1980s that uses restriction fragment length polymorphisms, then move on to the more modern version that uses short tandem repeats for identification.

15.1 Older DNA fingerprinting uses RFLPs

The first method of DNA fingerprinting was developed in the late 1970s to mid-1980s by Alec Jeffreys. He used probes to identify hypervariable minisatellites (tandem-repetitive regions of DNA) "to produce somatically stable DNA 'fingerprints' which are completely specific to an individual (or to his or her identical twin) and can be applied directly to problems of human identification, including parenthood testing" (Jeffreys et al., 1985). The technique utilizes restriction fragment length polymorphisms, or **RFLPs**.

Imagine you are a police detective who finds a large blood splatter at a crime scene. You collect a bit of the residue because you know there is DNA in blood. Be careful here, though, because the DNA is not in mature red blood cells, which have no nuclei and therefore lack DNA. (Other sources of crime scene DNA include semen, bone, skin, cells sloughed in the saliva, urine, feces, and cells at the root of hair follicles.) Back in the laboratory (in this example, you are not only a detective but also a laboratory technician), you are able to isolate cells from the sample, after which you lyse them and collect the DNA. You have now isolated the genome of the person from whom the blood originated. You then expose the DNA to a set of restriction enzymes.

Consider a typical human somatic cell with 23 pairs of chromosomes. Upon exposure to the restriction enzymes, these chromosomes will be chopped into very small fragments—small enough to run on an

Biotechnology and its Applications. https://doi.org/10.1016/B978-0-12-817726-6.00015-0

agarose gel. The fragments will represent a vast array of sizes so that, instead of bands, they would appear as a smear on the gel if you were to visualize them. In terms of restriction cuts and resulting fragment sizes (see Chapter 9), a smear by itself is of little use. However, probes can be used to further characterize the DNA in the smear. As with qPCR (see Chapter 10), the probes used in this technique are polynucleotides and pair up with specific DNA sequences—the core sequences of the minisatellites described by Jeffreys. The binding of one of these probes to its complementary sequence is known as **hybridization**.

Consider two suspects. Suspect 1 has three specific cut sites for a particular restriction enzyme within a hypothetical stretch of DNA, but in suspect 2 the equivalent stretch of DNA contains a mutation in one of the recognition sequences, so it will no longer be cut at that point (Figure 15.1A). Looking at the gel for this particular stretch of DNA for the two suspects (Figure 15.1B), the lane corresponding to suspect 1 would contain four fragments after exposing the sequence to the specific enzyme, while the lane corresponding to suspect 2 would contain only three. Note that there is not a single, well-defined, and predicted pattern of bands for the human genome but that every individual has their own characteristic pattern of bands.

Because the entire genome is exposed to the restriction enzyme, the resulting gel will contain a smear rather than distinct bands. After performing gel electrophoresis to obtain the smear, the genomic DNA is then blotted onto a membrane that is placed on top of the gel. Once bound to the membrane, the DNA is more accessible for further processing by means such as hybridization. The probe used in this technique is complementary to a specific DNA sequence, and it binds at the same genomic location in each of the two samples—but it will appear in different locations on the gel because of the selective degradation of the genomic DNA that has already taken place (Figure 15.1C). For a single band, it's a matter of whether the band appears high or low in the gel. More than one probe is used in practice, and it will bind to more than one location in the genome. The result is not a single band but rather a pattern of bands that is unique to each individual.

The probes used are radiolabeled, meaning they contain radioactive isotopes. To visualize them after the hybridization, the membrane is washed to remove unbound probe and then placed in a cassette with unexposed x-ray film. Once exposed, the film is developed to reveal the characteristic series of bands from each sample.

Consider a murder scene. The way the process takes place in a police crime lab is that the blood is rehydrated if necessary and the cells isolated. The cells are lysed, and the DNA is isolated, digested, and run out on a gel. It is then blotted from the gel onto a membrane, which will be exposed to radiolabeled probes. Blotting is performed by placing the membrane, which is a paper-thin piece of (typically) nylon, on top of the gel, under a weighted stack of paper towels. The paper towels will wick the fluid from the gel like a sponge. The DNA in the gel will also be wicked, but it will be trapped in the membrane and not reach the paper towels. The DNA-containing membrane is then washed and exposed to x-ray film. In the crime lab, the samples that undergo this process include not only evidence taken from the crime scene but also samples taken from individuals who may be suspects.

In our example, we will take blood or cell samples from suspects 1 and 2, and perhaps from the crime scene investigator who collected the original sample, and certainly from the victim. One cannot assume that all of the blood in a crime scene came from the victim; perhaps there was a struggle or an accident whereby the murderer shed (blood) cells. The DNA from each of these samples will yield RFLPs that produce a characteristic pattern upon hybridization (Figure 15.2). If the set of bands from the crime scene sample matches up with those originating from one of the tested individuals, an identification will have been made.

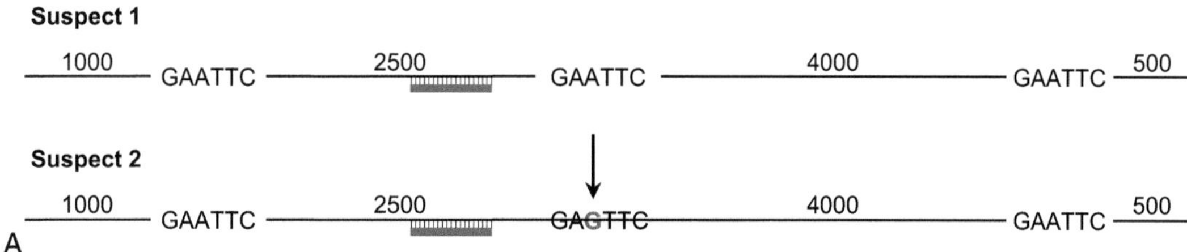

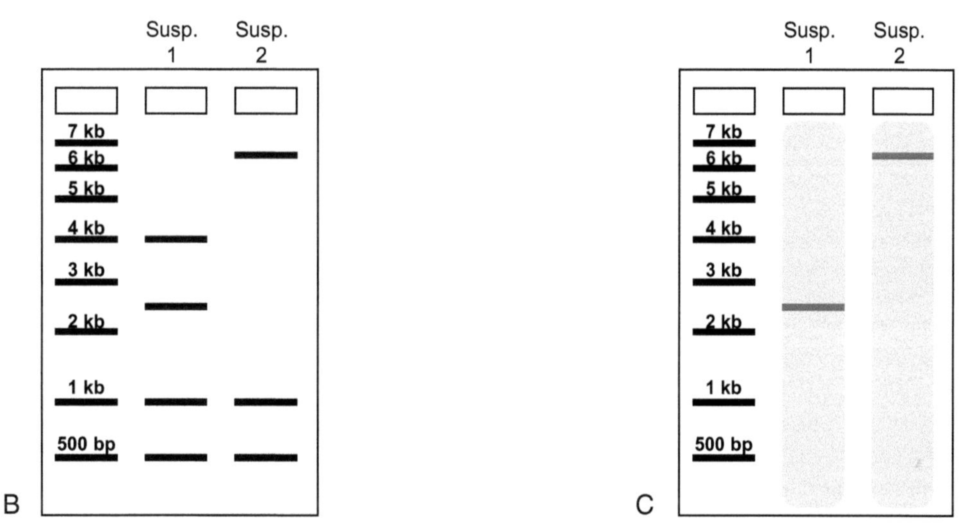

FIGURE 15.1

(A) Consider an 8000-bp stretch of genomic DNA from two suspects. In the DNA from suspect 1, there are three recognition sites for a certain restriction enzyme. In suspect 2, there has been a mutation in one of the recognition sites *(arrow)* so that there are only two cut sites for the enzyme. (Fragment lengths, in bp, are shown by numbers over each line segment.) (B) When the stretches of DNA are cut with the enzyme and the resulting fragments are run on a gel, the difference between the banding patterns for suspects 1 and 2 can be seen: the 2500- and 4000-bp fragments are apparently combined in suspect 2 due to the mutated cut site. (C) However, we must consider that the entire genome, not just an isolated 8000-bp fragment, is present in each DNA sample. After cutting the 3.08×10^9 bp genome with the restriction enzyme, there will be fragments of so many different lengths that the gel will appear as a smear (shown as *gray areas*). Exposure of the fragments to a probe (shown in *red*) allows for hybridization of the probe to a specific area of the genome. Refer back to panel (A) to convince yourself that, although the probe binds to the same place in each genome, the banding pattern will yield a different result for each suspect. When enough different probes are used, a distinct DNA fingerprint is revealed for each individual.

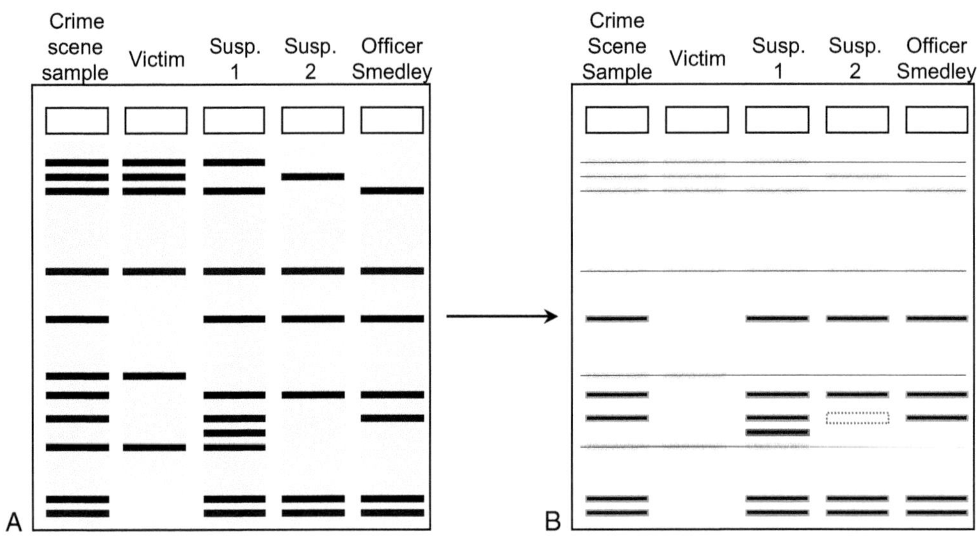

FIGURE 15.2

(A) RFLP pattern from a hypothetical DNA fingerprinting analysis of crime scene blood samples and DNA samples obtained from two suspects and the crime scene investigator, Officer Smedley. (B) To solve, eliminate all bands that are present in both the crime scene sample and the victim's sample. The perpetrator must have all the remaining bands and only those bands. Note that suspect 1 has an extra band not reflected in the crime scene sample and that suspect 2 is missing a band (both shown in *red*). From the data, we can conclude that Officer Smedley is the only one who satisfies the criteria. He is either the killer (which we would hope is very doubtful), or he has shoddy evidence collection techniques. Either way, the suspects have been exonerated.

15.1.1 RFLPs used for paternity testing

In this example, we will consider a single RFLP probe. Between two wild-type minisatellites, there is an enzymatic cut site that will produce two fragments that will be hybridized by the probe (Figure 15.3A and B). However, suppose an ancestor of a certain population from a remote island in the Philippines underwent a mutation that eliminated the central cut site for that minisatellite; descendants of that Filipino ancestor might carry the minisatellite that will produce only a single band upon hybridization (Figure 15.3A and B). We shall denote the wild-type version as the *A* form and the mutated minisatellite as the *a* form. If a woman from, say, Wyoming was homozygous for the wild-type form (*AA*), and she married a man who was homozygous for the Filipino mutation (*aa*), all their children would carry both alleles for that minisatellite, meaning they would be heterozygotes (*Aa*) for that allele. Their banding patterns would be a combination of the maternal and paternal patterns (Figure 15.3B, lane 3). We can see from the table that *AA* × *aa* yields 100% *Aa*:

		Mother	
		A	A
Father	a	Aa	Aa
	a	Aa	Aa

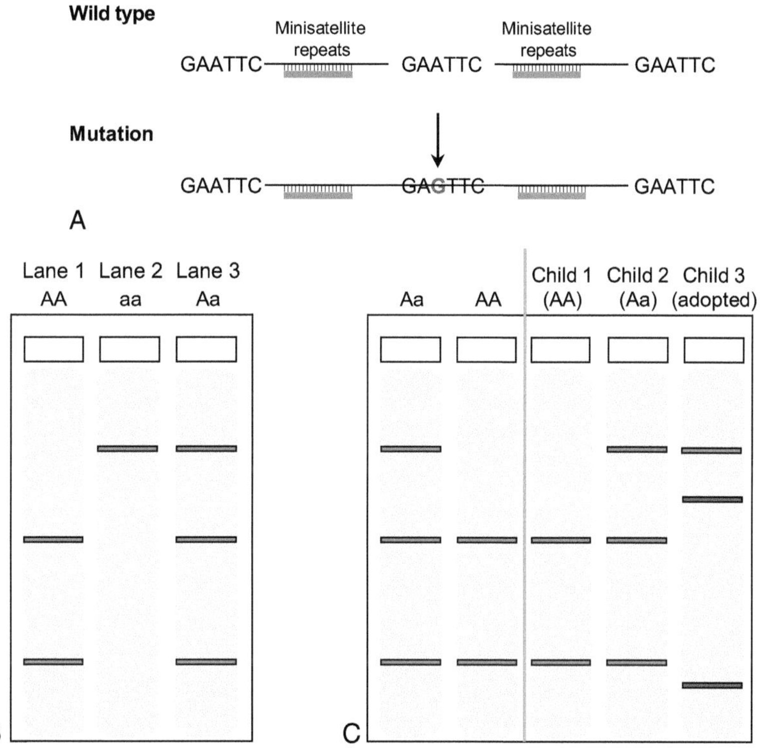

FIGURE 15.3

Example of how RFLPs can be used to establish paternity. (A) A wild-type minisatellite undergoes a mutation that will alter the resulting RFLPs. (B) Banding patterns of two homozygotes for the wild-type (*AA*) and mutated (*aa*) minisatellites. If these two people were to reproduce, all of their children would display the heterozygote banding pattern (*Aa*). (C) Following the example in the text, we have one of the heterozygotes from part (B) mating with another wild-type homozygote (the parents are shown to the left of the *orange line*). The only possible banding patterns are the ones shown for child 1 and 2: *AA* and *Aa*. In the right-hand lane, we can see from the appearance of new bands (red) that the third child is not the offspring of both parents.

Now suppose it is years later and a daughter from the couple's family marries another homozygous (*AA*) American. We could expect that the children from this union would either be homozygous (*AA*) or heterozygous (*Aa*):

		Mother	
		A	a
Father	A	AA	Aa
	A	AA	Aa

So *AA* × *Aa* yields 50% *AA* and 50% *Aa*; the children from this union should have either the *AA* (two-band) or *Aa* (three-band) pattern on the gel (Figure 15.3C). If the mother and father adopted a child and it displayed the banding pattern shown in the final lane of the figure, we could deduce that the child came from different parents. Perhaps the child came from the Philippines and was descended from the (*aa*) ancestor whose family line later suffered a second mutation in that minisatellite to introduce a new, differently located cut site (because the sum of the weights of the red bands in the figure is equal to the weight of the *a* band). We see the *a* banding pattern, but we see two new bands appear for the probed minisatellite. (We might consider the child to have the alleles *ab*.)

From this discussion, we can see that RFLP banding patterns can be inherited—so they can be used not only to show unique identity but also to confirm familial identity. The technique has also been used on a broad scale to show anthropological identity or to indicate migration patterns of ancient peoples.

15.2 Newer DNA fingerprinting uses STRs

The use of RFLPs is the older of the two techniques described here for DNA fingerprinting. The RFLP protocol requires a relatively large amount of DNA (>25 ng), which must be relatively undamaged. The reason that the DNA must be fairly intact is that fairly large amounts of the probe must bind for visualization to be possible. If the DNA sample has been in a tomb for hundreds (or thousands) of years—ample time for DNA damage from ultraviolet light, oxidation, or spontaneous deamination—the quantity of undamaged fragments (or the ratio of undamaged to damaged DNA) may be too low to obtain a reliable band pattern upon hybridization. What used to be a fragment of 1000 bp may have degraded so that the sample would produce many fragments of under 100 bp each or fragments of every size between 1 and 1000. There would be no way of determining whether these fragment lengths resulted from restriction enzyme cuts or inherent DNA damage.

The requirement for relatively undisturbed DNA is a significant limitation. One could not identify, say, the last Russian czar from a collection of burned bones using RFLP technology. However, identification of the remains of Czar Nicholas II were indeed made using a more modern technology, which involves **short tandem repeats (STRs)**. STRs, sometimes referred to as microsatellites, are 2- to 5-bp DNA sequences that are repeated several times in succession. For example, "GATAGATAGATAGATAGATAGATAGATAGATA" is an example of repeated GATA sequences, which is one of the main STR markers used for DNA fingerprinting. STRs occur throughout the genome. Although specific repeated sequences are common to all humans, the numbers of repeats can differ markedly between individuals. Consider a specific genomic locus that contains the STR D7S820, which resides on chromosome 7 and contains repeats of the tetramer GATA. You got half of your genome from your mother and half of your genome from your father, and each of them donated a short tandem repeat at this locus; you might find that, at the D7S820 locus, your genome contains alleles of perhaps 6 and 11 GATA repeats, whereas your friend's DNA shows 9 and 14 repeats at the same locus. So, even though you and your friend both possess the D7S820 STR locus, the lengths of your D7S820 alleles differ. In general, STR sequences can range from 3 to 21 repeats (some might argue a higher range), depending on the specific STR locus. For D7S820, the range is 6 to 14 repeats.

Given that we have two alleles for every genomic feature, including STR locations, how could differing repeat lengths yield any type of valuable information? The striking feature of STRs is that they are flanked by known sequences. These flanking sequences are the same for you as they are for your friend, or even the author of this book, even though the lengths of our STRs are different. We can use PCR to amplify the regions between these known sequences—meaning we can amplify the STRs—to levels that are sufficient for analysis, allowing us to determine the two lengths of a given STR that are present in a given genome. Odds are that the two lengths will be different for you, your friend, and the author of this book. However, because the number of repeats in an STR is a finite number, there is a finite number of combinations for their lengths (e.g., a range of 3–21 yields 190 combinations), so there is a small probability that you will share the same pair of STR lengths as another given individual for a given location in the genome. This possibility is offset by the fact that there are thousands of STR locations in the genome, and most of these sites have their own identifying flanking sequences.

The Combined DNA Index System (**CODIS**) is a database maintained by the United States Federal Bureau of Investigation that has records of core STR markers for over 1.9×10^7 individuals, and the number of records is steadily rising. In the United States, for STR evidence to be admissible at trial, it must match up with 13 agreed loci (Figure 15.4). In the United Kingdom, 10 STR loci are typically used for identification purposes (using the National DNA Database).

Consider D7S820, which can have 6 to 14 repeats. There are $(14\text{-}6+1)[(14\text{-}6+1)+1]/2 = 45$ combinations of allele lengths possible. (The number of combinations of n objects taken two at a time = $n(n+1)/2$ when order is not important.) We use this formula instead of 14^2 because, for example, 6 and 11 repeats would look the same as 11 and 6 repeats on a gel.

For the sake of a quick calculation, let's use D7S820 as a standard. If we assume that all STRs have the same number of possible combinations, then a DNA fingerprint for a given person, using 13 STRs, would be unique for one out of $(45)^{13} = 3.10 \times 10^{21}$ individuals, a specificity greater than one encompassing the entire population of the earth! The actual specificity is much lower, however, because allele lengths are not distributed randomly over a range, meaning some numbers of repeats are more likely than others. For the D7S820 locus, sizes of 10 and 11 repeats are the most common (at least for a population sample taken in northeastern Brazil).

Compared with the RFLP-based DNA fingerprinting technique, the STR system is far more sensitive. Only ~1 ng of DNA is needed, and the integrity of the DNA is not as much of an issue. We use PCR to amplify the STR regions, so as long as a primer can bind to one unbroken strand of DNA, that strand will be amplified exponentially and allow for accurate analysis. With RFLPs, no amplification takes place, so only the DNA that is collected can be visualized, and if it is damaged, it will yield poor or confounded results.

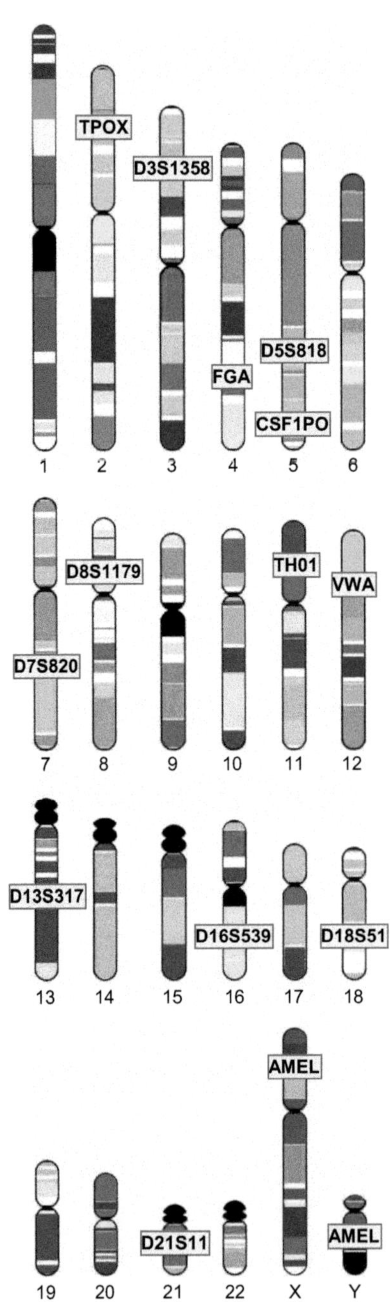

FIGURE 15.4

Names and relative positions of the 13 core Combined DNA Index System (CODIS) single tandem repeat loci on human chromosomes. Note that although it is not one of the core loci, AMEL is included here because it is often used to determine gender. (Relative positions taken from http://www.cstl.nist.gov/strbase/fbicore.htm.)

15.3 Summary

In this chapter, we learned about two different methods for DNA fingerprinting. The older method uses RFLPs and is based on minisatellite DNA that contains several repeats of specific DNA sequences. We saw that the method relies upon cleavage of genomic DNA with restriction enzymes, followed by hybridization with labeled probes. We then moved on to STR technology, which uses PCR to amplify microsatellites (the STRs) via their fixed flanking sequences. With this method, no DNA cleavage is required, and we can perform the technique with far less starting DNA.

In the next chapter, we will shift gears as we move into the world of fermentation. This chemical process is used by cells to produce interesting biochemical products. It has been harnessed by humans to entice cells to make products ranging from beer to biofuels.

Related reading

Butler, J.M., 2007. Short tandem repeat typing technologies used in human identity testing. Biotechniques 43 (4), ii–v.

Ferreira da Silva, L.A., Pimentel, B.J., Almeida de Azevedo, D., Pereira da Silva, E.N., Silva dos Santos, S., 2002. Allele frequencies of nine STR loci—D16S539, D7S820, D13S317, CSF1PO, TPOX, TH01, F13A01, FESFPS and vWA—in the population from Alagoas, northeastern Brazil. Forensic Sci. Int. 130, 187–188.

Jeffreys, A.J., Wilson, V., Thein, S.L., 1985. Individual-specific 'fingerprints' of human DNA. Nature 316, 76–79. STRBase (SRD-130), National Institutes of Standards and Technology, Biotechnology Division. http://www.cstl.nist.gov/strbase/intro.htm.

Seringhaus, M., 2009. The evolution of DNA databases: Expansion, familial Search, and the need for Reform. DePaul University College of Law 9th Annual Ciplit Symposium & 12th Annual Niro Distinguished Lecture, October, 2009, 3. (46 pages).

Questions

1. Name two advantages of using STRs rather than RFLPs for DNA fingerprinting.
2. What is the general range for the number of repeats in an STR?
3. If a certain STR can range from 8 to 13 repeats per allele, how many unique combinations of this STR are possible?
4. You and the king of Sweden both have an STR that involves the base sequence GATA, yet this sequence can potentially be used to tell you apart. If the sequences are the same, how can they be used to distinguish the two of you?
5. Why wouldn't a microinjected plasmid appear in DNA fingerprinting? What if the injection had been performed on an oocyte that then developed into a complete organism?
6. Describe some differences between DNA fingerprinting and DNA footprinting.

7. The story presented in Section 15.1.1 continues. About 2 years after the family adopted their third child, the wife got pregnant. It was a surprise to the family because the mother believed that she could not get pregnant again, which is why they adopted the third child. The RFLP patterns for the entire family are shown below:

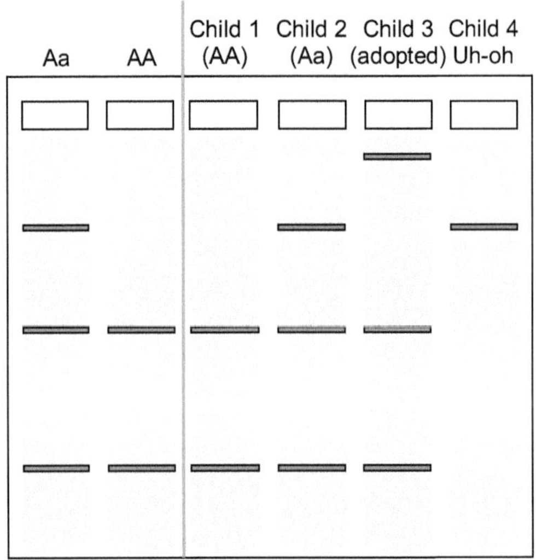

   a. We know that the mother is the true mother of the child. Could her husband be the father of the fourth child? (Refer back to the text to refresh yourself on the complete situation.) Justify your answer.
   b. Comment on the genetics of the fourth child.
   c. From your answers to parts a and b, give a likely scenario to explain the genetics of the fourth child.
   d. The banding pattern for the adopted child has been changed for this problem. Offer an explanation to what happened at the molecular level (Figure 15.3 should help).
8. Suppose that a DNA fingerprinting analysis is to be performed on a blood sample. Where does the DNA for the analysis come from?
9. Why is a larger sample needed to perform the RFLP form of DNA fingerprinting versus the STR method?
10. Why can you **not** use the RFLP method of DNA fingerprinting with moderately damaged DNA? What would your result look like?
11. Why **can** you use the STR method of DNA fingerprinting with damaged DNA?
12. Suppose there were only 7 combinations of numbers that could represent any given STR in any given person. What are the odds that an accurate STR-type DNA fingerprinting analysis could identify the wrong person in a US court of law?
13. Why is it that the range for the number of repeats in a given STR might be 6 to 14, but in remote parts of the world, the range is only 10 to 11?
14. Consider the STR named "THO1," which has the sequence AATG. One of the following is a good primer to use with this STR. Which one? Justify your answer.
    - AATG
    - $(AATG)_5$
    - CATT

- TTAC
- $(TTAC)_5$
- $(CATT)_5$
- GTGGGCTGAAAAGCTCCCGATTAT.

15. You decide to intern over the summer for your hometown police department. On your very first day, you are thrown in with Detective Park, who is investigating a murder. At the crime scene, you find a grisly slashfest has occurred. The detective asks you to collect biological samples from the victim and from a bloody switchblade that was found in the bushes. You do so, and then immediately quit the internship (after you finish vomiting on the lawn). You return to your school in the fall, knowing police work is not for you.

A couple of weeks later, you are called by the police to explain your collection techniques and the findings from the analysis of samples taken from the crime scene and three suspects. The detective messages a copy of the data to you (below) and asks whether you believe any of the three suspects should be arrested for the murder.

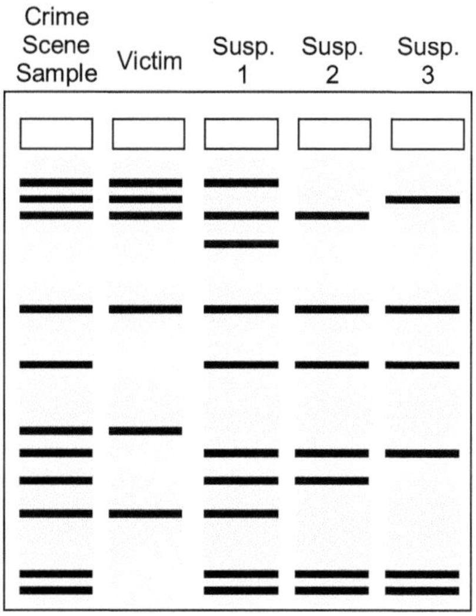

Crime Scene Sample | Victim | Susp. 1 | Susp. 2 | Susp. 3

a. Do you recommend that any of the suspects be detained? If so, which one(s)? Give a reason why or why not *for each suspect.*

b. Detective Park decides that, as a mere undergraduate, you could not possibly have done the experiment right. He obtains the original samples and performs another RFLP analysis using ethidium bromide to visualize the DNA. Draw a gel to represent what he will see for the sample taken from the victim.

16. Consider the STR D7S820, which has the tetrameric repeat sequence GATA. Suppose you used PCR to analyze this STR in a sample of DNA obtained from Officer Smedley using the primer set $[GATA]_5$ and a 20-mer of the complementary sequence. Officer Smedley has alleles containing 4 and 11 repeats. Draw a gel to indicate the results of the PCR. Include the size of each band.

17. a. If a murder scene is discovered 8 years after the fact, can DNA fingerprinting using RFLPs be used for analyzing DNA found at the crime scene? Why or why not?

b. Consider the crime scene from part *a.* Could STRs be used for DNA fingerprinting? Would this method be preferred over RFLPs? Why or why not?

CHAPTER

Fermentation, beer, and biofuels 16

Chapter outline

The harnessing of cells for the production of beer and biofuels involves processes that are (perhaps) surprisingly similar. Starting with glucose, different cell types can create different metabolites, and the metabolites they produce often depend on environmental conditions. The main environmental condition of interest in this chapter is anaerobic, and the metabolites of interest are the result of fermentation pathways.

In this chapter, we will learn two different methods of glucose breakdown. From there we will examine fermentation—what it is and what it can produce. We will discuss the production of ethanol and show how ethanol production is enticed from yeast in the production of beer. We will also discuss ethanol production for fuels. But the conversion of glucose to ethanol is not the only pathway of interest here. The production of glucose from cellulosic waste will be introduced, as will the production of alternative fuels from glucose feed stocks (and gene editing).

Before we can fully appreciate the applications of beer production and biofuels, the normal pathways of cellular metabolism must first be discussed.

Biotechnology and its Applications. https://doi.org/10.1016/B978-0-12-817726-6.00016-2

16.1 Glycolysis

Glycolysis is the cellular breakdown of the sugar glucose for energy. In-depth discussions of the glycolytic pathway can be found in any good biochemistry textbook and will not be included here; however, there are certain details that we should appreciate because they are important for the understanding of biotechnological applications that utilize cellular respiration.

16.1.1 The Embden-Meyerhof pathway

Glucose is the preferred energy source in most organisms. Although metabolism of many different sugars can be used for energy, the pathways involved all seem to converge on the glycolytic pathway at some point. The pathway shown in Figure 16.1, known as the Embden-Meyerhof pathway, is the glycolytic pathway used by humans (and yeast). It takes place in two phases: the investment phase and the payoff phase.

The investment phase refers to the steps of glycolysis that require energy input. The energy is obtained through the hydrolysis of a high-energy phosphate in an adenosine triphosphate (ATP) molecule. The investment phase includes the initial steps, the first of which is the conversion of glucose into glucose 6-phosphate, which can be written as:

Glucose + ATP → Glucose 6-phosphate + ADP

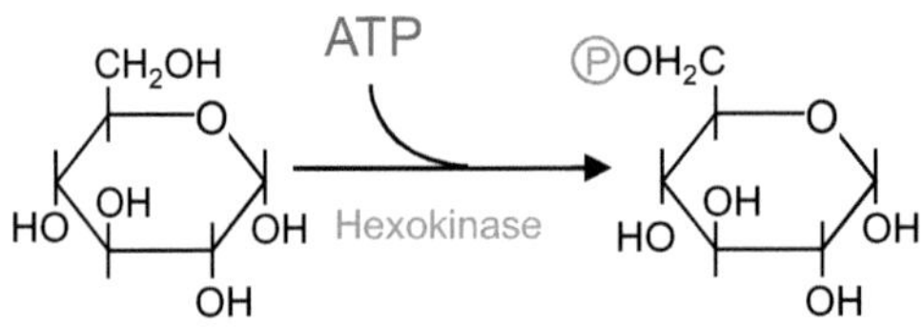

CH_2OH ATP ⓅOH_2C O OH HO OH Hexokinase OH HO OH OH OH

Notice that the ATP serves two purposes: it supplies the phosphate that is attached to the initial glucose molecule, and it supplies the energy needed to drive the reaction forward. After this initial step, the glucose of glucose 6-phosphate undergoes isomerization from a six-carbon ring into a five-carbon ring, yielding fructose 6-phosphate:

Glucose 6-phosphate → Fructose 6-phosphate

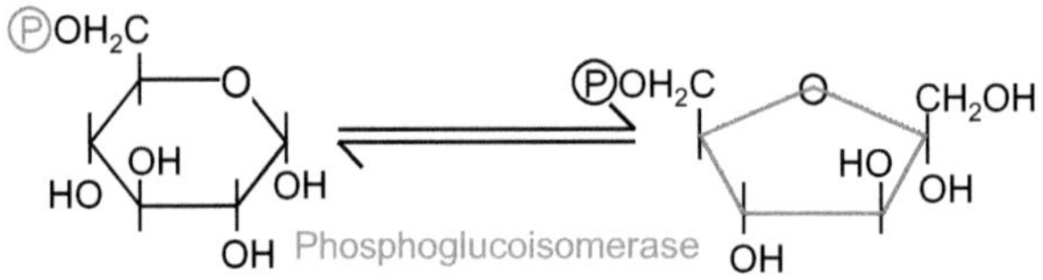

ⓅOH_2C O ⓅOH_2C O CH_2OH OH HO OH HO OH OH Phosphoglucoisomerase OH

In the next step, another ATP is invested to yield a fructose to which two phosphates have been attached, one on carbon 1 (introduced in this step) and one on carbon 6:

Fructose 6-phosphate + ATP → Fructose 1,6-bisphosphate + ADP

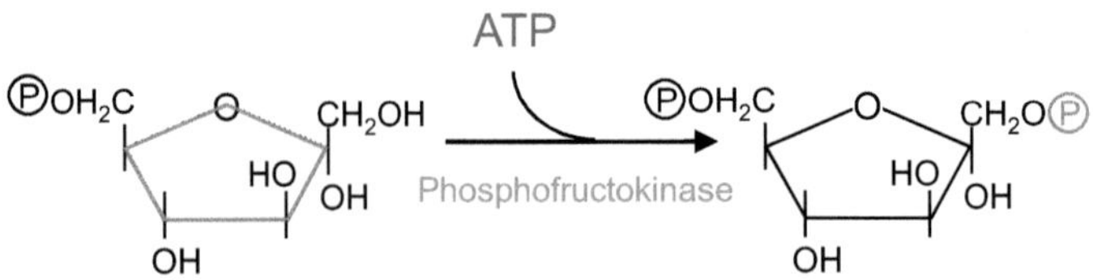

ATP ⓅOH_2C O CH_2OH ⓅOH_2C O CH_2OⓅ HO OH Phosphofructokinase HO OH OH OH

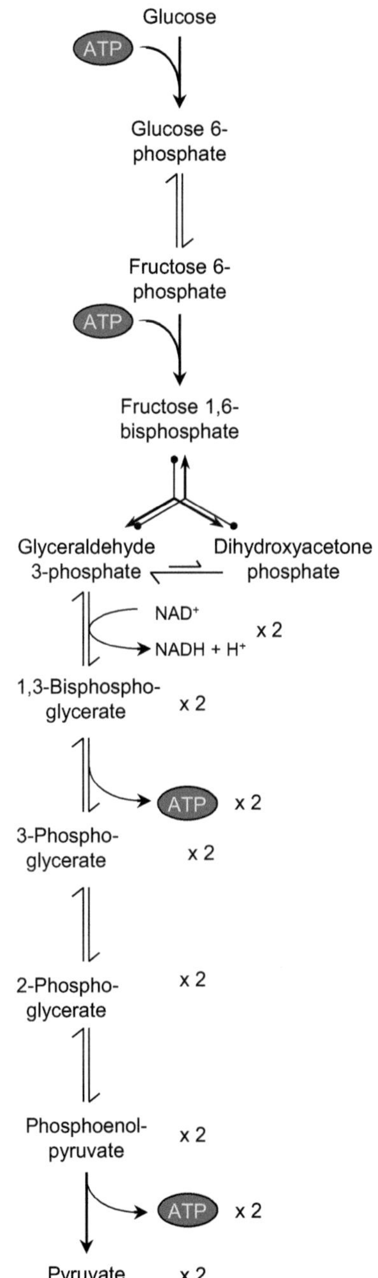

FIGURE 16.1

The Embden-Meyerhof pathway of glycolysis. The overall adenosine triphosphate (ATP) yield is 2 moles of ATP per mole of glucose. Specific steps involving ATP are shown.

If the purpose of glycolysis is to yield energy in the form of ATP for the cell and the initial steps require the hydrolysis of two ATP molecules, why would the process ever occur? The answer lies in the fact that reactions can be coupled. Despite the investment of two ATP molecules in the first phase, the yield of the payoff phase will be four ATPs, leaving the cell with a net gain of two ATPs for the conversion of one glucose molecule into two molecules of pyruvate, as we shall see.

The payoff phase of glycolysis has two ATP-yielding reactions. The first is the conversion of 1,3-bisphosphoglycerate—which can be viewed as a deprotonated glyceric acid with two high-energy phosphates attached—into 3-phosphoglycerate, which only holds one high-energy phosphate. The energy from this cleavage of a high-energy phosphate is used to form an ATP molecule:

1,3-Bisphosphoglycerate + ADP → 3-Phosphoglycerate + ATP

Further molecular conversions occur to set up removal of the final high-energy phosphate to produce another molecule of ATP. The reaction is:

Phosphoenolpyruvate + ADP → Pyruvate + ATP

From this cursory introduction to glycolysis, it may appear that two ATP molecules were invested to yield a payoff of two ATP molecules. However, notice from Figure 16.1 that the 6-carbon molecule fructose 1,6-bisphosphate is split into *two* 3-carbon molecules, dihydroxyacetone phosphate and glyceraldehyde 3-phosphate; each of these 3-carbon products progresses separately down the remainder of the glycolytic pathway to yield two ATPs and one pyruvate, for a total of four ATPs and two pyruvates. The net yield is thus two ATP and two pyruvates (and two NADH).

The above discussion by no means covers all of glycolysis. Enzyme names, kinetics, and individual reaction mechanisms are all important. The fate of pyruvate is also a key concern. For instance, in the presence of oxygen, not only can additional energy be extracted from the pyruvate molecules but also the two molecules of NADH produced in forming 1,3-bisphosphoglycerate can themselves be used to produce more ATP. Metabolism in the presence of oxygen is known as **aerobic respiration** and in the absence of oxygen is known as **anaerobic respiration**. Under anaerobic conditions, the pyruvate molecules will undergo **fermentation** to form products such as lactate in humans or ethanol in yeast.

Aerobic respiration yields far more energy than anaerobic respiration. The pyruvate is broken down into acetyl CoA, which can then enter the citric acid cycle (also known as the Krebs cycle or the TCA cycle) to produce carbon dioxide plus more ATP (or the energy-equivalent GTP), more NADH, and $FADH_2$. These latter two molecules can then enter the electron transport chain to eventually yield more

ATP molecules while converting oxygen into water. The difference in energy production from one mole of glucose is a net of 32 versus two moles of ATP for aerobic versus anaerobic respiration, which is why aerobic respiration is the preferred form of glucose metabolism in our bodies: the energy needs of the body can be met with a smaller amount of starting material. However, when the biotechnologist harnesses microbes to make fermentation products, oxygen supply will need to be eliminated to force the organisms into anaerobic respiration.

16.1.2 The Entner-Doudoroff pathway

The Embden-Meyerhof pathway is not the only glycolytic pathway used in nature. A second pathway, the Entner-Doudoroff pathway, is used by certain prokaryotes, such as *Zymomonas* and *Pseudomonas*, to break a molecule of glucose into two molecules of pyruvate. Although this pathway begins with the same conversion of glucose to glucose 6-phosphate, it is distinctly different from the Embden-Meyerhof pathway in the next three steps, which produce 2-keto-3-deoxygluconate (KDPG) (Figure 16.2). From here, KDPG, a 6-carbon molecule, is broken down to the 3-carbon molecules glyceraldehyde 3-phosphate (G3P) and pyruvate, two molecules we have already seen:

2-Keto-3-deoxyphosphogluconate → Glyceraldehyde 3-phosphate + Pyruvate

$$\begin{array}{c} COO^- \\ | \\ C{=}O \\ | \\ CH_2 \\ | \\ H{-}C{-}OH \\ | \\ H{-}C{-}OH \\ | \\ CH_2O\text{Ⓟ} \end{array} \xrightarrow{\text{KDPG Aldolase}} \begin{array}{c} COO^- \\ | \\ C{=}O \\ | \\ CH_3 \end{array} + \begin{array}{c} C{=}O \\ | \\ H{-}COH \\ | \\ CH_2O\text{Ⓟ} \end{array}$$

The G3P is converted to pyruvate using the same steps as in the Embden-Meyerhof pathway. The net result of the Entner-Doudoroff pathway is that one glucose molecule is broken into two pyruvate molecules with a net yield of one ATP (plus one NADH and one NADPH). The energy investment is cut in half because only one step uses ATP for phosphorylation. The lower ATP yield is due to the direct production of one pyruvate molecule from the 6-carbon intermediate KDPG without producing an ATP. This means that only one molecule of G3P is produced per glucose, so the pathway from G3P → → → pyruvate will only be utilized once per glucose molecule, thereby cutting the number of ATPs produced in the payoff phase in half.

The difference in ATP production between the two pathways has implications in terms of glucose utilization. Suppose a certain number of cells need a certain amount of energy from ATP to stay alive, and to get that energy they will burn glucose. In an anaerobic environment, cells that utilize the traditional (Embden-Meyerhof) pathway will produce two moles of ATP per mole of glucose, whereas cells using the alternative (Entner-Doudoroff) pathway will produce one mole of ATP per mole of glucose. Given that the energy needs of the two cultures are the same, it would take twice as much glucose to yield the same required amount of energy using the alternative pathway. The result is that cells using the alternative pathway, such as *Zymomonas*, will break down glucose at a faster rate, thereby

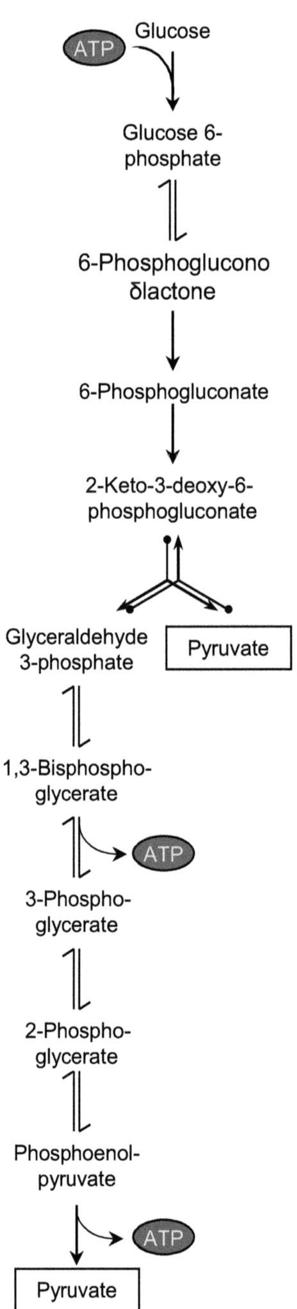

FIGURE 16.2

The Entner-Doudoroff pathway of glycolysis. The overall yield is 1 mole of adenosine triphosphate per mole of glucose. Two moles of pyruvate are also generated. Reactions involving NADH (and NADPH) have been omitted for clarity.

producing pyruvate at a faster rate. The rate of pyruvate production will soon become very important in the context of beer production, but first we must understand fermentation.

16.2 Fermentation

What can be done with the pyruvate molecules produced by glycolysis? Under anaerobic conditions, human cells convert pyruvate into lactate. Certain microbes, on the other hand, can convert pyruvate into acetyl CoA, which serves as a gateway to many other potential fermentation products; for instance, it can become phosphorylated and then used to yield acetate. Acetate is the conjugate base of acetic acid, which is found in the vinegar you might put on your salad. Acetyl CoA can also be converted into acetaldehyde, which can be further reacted to produce ethanol, another fermentation product with commercial value as a biofuel. Fermentation can also be used by certain microbes to produce acetone, a solvent commonly used in fingernail polish remover. From acetone, there is a chemical pathway to produce isopropanol, also known as rubbing alcohol.

Fermentation processes are performed by cells to extract energy from a starting material such as glucose. They do not consume oxygen—hence the extrapolation by some people that fermentation must occur in oxygen-free environments. This assumption is not strictly true: complete absence of oxygen is not a strict requirement for fermentation to occur, but oxygen is not involved in the chemical reactions. The hallmarks of fermentation processes are:

- Energy is produced.
- Oxygen is not consumed.
- The NADH/NAD^+ ratio is unchanged by the process.

Consider the fermentation of glucose into lactic acid in humans (Figure 16.3). Two molecules of pyruvate are formed by the breakdown of one molecule of glucose via glycolysis, and under anaerobic conditions the pyruvate is converted to lactate. This final step removes the NADH that was produced earlier in the glycolytic pathway. In the presence of oxygen, the NADH would be able to enter the electron transport chain and would be indirectly responsible for the hydrogenation of oxygen to form

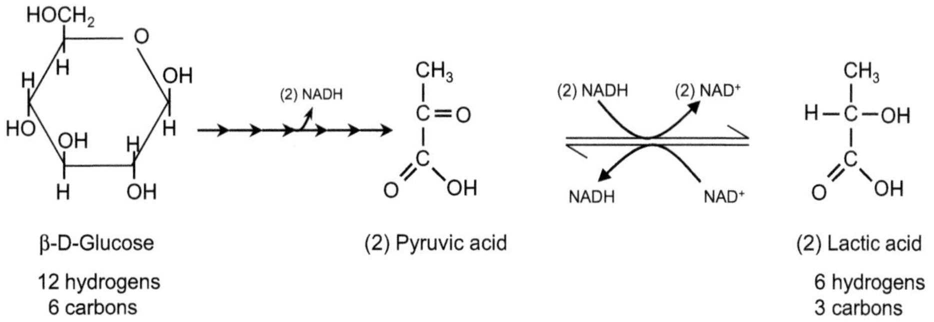

β-D-Glucose
12 hydrogens
6 carbons

(2) NADH

(2) Pyruvic acid

(2) NADH (2) NAD^+

NADH NAD^+

(2) Lactic acid
6 hydrogens
3 carbons

FIGURE 16.3

Fermentation of glucose into two lactic acid molecules. Notice that the net number of NADH=0.

water. In the absence of oxygen, the electron transport chain will be halted, so the NADH will not be used to produce energy. If left unchecked, the concentration of NADH would increase and turn off certain glycolytic enzymes, meaning the breakdown of glucose to obtain energy would be halted. To prevent this potentially deadly effect, NADH is siphoned off during fermentation to produce lactate. Notice that the conversion of pyruvate to lactate will alter the NADH/NAD^+ ratio (refer to the three hallmarks of fermentation), so this single step should not be considered fermentation; rather, the conversion of glucose into two molecules of lactate is a fermentation process. The NADH/NAD^+ ratio is unchanged in the process of converting one molecule of glucose into two molecules of lactate.

In aerobic respiration, NADH is an energy source. It passes an electron pair, via a hydride ion, into a chain whose members serve as a series of electron carriers. As the electrons are passed from one set of carriers to another, changes in free energy are harnessed to pump protons out of the mitochondrion to create a charge gradient across the inner mitochondrial membrane. The gradient eventually becomes large enough to drive protons back into the mitochondrion via a proton ATPase (see Chapter 2), but the ATPase is driven in reverse so that it forms ATP rather than hydrolyzing it. Because NADH can serve as an energy source, the cell uses the concentration of NADH (monitored by the ratio [NADH]/[NAD^+]) to help determine its own energy state. If there is an excess of NADH, the cell behaves in a way consistent with being in an energy-rich state because all of its energy needs are being met. Energy-producing processes such as glycolysis and the citric acid cycle are inhibited, and the cell will resort to alternative pathways designed to store energy.

In humans, fermentation is the production of lactic acid from glucose. As already mentioned, other organisms produce far different products via fermentation: ethanol, acetone, isopropanol, and butanol are all fermentation products. Several of these pathways are shown in Figure 16.4. The biotechnologist can harness microorganisms to produce each of these products, even on an industrial scale. For instance, yeast can be used to produce ethanol as part of the beer-making procedure.

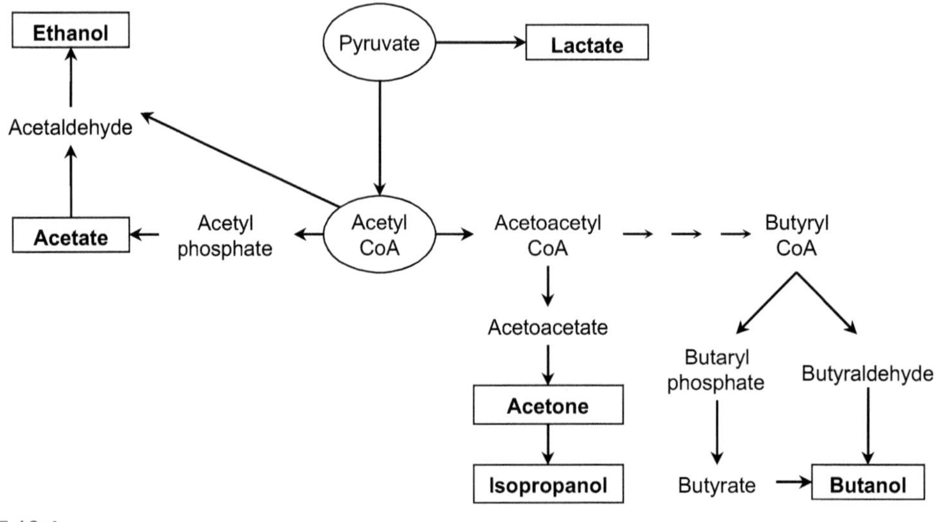

FIGURE 16.4

Rudimentary metabolic map of some common fermentation products (shown in boxes).

16.3 The production of beer

The production of beer is perhaps the oldest biotechnology known to humans. Beer has been around for thousands of years: ancient evidence of its existence can be found on a Sumerian tablet from 4000 BCE and again in a 4000-year-old Sumerian poem written to honor their patron goddess of brewing. Beer is mentioned in sonnets, in science books, and in religious texts—so evidently there is some kind of allure. Now, as budding biotechnologists, we will be able to appreciate beer for how it incorporates fermentation, cell respiration, and cell culture into a product with mass appeal.

The process of beer production uses fermentation, carried out by yeast, to produce ethanol. The fermentation uses carbohydrates found in grains as its primary fuel source. The process is carried out by yeast glycolytic enzymes using the Embden-Meyerhof pathway (Figure 16.1).

Keep in mind that grains are seeds. There's a reason why seeds are used: they contain dense energy stores and nutrients for the plant embryo that will develop from them, much as the yolk of an egg serves a developing chicken embryo. One of the most common seeds used for producing beer is barley, although other grains, such as wheat, can also be used.

16.3.1 Malt

The first step in the brewing process utilizes the seeds' own enzymes to break down the complex carbohydrates (starches) that are stored within. The products will be simple sugars, and the process is known as **malting** (Figure 16.5). The seeds are allowed to germinate because, in normal development, the growing young plant would need to access these complex carbohydrates for its own energy. The energy needs of a developing organism are relatively large, so the early plant embryo produces the enzymes needed to break down the complex carbohydrates stored in the endosperm into simple sugars. The brew master needs this production of simple sugars because they will be used later as fuel for yeast.

FIGURE 16.5

Malted barley.

FIGURE 16.6
Mashing.

If germination were to be allowed to continue to completion, the developing plant would use up the sugars for its own energy needs. The process is allowed to progress long enough for the embryo to produce the starch-degrading enzymes—but once the enzymes are present, germination can be halted because only the enzymes are needed to produce the simple sugars. (Recall that enzymes catalyze chemical reactions; a living organism is not needed for an enzyme to function.) Heat is used to halt germination. The sugars that will be produced in the next step include glucose, fructose, and the disaccharide maltose (which is made up of two glucose molecules). Maltose is typically the sugar of interest for beer production.

16.3.2 Wort

The product of malting—malt—is used to make a nutrient broth, similar in principle to the Luria-Bertani broth used for *Escherichia coli* growth. This nutrient broth, known as **wort**, is used as a culture medium for yeast. It is produced by adding water to the malt, which is then mashed (Figure 16.6). (*Sour mash* is slightly different: microbes are permitted to partially degrade the mash, which acidifies it, giving the resulting product a sour taste.) The malt enzymes then catalyze reactions in which maltose, glucose, and other sugars are formed. After this maturation, the mash is a slurry that contains seed debris, proteins, and both simple and complex carbohydrates (Figure 16.7). The solid portion is separated out by gravity, perhaps by using a mash tun, which uses a porous support bed to hold the solid grain remnants as the liquid slowly flows through.

The collected liquid is then boiled with hops (Figure 16.8). Hops give a characteristic bitter taste to the beer, but historically they were also used because of their antiseptic qualities: beer used to be transported around the world on ships, and the presence of a hop extract in the beer would prevent microbes from metabolizing it during transit. (As a side note, Indian pale ales—IPAs—have a fairly high alcohol content for a similar reason. Made in India, they once had to be shipped around the Horn of Africa to get to their final destinations, and the trip could take months. The higher alcohol content served as a preservative.) Hops are also used to lend a distinctive aroma to the final product. After the wort has been boiled with the hops, it is cooled and aerated. Aeration is necessary to enrich this nutrient broth with oxygen for the next step.

FIGURE 16.7

Mature mash.

FIGURE 16.8

Dried leaf hops.

16.3.3 Yeast cultures

The third step is to prepare the raw beer. A yeast culture is established in the aerated wort. Two-stage growth will be employed: first aerobic growth and then a switch to anaerobic conditions. The purpose of the aerobic growth is to allow the yeast population to increase rapidly. With oxygen present, the yeast can extract more energy per sugar molecule, which means its energy requirements are more easily met, allowing the cells to proceed through the cell cycle unhindered. The wort contains many simple sugars from the breakdown of starches. When the yeast cells are first added to the aerated wort, there is much room for expansion (recall contact inhibition), plenty of nutrients, and plenty of oxygen. These conditions allow them to grow and divide very quickly. The final products of aerobic metabolism of the sugars are water and carbon dioxide, which is the source of carbonation in beer. As the growth curve for the yeast culture enters the late log phase, when nutrients and room for expansion are becoming limiting factors, the culture is capped to prevent further access to oxygen. The culture will continue to grow and metabolize the sugars until all the oxygen is consumed, at which point the microbes will switch to anaerobic metabolism and utilize a fermentative pathway. The remaining sugars will be broken down into carbon dioxide and ethanol.

In Chapter 5, we discussed the growth curve, which certainly applies to yeast cultures in beer brewing. Within the growth curve, the beer maker must try to minimize the lag phase, which is critical to brewing: the longer the lag time, the greater the risk of contamination via microbial infection. Brewers will often make a starter yeast culture to minimize the lag phase (Figure 16.9). A starter culture is like a mini-beer that is used to increase the pitch rate (of yeast into the wort). The sugar selection for the starter culture is critical; if anything other than dry malt extract is used as the substrate, the lag time will increase despite an increase in final yeast numbers because the yeast will need to produce enzymes to metabolize different sugar types. For example, if your starter culture uses sucrose, the yeast cells will need time to produce the necessary enzymes for degrading maltose when you transfer them into the wort. If the starter culture is made with dry malt extract, there will be no need for the yeast to make new sugar-metabolizing enzymes when they are transferred.

That is the rudimentary process for making beer; many more intricacies can be employed to alter the color, clarity, foam, bitterness, carbon dioxide content, overall taste, aroma, and even the texture of the final product. For instance, the amount of foam (head) formed after pouring the beer

FIGURE 16.9

A mature yeast starter culture.

is related to the amount of protein in the liquid; this can be controlled during the malting process. The addition of a protease such as shortens the polypeptides and makes them less able to form a large, stable layer around bubbles as they form, which translates to less head on the beer. Although a good amount of foam is considered a positive thing, too much protein in the beer will result in cloudiness, especially when the beer is cold. This is related to proteins partially coming out of solution, to the point at which they become visible. Letting the proteases work too long can result in flat or foamless beer, but not letting them work long enough will give you cloudy beer. The amount of foam in the final product can also be controlled by regulated venting of the carbon dioxide gas during the growth of the yeast culture. Additional carbonation can also be added via an in-line carbonator. (Notice the reappearance of our old friend, papain. While papain is used here to produce a clear, amber beer, it is also routinely used in establishing primary cell cultures from explants (Chapter 7). Papain exists in the leaves, roots, and fruit of the papaya plant.)

16.3.4 Skunky beer

Let us now briefly revisit the Entner-Doudoroff pathway (Figure 16.2), the glycolytic pathway used by certain microbes such as *Zymomonas*. Recall that, in this pathway, for every glucose molecule and investment of one ATP, there is a yield of two ATP and two pyruvate molecules. This is in contrast to the Embden-Meyerhof pathway used by humans and yeast, which requires an investment of two ATP to yield four ATP and two pyruvate molecules per glucose. The net ATP yields are one ATP for the Entner-Doudoroff pathway and two ATP for the Embden-Meyerhof pathway. In section 16.1.2, we considered two imaginary cell cultures that had identical energy requirements for the maintenance of normal functions; let us now suppose that amount is 10 (arbitrary) units of ATP equivalents per second. To get that amount of energy, the cell would have to use $10/2=5$ units of glucose if it used the Embden-Meyerhof pathway for glycolysis, while it would require twice as much glucose—10 units—if it used the Entner-Doudoroff pathway, assuming anaerobic respiration. The point here is that the cell using the Entner-Doudoroff pathway would have to burn twice as much glucose just to stay alive, which implies that glucose would be metabolized twice as quickly to produce the necessary energy for survival. One might wonder whether this pathway could be applied to make beer production more efficient. Instead of using yeast, could *Zymomonas* be used to ferment the sugars in wort into ethanol? *Zymomonas* would break down the sugars faster than yeast; therefore, the rate of ethanol production should be increased. From a business standpoint, this one fact might justify a switch from yeast to *Zymomonas*. The problem is that *Zymomonas* will produce other side products, such as acetaldehyde—a two-carbon aldehyde referred to in the brewing industry as ethanal. A beer drinker might care about this because acetaldehyde, a molecule produced during ethanol metabolism in the liver, contributes to the ill effects of a hangover. Having this molecule present in the beer would help ensure that the beer would provide a more significant hangover to the consumer, which is not a great marketing ploy. In addition to acetaldehyde, glucose metabolism in *Zymomonas* yields hydrogen sulfide, also known as sewer gas. *Zymomonas* happens to be the number one contaminant responsible for beer spoilage, imparting an undesirable taste and smell to the product. Returning to our business model, not only is *Zymomonas* bad because it produces a poor product, but it will outcompete the yeast for glucose by metabolizing it more quickly, resulting in a cell culture made primarily of *Zymomonas* and not yeast. Not only should *Zymomonas* not be added to the wort, but active steps should be taken to avoid its introduction as a contaminant.

Zymomonas contamination is often the result of scale-up. When an individual goes from a home-brew to an industrial scale of production, adequate adjustments must be made for cleaning the equipment because the probability of contamination is increased. During the brewing process, a layer of matter will collect on the bottom of the casks; *Zymomonas* lives easily in that layer because the sediment is energy rich. Once the *Zymomonas* culture is established, it is very hard to kill. To prevent contamination, much of the brewing equipment must undergo routine sterilization.

16.4 Fermentation to produce biofuels

16.4.1 Ethanol: a biofuel with problems

In addition to being used as a beverage ingredient, ethanol can be used for energy. In the United States, the government mandates that most gasolines must contain ethanol. China had agreed to follow suit in 2020 but suspended its plans to do so partially because of a decline in their corn stocks. The purpose of using ethanol in gasoline is not to reduce greenhouse gases like carbon dioxide—in fact, ethanol combustion releases about the same amount of greenhouse gases per carbon as gasoline. Ethanol is used as a means to stretch existing oil and gas supplies and for corn-producing countries such as the United States to reduce their dependence on foreign oil.

The ethanol used in gasoline is a bioproduct, produced by yeast fermentation of sugars derived from corn starches. The process is strikingly similar to the production of the ethanol in beer: a mash is produced, corn starches are broken down into simple sugars via added enzymes (α-amylase), and yeast fermentation converts the sugars into ethanol. The fermented corn mash is even called "beer." After the fermentation, the ethanol concentration is 8% to 12%; subsequent concentration via distillation will bring the purity up to 92% to 95%, and most of the remaining water will be removed from the vapor phase by molecular adsorbents to yield more than 99% ethanol.

In addition to not reducing greenhouse gases during combustion, ethanolic fuels come up short in terms of environmental friendliness because the production of ethanol involves the use of fossil fuels. Corn must be harvested, typically using tractors, and then transported to the processing facility by trucks burning diesel fuel. Heat for cooking the corn mash and distillation is generated by coal, oil, or natural gas. Fossil fuels must also be used in transporting the ethanol product to facilities that mix it with gasoline. Other problems that distance ethanol from the status of a perfect biofuel include its heat of combustion, for which gasoline automobiles are not optimized, and the fact that the corn could instead be used as a food supply. The 2008 food riots in Haiti, Egypt, and Indonesia—and many other countries—can be traced back to the decreased food supplies that resulted from corn crops being earmarked for ethanol production in the United States, which drove up the price of corn.

Some well-meaning environmentalists have argued that ethanol should be used in a pure form as a fuel source for cars, but one major problem with an ethanol-based system lies in the transport of the ethanol. Compare ethanol with oil: oil can be transported via pipelines, but ethanol cannot. Ethanol mixes readily with water, including the water in the air we breathe. Humidity presents a problem: the supplier must spend money and energy to transport the unwanted water that is mixed with the ethanol product. Even worse, when you put this ethanol from the pipeline into your car, upon combustion a significant amount of the energy will be used to heat the water in the fuel instead of being used to make the car move. Removal of the water at this point would be economically infeasible, requiring another distillation step that would require more fuel energy.

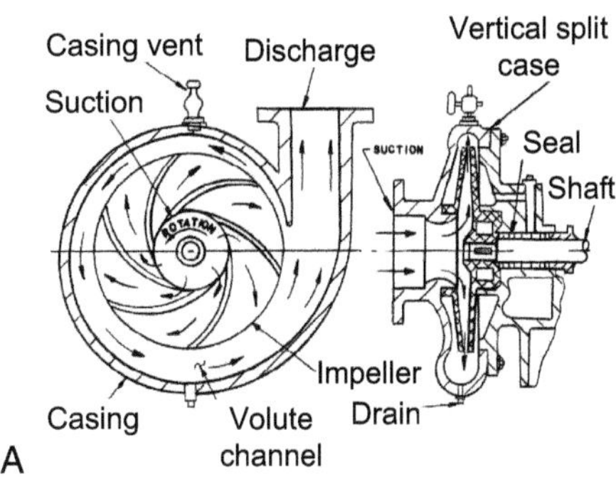

FIGURE 16.10

(A) Schematic of a centrifugal pump. (B) Centrifugal pump in use.

(A) from http://navalfacilities.tpub.com/mo230/mo2300161.htm.

Another major problem with using pure ethanol as a fuel for the mass automobile market is that it cannot be dispensed through traditional pumping systems. One reason for this is that ethanol vapor would erode metal components in conventional pumps, making small cavities. This effect is known as **cavitation**. Given the high volatility of ethanol, this is a significant potential problem. An alternative type of pump could be used, such as a centrifugal pump (Figure 16.10). This type of pump operates like a fan, spinning between 1750 and 3500 RPM. However, the spinning requires energy, which must be weighed against the energy benefits of the ethanol fuel. Another common type of pump to consider is the positive displacement pump, which works like a piston to force liquid through tubes via pressure. If there is vapor in the liquid (as there will be in the case of ethanol), then part of the energy used to move the liquid through the tubing is going to be wasted on compressing the vapor. Once again, the overall energy yield of the ethanol fuel will be lowered at the pumping site. One way to get around this problem is to cool the ethanol so that less of it is in the gas phase, meaning there is less vapor to compress with the positive displacement pump. The problem with this solution is that the cooling process will again use energy and cost money.

One undeniable advantage of using ethanol as a fuel is that it can originate from a renewable source. However, one must consider the ramifications of messing with the food supply. The planet can only support a finite number of people—on the order of 1×10^{10}. This number is based, in part, on the amount of food that can be produced. If some of this food is instead used to produce energy, the total number of people that can be supported will be reduced.

16.4.2 Biobutanol

Some of the end products in the biochemical fermentative pathways are ethanol, acetone, isopropanol, acetate, butyric acid, and butanol. Of these, butanol might serve as a biofuel with properties superior to ethanol. Butanol is a four-carbon molecule, as opposed to the two-carbon ethanol, which makes butanol less volatile. Butanol is also far less hygroscopic, meaning it will pull far less water out of the air. Butanol and water don't mix very well due to butanol's longer carbon chain, which makes it less hydrophilic than ethanol. This has significant implications, such as with regard to transporting butanol via pipelines. Also, because the longer carbon chain renders butanol less volatile, conventional pumps can be used to dispense the fuel. The alternative pumping systems just described can also be used: less vapor will be associated with the butanol than the ethanol, so the problem of spending energy to compress butanol in the gas phase is greatly reduced. Inside the engine, a reduction in the amount of water in the fuel means a greater amount of energy will be available to make the vehicle move because energy will not be wasted in heating the water. Finally, every mole of butanol contains more chemical potential energy than a mole of ethanol because of the two extra hydrocarbons that can be oxidized; because of the relative density of the two compounds, butanol carries more energy per unit volume than ethanol. Specific parametric values are given in Table 16.1. In short, butanol would be a more attractive biofuel than ethanol because of its higher energy potential and the possibility of using traditional pumps and transport systems.

Notice that the pathways shown in Figure 16.4 provide for many products in addition to butanol. If all we were interested in was the production of butanol, then these alternative pathways would serve to reduce the total yield of our desired product. One approach aimed at increasing butanol yield is to prune this reaction tree. For instance, by knocking out the gene that codes for the enzyme acetate kinase, the pathway from glucose to acetate will be eliminated, and theoretically this should increase the use of all

Table 16.1 In comparing butanol with ethanol, one can see that butanol carries more energy per unit volume, is less volatile, and can be mixed with a greater amount of air for combustion

	Butanol	Ethanol
Energy content (kBTU / gal)	110	84
Vapor pressure (PSI @ 100°F [37.8°C])	0.33	2.0
Useful air / fuel ratio	11.1	9.0

kBTU, Kilo-British thermal unit; *PSI,* pounds per square inch.

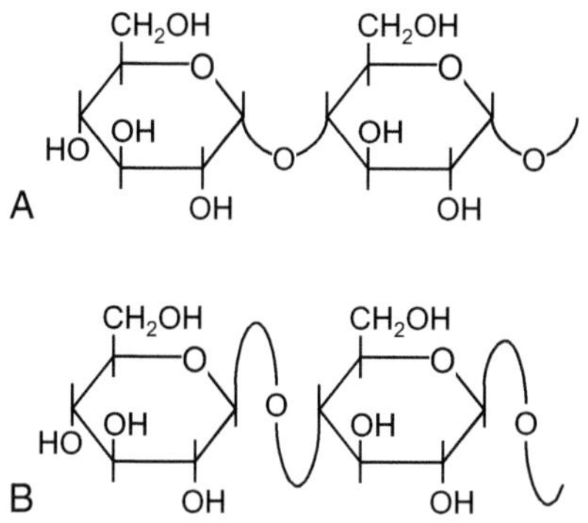

FIGURE 16.11

The structures of two polymers of glucose: starch and cellulose. (A) For starch, the glucose repeating units are connected by (α1→4) linkages. (B) For cellulose, the glucose repeating units are connected by (β1→4) linkages. Humans do not produce an enzyme to help break these bonds.

of the other pathways in the tree. Because the cell is not spending time, energy, and resources on the production of acetate, the potential for greater butanol production should be realized. Genes can be knocked out via a gene editing technology such as a CRISPR-Cas system (see Chapter 14).

Additional branches could also be trimmed, and this has been achieved in the laboratory with some success. For example, a significant increase in the amount of butyrate produced by *Clostridium* has been achieved by knocking out the gene coding for phosphotransacetylase (plus two other mutations). Although the results are very encouraging, the butanol yield from these microbes is still not sufficient for industrial-scale use. A yield of 12% from glucose would make this method of butanol production economically feasible on a large scale. However, improvements in microbial butanol yields are being achieved little by little. A review from 2020 showed microbial yields of biobutanol have been improved to the order of 1%, although new strains of Clostridia have improved tolerance and production of butanol to values of 1.9% to 2.1% *(w/v)*.

16.4.3 Cellulose

There is another potential energy source—cellulose—that is being investigated as a feed stock for biofuel production. Cellulose is found in grasses, paper, and plant husks, all of which are thrown away by humans as waste materials. Like starch, it is a polymer of glucose molecules; the major difference between the two polymers lies in how the glucose molecules are linked together. Starch uses α-acetal linkages, whereas cellulose uses β-acetal linkages (Figure 16.11). Although humans can digest starch, cellulose is useless to us as a food source because we lack the enzyme to break down the glc(β1→4)glc (β-acetal) linkages—an enzyme called cellulase. By incorporating a gene coding for cellulase into the genome of certain microbes, some biotechnologists are making strides toward converting what was once considered a waste material into a feedstock for biofuel production. By allowing microbes to break down cellulose to glucose, greater potential would exist for producing biofuels such as butanol.

16.5 Summary

In this chapter, we learned about the Embden-Meyerhof and the Entner-Doudoroff pathways of glycolysis. These pathways have direct implications for the production of pyruvate and the products of fermentation that may follow. We then identified properties common to all fermentation processes and saw how ethanol and other products can be produced. We looked at beer brewing—one of the oldest biotechnologies—and biofuel production, a relatively recent biotechnology involving the same fermentation product, ethanol. We ended by discussing butanol as an alternative biofuel and saw several advantages it presents versus ethanol. We also took a brief look at the potential of cellulose as an upstream source of glucose.

In the next chapter, we will shift gears back to mammalian (human) cells as we examine stem cells and their use in tissue engineering and regenerative medicine.

Acknowledgment

Special thanks to Graham Satterwhite, beer home-brewing expert.

Related reading

Farrell, A.E., Plevin, R.J., Turner, B.T., Jones, A.D., O'Hare, M., Kammen, D.M., 2006. Ethanol can contribute to energy and environmental goals. Science 311, 506–508.

Lehmann, D., Hönicke, D., Ehrenreich, A., Schmidt, M., Weuster-Botz, D., Bahl, H., Lütke-Eversloh, T., 2012. Modifying the product pattern of *Clostridium acetobutylicum*: physiological effects of disrupting the acetate and acetone formation pathways. Appl. Microbiol. Biotechnol. 94, 743–754.

Liew, F.M., Köpke, M., Simpson, S.D., 2013. Gas fermentation for commercial biofuels production. In: Fang, Z. (Ed.), Liquid, Gaseous and Solid Biofuels—Conversion Techniques (Chapter 5), IntechOpen, London, UK.

Mosier, N.S., Ileleji, K., How fuel ethanol is made from corn. Purdue Extension Bioenergy Series (ID-328) http://www.ces.purdue.edu/bioenergy.

Veza, I., Said, M.F.M, Latiff, Z.A., 2021. Recent advances in butanol production by acetone-butanol-ethanol (ABE) fermentation. Biomass and Bioenergy, 144, 105919.

Questions

1. Consider the Embden-Meyerhof and the Entner-Doudoroff pathways of glycolysis.
   a. How many moles of pyruvate will be produced by each pathway when starting with one mole of glucose?
   b. How many moles of ATP will be produced by each pathway when starting with one mole of glucose?
2. a. What are the investment steps in the Embden-Meyerhof pathway of glycolysis (used by yeast)?
   b. What are the payoff steps in the Embden-Meyerhof pathway of glycolysis (used by yeast)?
3. Explain how it is that one mole of fructose 1,6-bisphosphate is responsible for the production of *two* moles of 1,3-bisphosphoglycerate in humans.

4. Suppose that you have two vats of identical glucose-containing solutions stored in your shed. They each become contaminated on Monday, one with a microorganism that utilizes the Embden-Meyerhof pathway of glycolysis and the other with a microorganism that utilizes the Entner-Doudoroff pathway. On Tuesday, unaware that the solutions are contaminated, you sneak into your shed to enjoy a sweet beverage. Ugh! Neither solution tastes like you expected. Compare the tastes of the drinks in terms of sweetness.
5. What are the three defining characteristics of fermentation?
6. a. What will happen if we put a small sample of yeast into a vat with wort and seal the vat, so no oxygen is available? Draw the growth curve, starting from the initial addition of yeast.
   b. What will happen if we put a small sample of yeast into a vat with wort, allow it to grow to the mid-log phase, then seal the vat so no oxygen is available? Draw the growth curve, starting from the initial addition of yeast.
7. You have two starter cultures of yeast. The first was grown in a solution containing glucose as the energy source, and the second was grown in a solution containing malt extract. Compare and contrast the growth curves of the two cultures after each is added to wort for beer brewing.
8. What is the purpose of each environmental condition when growing alcohol-making bacteria first in an aerobic and then in an anaerobic environment?
9. Consider the process of beer making. If, in making the malt, we let the process proceed for three times longer than we should, predict the effect on the final beer product.
10. "Ethanol derived from corn and used in gasoline is good for the environment because it reduces CO_2 emissions." Give three reasons why this is not the case.
11. Name two advantages of using butanol over ethanol as a fuel source.
12. What would happen if, instead of producing lactic acid, a human produced ethanol as a fermentation product?
13. a. Does the human body produce ethanol via fermentation?
    b. Can *Clostridia* produce lactic acid via fermentation?
    c. Can any living organism produce the powerful solvent acetone?
14. a. Why can yeast not produce beer that is 50% alcohol?
    b. Yeast are used in producing liquors that contain over 50% alcohol. How is this possible?
15. Name two living organisms (not including humans!) positively involved in the production of beer for human consumption.
16. Name one reason why positive displacement pumps are not an economically feasible method for pumping ethanol.
17. To combat the problem you named in question 16, we could use a centrifugal pump. Name one reason why centrifugal pumps are not an economically feasible method for pumping ethanol.
18. Suppose you were making beer in your closet. Predict what effect each of the following would have on your final product, and give a brief reason to support your answer:
    a. You quadrupled your malting time.
    b. You boiled your wort with tea instead of hops.
    c. There was a small hole in your tun, which allowed oxygen to enter during the fermentation step.
    d. You used roasted wheat instead of barley.
19. Would *Zymomonas* be a better choice than yeast for the biological production of ethanol for chemical processes? Why or why not?
20. What is cavitation and what does it have to do with ethanol?

CHAPTER

17 Stem cells, tissue engineering, and regenerative medicine

Chapter outline

In the first half of this chapter, we will examine stem cells from two philosophical points of view. Although a definition that encompasses both views is readily available, we will need to learn a little embryology to be able to appreciate what it means. After understanding what stem cells are, we will turn our attention to what must be done with them to allow us to use them in a controlled manner.

In the second half, we will learn about tissue engineering and regenerative medicine, which use cells (or scaffolds populated with cells) to restore, maintain, or improve tissues or organs. Special consideration will be given to bioreactors, which provide the environments for cell growth, and polymeric scaffolds, which provide the supports for cellular attachment, proliferation, and communication. We will end with a specific example of a tissue engineering application: engineered artery.

17.1 Stem cells

17.1.1 What is a stem cell?

We begin this chapter with the definition of a stem cell. There is some controversy over this definition, which varies between laboratories. However, for the purposes of this text, we will define a stem cell as being a cell that has the potential to be differentiated down any of the three germ cell lineages.

Biotechnology and its Applications. https://doi.org/10.1016/B978-0-12-817726-6.00017-4

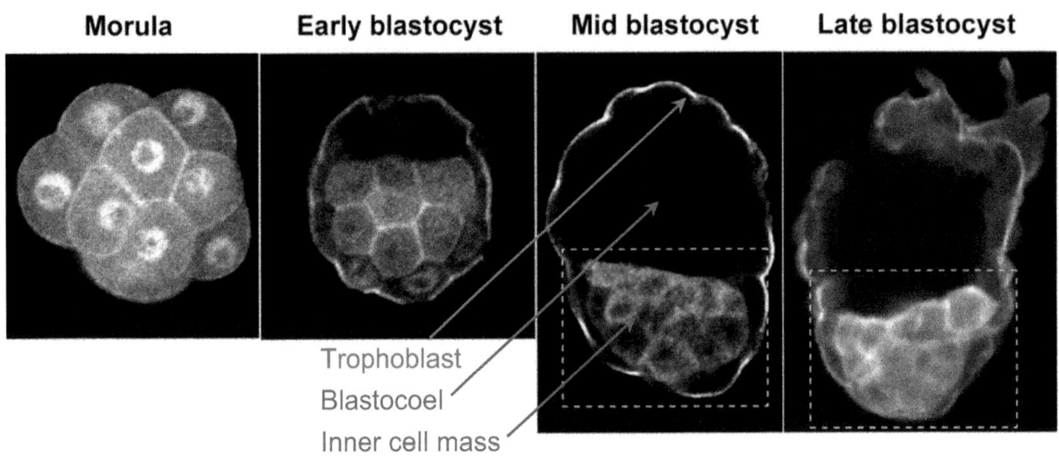

FIGURE 17.1

Development of the mouse blastocyst. The layers of the blastocyst are indicated in color in the third panel.

From: Saiz N, Grabarek JB, Sabherwal N, Papalopulu N, Plusa B. Atypical protein kinase C couples cell sorting with primitive endoderm maturation in the mouse blastocyst. Development. (2013) 140: 4311-4322. doi: 10.1242/dev.093922. Epub 2013 Sep 25.

Germ cell lineages are related to embryo development. All the cells in your body have basically the same DNA (not counting mutations), so it is reasonable to assume that your bone cells have the same DNA as your liver cells, which have the same DNA as your nerve cells. These three cell types behave very differently, but they still have the same DNA. The reason for their common genomes is that they all came from the same fertilized egg. Given that all the cells in your body were derived from this one cell, it is reasonably straightforward to see why they all have the same DNA, for the most part. After an egg cell has been fertilized, it divides into two cells, which divide into four, which divide into eight, and so on (the cells will then continue to divide asynchronously, so the number of cells in the growing mass will not necessarily be a power of two). At approximately 96 hours after conception, there will be a solid ball of about 32 cells, known as a **morula** (from the Latin word *morus*, meaning "mulberry" because of its appearance). From this point, the cells continue to divide, developing into a **blastocyst**, which is a hollow sphere of about 150 cells (Figure 17.1). The outer layer of the blastocyst, known as the **trophoblast**, surrounds a fluid-filled cavity known as the **blastocoel** (ˋblas-tə-sēl) and an **inner cell mass**. It is the cells of this inner cell mass that will develop into the embryo and further on into the fetus.

The blastocoel continues to develop in a process known as **gastrulation**; by 15 or 16 days after fertilization, the **gastrula** has been formed (Figure 17.2). The blastocoel has developed into an embryoblast, with the cells of the inner cell mass developing into a two-layered structure called the bilaminar disc. The bilaminar disc separates two hollow cavities, the amniotic cavity and the yolk sac. After further maturation, an indentation known as the primitive streak forms in the upper layer of the bilaminar disc (the epiblast). The primitive streak develops into the primitive groove, which will serve as a conduit for cells migrating from the epiblast. The cells of the bottom layer of the bilaminar disk (the hypoblast) are replaced by migrating cells to form the **endoderm**. Cells in this layer will go on to form the endocrine glands and the linings of the digestive and respiratory tracts. At this point, the layers of the bilaminar disc are separated, creating space for more migrating cells which will make up the **mesoderm**. Cells from this layer will form muscle (including heart), bone, and cartilage. The epiblast layer is now referred to as the **ectoderm**, and its cells will go on to form skin and nerve tissue.

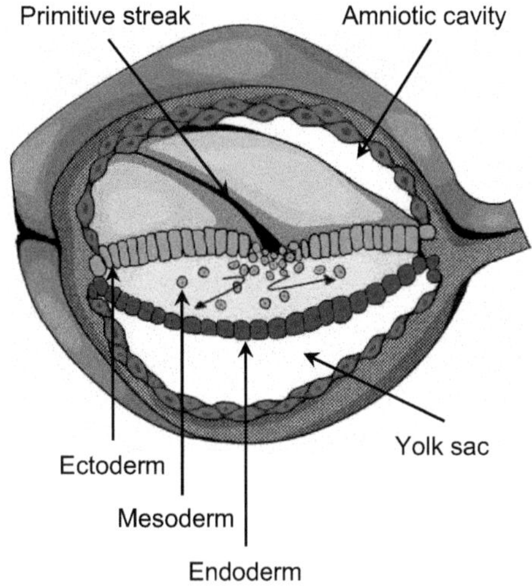

FIGURE 17.2

Cross section of a gastrula at 15 to 16 days of development. The trophoblast has matured into an embryoblast, having two cavities (the amniotic cavity and the yolk sac) separated by a bilaminar disk. After further maturation, the primitive streak (groove) forms, which will serve as a conduit for cells from the top layer of the bilaminar disc to migrate. The cells of the bottom layer of the bilaminar disk are replaced by migrating cells to form the endoderm *(blue)*. At this point, there is a separation between the layers of the bilaminar disc, which will be partially filled by cells of the mesoderm *(pink)*. The top layer is now known as the ectoderm *(green)*.

Drawing courtesy of Dustin Godbey, Lusher Charter School.

Returning to the definition of the stem cell, it should now be clear that "the three germ cell lineages" refers to cells of endodermal, mesodermal, and ectodermal origin. On a smaller scale, a mesoderm progenitor cell can differentiate to produce muscle cells (skeletal, smooth, or cardiac muscle), cartilage cells (chondrocytes), and bone cells (osteoblasts, osteoclasts, and osteocytes) because the cell types all belong to the same germ cell lineage: the mesoderm. After a stem cell has differentiated into a specific germ cell lineage, it is relatively difficult to change its differentiation pathway to yield a cell of a different germ cell lineage.

17.1.2 Potential

A fertilized egg can develop into a complete organism. After the initial division into two cells, if the cells are separated, they can also each develop into a complete organism. Even at the eight-cell stage, each of the cells has the potential to form a complete organism on its own. In other words, the potential of each of those cells is that they can make a complete (total) organism. The term used to describe this potential is **totipotent**. At or shortly after the eight-cell stage, the cells begin to lose this ability, but they still retain the ability to become members of any of the three germ cell types. This plurality of potential is the basis for the term **pluripotent**. It is the cells of the inner cell mass in the blastocyst that give rise to embryonic stem cells. As the inner cell mass develops into the gastrula, the cells commit to one of the three germ cell lineages. A mesodermal cell, for example, can be expected to produce skeletal muscle, smooth muscle, heart muscle, chondrocyte, osteoblast, osteoclast. ... Such a mesodermal cell has the potential to become any of multiple types of cells under the same germ cell lineage; it is said to be **multipotent**. These levels of cellular potential are laid out in Figure 17.3.

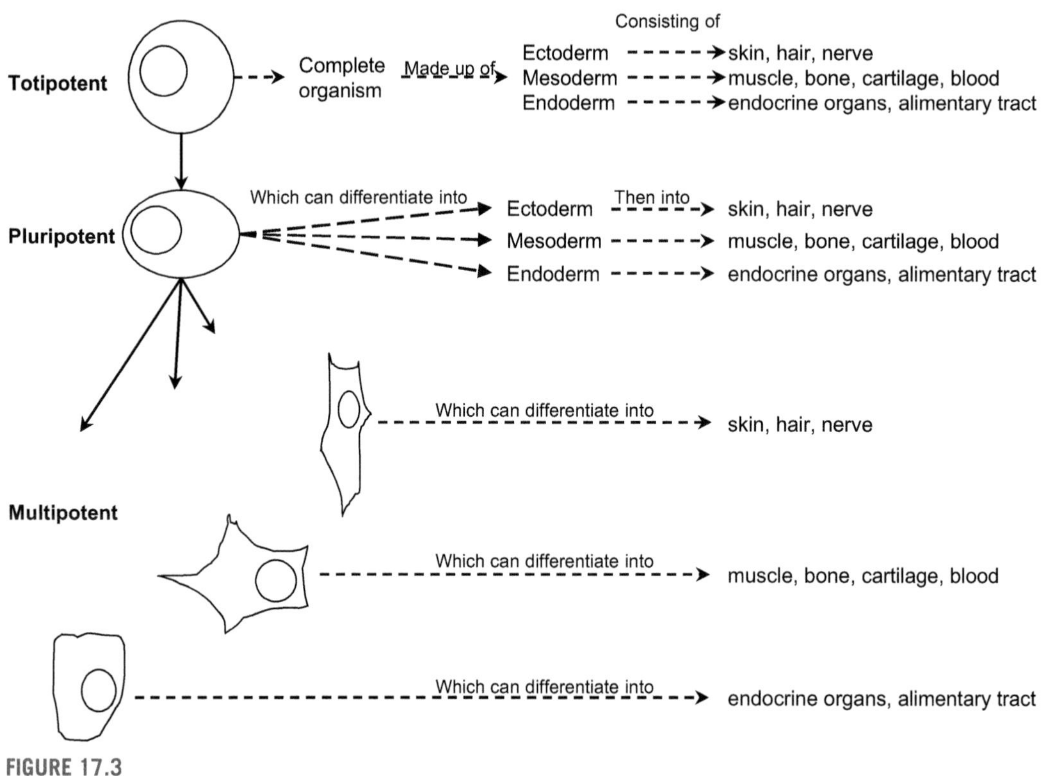

FIGURE 17.3

Levels of cellular potential, and some of the cell types that can differentiate from the cellular progenitors.

Mesoderm is also called mesenchyme, and cells from the mesenchyme are also called mesenchymal cells. This brings us to the mesenchymal stromal cell (MSC), a type of cell that has been studied intensely for its possible applications in tissue engineering and regenerative medicine. Historically, "MSC" used to stand for "mesenchymal stem cell." However, as the nature of stem cells became better understood and definitions were adjusted, it was realized that MSCs did not fit the definition of a stem cell because they were already committed to a particular (mesodermal) lineage; at the time, they could not be driven down the other two germ cell lineages. Multipotent cells are not stem cells, so MSCs could no longer be referred to as such. But the abbreviation "MSC" was already well accepted in the scientific community and could not simply disappear. For a while, the term "MSC" was no longer regarded as an acronym, although it still referred to this cell type. Eventually, it was accepted as standing for "mesenchymal stromal cell." This is an interesting example of one of the problems associated with a rapidly developing field: the language associated with the field must also develop. New words must be invented and definitions sometimes shifted. No matter the definition used by a specific laboratory for "stem cell," an MSC could be universally called a mesenchymal stromal cell.

(Note that following the adoption of "MSC" to mean "mesenchymal stromal cell," the science progressed to the point of being able to drive the cells into each of the three germ cell lineages. At that point, MSCs could again be referred to as stem cells.)

17.1.3 An alternate view of stem cells

Every somatic cell in your body has basically the same DNA, notwithstanding things like mutations and recombinations. (A **somatic cell** is any body cell that is not a germ cell. **Germ cells** are reproductive cells—sperm and eggs—which have only half as many chromosomes as somatic cells.) From your bones to your liver to your nerves, the genomes of each cell type are basically the same, despite the cells looking and behaving very differently (and belonging to different germ cell lineages). What makes a bone cell a bone cell as opposed to a liver cell lies largely in the set of genes it expresses. In theory, if a bone cell could turn off the bone-specific genes and begin to express the genes necessary to be a liver cell, then it could become a liver cell—or any other cell type, for that matter. This would make the original bone cell a stem cell, would it not? Your bone cells, however, do not have this ability. This brings us to an alternative view of what it means to be "stem": rather than a stem cell being a particular cell type that we can pick out of a body, hold, point to, or identify via a set of surface markers, many people believe that "stem" refers to a property that a cell may or may not have. Certain cells would then be said to have a stem property. Our original definition of a stem cell would not change using this alternative view. Thinking of "stem" as a property makes the logic tree less convoluted as more and more cell types have been driven into alternate lineages via the control of gene expression.

17.1.4 Using stem cells

Now that we have a definition for stem cell, we can turn our attention to stem cell differentiation. One of the ideas for stem cell therapy entails the cells being grown en masse in an undifferentiated state, prior to the need for an engineered clinical application. Suppose a patient comes into the clinic in need of a new liver. Currently, many patients will die waiting for a donor liver because the wait can last for years; the number of organs available for transplant is low, and it is difficult to find a tissue-type match from the limited supply. One goal of stem cell therapy is to have a large supply of stem cells that can be differentiated into whatever tissue or organ a patient may need—be it liver, kidney, bone, or anything else. The popular press often misrepresents the true intent of these therapies, focusing on totipotent cells, which are not used for stem cell therapy. Similarly, embryonic stem cells are the focus of moral objections to stem cell therapy. Developing embryos are not the only source of stem cells; adults have a supply of stem cells distributed throughout their bodies. Although bone marrow and amniotic fluid have both been used to produce adult stem cell cultures, stem cells can be found throughout the body, albeit in very low numbers; a current challenge is the isolation of these cells, which make up a very low percentage of the population in even a stem cell–rich tissue. There is no universal stem cell marker, although there are combinations of certain markers (membrane proteins, glycoproteins, and the like) that appear in combinations that render stem cells distinct from other cell types.

Let us now return to the concept of a somatic cell being reprogrammed to become a different cell type. This has, in fact, been achieved with mouse fibroblasts. In their landmark paper from 2006, Takahashi and Yamanaka were able to generate pluripotent cells from embryonic and adult fibroblasts via the addition of four growth factors (Oct3/4, Sox2, c-Myc, and Klf4) to the growth medium. These modified cells were injected into developing blastocysts, where they took part in embryonic development. The cells were said to have been induced into pluripotency, hence the term **induced pluripotent stem cells** (iPSCs). For this breakthrough, Shinya Yamanaka, along with John Gurdon, who also worked with the reprogramming of cells to take on the stem character, won the Nobel Prize in Physiology or Medicine in 2012. iPSCs have been a boon to the fields of tissue engineering and regenerative medicine because they are a replenishable source of otherwise scarce stem cells, and they are derived from adult cells, which eliminates many of the moral objections to stem cell research.

One problem with stem cell culture is how to maintain the cells in an undifferentiated state. Many times, teratomas will form in the culture vessel. A **teratoma** is a tissue mass that contains differentiated cells from all three germ lineages. Teratomas can have teeth, hair, muscle, bone—you name it. Some can even resemble complete organisms. They are not functioning organisms, though; they are merely a complex amalgamation of cells. They are not a good source of tissues (or even cells) because of the array of differentiation found within. In terms of stem cell culture, a teratoma can be considered a contaminant. They sometimes appear on the culture plate as uncharacteristic lumps on the monolayer, and they can grow and take on odd visual characteristics if allowed to develop. Teratomas have also developed during in vivo research, occasionally forming after the transplantation of iPSCs just under the skin.

One might ask whether an adult stem cell or an iPSC has less potential to divide than an embryonic stem cell. These types of stem cells all display a characteristic that is common to stem cells: they can renew themselves through cell division, even if they've been inactive for a long time. Stem cells do not have a Hayflick limit: they are immortal. However, do not confuse this with being transformed. There is a fine line between a stem cell and a cancer cell, and there is some concern in the field about cultured stem cells transforming into cancer cells after implantation. Like cancer cells, stem cells can renew their telomeres during cell division, although they achieve this through a different, telomerase-independent pathway.

17.2 Tissue engineering and regenerative medicine

After the initial stem cells have been obtained, their numbers will often be expanded through culturing, often for the purposes of tissue engineering or regenerative medicine. Most tissue engineering and regenerative medicine applications involve the implantation of cells, often housed in a porous scaffold, into a host (Figure 17.4). The difference between the two fields lies mainly in where cell proliferation within the scaffolds occurs. If scaffolds are seeded with cells that are allowed to proliferate in the laboratory prior to implantation, the application is an example of tissue engineering. If the scaffolds are implanted and development of the tissue occurs within the body, the approach is an application of regenerative medicine. In both cases, the scaffolds mature inside a bioreactor; if the bioreactor is an apparatus, we are referring to tissue engineering, and if it's a body, we are referring to regenerative medicine. The cartilage cells seeded onto the scaffold in Figure 17.4 were allowed to incubate for 1 week before implantation, so that experiment was an example of tissue engineering.

The goal of tissue engineering is the restoration of tissue structure or function using biological components. Certain applications of tissue engineering, such as engineered skin, have already been realized in the clinical setting. For burn victims, the application of engineered skin to wound sites helps to prevent dehydration and infection. During the early days of this technology, the skin was created from an initial biopsy of skin cells, often from the foreskins of male babies following circumcision. More recent research uses cells from the patient's own body. From one biopsy, an explant culture can be established and expanded to an area that would cover a tennis court. (No, not in a single sheet!) Recall from Chapter 7 that an explant involves a small amount of tissue, mechanically and/or chemically treated to break down extracellular matrices to free individual cells. The resulting mixture is then plated into a bioreactor to allow the cells to adhere to a surface and multiply. After a culture has been established, selective media can be used to ensure that only the cell type of interest can grow. Culture expansion is when the cells are passed from one flask or bioreactor into multiple flasks (bioreactors), thus expanding the total surface area of that particular culture.

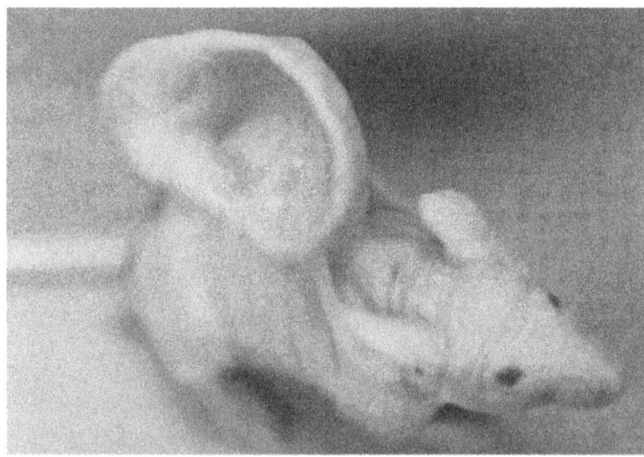

FIGURE 17.4

In the early days of tissue engineering, Charles Vacanti seeded biodegradable, porous scaffolds in the shape of an ear with cartilage cells from a cow and implanted the constructs under the skin of athymic mice. Although these particular constructs were not to be used for implantation onto a human, they did demonstrate the feasibility of growing cells on constructs to produce tissues.

Engineered vaginal tissue has also made it to clinical trials. The methods employed for the underlying research were very similar to those described earlier: biodegradable scaffolds were seeded with epithelial cells, and the constructs were cultured in a perfusion bioreactor prior to implantation. In the rabbit model, these constructs were implanted as a total vaginal replacement and were successfully accepted and integrated into the host tissues. After several months, the histologic characteristics of the constructs were similar to those of normal vaginal tissue.

Arterial tissue has been engineered in multiple settings. Figure 17.5 shows a construct that was seeded with a mix of endothelial, smooth muscle, and fibroblast cells. Tissue-engineered arteries must necessarily employ more than one cell type. Niklason et al. produced engineered arteries using pulsatile media flow to induce migration and orientation of smooth muscle cells seeded into scaffolds. After incubation, a second layer of cells—this time, endothelial cells—was seeded onto the luminal side of the vessel construct. The problem of vessel rupture due to poor mechanical properties was addressed by this approach, purportedly due to the smooth muscle cells and the matrix proteins they produced.

A different tissue-engineered vascular model used three cell types. In this model, the support for the construct was made not from a degradable synthetic scaffold, but from dehydrated fibroblasts wrapped temporarily around an inert tubular support; a sheet of smooth muscle cells was layered around the support and covered with a sheet of fibroblasts to provide an external structural layer. The constructs were allowed to incubate for at least 8 weeks, after which a layer of endothelial cells was seeded onto the dehydrated fibroblasts on the luminal side of the constructs. The constructs had sufficient mechanical strength to handle the loads generated by flowing blood, probably due to the adventitial layer of externally-seeded fibroblasts and the matrix they produced.

This section is by no means an exhaustive list of tissue engineering and regenerative medicine applications. Virtually every tissue and most organs of the body have been under investigation for eventual clinical applications. Further discussions can be found in readily available textbooks on the subject.

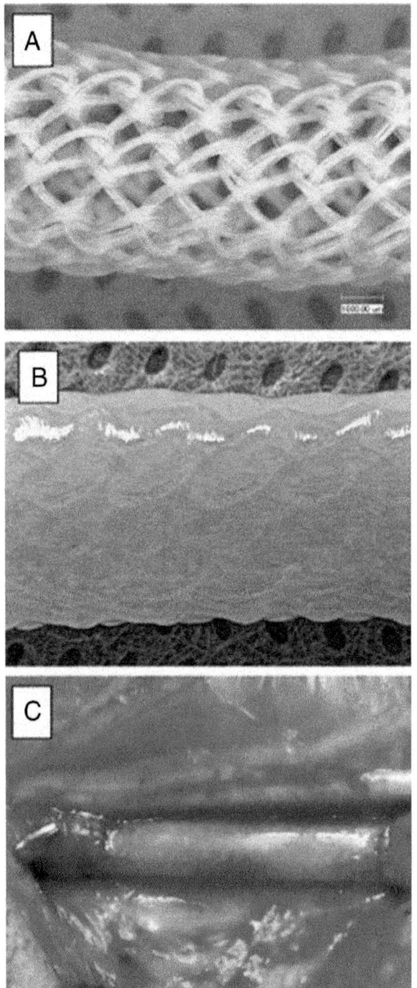

FIGURE 17.5

Example of engineered artery. Biodegradable, macroporous mesh (pore size ~1 mm) (A) before and (B) after embedding into a fibrin/cell matrix. (C) Fibrin-based vascular graft after implantation in the arterial circulation (ovine carotid model).

From Jockenhoevel and Flanagan (2011).

17.2.1 Bioreactors

Bioreactors can range from very simple to very complex. Even a cell culture dish or flask made of treated polystyrene and placed in an incubator with controlled temperature and humidity can count as a bioreactor (Figure 17.6). The polystyrene can be coated with collagen or poly(lysine), or it can be treated electronically via corona discharge, to provide a charged culture surface for cell adhesion. Culture vessels for simple systems are not airtight, so oxygen–carbon dioxide gas exchange can occur.

FIGURE 17.6

A bioreactor consisting of an incubator containing polystyrene tissue culture dishes and flasks.

17.2.1.1 Incubators

The incubator is an important part of the bioreactor system. The standard incubator participates in controlling three parameters: the temperature, the osmolarity, and the pH of cell media.

17.2.1.1.1 Temperature

Temperature is controlled via a heater and a simple thermostat. Just as you get uncomfortable if the temperature gets too hot or too cold, cells too are affected by suboptimal temperatures. If they are too cold, their metabolism and other cellular processes will slow, in part because energy has been removed from the system and chemical reaction rates will be affected. If the temperature is too high, in addition to experiencing altered reaction kinetics and protein folding, the cell may begin to transcribe and translate a new set of genes that encode products known as heat shock proteins. The optimal temperature for cells differs depending on the organism. For example, human cells are kept at 37°C in the incubator, the same as normal human body temperature, while yeast cells are kept at 30°C.

17.2.1.1.2 Osmolarity

If cell medium were to evaporate, water would leave the liquid and enter the gas phase, while salts, sugars, and proteins would be left behind, resulting in an increased osmolarity of the medium. The way to prevent this is to prevent evaporative water loss. Although sealing the culture vessel would prevent water loss, it would also affect gas exchange, which would result in altered pH, the induction of hypoxia, an eventual switch to anaerobic/fermentative metabolism, and eventually cell death. A safer way to prevent water loss is by increasing the partial pressure of water over the medium to the saturation point. Although the value for the amount of water needed to reach saturation in an incubator of a given size and temperature can be calculated, such calculations are not necessary: saturation can be achieved simply by putting a pan of water inside the incubator. As long as water is kept in the pan, the water levels in the atmosphere of the incubator will remain at saturation. Evaporation from cell media will be slowed or stopped, thus preserving the osmolarity of the media. (Of course, a pan of water in a nonsterile 37°C environment will eventually allow the growth of microorganisms such as algae, so the pan should be cleaned regularly, and an algaecide should be mixed into the water.)

17.2.1.1.3 pH

Recall the reaction:

$$CO_2 + HOH \rightleftarrows H_2CO_3 \rightleftarrows HCO_3^- + H^+$$

This reaction is used in incubators to control the pH of most cell media. First, CO_2 is pumped into the incubator and maintained at a concentration of 5%. This amount is much higher than the concentration of CO_2 in the air (~0.039%), so the reaction will be driven to the right in a controlled fashion (because there is plenty of water in the humidified atmosphere of the incubator). The pH of the cell medium is buffered by including sodium bicarbonate (~0.37% [w/v]), and the medium is typically titrated to a pH of 7.2 to 7.4 before use. The bicarbonate will serve as a buffer to hold the pH of the medium fairly constant; however, as the cells metabolize the nutrients in the medium, they will produce their own CO_2 and eventually lower the pH over several days. It is therefore generally accepted that mammalian cell culture media should be changed twice per week.

17.2.1.2 Static and dynamic cultures

Tissue culture plates and flasks come in various sizes, with growing areas ranging from less than 1 cm^2 to 225 cm^2 and larger (Figure 17.7). The size of a tissue culture flask is limited by the size of the incubator in which it will be placed. To help maximize the amount of growing area without requiring the purchase of enormous incubators, improvements to flask design have been made; for example, some tissue culture flasks have multiple layers of growing surface within the same outer shell to increase growing surface area per flask volume. Three-dimensional growing surfaces, in the form of scaffolds, also exist and allow for the culture of a greater number of cells in a given volume. Three-dimensional scaffolds also allow more complex culturing and can be used for producing engineered tissues containing more than one type of cell.

Cells growing in tissue culture plates or flasks, or on scaffolds suspended in a bottle of medium, are all examples of **static culture**: there is no stirring or flow involved. A more complex bioreactor utilizes **dynamic culture**. In such cultures, the media may be stirred slowly (Figure 17.8). Dynamic culture is used for cultures of nonadherent cells but is also often applied to cells growing on three-dimensional scaffolds. The rate of mixing can be controlled by controlling the RPM of the stir bar. Mixing facilitates better gas exchange, because pockets with high waste concentrations are not allowed to develop around the cells. Mixing also allows nutrient and oxygen concentrations to be evenly distributed throughout the culture medium.

Another example of a dynamic culture system is the perfusion bioreactor, a system utilizing a pump that allows medium to flow over or through culture surfaces. Incorporating flow into the culture conditions is used in both two-dimensional and three-dimensional cultures. In two-dimensional cultures,

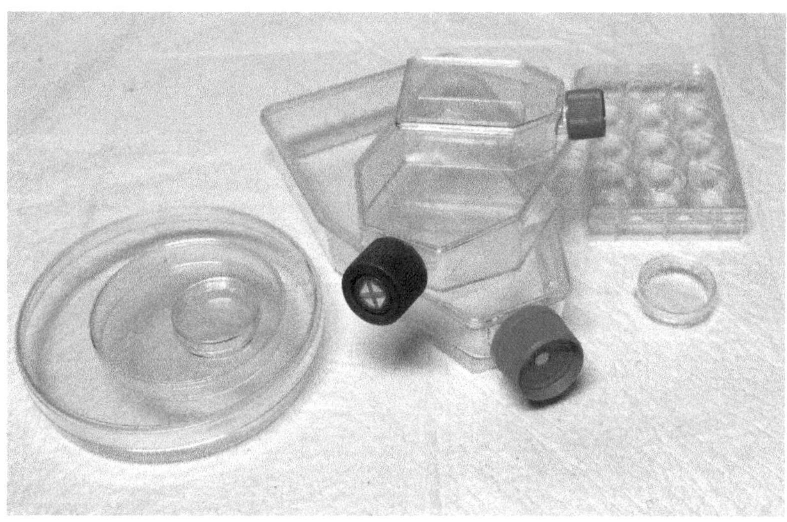

FIGURE 17.7

Tissue culture plates and flasks come in many sizes.

FIGURE 17.8

A spinner flask, an example of a dynamic culture bioreactor.

cells grow on an interior surface of a chamber that has an inlet and outlet allowing medium to pass over them (Figure 17.9). The three-dimensional system is very similar, except that medium is either pumped through the cell-containing scaffolds or is dripped on top of them and allowed to move through via gravity flow (Figure 17.10). These bioreactors are usually closed systems, so the medium passes over the culture surfaces several times before it is replaced.

The advantages of having medium flow over the cells stem from controlled and directional shear forces, which aid in the differentiation of pluripotent and multipotent cells. MSCs grown in static

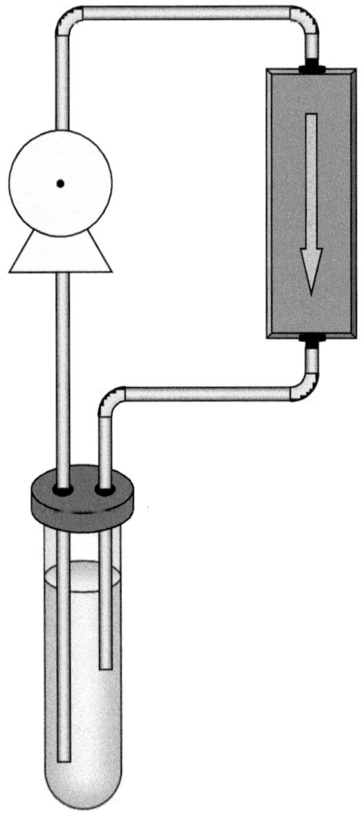

FIGURE 17.9

Scheme for a two-dimensional perfusion bioreactor which uses a pump to bathe cells growing on a surface.

FIGURE 17.10

A three-dimensional perfusion bioreactor set up in an incubator. Cells are seeded into scaffolds which are placed in perfusion cassettes (top of figure). A pump is located on the left-hand side of the figure. Bottles containing medium *(pink)* serve as a source *(right)* and a sink *(left)* for the circulating medium.

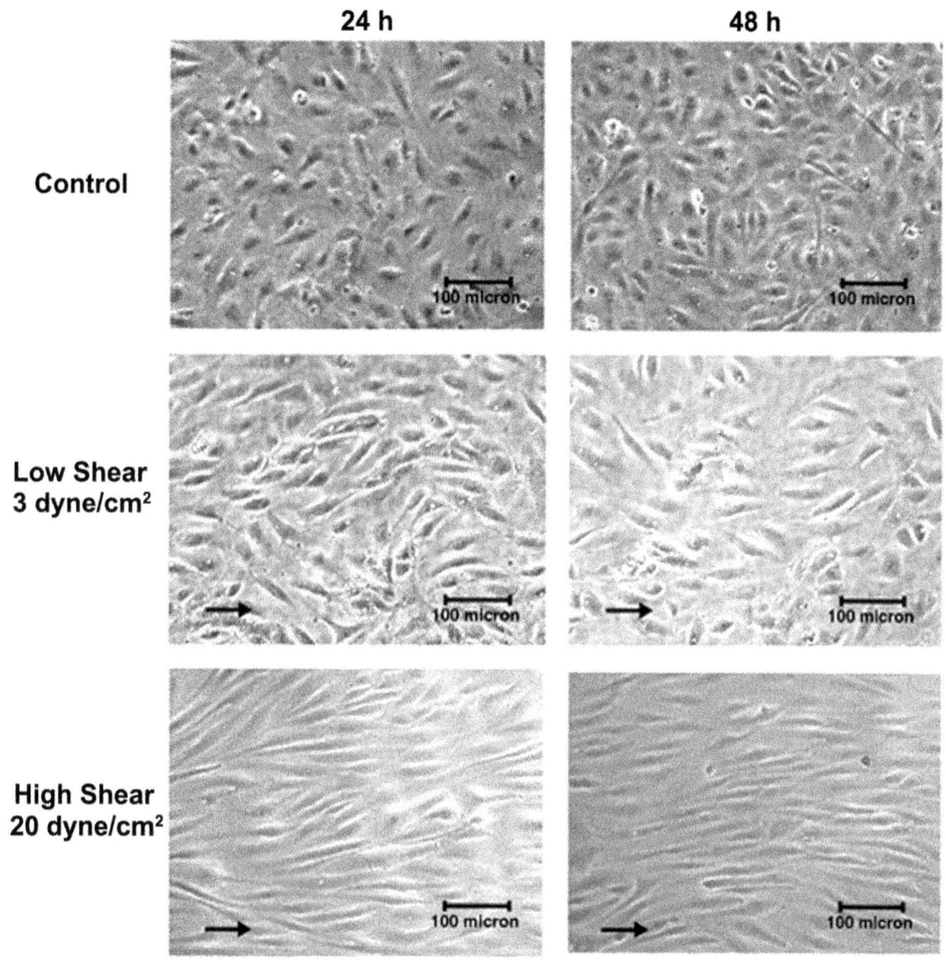

FIGURE 17.11

Endothelial cells react to shear stress that has been introduced by laminar fluid flow. *Arrows* indicate direction of flow.

From McCormick et al., 2012.

culture with no growth or differentiation factors will either not differentiate or, more likely, will differentiate along multiple pathways to produce a heterogeneous culture. MSCs grown in the presence of shear stress tend to differentiate into cell types that provide structure, such as bone cells. Muscle cells also tend to line up in a parallel fashion when exposed to shear (Figure 17.11). Recall that a tissue is more complex than just multiple cells of the same type growing on the same particular scaffold; it is a collection of cells or cell types that interact and function as a unit. Including the parameter of shear is useful for creating tissues to mimic those that are exposed to shear in their normal functioning.

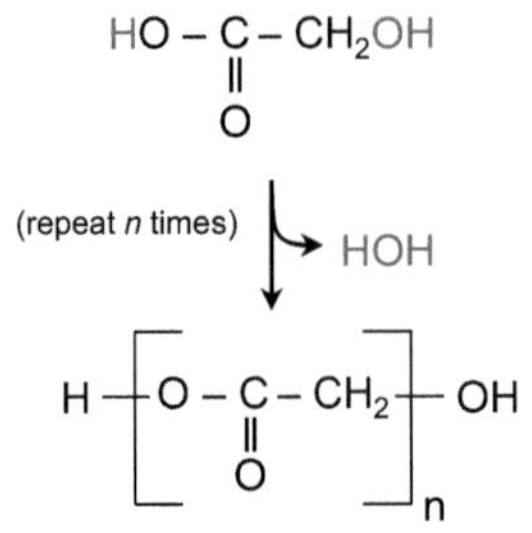

$$HO-\underset{\underset{O}{\|}}{C}-CH_2OH$$

(repeat *n* times) $\downarrow$ HOH

$$H-\left[O-\underset{\underset{O}{\|}}{C}-CH_2\right]_n-OH$$

FIGURE 17.12

Glycolic acid is polymerized to form poly(glycolic acid).

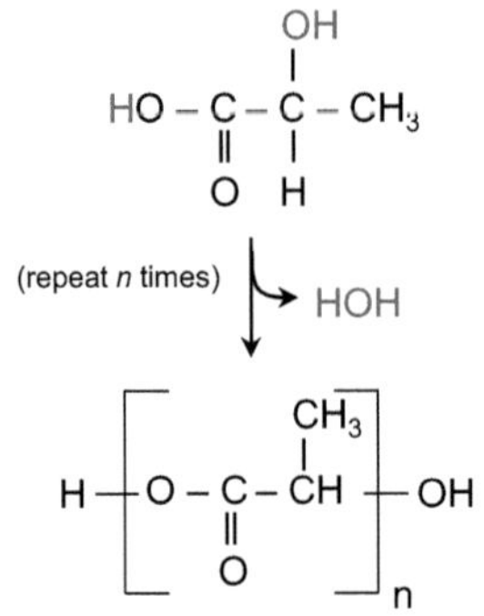

$$HO-\underset{\underset{O}{\|}}{C}-\underset{\underset{H}{|}}{\overset{\overset{OH}{|}}{C}}-CH_3$$

(repeat *n* times) $\downarrow$ HOH

$$H-\left[O-\underset{\underset{O}{\|}}{C}-\overset{\overset{CH_3}{|}}{CH}\right]_n-OH$$

FIGURE 17.13

Lactic acid is polymerized to form poly(lactic acid).

17.2.2 Polymeric scaffolds

Supports that are used for three-dimensional cultures are known as scaffolds; most scaffolds currently in use for tissue engineering and regenerative medicine are made of biodegradable and biocompatible materials. As mentioned earlier, **biodegradable** means the material will be broken down under physiological conditions, such as when a polymer is hydrolyzed (broken apart by water), and **biocompatible** means the original material and its degradation products of the are not toxic.

17.2.2.1 Homopolymers

A **homopolymer** is a polymer made from many copies of a single repeating unit. For example, a number of glycolic acid molecules can be conjugated as shown in Figure 17.12 to create the homopolymer poly(glycolic acid). Lactic acid can also be used to make a homopolymer—in this case, poly(lactic acid) (Figure 17.13). Poly(glycolic acid) and poly(lactic acid) are typically abbreviated PGA and PLA, respectively. During the polymerization of either of these molecules, every polymerization step will have an H and an OH come off as water (shown in red in both figures); this implies the reaction could go backward so that the polymers could be broken apart by water, a process known as **hydrolysis**.

The consistency of PGA is like felt (Figure 17.14). It is biodegradable via hydrolysis, and glycolic acid is biocompatible (as are oligomers of glycolic acid). When PGA is implanted into an animal, it is

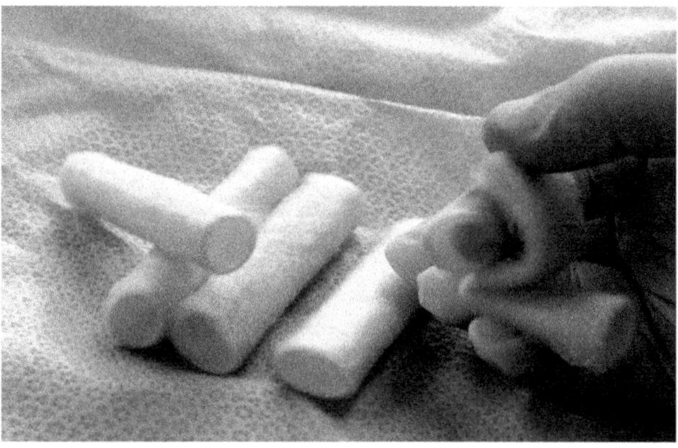

FIGURE 17.14

Poly(glycolic acid) is a flexible polymer that can be spun into different geometries, such as sheets or open cylinders. The polymers shown here could serve as scaffolds for engineering arteries.

being placed into an aqueous environment, which means it will degrade over time. This fits well with an overarching goal of tissue engineering: after the construct is seeded with cells and implanted, it will eventually degrade to leave only a cell-derived tissue: cells and extracellular matrix, with no scaffold material remaining. As the cells grow and proliferate, they will secrete their own extracellular matrix. The perfect system would have the rate of extracellular matrix construction occur at the same rate as PGA degradation to keep the mechanical properties of the construct constant as the construct matures into a tissue. The perfect final product would have cells in the same orientation, with the same distribution of cell types and the same amount of extracellular matrix as found in the normal tissue (or organ).

A molecule of lactic acid contains a carboxylic acid and an alcohol side group that can be used for polymerization (Figure 17.13). As with glycolic acid, polymerization of two lactic acid units causes the release of a water molecule. This implies that PLA is hydrolysable (and therefore biodegradable). PLA has different properties than PGA. Whereas PGA is like a felt, being easily shaped or deformed, PLA is relatively stiff but more able to hold its shape under compressive forces, making it more suitable for applications that require load bearing. Sometimes PGA scaffolds are formed into the desired shape and then coated with PLA to help the construct retain its shape.

17.2.2.2 Copolymers

Not all polymers are homopolymers. If two or more different repeating units are used to make the polymer, the product is said to be a **copolymer**. There are many different possible arrangements of the repeating units—too many to describe here. However, some rudimentary architectures are shown in Figure 17.15.

The repeating units of a copolymer can come together in a random fashion, forming a **random copolymer**. However, important structural characteristics can emerge if the repeating units are joined in nonrandom ways. If each repeating unit is repeated in a regular, alternating fashion, the copolymer is an **alternating copolymer**. Quite often, small blocks of homopolymers can be covalently linked to produce **block copolymers**. These are given more descriptive names to indicate how many blocks were joined, such as di-block or tri-block copolymers. Polymers can also be branched. One method used to

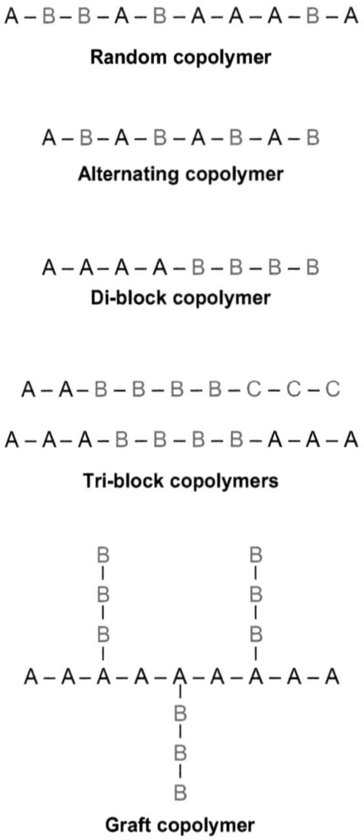

A – B – B – A – B – A – A – A – B – A

Random copolymer

A – B – A – B – A – B – A – B

Alternating copolymer

A – A – A – A – B – B – B – B

Di-block copolymer

A – A – B – B – B – B – C – C – C

A – A – A – B – B – B – B – A – A – A

Tri-block copolymers

```
        B               B
        |               |
        B               B
        |               |
        B               B
        |               |
A – A – A – A – A – A – A – A – A
                |
                B
                |
                B
                |
                B
```

Graft copolymer

FIGURE 17.15

Examples of copolymer architectures.

achieve branching is to take one polymer and graft it onto the chain of another to form a **graft copolymer**. Graft copolymers are generally made up of homopolymers grafted onto homopolymers.

An example of a copolymer used in tissue engineering is poly(lactic-co-glycolic acid) (PLGA), a copolymer of the two homopolymers we have already mentioned, PLA and PGA. Different ratios of PLA and PGA can be used to yield PLGA constructs with different properties, such as predetermined compressive strengths and degradation rates. Stiffer than PLA but more pliable than PGA, PLGA has been used in several applications, such as for bone and cartilage engineering.

17.2.3 Bringing it all together: a tissue engineering application

Let's return to the example of engineered blood vessels. A naïve view of blood vessels is that they are a series of tubes that run throughout the body for the purpose of circulating blood cells, which carry oxygen, to all of the tissues. However, replacing a vessel such as the radial artery with a piece of plastic tubing would not be sufficient to replace the function of the artery. Although the plastic tubing would prevent the loss of blood through its walls and could be made pliable enough to allow for movement of the host's arm, it would not allow for the leakage of immune cells or macromolecules, as may be

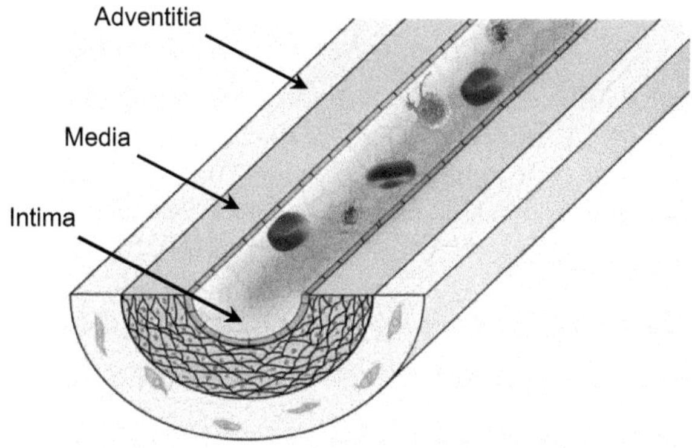

FIGURE 17.16

Structure of an artery, with three layers as indicated. The intimal layer is composed of endothelial cells, and the medial layer is composed of vascular smooth muscle cells. The adventitia is structural, composed mainly of connective tissue made by the cells within.

needed by the body along the site of the implant. The foreign material of the plastic might also initiate a foreign body response that would be deleterious to the host. New blood vessels sprouting from the implant would likewise not be possible, and gas exchange in the area of the implant would be impeded. A superior alternative would be the seeding of a porous scaffold with cells that could eventually proliferate to create arterial tissue.

The structure of an artery is relatively simple. As shown in Figure 17.16, there are three distinct layers to the vessel: the **tunica intima**, the **tunica media**, and the **tunica adventitia**. "Tunica" means "coat" (as in "tunic"); this part of the terms is sometimes dropped to yield *intima*, *media*, and *adventitia*. The intima is the layer that surrounds the lumen of the artery (it is "intimate" with the blood); it is composed of endothelial cells. The medial layer is composed of smooth muscle cells, and the adventitia is composed mainly of connective tissue such as collagen. Fibroblasts, immune and inflammatory cells, and stem/progenitor cells can be found within the adventitia, which also serves to attach the artery to another tissue or organ. Larger arteries will also have their own blood supply (the *vasa vasorum*) to feed the muscle cells in the media and allow immune cells access to the intima.

An engineered construct of an artery can be made starting with a porous scaffold. Pores allow cells to migrate throughout the volume of the scaffold. The percentage of a given volume of scaffold that is made up of pores is described as its **porosity** and has a direct effect on both the strength of the scaffold and how easily cells can move into and throughout the scaffold interior. In the absence of cells, as a biodegradable scaffold is hydrolyzed its porosity will increase. In the presence of cells, however, the deposition of extracellular matrix should make up for the loss of scaffold material. An ideal construct would have a supporting scaffold polymer that degrades at the same rate that the cells establish extracellular matrix.

A specific application of engineered artery starts with tubular PGA scaffolds being seeded with smooth muscle cells, followed by 8 weeks in culture in a specialized perfusion bioreactor. The incubation allows the smooth muscle cells to migrate into the interior of the PGA matrix and provide a smooth surface for a second seeding of cells onto the construct. This second seeding will deliver endothelial cells only to the luminal surface of the construct. After an additional culturing period, the process will produce intact vessels. This particular application has been used to produce arteries that were successfully implanted into pigs.

Although the production of the engineered arterial constructs containing more than one cell type was an important procedural step at the time (1999), the striking feature of that piece of work was that the bioreactor utilized pulsatile flow of cell medium to produce constructs that mimicked native artery more closely. The approach has since undergone over two decades of development, with improvements being made to both chemical and physical aspects of the culturing system and the way in which cells are delivered to the scaffold. An important lesson from this application is that a good bioreactor will necessarily mimic the conditions of the eventual implantation site. One should be aware of as many parameters as possible, both chemical and physical, to allow constructs to develop with the desired complexity of form and function.

17.2.4 Another application: gene editing

In Chapter 14, we learned several biotechnologies for gene editing but without knowing about progenitor cells, tissue engineering, and regenerative medicine, it was difficult to give a good example. We are now ready for such an example.

Gene editing cannot be used to edit the genomes of every cell in an adult human. The approach is more aligned with single or multiple cells, especially when we consider issues such as delivery and off-target effects. If an adult (or juvenile) human is to be treated, some of their cells will need to be isolated; the cells may be pluripotent, multipotent, or differentiated, but a common theme is to use or produce stem cells (such as iPSCs). After isolation, the cellular genomes are edited and then screened to isolate only those cells in which editing has been successful. (Some cells may have undergone no changes, some may have only eliminated the function of the target gene via nonhomologous end joining, and only a very low percentage may have had an insertion of the gene of interest if an insertion was the goal of the editing.) The screened cells are then reimplanted into the host in either a pluripotent or a differentiated form. Differentiation of the cells prior to implantation gives the investigator a chance to verify that the expression of the edited gene is at a clinically relevant level, while implantation of undifferentiated cells may allow for differentiation into multiple cell types in whatever proportions the body needs. The implantation itself can occur in the form of injection of cells into a given organ or tissue, injection of encapsulated cells, or implantation of a scaffold that has been seeded with the cells.

17.3 Summary

In this chapter, we learned about stem cells, tissue engineering, and regenerative medicine. We learned that stem cells are cells that can be driven down any of the three germ lineages (ectodermal, mesodermal, endodermal), and we learned what germ lineages are. We also saw that "stem" can mean a cell type, or it can be used to describe a property that cells have (or can be given, as in the creation of iPSCs). We also saw that stem cells might be the key component to producing engineered tissues or organs on demand.

In the second half of the chapter, we learned the difference between tissue engineering and regenerative medicine. We discussed how cells are obtained and expanded, paying specific attention to how parameters affecting growth and development are controlled via bioreactors. Because tissue engineering and regenerative medicine usually involve the seeding of cells onto porous, biodegradable, biocompatible scaffolds, we spent time examining different polymers that are used as scaffold material.

In the next chapter, we will turn back to applications that involve genetic manipulation of cells: transgenics. The agricultural applications of transgenics have given us ways to increase crop yields and produce foods that can be processed using less energy. One specific application has even been credited with saving the lives of millions of people.

Related reading

Atala, A., Kasper, F.K., Mikos, A.G., 2012. Engineering complex tissues. Sci. Transl. Med. 4, 160rv12.

Battula, V.L., Bareiss, P.M., Treml, S., Conrad, S., Albert, I., Hojak, S., et al., 2007. Human placenta and bone marrow derived MSC cultured in serum-free, b-FGF-containing medium express cell surface frizzled-9 and SSEA-4 and give rise to multilineage differentiation. Differentiation 75, 279–291.

Cao, Y., Vacanti, J.P., Paige, K.T., Upton, J., Vacanti, C.A., 1997. Transplantation of chondrocytes utilizing a polymer-cell construct to produce tissue-engineered cartilage in the shape of a human ear. Plast. Reconstr. Surg. 100, 297–302.

Dahl, S.L., Kypson, A.P., Lawson, J.H., Blum, J.L., Strader, J.T., Li, Y., et al., 2011. Readily available tissue-engineered vascular grafts. Sci. Transl. Med. 3, 68ra9.

De Filippo, R.E., Bishop, C.E., Filho, L.F., Yoo, J.J., Atala, A., 2008. Tissue engineering a complete vaginal replacement from a small biopsy of autologous tissue. Transplantation 86, 208–214.

Godbey, W.T., Atala, A., 2002. In vitro systems for tissue engineering. Ann. NY Acad. Sci. 961, 10–26.

Jockenhoevel, S., Flanagan, T.C., 2011. Cardiovascular tissue engineering based on fibrin-gel-scaffolds. In: Eberli, D. (Ed.), Tissue Engineering for Tissue and Organ Regeneration. InTech. http://www.intechopen.com/books/tissue-engineering-for-tissue-and-organ-regeneration/cardiovascular-tissue-engineering-based-on-fibrin-gel-scaffolds.

Lee, A.A., Graham, D.A., Dela Cruz, S., Ratcliffe, A., Karlon, W.J., 2002. Fluid shear stress-induced alignment of cultured vascular smooth muscle cells. J. Biomech. Eng. 124, 37–43.

Lee, E.J., Vunjak-Novakovic, G., Wang, Y., Niklason, L.E., 2009. A biocompatible endothelial cell delivery system for in vitro tissue engineering. Cell Transplant. 18, 731–743.

Majesky, M.W., Dong, X.R., Hoglund, V., Mahoney Jr., W.M., Daum, G., 2011. The adventitia: a dynamic interface containing resident progenitor cells. Arterioscler. Thromb. Vasc. Biol. 31, 1530–1539.

McCormick, S.M., Seil, J.T., Smith, D.S., Tan, F., Loth F., 2012. Transitional Flow in a Cylindrical Flow Chamber for Studies at the Cellular Level. Cardiovasc. Eng. Technol. 3:439-449. doi: 10.1007/s13239-012-0107-5. Epub 2012 Sep 11.

Moran, J.M., Pazzano, D., Bonassar, L.J., 2003. Characterization of polylactic acid-polyglycolic acid composites for cartilage tissue engineering. Tissue Eng. 9, 63–70.

NIH Stem Cell Information Home Page. In Stem Cell Information [World Wide Web site]. Bethesda, MD: National Institutes of Health, U.S. Department of Health and Human Services, 2016 (Last modified March 13, 2020) [cited December 29, 2020] Available at: <//stemcells.nih.gov/info/basics.htm> .

Niklason, L.E., Gao, J., Abbott, W.M., Hirschi, K.K., Houser, S., Marini, R., et al., 1999. Functional arteries grown in vitro. Science 284, 489–493.

The Nobel Prize in Physiology or Medicine: 2012 Press Release. 8 October 2012. Nobel Media AB. http://www.nobelprize.org/nobel_prizes/medicine/laureates/2012/press.html.

Panta, W., Imsoonthornruksa, S., Yoisungnern, T., Suksaweang, S., Ketudat-Cairns, M., Parnpai, R., 2019. Enhanced hepatogenic differentiation of human Wharton's jelly-derived mesenchymal stem cells by using three-step protocol. Int. J. Mol. Sci. 20 (12), 3016.

Solan, A., Dahl, S.L., Niklason, L.E., 2009. Effects of mechanical stretch on collagen and cross-linking in engineered blood vessels. Cell Transplant. 18, 915–921.

Takahashi, K., Yamanaka, S., 2006. Induction of pluripotent stem cells from mouse embryonic and adult fibroblast cultures by defined factors. Cell 126, 663–676.

Ye, L., Sun, L.X., Wu, M.H., Wang, J., Ding, X., Shi, H., et al., 2018. A simple system for differentiation of functional intestinal stem cell-like cells from bone marrow mesenchymal stem cells. Mol. Ther. Nucleic Acids 13, 110–120.

Questions

1. Describe the difference between multipotent, pluripotent, and totipotent cells.
2. **a.** Are the cells that go on to create a teratoma totipotent? Why or why not?
   b. Are the cells that go on to create a teratoma stem cells? Why or why not?
3. Are totipotent cells stem cells?
4. Suppose Joe has some MSCs in his laboratory. Are they stem cells? Give an example of something Joe could do to convince you that they are.
5. Consider the fact that the liver can regenerate itself to some extent. Does this mean liver cells are stem cells? Does it mean liver cells are multipotent?
6. Stem cells are immortal, which means they are able to circumvent the Hayflick limit. Does this mean stem cells are cancer cells? Why or why not?
7. Which is the most remarkable? Explain your answer.
   a. A hematoprogenitor cell becoming a muscle cell
   b. An embryonic stem cell becoming a kidney cell
   c. An MSC becoming a liver cell
   d. An adult stem cell becoming a nerve cell
8. Which is the most remarkable? Explain your answer.
   a. A totipotent cell becoming/producing a multipotent cell
   b. A multipotent cell becoming/producing a pluripotent cell
   c. A pluripotent cell becoming/producing a multipotent cell
   d. A totipotent cell becoming/producing a pluripotent cell
9. True or false? Give a reason for every "false" answer:
   a. Pluripotent cells are stem cells.
   b. MSCs are stem cells of mesenchymal (mesodermal) lineage.
   c. A stem cell is a cell that can, for example, become a muscle cell, then become a nerve cell.
   d. A pluripotent cell can become an entire organism.
   e. Blood cells belong to a mesenchymal lineage (i.e., they originate from mesoderm).
   f. A multipotent cell from the ectoderm could be differentiated into skin or nerve cells.
   g. Embryonic stem cells have been isolated from blastocysts.

h. Scientists have isolated totipotent cells from bone marrow.
i. A cell that produces a teratoma is (was) a stem cell.
j. A cell that produces a teratoma is (was) a totipotent cell.

10. Why doesn't the water in a cell incubator need to be carefully measured?
11. If the water in a cell culture incubator pan completely evaporates, what will happen?
12. How is the pH of cell cultures controlled by cell media? How does the incubator play a role in this process?
13. What is the difference between static and dynamic cell culture?
14. What is a perfusion bioreactor? Why might a biotechnologist want to use one?
15. Name one major problem with growing cells seeded into a 10 mm × 10 mm × 3 mm scaffold via a static culture.
16. Name two homopolymers used for tissue engineering.
17. Name two properties a good scaffold for tissue engineering should have.
18. What is the difference between tissue engineering and regenerative medicine?
19. What is the difference between an alternating copolymer and a di-block copolymer? Name a property each might have that would make it a desirable choice for a tissue engineering application.
20. Name three reasons why the porosity of a polymeric scaffold might be important.
21. How do the cells of the tunica intima in an artery obtain the oxygen they need to stay alive?
22. Stress can cause high blood pressure. Use your knowledge of arterial structure to speculate what is going on.

CHAPTER

Transgenics and genetically modified organisms in agriculture 18

Chapter outline

Biotechnologists think big. Even though bioresearch is focused on the cellular, subcellular, and molecular levels, the driving forces behind it are quite often concerned with much grander scales. Human health is one example of a grand-scale application. We have seen how biotechnology is being applied in the medical field: the use of DNA- and RNA-based drugs and the growth of tissues and organs to restore function via tissue engineering and regenerative medicine. The energy sector is another example of a grand scale. There is a direct correlation between the human development index of a country and its total primary energy demand. Even for underdeveloped countries, energy requirements are ever-increasing. The availability of energy affects nearly every person on an everyday basis. We have seen how biotechnology is being used in an attempt to increase the availability of fuels via the manipulation of living organisms. The food supply can also be regarded as a grand-scale application. As the population of the earth continues to increase, the potential for world food shortages becomes more imminent. This chapter will look at some applications of biotechnology to agricultural food supplies.

Transgenics are genetically modified organisms (GMOs). The term also refers to the transfer of genetic material into a cell or organism, with the gene being either one from a different species, or a modified form of a gene from the same species created by genetic engineering techniques. In this chapter we will see several classic examples of transgenic foods, including potatoes, soybeans, tomatoes, and rice. Particular attention will be paid to the science that went into producing some of these agricultural products, with the aim of showing that although caution should be exercised regarding products that are meant to be toxic, genetically modified foods are not inherently bad. By the end of the chapter,

Biotechnology and its Applications. https://doi.org/10.1016/B978-0-12-817726-6.00018-6

FIGURE 18.1

Certain bacteria produce proteins that serve as nucleation sites for ice crystal formation. Spraying plants with ice-minus bacteria can inhibit the formation of ice crystals to a small degree.

Photo: George Hodan, from http://www.publicdomainpictures.net

you should be aware that organic foods are not automatically safe and that GMO foods could hold the solution to future food shortage problems.

We will also take a brief look at business aspects related to biotechnology—from nonscientific issues that have led to the demise of sound biotechnology companies, to ways that companies can protect their technologies through biological security. We will begin, though, with a classic example of a transgenic organism being used to increase crop yields: ice-minus bacteria.

18.1 Ice-minus bacteria

The first experiments involving transgenics, as applied to plants, did not produce transgenic plants at all; they utilized genetically modified bacteria that were sprayed onto plants. The bacteria were *Pseudomonas syringae*, also known as ice-forming bacteria, which contain a protein in their outer membranes that serves as a nucleation site for water as it freezes. When the temperature drops down to around 0°C and frost forms, it will form first on and around these proteins on the bacteria. The bacteria reside all over the plant, which means the plant will be covered with frost and will die or suffer extensive damage (Figure 18.1).

Experiments were performed to knock out the gene coding for the nucleation site protein. **Knockout** refers to the technique of mutating a specific gene in the genome by targeted recombination, with the intention of rendering the gene useless and thereby knocking out the expression of the original, functioning protein. By knocking out the gene that codes for this membrane protein, a strain of *P. syringae* known as **ice-minus** was created. These bacteria were then sprayed onto crops to help retard the formation of frost crystals.

18.2 Bt plants

"Bt" in this case stands for *Bacillus thuringiensis,* a type of gram-positive bacterium that can be found in soil. *B. thuringiensis* is known to produce a toxin to aid in its own survival. Biotechnologists have

FIGURE 18.2

(A) Larvae of the lesser cornstalk borer extensively damaged the leaves of this unprotected peanut plant. (B) After only a few bites of peanut leaves with built-in Bt protection, larvae of the lesser cornstalk borer crawled off the leaf and died.

Photos by Herb Pilcher

been successful in incorporating the gene for this toxin into many plant species, including cotton, corn, soybeans, peanuts, and potatoes (Figure 18.2). These modified plants are known as **Bt plants**, and they express this gene in sufficient quantities to form crystals of the toxin. Caterpillars, beetles, and other similar pests that consume these crystals while eating the plants will die. The crystals are harmless to humans and birds. In this way, farmers have been able to grow crops that produce their own pesticides without the need to spray chemicals, thus allowing the crops to be classified as "organic." Advantages of Bt plants include better crop yields because plants are not destroyed by pests and cheaper crop production because the expense of spraying is not needed. By the year 2000, over 50% of the soybean crops in the United States were Bt plants.

Just as bacteria can develop resistance to certain drugs, insects can develop resistance to certain pesticides. As is also a common response in evolution, different strains of *B. thuringiensis* have developed different forms of toxin for protection. Biotechnologists have identified and isolated genes for

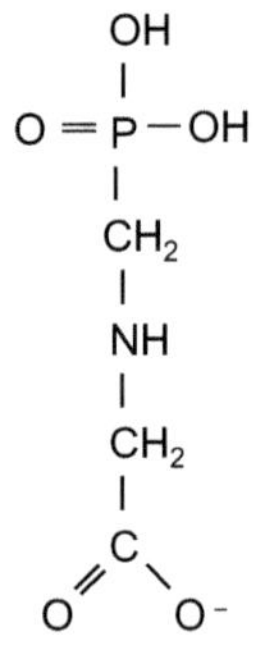

OH
|
O = P − OH
|
CH_2
|
NH
|
CH_2
|
C
O O−

FIGURE 18.3

The structure of glyphosate, the active ingredient in Roundup.

these different toxins and have incorporated genes for multiple toxin forms into single plants in a process known as pyramiding, to produce **pyramided plants**.

In the United States, the first Bt plants sold in supermarkets were NewLeaf potatoes, produced by Monsanto and distributed as early as 1996. Although the first patent for Bt technology was filed in 1988, a legal fight between Mycogen Plant Seeds, Inc. (an affiliate of Dow Agrosciences, LLC) and Monsanto delayed the awarding of patent protection until 2005. The patent was ultimately awarded to Dow, and Monsanto suspended its Bt potato program, which had expanded to include several varieties of potatoes.

18.3 Herbicide resistance

Broad-spectrum herbicides are chemicals that kill virtually all plants. Imagine if a plant were developed that was resistant to a broad-spectrum herbicide: crops of this plant could be sprayed with the herbicide, and all weeds and competing plants would die while the desired crop survived. Glyphosate is the active ingredient in most broad-spectrum herbicides, including a product known as Roundup; it has the structure shown in Figure 18.3. Glyphosate acts to inhibit the activity of 3-enol-pyruvylshikimate-5-phosphate synthase (EPSP synthase), which is involved in the biosynthesis of aromatic amino acids, tetrahydrofolate, ubiquinone, and vitamin K. These particular pathways are not used in mammals, fish, birds, or insects, so EPSP synthase presents an attractive target for killing plants.

With glyphosate being such an effective herbicide, engineering plants resistant to its action became the goal of the above biotechnology for enhanced crop production. One approach to obtaining plants with glyphosate resistance is to have them make more EPSP synthase than an application of glyphosate could inhibit. This can be achieved by inserting additional copies of the EPSP synthase gene into the plant genome, along with multiple enhancers in front of the inserted genes. Another approach is to give the plants a gene coding for a slightly different EPSP synthase—one that is resistant to glyphosate. The result in this case was the **Roundup Ready plants** produced by Monsanto (which also produces Roundup). A specific example of a Roundup Ready plant is Monsanto's genetically modified soya plant.

Questions have arisen as to whether the food produced from Roundup Ready plants is affected by exposure of the plants to glyphosate. One line of reasoning is that herbicides and pest control agents that are sprayed onto crops must gain Food and Drug Administration (FDA) approval prior to use with the food supply. Once these agents have been deemed safe, a plant that has been exposed to them is supposedly safe for consumption if it has been properly washed. A counterexample to this reasoning is DDT, which is a very effective insecticide that is still used in parts of the world to control mosquitoes to combat malaria. One problem with DDT is that it is not metabolized very quickly, so prolonged consumption of food that has been sprayed with it can lead to a buildup of the chemical in fat cells, eventually leading to toxic effects. And the presence of DDT is not limited to plants: it will be present in birds that consume the crops and in fish (due to runoff of rainwater into rivers and lakes). These multiple points of entry into the food chain led to DDT levels that eventually adversely affected humans, leading to premature births and low birth weights. This counterexample provides a vivid illustration that governmental agencies are not infallible, despite their best testing efforts. However, this is only a singular counterexample and is not intended to cause one to doubt every single pesticide in use today. The debate around whether to eat only organic produce cannot be easily resolved because arguments against existing pesticides and herbicides are often based on predicted results that have not been realized.

It should be noted that glyphosate was classified as "probably carcinogenic to humans" in 2015 based on in vivo animal testing and in vitro evidence of chromosomal damage in human cells; however, the word "probably" indicates that evidence for a link with cancer is not conclusive. In 2018 a study involving 54,251 licensed pesticide applicators concluded that there was no apparent association between glyphosate and any solid tumors or lymphoid malignancies overall, including non-Hodgkin lymphoma and its subtypes, although there was some evidence of an increased risk of acute myeloid leukemia among individuals with the highest exposure—an observation that requires confirmation. The finding regarding non-Hodgkin lymphoma is especially interesting because a groundskeeper in the United States was awarded $289 million in 2018 for his claim that exposure to Roundup caused him to develop non-Hodgkin lymphoma. He won the award because a jury was convinced—but juries do not decide science. The scientific community has yet to prove whether glyphosate causes cancer in humans. However, effects beyond cancer and beyond humans should also be considered. There is recent evidence that glyphosate affects larval development in honeybees, and this could ultimately affect humans because pollinators are important to the agriculture industry and the food chain.

The only difference between organic foods and regular crops is whether or not pesticides or herbicides have been used during cultivation. A problem with organic crops is that, by necessity, they cost more to produce. Without pesticides, there will be increased crop loss due to pests, and the monetary value of crops lost is expected to exceed the amount that would have been spent on pesticides. As a result, the price of the produce will be inherently higher if the grower is to break even or realize a profit.

18.4 Tomatoes

18.4.1 The Flavr Savr tomato

Most adults like tomatoes, and there is nothing quite like a homegrown tomato picked ripe. The problem is that finding a fresh, ripe tomato at the grocery store is very difficult because the tomatoes sold at most grocery stores are mass produced on farms and then shipped long distances. Ripe tomatoes are kind of squishy; if they are loaded into crates and transported hundreds of miles in

Box 18.1 Antisense technology

As an aside, let us briefly look at antisense technology. We already know from the central dogma that transcription of DNA produces RNA, which can then be translated into a protein. Antisense technology utilizes the insertion of a new "gene" into the genome, complete with a promoter and "exon." The RNA product of transcription of this inserted sequence will be complementary (antisense) to the mRNA sequence of the targeted gene. The result is a double-stranded RNA sequence that will not undergo translation but rather will be degraded by the cell. If you have read Chapter 13, you probably have a better idea of how this technology works than the Calgene inventors did in 1987, when they were producing their first genetically modified tomato plants with suppressed polygalacturonase.

trucks, they will be bruised, ruptured, or completely squashed during the trip. To combat this problem, mass-produced tomatoes are picked before they are ripe; green tomatoes are hard, so they can be packed into crates and shipped. Prior to being stocked on a grocery store shelf, they will be exposed to ethylene gas to finish the ripening process on site. Over time, consumers have become used to mediocre tomatoes and are generally satisfied with what is offered in the grocery store—until they have the good fortune to eat a homegrown tomato, which makes them wonder why they can't get such good tomatoes in the store. Researchers at Calgene decided that, if they used "antisense technology" (see Box 18.1 and Chapter 13), they could target and suppress the gene encoding polygalacturonase (PG). PG is an enzyme that digests proteins in the cell wall. The idea was that by knocking down expression of this gene, there would be a reduction in protein digestion in the cell wall—specifically, the type of protein digestion responsible for the softening of tomatoes during ripening. The hypothesis was that all of the ripening processes would still occur, yielding a great-tasting tomato that was still firm enough for shipping.

There was one problem: the product didn't meet expectations. The idea was good, but the product wasn't satisfactory because the company went about the process in the wrong way. First of all, the tomatoes were bland. Second, they still bruised easily. So, the product was a tomato that still couldn't be shipped when it was ripe, and even if it could be shipped, it didn't taste good. Third, the tomatoes cost about twice as much to produce because the plants yielded about half as many tomatoes per vine as normal tomato plants; half the tomatoes equals double the cost per fruit.

One reason for the lower production numbers per vine was a matter of logistics. Calgene was a California company, and research and development on the Flavr Savr tomatoes took place in California. However, when the project progressed to mass production, the fields used for the scale-up were in Florida. The environment at the Florida farms was more humid, and the soil was sandier (Figure 18.4). Sandy soil caused the root systems to develop differently so that the plants were not as hardy, and the higher humidity allowed for a greater amount of fungus in the air. Fungal infections were a problem for the Flavr Savr crops and again raised the cost of production.

Further problems for the product stemmed from public opinion. Around this time, American attitudes toward genetically modified foods were not favorable: the general public perception was that if somebody ate one of these "Frankenfoods," terrible things could happen to the consumer.

At this point, we have a product that doesn't taste as good as what is already on the shelf, costs more to produce, and has been rejected by the public. As business plans go, it's not great. Meanwhile, there were many legal disputes between Calgene and Monsanto regarding the patent rights: Monsanto claimed to have patent protection for the process of genetically modifying plants, whereas Calgene claimed that such protection only applied to the GMOs Monsanto had actually produced, as opposed to all genetically modified plants. Calgene argued that they were suppressing PG, which was not covered by Monsanto's patent. Of course, the matter went to court, and nothing goes through the courts quickly.

FIGURE 18.4

Not all dirt is the same. Soil from Florida tends to be sandier than soil from southern California. Shown are samples of soil from numerous locations.

Photo: Sean Brady, from https://cen.acs.org/articles/96/i8/Genetic-screen-soil-microbes-uncovers.html

So while Calgene was paying massive legal bills to try to fight Monsanto, its product was not doing very well in the marketplace. The death blow came when Monsanto bought Calgene, and the Flavr Savr tomato was no longer pursued.

That, however, is not the end of the story for genetically modified tomatoes. …

18.4.2 Safeway Double Concentrated Tomato Purée

Another company, Zeneca (which went on to become AstraZeneca), based in the United Kingdom, was also working with GM tomatoes. The Zeneca product found greater success than the Calgene tomatoes, in part because of better business decisions, plus a public perception of GM foods that was more favorable in Great Britain than in America.

FIGURE 18.5

Shelf edge label for genetically modified tomato puree sold at Sainsbury's in the United Kingdom.

Zeneca aimed its tomato project at developing a bulkier product. The ripening process was not to be altered; the tomato had a lower water content, so by its very nature, it was firmer and therefore easier to ship, whether green or ripe. The lower water content also meant the product had greater viscosity, which made it more suited for purées and soups. The tomatoes in fact tasted good and did make it to market.

Business acumen came into play through marketing. Zeneca was aware that the consumer who buys a tomato from the store expects a certain product and that these tomatoes were not that product. However, the consumer who buys purée just expects something red and gloppy that tastes good. The GM tomatoes were therefore used for sauces and purées.

The tomatoes were grown and developed in California around 1994. By 1996 they were being sold at certain locations of Safeway and Sainsbury's, two UK grocery store chains (Figure 18.5). At Safeway they were marketed as Safeway Double Concentrated Tomato Purée. It was a more viscous product, so the production costs were lower: normally, to make a tomato purée, the tomatoes are picked and chopped up, then stewed and reduced—simmered to boil away some of the water content—before packaging. Because the Zeneca product had a lower water content in the first place, the amount of boiling time was reduced, so the amount of energy needed for processing was also lower. These lower costs presented an opportunity to the company: the purée could be marketed in the same sized cans as the competitors' products but at a cheaper price, or it could be sold in larger cans at the same price. The company realized that offering the product as a store brand with a cheaper price might create the perception that the product was of lower quality. However, by making the can larger but charging the price already accepted by the public, the company could create the perception of great value.

By 1999 the Safeway GM tomato purée had 60% of the UK market share. Although that sounds like a great success, consider this question: if as recently as 1999 the Safeway product was selling so well, why are we all not eating GM tomato purée now? The answer is that something went wrong. First of all, there was a problem with public perception. The grocery stores did not try to hide the fact that they were selling a genetically modified product; in fact, customers visiting the store were presented with a flyer that touted the new product. "Hey, we've got genetically modified foods that taste great; this is science at its best! It tastes great *and* it gives you better value. Try some today!" People did try it and realized it was a quality product with better value. However, around that time, news of the failed Flavr Savr came out and, as a result, public opinion changed: "I've heard about these tomatoes. They're awful!" Public perception dictated that sales would go down. At that point the market share for Safeway Double Concentrated Tomato Purée went from 60% to just 25% better than normal tomato purées. It was still great that the GM product sales were outpacing those of the traditional products, but the GM tomatoes were grown on small farms but their competitors came from large conglomerates, so a large competitive advantage was required for the small-farm product to compete successfully. The smaller advantage that existed after the Flavr Savr debacle represented a turning point for both the product and the company.

Around the same time, not only had public perception of GM foods started to change in the United Kingdom, but the public also began to regard store brands as being of lower quality. Because these tomatoes were being sold under the Safeway moniker, people started to assume they were inferior, and this further contributed to decreasing market share. Because of these two unfortunate events—neither of which had anything to do with the reality of the fine tomato product in the cans—not enough profit was generated to sustain the venture.

Entrepreneurs, take heed: public perception changes all the time, and factors completely unrelated to a given product can bring down that product or even the entire company. A moral to this story is that if you're forming a new biotech company, make sure you invest in a good business staff. You might perform the best science in the world, but if you can't persuade people to buy into it or the product the company will fail. Some people would suggest that over 50% of your budget should be spent on marketing and sales.

18.5 Rice

Rice is a good focus for genetic modification because over 3 billion people in the world use it as a staple. If we wanted to affect world food supplies, modifying rice would be a great way to go about it.

18.5.1 Miracle Rice

In this example of biotechnology being used to address the food supply, we are not necessarily referring to a transgenic organism. Selective breeding was used to create the final product, but neither the delivery of genes from different species nor the delivery of specifically engineered genes was employed. Still, Miracle Rice represents one of the greatest success stories in the world of biotechnology.

In 1960, the International Rice Research Institute (IRRI) was set up in the Philippines. A team of agricultural biotechnologists collaborated with plant breeders from all over the world in an attempt to produce rice crops with higher yields and better disease resistance. One of the problems that affected yields was **lodging**, which is when the rice stalks cannot support the weight of heavy, seed-bearing heads and they fall over. The IRRI researchers investigated over 10,000 varieties of rice from around the world. One idea to increase crop yields was to cross high yield varieties with dwarf plants to create short, high yield variants that would not lodge. They were also able to accelerate their research by incorporating a technique known as **shuttle breeding**, whereby crops are grown in different regions during the summer and winter, which effectively doubled the number of crops they could investigate in a single year.

Two years later, 38 investigational crosses were completed. Of these novel variants, the eighth one showed special promise; it had been produced by a cross between Dee-geo-woo-gen (DGWG) – a dwarf variety from Taiwan known for high-yield – and Peta – a tall Indonesian rice plant known for heartiness, quick growing times, disease resistance, and seed dormancy. This cross only produced 130 seeds, but these seeds were the genesis of one of the most famous breeds of rice ever produced: the **IR8**. The cultivation and selection process that ensued is worth a closer look:

- The first generation (F1) was the result of potting the 130 seeds obtained from the initial IR8 cross. All of these plants were tall, but that did not mean the experiment was a failure.
- Seeds from the F1 plants were isolated and planted in a field to yield second-generation (F2) plants. By this point there were around 10,000 of them. The striking feature of the F2 generation

was that roughly one-quarter of the plants were short, which meant dwarfism was a recessive trait controlled by a single gene from the DGWG parent. The knowledge that only a single gene was involved made the process of producing a variety with the desired traits much more straightforward.

- The tall F2 plants were discarded, and seeds from the short, early maturing plants were used to produce the F3 generation. In this step, these seeds were planted in a special nursery which allowed the researchers to expose the developing plants to the fungus responsible for the plant disease known as "rice blast". Only plants that were resistant to rice blast were retained.
- The F4 generation started with seeds taken from the most promising F3 plants; 298 F3 plants were selected, and seeds from each one were sewn into individual "pedigree rows" in the blast nursery. The F4 plants were exposed to rice blast like in the F3 generation, and the ones most susceptible to the disease were removed.
- From that F4 generation, three rows were especially noteworthy: rows 36, 246, and 288. In row 288, the third plant in was especially striking. Carrying the designation *IR8-288-3*, this plant produced progeny that went on to feed millions. Its name was shortened to "IR8", but it gained a more spectacular name after its performance was noted in numerous performance trials.

The IR8 seed line eventually produced plants that were 100 to 120 cm tall, with strong stems that supported large heads without lodging. The plants needed only 130 days to mature compared with up to 170 days for traditional varieties or rice. In 1965, the IRRI performed yield trials on 23 of their new rice breeds. The top three yields all came from derivatives of the eighth cross. These three varieties were assessed further in the 1966 international yield trials and the All India Coordinated Rice Improvement Project at Hyderabad, for which data are shown in Table 18.1. When examining the table, compare the yields of one very special variety - IR8-288-3 - to the yields of typical rice crops around the world.

The IRRI scientists were interested in more than just days to maturity, yield, and degree of lodging. They also observed incidence of disease, seed dormancy, and milled rice yields plus the gelatinization temperature and amylose content of the rice starch. It was not enough to produce tons of rice; the product had to have properties that millers could work with and consumers would accept.

Early IR8 was not perfect. It had a chalky grain that detracted from the market appearance of the polished rice, and there was substantial breakage during the milling process. In addition, the amylose content of the starch was too high for many Asian consumers—not that Asian consumers were measuring the amylose content of their rice, but a high amylose content would cause the rice to harden after cooking, preventing the desired soft gel consistency of common dishes such as mochi, puto, and tteok. Moreover, IR8 was found to be susceptible to bacterial blight and some particular types of the rice blast disease. However, much effort was later put into improving the quality of the IR8 grain by reducing the amylose content and improving the chalky texture. The IRRI achieved success on these fronts.

In 1966 a large portion of Asia was going through a period of drought, which could have led to widespread famine. The IRRI responded by making IR8 freely available to the world. Over 2300 farmers came to IRRI by any means possible, to get 2-kg bags of seed. The people took buses, rode bicycles, or even walked many miles to get starters for new crops that would ultimately save the lives of many thousands, if not millions, of people throughout Southeast Asia.

IR8, the product from the 1960s, eventually became known as Miracle Rice, from a newspaper headline—MARCOS GETS MIRACLE RICE—published in Manila when President Ferdinand Marcos met with IRRI officials and received IR8-288-3. The product was very successful, producing

Table 18.1 (A) Yields for the top three producers of the F4 generation of IR8 (the eighth cross), grown at the IRRI in the Philippines (from Chandler, 1982). (B) Yields obtained for IR8-288-3 grown in large-scale rice trials in 1965 and 1966 (from Chandler, 1982). (C) Average yields for selected countries in 1965.

(A) Initial F4 production	**Production amounts (kg/ha)**	
IR8-246	6104	
IR8-288-3	6060	
IR8-36	6047	
(B) Further IR8-288-3 yields	**Country**	**Production amounts (kg/ha)**
	Thailand	6031
	Malaysia	6600
	CRRI India	7034
	Hyderabad (India)	7753
	Mexico	8000
	Bangladesh (3 sites)	6710–8200
	Pakistan	10,248
(C) Average paddy yields for selected countries in 1965		
	Philippines	1310
	Thailand	1840
	Mexico	2740
	Pakistan	1420

From IRRI database, USDA data: http://ricestat.irri.org:8080/wrsv3/entrypoint.htm, Accessed January 21, 2020

several times the yield of traditional rice with a growing time 1 month shorter. The strain was so successful that it was planted throughout Southeast Asia. The problem with the crop's overwhelming success was that a **monoculture** was established—that is, all of the plants were of the same strain. This lack of genetic diversity made rice crops vulnerable to many kinds of threats: attacks from viruses, microbes, or insects, or exposure to adverse weather conditions. For example, an insect known as the brown planthopper developed quite a taste for the plants. When the pests set up residence, farmers either had to combat the infestation with pesticides or go hungry; as a result, an entire area of the world became dependent on pesticides. Farmers eventually drifted away from planting Miracle Rice.

18.5.2 Golden Rice

In 2013 the World Health Organization estimated that 250 million preschool children had from vitamin A (retinoic acid) deficiency. β-Carotene is a precursor of vitamin A (Figure 18.6) and is associated with orange foods like carrots, pumpkins, and sweet potatoes but is also found in non-orange foods like spinach and kale. Vitamin A deficiency can be much more serious than simply making you have trouble seeing at night; severe cases can be fatal. An estimated 250,000 to 500,000 children lose their sight each year as a result of vitamin A deficiency, and half of them within one year of going blind. Because so

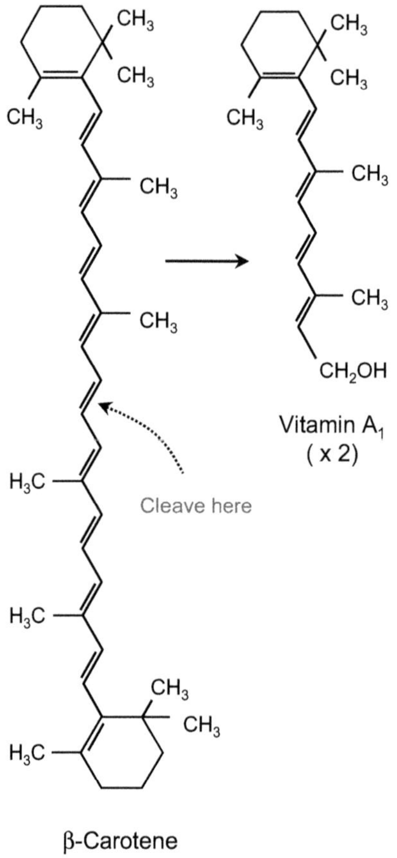

CH_3 CH_3 CH_3 CH_3 CH_3 Cleave here H_3C H_3C H_3C CH_3 CH_3 β-Carotene

CH_3 CH_3 CH_3 CH_3 CH_3 CH_2OH Vitamin A_1 (x 2)

FIGURE 18.6

One molecule of β-carotene can be cleaved to produce two molecules of vitamin A_1 (retinol), a member of the vitamin A family.

many people consume rice—especially in the countries most afflicted with rampant vitamin A deficiency—rice has been chosen to combat the problem via genetic modification.

To create the first generation of Golden Rice, gene delivery was used to introduce an entire biosynthetic pathway into the plant cells, using promoters specific to the endosperm. The endosperm was chosen because it is the part of the rice grain that is eaten, and the endosperm plastids are the sites where geranylgeranyl diphosphate is formed. The genes inserted into the rice genome, originating from daffodil and the bacterium *Erwinia uredovora*, caused the cells to produce enzymes that convert geranylgeranyl diphosphate into lycopene (Figure 18.7). The plant's native enzymes can convert lycopene into β-carotene, and the β-carotene is then converted to vitamin A in the animal gut.

The second generation of Golden Rice replaced the daffodil gene (*psy*, encoding phytoene synthase) with the *psy* gene from maize to yield a plant that produced up to 23 times more carotenoids (including α- and β-carotene, β-cryptoxanthin, zeaxanthin, and lutein), and preferentially produced α- and β-carotenes over the other carotenoids.

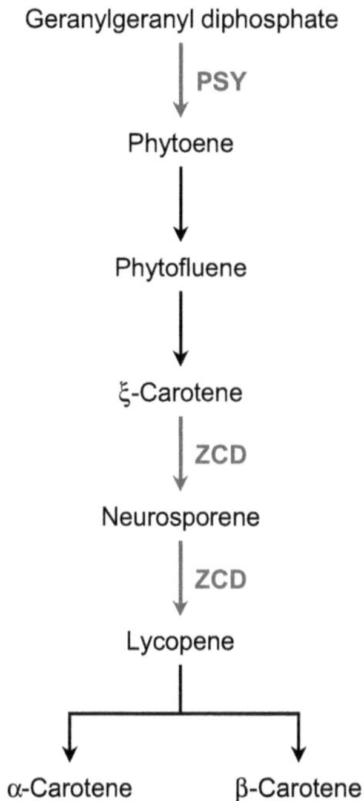

FIGURE 18.7

The biochemical pathway that is used by GR_2E (Golden Rice) to convert geronylgeronyl diphosphate to β-carotene. Bioengineers inserted two genes, encoding phytoene synthase (PSY) and zeta-carotene desaturase (ZCD), into the rice genome to achieve β-carotene production. Bioengineered steps utilizing the inserted genes are indicated with *red arrows* in the pathway. Once ingested, β-carotene is converted into vitamin A in the gut.

Golden Rice received regulatory approval for field trials in the Philippines in 2012. The FDA served as a consultant for the IRRI and stated in 2018 that it had no questions concerning human or animal food derived from Golden Rice (specifically, the cultivar known as GR2E). The FDA was satisfied with the IRRI's claims regarding the safety and nutritional assessment of the product, although the letter noted that the rice was not presently intended for human or animal food uses in the United States. It did, however, acknowledge that the rice could enter the US food supply via imports from countries of production. Interestingly, the FDA pointed out that the amount of β-carotene in the product was too low to warrant a nutrient content claim in its labeling.

The aim of Golden Rice was to attack the issue of widespread vitamin A deficiency by having the rice produce a vitamin A precursor, but the project ran into problems. Once again, it was not the science that restricted the progress of this technology to the consumer, but a societal issue. With the Safeway tomato purée, it was public perception; with the Flavr Savr tomato, it was legal issues; with Golden Rice, the problem has been patent issues.

A company that creates a new product needs to recoup its development costs. It takes a lot of money to develop a new product, whether it is a drug, a device, or a GM food. In the case of a GM food such as rice, although the company must recoup its investment, the product is nevertheless intended for developing countries that cannot afford to pay exorbitant sums for it. Ethical issues thus arise: how can poor or developing countries be given access to a technology that cost so much to develop?

If you pursue a career in biotechnology with the idea that you're going to save a large number of lives globally, consider the people you are intending to save and whether they have the means to pay for your product. The altruists among you may be asking why we can't develop solutions and just give them to the consumer? Again, keep in mind that you must recoup enough money to keep your company running so it can continue to do good work and that you will also need to pay back the investors who gave you the funds to develop your product in the first place. A serious issue can arise when the people you're servicing do not have the money to pay for your final product.

Sometimes a government will step in and work with the biotechnology company to help offset the cost of development and production to help its own people. Other times, however, another group might attempt to steal the technology so that they can sell it more cheaply (because they do not have to pay the original costs associated with research). In the next section, we will examine two methods used to try to prevent technology theft.

18.6 Terminators and traitors

There are a couple of ways in which companies try to get around the issue of getting payment for a newly developed product. Suppose Suki grows a special strain of tomato that is exceedingly good and easily shipped and that she sells a couple of these tomatoes to Rachel. Rachel eats one and says, "Wow, these really are great!" She plants seeds from the other tomato and grows her own vines, eventually starting her own company to sell tomatoes that are really the product of Suki and her company. One might argue that it would be problematic for Rachel to form a company to sell tomatoes that were developed by Suki. However, what if Rachel simply continues to grow the tomatoes for her own use without selling them? What if she grows enough to give to all of her neighbors? After all, she purchased the tomatoes and everything in them, including the seeds, so why not grow them? Agriculture companies are aware of this potential situation and have developed ways to deal with the potential loss of profits.

18.6.1 Terminators

Terminator technology is a way of engineering plants to produce seeds that cannot be replicated beyond the F1 generation. Suppose you are a farmer who wants to grow some great new GM plants that are resistant to all pests, so you don't need to spray your crops. The marketing potential is great: people like the idea of eating food that has never been exposed to pesticides. So you purchase seeds from the biotech company, plant them, and later harvest a great crop. Having sold the crop and made a good profit, you want to produce the same crop the following year. However, the company that produced the original seeds used terminator technology, so after harvesting, you cannot go and plant seeds from the plants you just grew—you have to go back to the company and buy more seeds and must do so every single year.

Terminator technology is great for the inventing company because the farmer must repurchase seeds every year, but it is not good for the farmer. Many farmers around the planet live in poverty, growing crops for their own survival and hoping they can take enough surplus to market to help pay for farm upkeep and other expenses they will incur over the coming year. These people often cannot bear the added expense of repurchasing seed from the company every single year.

One of the fears associated with terminator technology is this: what if the molecular principle you used to render the seeds sterile spreads to other crops—perhaps via a virus, bacterium, or insect? This could potentially render all other crops in the country sterile, so that after one season we would not be able to grow any more food because we would have no more viable seeds. Even if we had seeds stored in a seed bank, it would only be a matter of time until plants from those seeds would also be infected; the agriculture industry would be completely shut down. Although this is not a likely scenario, it is still a concern associated with the technology.

18.6.2 Traitors

Traitor technology, also called **GURT** (an acronym for genetic use restriction technology), involves a genetic switch that is turned on and off by the use of a chemical additive. For instance, a company might sell you seeds for a crop containing a gene that kills caterpillars but that only expresses the gene if you spray the plants with the company's magic potion. The result is that you can grow your crops and you won't need to buy seeds every year, but you will have to keep coming back to the company for the spray.

There are two problems with traitor technology. One is the same problem associated with terminator technology: poor farmers are tied to the company, in that they will have to pay the company every year to be able to grow these crops. The other problem—this time a problem for the company—is that the spray could be analyzed and reverse engineered by a third party who can then manufacture and sell it at a greatly reduced price. Suppose you own the biotech company and that you sell the spray for $10 per unit to try to recoup your research and development costs. The third-party company might be able to sell the same spray for $5 per unit because t does not have to account for the same development costs that you incurred while developing your initial product. Eventually, customers start to figure out that this second spray works just as well for half the price and stop purchasing your spray. Who wins in this situation? Is it the farmer because he has found out a way to pay half as much for the spray? Does the third-party company win because it is stealing money from the original company that developed the technology? Perhaps no one wins; the poor farmer is still being forced to either pay $10 here or $5 there when he used to pay nothing, and the original company is unable to recoup its original investment costs, so it goes bankrupt and is no longer able to invent great products or help humankind.

18.7 Summary

In this chapter we saw specific examples of biotechnology as applied to agriculture. As with many of the other applications covered in this book, biotechnologists used or targeted a specific gene to improve some aspect of an overall system. Bacteria were modified to help prevent the formation of ice crystals on plants, or plants were given the ability to produce crystals that were toxic to caterpillars and other pests.

An interesting model of an engineered system was seen with the glyphosate example. This powerful herbicide was created to kill virtually every plant, and then special plants were designed to be resistant to its toxicity. This led to great control over the plants that could grow in a farmer's fields.

We then moved on to two genetically modified tomatoes: the Flavr Savr and the tomatoes used for Safeway Double Concentrated Tomato Purée. These products were presented for reasons beyond the science of genetic modification: they were examples of engineers identifying an aspect of a system that could be improved for higher product quality or a product that required less processing after harvesting. They were also examples of how nonscientific issues can cause very real problems for biotechnology companies.

Next, we covered two products of the IRRI: Miracle Rice and Golden Rice. Miracle Rice was the product of engineering, although the gene transfers were carried out by selective breeding. The result was one of the most successful engineered foods in history, even though its success did not endure for all of eternity. (Then again, what technological success does? The world will always adapt and present new challenges.) Golden Rice is a current example of a transgenic food aimed at addressing a specific nutritional issue seen in underdeveloped parts of the world, and its story is still developing before our eyes.

We ended the chapter by looking at two forms of biosecurity: terminator and traitor technology. These technologies are used by companies to try to protect their agricultural products from being used by unauthorized individuals. (Unauthorized use and sale of a technology is an example of how the world adapts and presents new challenges.) We also considered the reality of money and its necessity in terms of the life of a business and discussed the paradox of technologies that are intended to help the poor but are too expensive for them to use.

In the next chapter, we will look more closely at the business side of biotechnology. Specifically, we will learn about patents and licenses: their purpose, what they mean, and how they are obtained.

Related reading

Alibhai, M.F., Stallings, W.C., 2001. Closing down on glyphosate inhibition—with a new structure for drug discovery. Proc. Natl. Acad. Sci. U.S.A. 98, 2944–2946.

Amrhein, N., Deus, B., Gehrke, P., Steinrücken, H.,C., 1980. The site of the inhibition of the shikimate pathway by glyphosate: II. Interference of glyphosate with chorismate formation in vivo and in vitro. Plant Physiol 66, 830–834.

Andreotti, G., Koutros, S., Hofmann, J.N., Sandler, D.P., Lubin, J.H., Lynch, C.F., et al., 2018. Glyphosate use and cancer incidence in the agricultural health study. J. Natl. Cancer Inst. 110, 509–516.

Annex, I.V., Approval Registry for Field Testing of Regulated Articles. Bureau of Plant Industry, Department of Agriculture, Republic of Philippines. http://biotech.da.gov.ph/upload/annexIV.pdf.

Arto, I., Capellán-Pérez, I., Lago, R., Bueno, G., Bermejo, R., 2016. The energy requirements of a developed world. Energy Sustain. Dev. 33, 1–13.

Chandler Jr., R.F., 1982. An Adventure in Applied Science: A History of the International Rice Research Institute. International Rice Research Institute, Manila, Philippines.

Galli-Taliadoros, L.A., Sedgwick, J.D., Wood, S.A., Körner, H., 1995. Gene knock-out technology: a methodological overview for the interested novice. J. Immunol. Methods 181, 1–15.

Ganzel, B., 2007. Farming in the 1950s and 1960s: The Green Revolution: "Miracle Rice." Wessels Living History Farm, York. Nebraska, USA (Accessed February, 2014) http://www.livinghistoryfarm.org/farminginthe50s/crops_17.html.

Gruys, K.J., Sikorski, J.A., 1999. Inhibitors of Tryptophan, Phenylalanine and Tyrosine Biosynthesis as Herbicides. Dekker, New York.

Guyton, K.Z., Loomis, D., Grosse, Y., El Ghissassi, F., Benbrahim-Tallaa, L., Guha, N., et al., 2015. Carcinogenicity of tetrachlorvinphos, parathion, malathion, diazinon, and glyphosate. Lancet Oncol. 16, 490–491.

Haslam, E., 1993. Shikimic Acid: Metabolism and Metabolites. Wiley, New York.

Nayar, A., 2011. Grants Aim to Fight Malnutrition. . https://doi.org/10.1038/news.2011.233. http://www.nature.com/news/2011/110414/full/news.2011.233.html.

Paine, J.A., Shipton, C.A., Chaggar, S., Howells, R.M., Kennedy, M.J., Vernon, G., et al., 2005. Improving the nutritional value of Golden Rice through increased pro-vitamin A content. Nat. Biotechnol. 23, 482–487.

Suszkiw, J., 1999. Tifton, Georgia: a peanut pest showdown. Agric. Res. Magazine 47, 9.

Vázquez, D.E., Ilina, N., Pagano, E.A., Zavala, J.A., Farina, W.M., 2018. Glyphosate affects the larval development of honey bees depending on the susceptibility of colonies. PloS One 13, e0205074.

Ye, X., Al-Babili, S., Klöti, A., Zhang, J., Lucca, P., Beyer, P., et al., 2000. Engineering the provitamin A (beta-carotene) biosynthetic pathway into (carotenoid-free) rice endosperm. Science 287, 303–305.

Zhao, J.Z., Cao, J., Li, Y., Collins, H.L., Roush, R.T., Earle, E.D., et al., 2003. Transgenic plants expressing two *Bacillus thuringiensis* toxins delay insect resistance evolution. Nat. Biotechnol. 21, 1493–1497.

Questions

1. The Safeway Double Concentrated Tomato Purée (circle all that are true):
   - **a.** Was based on the concept that a tomato with less water content would be easier to ship when ripe and would also lower production costs because less water had to be removed to produce a purée
   - **b.** Never caught on in the United Kingdom because the Flavr Savr tomatoes in the United States tasted so bad
   - **c.** Was a great hit in the United Kingdom until people found out the tomatoes were genetically modified
   - **d.** Was marketed as a marvel of genetic engineering and as an example of how science could yield great values for the consumer
2. Circle all of the following that are true for Bt plants:
   - **a.** Bt plants are Roundup Ready.
   - **b.** Bt is an acronym for "Better than," meaning they have been genetically modified to produce larger, tastier produce than their non-GM counterparts.
   - **c.** Bt plants such as Soya are resistant to the actions of glyphosate because several copies of the gene for EPSP synthase have been transferred into the plant genome.
   - **d.** Bt plants contain a specific gene from a bacterium that prevents the formation of ice crystals on the plant during light frosts.
3. **a.** What was the idea behind the Flavr Savr tomato?
   - **b.** What was the idea behind the tomatoes used for Safeway Double Concentrated Tomato Purée?
4. "Golden Rice" gets its name from:
   - **a.** β-Carotene production
   - **b.** The high profit gleaned from the sale of this engineered food
   - **c.** The fact that it uses urea, a nitrogen-containing molecule found in urine, as a nitrogen source
   - **d.** Engineered production of B vitamins, which turn urine bright yellow

5. Why are the words “gene” and “exon” in quotation marks in Box 18.1?
6. To what does the “8” in “IR8” refer?
7. Referring to Table 18.1, we see that IR8-246 had the highest rice yield of the experimental cultivars, and the second and third place breeds were nearly identical in output. Yet IR8-288-3 was selected for further propagation. Speculate why this happened.
8. Miracle Rice suffered from rice blast, but the breeding process that eventually produced Miracle Rice specifically screened for plants resistant to this disease. Resolve this apparent contradiction.
9. Without regard to legal and safety issues, speculate on the ultimate effectiveness of Golden Rice in terms of reducing the number of deaths caused by vitamin A deficiency. Give reasons to support your answer.
10. Give three advantages of Miracle Rice over traditional rice, such as Peta.
11. Why would a biotechnologist care about the amylose content of rice?
12. If Miracle Rice was such a superior crop plant, why is it not being used today?
13. a. Calgene came up with an idea that would let tomatoes ripen without getting mushy. Explain the biochemical basis behind the idea.
    b. What could the idea from part a be used for?
14. Miracle Rice was a very good product that saved the lives of perhaps millions of people. It was good for the farmers, and it was good for the consumers. Why, then, was it not a good idea for every farmer to grow this strain of rice?
15. Name two ways that you could give a plant resistance to glyphosate.
16. Name five problems that led to the demise of the Flavr Savr tomato.
17. What are ice-minus bacteria? What is a commercial application of these bacteria?
18. Suppose a biotechnology company is using GURT to protect against unauthorized use of its plants. The technology is such that the company sells seeds of its engineered plants but also requires the grower to use a spray that is also sold by the biotechnology company.
    a. Suppose the spray only cost $1 per gallon to manufacture, and the company sells it for $20 per gallon to generate funds in a way that is not driven by greed. Give a key reason, related to existing debt, why the markup is so high.
    b. Suppose the business model was to pay off the existing debt from part a within 3 years. The company could have sold its product for a much cheaper price if it used a model that would pay off its debt within 10 years. Give a practical reason why the 10-year payoff structure was not chosen.
    c. Other reasons for product markup are not related to historical debt but rather are used to cover the costs of distribution. One example is the cost of packaging (bottling, in this case). Name three additional costs that must be covered.
19. Discuss one benefit and one concern regarding terminator technology.
20. In the second generation of Golden Rice development, the *psy* gene from maize replaced the *psy* gene from daffodils because the new gene resulted in 23 times more carotenoids. Speculate at least two reasons that might explain the improvement. (Your knowledge from Chapter 4 might help with this question.)
21. Is a Bt plant organic? Would you eat Bt soybeans right off of the vine in the field?
22. Was the Safeway Double Concentrated Tomato Purée safe? Why or why not?

CHAPTER

Patents and licenses

19

Chapter outline

We have covered the basic science and scientific applications of biotechnology, but there is much more to consider if an innovation is to make it out into the world to help improve the lives of people. In this chapter, we will look at legal and business issues related to technology development.

Early in the development process, an inventor might want to seek legal protections for a technology in the form of a patent. We will look at three types of patents: composition of matter, method of use, and medical devices. We will also discuss some of the expectations that must be met before a patent examiner will award patent protection.

Patents do not get products to market; they only offer protections. However, they do provide a public declaration of a new product or technology, thus allowing the world to consider ways in which it could be used. If a business wants to adopt or develop a patented product or technology for its own purposes, a license can be granted by the holder of the patent to permit such activities. A patent is a declaration, but a license is permission; we will examine licenses and how they can be structured to allow for product development. We will also be looking at the interplay between money, risk, options, and product development as companies and inventors come together to drive a product toward the marketplace.

19.1 Types of patents

Simply holding a patent is not enough to make you rich.

Getting a technology to market uses a variant of the Edison equation: the process is 1% inspiration and 99% perspiration. A patent application requires both. Getting to the point of filing a patent application is a good start, but it is not what actually gets a technology to market. A patent is the legal cover for something of value that you have created. It is the deed to your intellectual property, but it is still up to

Biotechnology and its Applications. https://doi.org/10.1016/B978-0-12-817726-6.00019-8

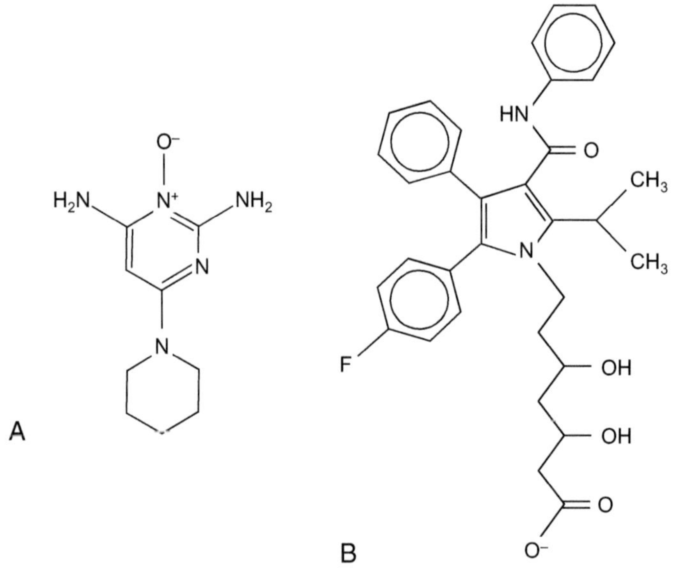

FIGURE 19.1

Chemical structures of (A) minoxidil (Rogaine) and (B) atorvastatin (Lipitor).

you to improve upon it. Just as you may buy some physical property in the form of land and decide to build a house, a parking lot, or a race track, it is also up to you to decide what to do with intellectual property to which you have acquired the deed in the form of patent protection.

The protections we most commonly deal with in the biotech community are **composition of matter** patents. A composition of matter patent is basically a chemistry patent on a unique molecule. It describes the molecule, including the structure that is going to be used in your product, such as $C_9H_{15}N_5O$ for Rogaine, or $C_{33}H_{34}FN_2O_5$ for Lipitor (Figure 19.1). The patent states that you have the molecule, what the molecule looks like; that the molecule is original; that you discovered, found, or created it; and that you make certain claims regarding that molecule such as method claims, which include ways of using the molecule to do something (such as growing hair or lowering blood cholesterol levels). These are **Method of Use** claims. Such claims should appear in the composition of matter patent application because, even with a novel composition of matter, your invention must be good for something in order to be awarded a patent. It is not enough just to say, "I have this original molecule"; you must state "I have this original molecule, and it does *A, B,* and *C*."

The person filing the patent will make claims as to some uses in the original application, but some uses of a molecule will be claimed later. Minoxidil (Rogaine) is a good example. The minoxidil patent was initially a composition of matter with claims for use as an antihypertensive (a drug that lowers blood pressure). However, after patients had been taking the blood pressure drug for a while, they started to report a side effect: they were growing more hair on their arms, legs, chest, back... and scalp. It did not take long for doctors to figure out that applying the drug externally to the head could cause directed hair growth and be used to combat hair loss. At this point the inventing company (Upjohn, now a division of Pfizer) filed a new patent application claiming an entirely new method of use for the old composition of matter. If the new use had been something closely related to hypertension, the company probably could not have patented it because it would not have been novel: it would have been obvious, and *patent claims must be non-obvious*. However, the switch from antihypertensive agent to hair restorative moved the drug's method of use far enough away from the original claims that the new use was freshly patentable. To be clear, while methods of use must appear in composition of matter patents, they can also stand alone as a separate category.

Another broad category of patents is **medical devices**. Medical devices are a large part of biotechnology, and the category will continue to expand as other forms of technology advance. For example, consider the field of imaging and the ability to steer probes and catheters through the body in real time: the probes and catheters are devices. As another example, luminescent chemicals can be thought of more as devices than as treatments. They could serve as diagnostic devices by providing a new way of getting inside the body for imaging. Perhaps they could allow a surgeon to find an exact location for excision. The luminescent chemicals are not being used to treat anything but as a tool to guide the surgeon performing the treatment.

The rules governing approval of medical device patents are broadly different from those surrounding composition of matter patents. Medical device patents are far more challenging: with a composition of matter application, you must somehow ultimately prove to a patent examiner that your material is different from materials that have come before—but devices are almost always built on technologies that already exist. Devices have much more of a history. Claims for medical devices often amount to "my invention is like this existing device, only different," so device claims are difficult to prepare and defend. Good legal counsel is needed to write a patent application to distinguish the device from all others. As with a composition of matter application, medical device patent applications also require methods of use claims. It is not enough to say that you have an original device; the device must be useful for something before patent protection can be awarded.

Consider another example. At one time, Tulane University had a certain medical device that was licensed and under development—an oxygen-sensing catheter that allowed the physician to go through the body looking for a spot where the oxygen saturation was at a specific level. It was a combination of two existing technologies: an oximeter and a steerable catheter to which the oximeter would be attached. The patent examiner agreed that Tulane had come up with a new use for the technologies, but he did *not* agree that Tulane had created a new device, the rationale being that anyone can take two existing products off the shelf and tape them together. Tulane was awarded protection on its method of use claims, but the patent office would not allow protection of the device itself because it was not sufficiently novel.

Box 19.1 Are patents secret?

Were any secrets revealed in the previous example? The answer is "No," by definition. "Patent" means "to make clear." The philosophy behind a patent is that the inventor announces a new compound, device, or use to the world and explains it so that somebody else in the field could reasonably expect to replicate the invention or use. In exchange for making the technology available for the benefit of the rest of society, the inventor(s) are given legal protection to ensure that nobody else can make money from the discovery without their approval, for a given period of time. The fact that a patent was awarded means that the details are publicly available. One could look up this or any other patent and read the history if so desired. A good place to start is the United States Patent and Trademark Office: https://www.uspto.gov/patents-application-process/search-patents.

19.2 Licenses

After one gets patent protection, the next step toward gaining some kind of monetary recognition is to license the technology. This is where the money is: getting a company to license the rights to use your patented technology.

The above idea is correct. (We are confining this discussion to the biotechnology industry.) Once you have been awarded patent protection, you can find a company that is already in the space and approach them about taking the license to your intellectual property. The patent does not necessarily have to be granted yet; it's possible to get license deals on pending patent applications. Usually, the licensee (the person taking the license) will want to put safeguards in the agreement stating that if the patent fails to be issued, certain things will happen, because the patent process can take years to complete. In fact, pre-award licensure is becoming more common because the process is so long, and the life of patent protection begins on the day you file for it, *not* the day it is granted. Your patent's life starts to dwindle the day you do the right thing and submit it to the Patent and Trademark Office (in the United States; office names vary by country). Filing starts your 20-year clock. Because of this, the earlier you can find a licensee, the better, if you are interested in licensing the technology to an outside party. Potential licensees understand this. Agreements must be made regarding what will happen if the inventor never gets a valid claim issued on the patent application.

Suppose you as an inventor have a medical device that is going to do all sorts of cool things. There is more patentability in this case, and you will have a market edge just by having a functioning prototype. However, from the licensing company's point of view, pursuing your technology would basically be licensing it based on nothing but vapor because nobody in the patent office has taken a look at the application and probably will not do so for another 6 months. If your device is novel and exciting, the company might go ahead with licensing it even without patent protection on day 1, but you can be certain they will write safeguards into the licensing agreement to cover the possibility that the technology is not awarded a patent in a timely manner.

Right now, the US and other patent authorities around the world are aligned in terms of patent life: 20 years from the date of filing. There are a couple of ways to extend this, but the amount of extra protection is usually minimal. For example, such an extension might be granted to make up for any extra time that was taken in issuing the original patent if you can show that the process took longer than normal, but the time will not be significant enough to make a material difference; you cannot expect the life to be extended for an additional 10 years. As a result, it is important to find licensing opportunities as early as you possibly can.

It is important to start talking to companies not only because they might license your technology but also because they will give you feedback. Talking to a company almost always entails a **confidentiality agreement** (also called a nondisclosure agreement) so that you can show them everything you've got and have a valuable discussion without the fear of having your idea stolen (Figure 19.2). A good confidentiality agreement is usually not very onerous to either party. Such agreements are routine, but they are very important, especially when dealing with patentable matter about which the two parties want to be able to talk freely; many companies will not talk to you at all without having such an agreement in place. There

NONDISCLOSURE AGREEMENT

This Nondisclosure Agreement (the "***Agreement***") is entered into as of the date set forth below by and between the individual or entity identified by signature below ("***Recipient***") and Company Name, Inc. ("***Company***"), having a place of business at 123 Somewhere Street, City, State.

WHEREAS Company has made discoveries or improvements relating to Technology X and/or its use for drug delivery or targeting, or its use as a medical therapy (collectively; "***The Developments***");

WHEREAS Company possesses proprietary and confidential information and materials related to The Developments;

WHEREAS Recipient desires to receive information or materials related to The Developments in order to evaluate a potential business relationship including, but not limited to, consulting, collaboration or provision of services to Company;

WHEREAS Recipient appreciates that Company has expended and continues to expend money and effort to establish a proprietary position with respect to The Developments it has made and that Company considers The Developments and information or materials pertaining thereto to be its confidential property; and

WHEREAS Company is willing to reveal or provide to Recipient information or materials relating to The Developments on a confidential basis.

NOW, THEREFORE, in consideration of the foregoing, and of the mutual covenants, terms and conditions hereinafter expressed, Recipient and Company agree as follows:

1. Confidential Information. "***Confidential Information***" means any information disclosed to Recipient by Company, either directly or indirectly in writing, orally or by inspection of tangible objects, ...

2. Non-use and Nondisclosure. Recipient agrees not to use any Confidential Information for any purpose except ... Recipient agrees not to disclose any Confidential Information to any third parties ...

3. Maintenance of Confidentiality. Recipient agrees that it shall take all reasonable measures to protect the secrecy of and avoid disclosure and unauthorized use of the Confidential Information. ...

FIGURE 19.2

A sample from a confidentiality agreement.

are different levels of confidentiality agreements, but they are all fairly simple and very valuable, especially for moving forward with some kind of collaboration, partnership, or transaction.

After you get a confidentiality agreement in place—generally a simple matter—you talk to the company and hear its feedback. Even if they are not interested in the technology, it will be good to hear why. You can learn a lot from an opportunity that did not develop. Why didn't it? How can you make it better for the next pitch? Consider that you might be pitching to someone who has been in the industry for over 20 years. If they are willing to sit down with you or even just have a phone call or respond to your emails, pay attention to what they say. It is tremendously useful to hear the insights of experienced people who are already established in the trade. Then, hopefully, during the next conversation you will be better informed and will be able to address the issues that were raised during the first set of conversations.

Once you have found a licensee and it looks like there may be a license there, you can write up an agreement. The internet is fabulous for finding things like template agreements or what royalty rates should be used in a given sector (e.g., rates in pharmaceuticals are different from those in diagnostics; rates in vaccines are different from those in devices). You can further break down devices into subclasses— for example, is it an orthopedic device, a surgical device...? You can perform a great deal of research on the web to get an idea of what a ballpark deal should look like.

19.3 After a license is granted

19.3.1 The inventor works with the licensee after the patent has been licensed

From the point of view of a university technology transfer office, it would be impossible to license technologies if the inventor were not to be a participant in future work with the technology because they know more about it than anybody else on earth. Inventors do not come up with patentable inventions every 5 minutes; they spend some huge proportion of their lives thinking about it every waking moment (and even some sleeping and dreaming ones). They pour their hearts and souls into the invention because it is something they are passionate about; they're not just looking for a new flavor of toothpaste. Perhaps they have a medical condition that affects themselves or somebody they know, and that has driven them to work tirelessly to come up with their invention.

Almost every licensee will want to have the involvement of the inventors, as opposed to a clean handoff. As an inventor, you work hard to get the patent, you work hard to get the license—but if the license is granted for January 1, it does not mean you are finished on January 2. The license agreement will include access to know-how (difficult to quantify but tremendously important) or a consulting arrangement with the inventor. The inventor has done a ton of work and the company will want access to him. That will simply be part of the deal. Getting a good licensee or collaborator with experience—somebody who will work hard to get the invention to market—is a tremendously valuable part in the education of the young inventor.

19.3.2 Remuneration

Once a licensure agreement has been made, things are contingent upon the eventual success of the product in the market. It is not the case that today you were able to license your invention so now you get a check for $1,000,000. How is remuneration typically structured?

There are multiple levels in a licensing agreement. You can ask for an **upfront fee**, meaning that on the day everybody signs the agreement, the company pays the inventor a lump sum of cash. As an inventor, you can (and should) also have a **point of development plan** identifying milestones the licensee must meet to ensure the invention is being aggressively developed. Recall the limited window of protection that is afforded by a patent. By licensing the patent to a company that did not pursue bringing the invention to market, the inventor would lose out on a potentially large amount of revenue. In the case of a potential drug, milestones might include when a leading formulation candidate is identified, when toxicology studies will be performed, and so on. With clinical trials, a target milestone might be having the first patient enrolled in a Phase I trial by a certain date. It can be written into the license agreement that if the company does not meet these milestones, the license comes back to the inventor, and you can put a monetary figure on it. The company must make a decision at each point: commit more resources to the project, or give it back to the licensor (the inventor)? Will it decide to enroll that first patient into the Phase I trial tomorrow and pay the inventor $100,000, or will it decide to terminate the license? If the company chooses to progress through each of the developmental points and the product goes to market, then a steady royalty stream will be established for the inventor.

Of the licensing-related income that Tulane University has collected in the past 20 years, about 90% has come from royalties from products that were actually sold on the market. All of the other items mentioned earlier—upfront fees, milestone fees, and so on—comprise only a tiny bit of the total package, despite being six- or seven-figure fees. The big payoff is in getting a product to market. There are rare cases in which a large upfront fee is paid, but if these products never make it to market (and the odds dictate that they will not), then the majority of potential revenue for the inventor is lost. A small upfront fee, combined with diligent effort between the inventor and the company, can eventually yield a royalty stream of well over $1,000,000 a year after a biotechnology product reaches the market.

The licensing structure should be focused on the possibility of success. For the inventor, getting a large upfront fee would come at the price of a reduction in control of the technology down the line. For instance, if a large upfront fee is paid, the company might not have to give the technology back if they do not develop it, or there might be a low royalty rate attached to the product if it reaches the market. A more effective long-term strategy is to have:

- a low upfront fee;
- relatively low (or absent) milestone fees;
- control over the development plan retained by the inventor;
- a larger royalty rate; and
- the ability for the inventor to take the technology back if the company does not continue to pursue it.

With this type of agreement, the ultimate success is measured in terms of the number of products that make it to market. It is also a good strategy because it motivates the inventor to remain highly involved in the development of the product. If the technology is licensed to a small company that does not have a lot of cash, asking for large upfront fees would end up taking a large portion of the cash that it would otherwise use for development of the product. When the end game is to get products to market, large upfront fees are self-defeating for both the inventor and the licensee.

19.3.3 If the license is released back to the inventor

If a product is not developed after it is licensed, the licensor (inventor) will want to get it back promptly because the patent clock is still ticking. If the licensee cannot develop the product, the inventor needs to be able to get it into the hands of somebody who can. You want to have a clean break with the licensee; this is achieved through termination language in the license agreement, spelling out that the licensee will have no further rights. If you can get the licensee to turn over their data—another negotiation point—this will help to make the product more licensable to the next company. You should keep a positive relationship with the licensing company throughout development because you need to know what is going on. You need to know why the company is not moving forward: what failed, and at what point did it happen?

If a subsequent company chooses to license the technology, they will be aware that the first license went sour. The subsequent license deal will be less lucrative for the inventor, in part because there will be less patent life left on the technology but also because there is now some doubt or risk associated with the product. It's tough to relicense something once it has failed, but if you do not have the rights, you can't even try. However, there are some things you can do that can help your position. Perhaps the science in related areas has improved so that the problems can be dealt with. The problems may be in formulation or how to make the drug more bioavailable. Solutions that did not exist 2 or 3 years ago when the first company gave the technology back may now be readily available.

19.4 Summary

In this chapter, we were introduced to the idea of a patent. We saw that a patent is legal protection that is offered in exchange for a public declaration of a new product or technology. We also discussed three types of patents:

- composition of matter
- method of use
- medical device

We then followed the progress of an idea from the patent stage to the marketplace and how licensing the idea is an essential step if an inventor wants to enlist outside help in getting a product to market. We discussed different ways in which inventors can get paid during the development process and ways that companies might protect themselves to prevent huge expenditures if an idea does not pan out. Although inventors and companies must work together for the common goal of getting a product to market, each must protect themselves so that successes and failures are shared in an equitable manner.

Acknowledgment

Special thanks to John Christie, Executive Director of Technology Transfer, Tulane University.

Questions

1. What should be the main goal of patenting a technology? Is it to get rich?
2. What should be the main goal of licensing a technology? Is it to get rich?
3. Name three types of patents and give an example of each.
4. True or false: When you patent a technology, you are granted legal secrecy for a period of time to allow you to develop the technology without outside competition. (Explain your answer.)
5. True or false: When you license your technology to a company, you can reasonably consider retiring to live a stress-free life. (Explain your answer.)

6. Suppose you have found a process by which you can attach a novel molecule to a constituent of butter to render the compound able to eradicate the AIDS virus, without altering the taste or texture of the butter. What type of patent would you file?
7. Suppose you have invented a medical device that takes an intravenous (IV) catheter and an ultrasound device and uses them together to monitor individual cells in the bloodstream and remove cell types that fit within a programmable set of parameters, such as sickle cells or microbes. Would this be patentable as a medical device? Given that ultrasound is already used to visualize blood and IV catheters are used to gain access to the blood, would the technology instead be patentable as a method of use? Explain your answer.
8. Name two problems associated with a large upfront fee.
9. You have come up with a novel drug that can be used to treat type I diabetes. You have filed for a patent, but it has not yet been awarded. Should you wait until the patent is awarded before pursuing a license with a company? Why or why not?
10. Suppose you want to talk about an idea for a new drug with Johnson & Johnson because they have masses of expertise in your area of research. You know that talking with them could save you years in development because they could give great guiding advice. On the other hand, they have legions of researchers who could steal your idea and develop it faster than you could ever hope to. What is the best course of action for you and your idea?
11. Suppose I file a patent application that describes my invention of the 10-blade shaving razor. Nobody has made a 10-blade shaving razor. Predict whether a patent will be awarded for my idea. Justify your answer.
12. Suppose I file for a patent on my invention. This invention is so sensitive that it can be influenced by sound to drive a needle across rotating cylinders of wax to create impressions that are proportional to both the frequency and amplitude of the sound waves. The patent application will be denied. Why?
13. If an inventor is to net great amounts of money, at what stage in the life of an invention would that happen?
14. When a company licenses a technology, milestones are often incorporated into the licensing agreement to protect the inventor.
    a. What is the inventor being protected from?
    b. What term is given for the milestone portion of the licensure agreement?
15. Suppose you have patented a technology and licensed it to a company and that five major milestones must be met as the product makes its way to market. The first two milestones were met, but now the company says it is no longer interested in developing your technology. Name two things you could do, given that the licensure agreement was written well and is being followed.
16. You have just designed a polymer that nobody has ever thought of before. Can you be awarded a patent based on that alone? Why or why not?
17. You have just produced a polymer that nobody has ever even thought of before. Can you be awarded a patent based on that alone? Why or why not?
18. You have just designed a novel polymer and explained all the great things it can be used for. Can you be awarded a patent based on that alone? Why or why not?
19. You have produced a polymer that nobody has ever thought of before, and you used it to build a 100-kg, full-sized car that is more impact resistant than any other car in existence. Can you

be awarded a patent based on that alone? Why or why not? (Keep in mind that cars have been around for a long time.)

20. You are a professor at a research university, and you have designed a new drug to treat Parkinson's disease. The university has been awarded a patent, with you named as the inventor. You want to take the drug further and have it licensed to a big pharmaceutical company, but all the companies say the drug is not far enough in development, so there is significant financial risk for them.

   a. Your university suggests that you form your own company, and they will license the drug to it. You are just a poorly paid researcher, so how can you pay for the licensing fee?

   b. If you were to form the company, what would be in the licensing deal for you? What would be in the deal for the university?

21. You patented a drug to treat uncontrollable laughter and licensed it to Novartis, which has developed the drug and now sells it under the name Kalmzitol. It has only been 10 years since the patent was filed, and your best friend has found that the drug can be used as a *very* effective cockroach bait and poison. She has looked up the original patent and is now able to make the drug on her own.

   a. Do you have any legal recourse against your friend?

   b. Your friend files for patent protection on her invention. What kind of patent will she file? Is it likely to be awarded?

   c. Does Novartis have any legal recourse against your friend?

Glossary

ABC transporters A class of membrane proteins that can harness the energy provided by adenosine triphosphate–binding cassettes to move hydrophobic molecules (such as drugs) out of the cell.

absorbance A measure of how much light energy is lost as it passes through a given medium. Absorbance can be calculated via Beer's law: Absorbance = (molar absorptivity) × (path length) × (concentration).

acid hydrolase A type of enzyme that catalyzes hydrolytic degradative reactions and is active only in acidic environments.

activator (protein) Aids in building the transcriptional machinery by binding to enhancers.

active site The portion of an enzyme where the catalyzed chemical reaction takes place.

active transporter A membrane transporter that requires energy, often via adenosine triphosphate hydrolysis, to perform its function.

adaptin A molecule that adds specificity to the interaction between certain membrane-bound receptors and clathrin.

adherent cells In cell culture, cells that attach to a substrate such as a culture dish or a three-dimensional scaffold as they grow, divide, or even migrate.

aerobic respiration The metabolism of a molecular feed source (such as glucose) into molecular products plus energy in the presence of oxygen.

agarose A sugar extracted from red algae, used to create porous networks for gel electrophoresis.

alkaline phosphatase An enzyme that removes the 5′ phosphates from DNA or RNA. It is used in genetic engineering to help prevent the formation of empty vectors.

allele The presence of a specific form of a gene. In humans, who have two copies of every gene (one from each parent), we refer to the two *alleles* of a given gene.

alternating copolymer A copolymer containing two different repeating units that are distributed in an alternating fashion in the molecule.

alternative lengthening of telomeres (ALT) A process that adds telomere repeat sequences to the ends of eukaryotic chromosomes without using telomerase.

amphipathic A molecule that has both hydrophilic and hydrophobic regions.

amplicon The specific DNA bases that will be copied (amplified) during polymerase chain reaction. The bases lie between and include the primer binding sites, and do not necessarily include an entire exon or gene.

amplification The process of increasing the amount of something. In genetic engineering, amplification refers to the attainment of many copies of DNA, typically plasmid DNA.

anaerobic respiration The metabolism of a molecular feed source (such as glucose) into molecular products plus energy in the absence of oxygen.

annealing In polymerase chain reaction, the attaching of primers to ssDNA.

antibiotic An antimicrobial drug.

antibiotic resistance marker A gene that serves as a selection marker by virtue of its product being a protein that lends resistance to a specific antibiotic. Often a component of plasmid DNA.

anticodon The portion of a tRNA molecule that pairs directly with a specific codon sequence of mRNA being held inside the ribosome.

antigen Any substance capable of invoking an immune response.

antiporter A membrane transporter that carries two or more molecules across the cell membrane at the same time but in opposite directions.

antiseptic Chemical applied to body surfaces to destroy or inhibit the growth of vegetative pathogens. Similar to a disinfectant but applied to living surfaces.

apoptosis Also known as "programmed cell death," this process of cellular suicide is a complex molecular cascade that leads to the cell dismantling its own cytoskeleton, chopping up its own genome, and packaging the remains into vesicles known as "apoptosomes."

apotransferrin A protein that binds with iron in the blood to create transferrin, which is used to transport iron to and into the cells of various organs.
autoclave An apparatus used to sterilize objects via a combination of heat and pressure.
back-biting termination A type of termination of polymerization whereby the leading edge of a growing polymer becomes bonded to the origin of the polymer, thus eliminating that leading edge of polymerization.
band-pass filter Light filter that allows wavelengths within a certain range to get through. Described by the wavelength at the center of that range and the total width of the range.
basolateral The side of a cell that faces away from the luminal surface.
batch culture Culture in which cells are given a set amount of medium and allowed to grow without medium changes.
bioactive Having a biological effect.
biocompatible The case in which a material and its degradation products are nontoxic.
biodegradable Capable of being broken down under physiological conditions, often by hydrolysis.
blastocoele The hollow cavity of a blastocyst.
blastocyst The hollow, approximately 150 cell structure that develops from a morula.
block copolymer Copolymer containing two or more subunits that are clusters of a single type of repeating unit. Block copolymers are often made by joining two or more homopolymers.
blunt ends The result of a restriction endonuclease that cuts both strands of double-stranded DNA without leaving an overhang.
brightness The readiness with which a fluorophore will absorb photon energy, multiplied by the probability that it will be released as a photon once absorbed. More precisely, the product of the extinction coefficient and the quantum yield.
Bt plant Transgenic plant that has integrated a gene encoding a toxin to agricultural pests. "Bt" stands for *Bacillus thuringiensis*, the microbe that was the original source of the gene.
cadherin One of a family of calcium-dependent adhesion proteins used by cells to attach to other cells.
cargo molecule A molecule that is carried within a separate, enclosed structure. Examples include DNA carried inside a viral head, and hydrophobic drugs or oils carried within micelles.
caveolin A protein involved in the formation of caveoli (pinocytotic vesicles).
cavitation The pitting and wearing away of solid surfaces (such as metal) as a result of rapid formation and collapse of vapor pockets.
cDNA See *copy DNA*.
centriole Formed from short microtubules and generally found in pairs within the centrosome, which acts as the spindle pole during mitosis.
centromere The region where sister chromatids are held together; the place where the kinetochore will form during mitosis.
channel A membrane protein that allows certain molecules to pass via passive diffusion, driven by the electrochemical gradient.
Chargaff's rules In relation to the genomic DNA of a given species, the percentage of DNA bases containing thymine roughly equals to the percentage of bases containing adenine, and the percentage of deoxycytosine roughly equals the percentage of deoxyguanine.
chiral A molecule is said to be chiral if its mirror image is not superimposable upon the molecule itself. Chirality is a necessary (but not sufficient) condition for optical activity.
cholesterol A steroid molecule often present in the plasma membranes of animal cells. It lends fluidity and rigidity to membranes, depending on temperature.
chromatid A condensed form of DNA that appears during the first portion of mitosis (prophase).
cistron A polynucleotide unit that codes for a polypeptide.
clathrin A protein that self-assembles into cages containing bits of the plasma membrane. Used in one form of endocytosis.

coding strand The strand of genomic DNA that has the same sequence as the primary RNA transcript (with the exception of T vs U bases).
CODIS Combined DNA Index System, a database of genetic information created by the United States Federal Bureau of Investigation.
codon A group of three mRNA bases that encodes an amino acid.
cold gas sterilization Sterilization via ethylene oxide. "Cold" refers to the fact that the sterilization can take place at room temperature.
colony A distinct mass of microorganisms such as bacteria.
colony-forming unit (CFU) A cell capable of multiplying and forming a colony on a solid medium (a culture plate).
competent (bacteria) Having an enhanced ability to take up exogenous DNA.
competitive inhibition Reduction in activity of an enzyme caused by the introduction of a molecule that can bind to the active site of the enzyme but is not acted on by it.
complex medium Medium for which the exact chemical makeup is not known.
composition of matter A type of patent that details a new molecule or compound and its uses. The patent will state that you have the molecule; what the molecule looks like; that the molecule is original; that you discovered, found, or created it; and that you make certain claims regarding uses for that molecule.
confidentiality agreement A means of allowing two parties to discuss a technology without fear of having the information stolen or used outside of the meeting. Also known as a nondisclosure agreement.
consensus sequence An accepted sequence of bases in a DNA element, determined by individually identifying the most common base found in each position of the sequence.
contact inhibition The property of untransformed somatic cells that halts their growth, replication, and migration upon contact with other cells (or the walls of a culture vessel) on all sides.
control An experimental sample with known parameters that should produce an expected result. Used to verify that an experimental procedure is working properly.
copolymer A polymer made up of two or more different compounds (repeating units).
copy DNA (cDNA) DNA that is generated via reverse transcription of RNA (often mRNA) molecules.
counterstain As used in the Gram stain, a second dye used to stain all cells for the purpose of making gram-negative cells visible.
coupled reaction Reaction in which the energy released by one process is used to drive a separate process.
CpG island Long stretch of DNA rich in unmethylated Cs and Gs, surrounding housekeeping genes.
CRISPR Acronym for "clustered regularly interspaced short palindromic repeats," which are DNA features targeted in a specific type of nuclease-mediated genome editing.
CRISPR-Cas Acronym for CRISPR-"CRISPR-associated system," which is the combination of palindromic repeats and nucleases used in a specific type of nuclease-mediated genome editing.
cryptic growth A short period of cell growth that takes place as nutrients from dead cells are consumed by starving cells. Sometimes seen at the end of the plateau phase of the growth curve.
cycle threshold (C_t) For a given sample undergoing real-time quantitative polymerase chain reaction (qPCR), this is the cycle number at which a minimum amount of fluorescence is first reliably detected.
cytokinesis The process whereby a binucleated cell pinches off into two daughter cells following mitosis.
cytotoxic Poisonous to cells.
decolorizer Used in the Gram stain to dissolve and remove dyes that are not firmly entrapped.
defined medium A cell culture medium for which every component is known, measured, and controlled.
degeneracy Term used to describe the fact that most amino acids are encoded by more than one codon sequence.
deletion A type of DNA mutation whereby a base is removed from the genome.
delta delta method ($2^{-\Delta\Delta}$) Mathematical model used in quantitative polymerase chain reaction to determine the fold difference in transcription of a gene of interest in treated versus untreated cells.
denature To unfold, as in a protein.

dendrimer A highly branched polymer containing a central core molecule and regularly spaced, well-controlled bifurcations.
dendriplex A gene delivery complex formed by the interaction of DNA with a dendrimer.
diauxic growth A growth curve characterized by two log phases corresponding to the culture utilizing two different carbon sources.
Dicer A protein involved with RNA interference that cleaves double-stranded RNA molecules.
disinfection The inactivation/destruction of vegetative pathogens, typically by liquid chemicals.
DNA ligase An enzyme that seals breaks in the phosphoribose backbone of DNA via the creation of phosphodiester bonds.
dry weight The weight of a cell pellet following centrifugation, washing, recentrifugation, and subsequent drying.
dynamic culture A culture of cells grown in an environment where medium is either stirred or is caused to flow across or through the cell support.
dynamin Protein that helps to pinch off an endocytotic vesicle from the plasma membrane to prevent leakage of cytoplasmic contents from the cell.
E-box A transcriptional element found within many enhancers. Has the general sequence CANNTG. The E-box has the ability to form a stem-loop which can be recognized by transcription factors.
ectoderm Germ cell lineage originating from the outer layer of the gastrula. Can develop into skin or nerve cells.
electrochemical gradient A combination of the difference in charge and the difference in concentration of a particular ion across two sides of a membrane.
electroporation The use of an electric field to create pores in cell membranes. Often used in the laboratory to introduce genetic material into cells.
elution The removal of adsorbed molecules from a material by application of a solvent.
E_m Emission wavelength: the wavelength of light produced in the greatest amount by a stimulated fluorophore.
empty vector A plasmid that has been sealed without an insert.
endocytosis The transport of material into a cell via invagination of the plasma membrane.
endoderm Germ cell lineage originating from the inner layer of the gastrula. Can develop into the endocrine glands, or cells lining the digestive and respiratory tracts.
endonuclease An enzyme that cuts polynucleotides at an interior position.
endotoxin Another name for the lipopolysaccharides that constitute the bacterial outer membrane.
enhancer A noncoding portion of a gene that aids in the building of the transcriptional machine.
enzyme A catalyst, typically a protein (although RNA enzymes called "ribozymes" also exist).
equity Percentage ownership of a company.
ethylene oxide A gas used for sterilization at relatively low temperatures.
eukaryotic A eukaryotic cell has DNA stored in a membrane-bound nucleus and utilizes membrane-bound organelles (*c.f. prokaryotic*).
E_x Excitation wavelength: the optimal wavelength of light absorbed by a fluorophore.
exon The coding portion(s) of a gene (*c.f. intron*).
exonuclease An enzyme that cuts polynucleotides from their 5′ or 3′ terminus.
expansion (culture) Increasing cell numbers by allowing a culture to grow until the vessel is well populated. The culture may then be passed into multiple other vessels and allowed to proliferate (*c.f. maintenance*).
explant A process whereby a primary cell culture is established via processing of whole tissue.
extinction coefficient (ε) A descriptor of how quickly light is absorbed by a fluorophore, standardized to a 1 M solution and a penetration distance of 1 cm into the liquid.
extracellular Sited outside of a cell.
F_{ab} The antigen-binding region of an antibody.
F-ATPase A proton transporter found in the inner mitochondrial membrane that can pump protons via adenosine triphosphate (ATP) hydrolysis. This pump typically works in reverse, however, producing (not using) ATP as protons run through the transporter, down their concentration gradient.

F_c The constant region of an antibody; also known as the fraction crystallizable region.

fermentation A process used by cells to produce products such as ethanol or lactic acid under anaerobic conditions. The chemical pathway is adherent to three rules: (1) energy is produced, (2) oxygen is not consumed, and (3) the NADH/NAD+ ratio is unchanged by the process.

flippase A type of phospholipid translocator that moves phospholipids, such as phosphatidylserine, to the opposite face of the plasma membrane in a specific fashion.

fluorescence resonance energy transfer (FRET) The phenomenon whereby the photons emitted by one fluorophore excite a second fluorophore.

fluorescent tag A fluorescent molecule used to identify another molecule or structure. The tag may be attached to its target covalently or noncovalently, but the attachment is more informative if it is highly specific. "Fluorescent" means the molecule will emit photons in the visible light spectrum.

frameshift mutation An insertion or deletion of a DNA base within a gene that causes the reading frame of that and all subsequent codons in the associated mRNA molecule to be shifted, resulting in a different (or no) message being encoded.

fusion protein A protein with two or more distinct domains, made by adjoining the exons for each at the genetic level. Often used in genetic engineering to create proteins with a fluorescent tag.

gastrula A hollow structure in early embryonic development, containing the three germ cell layers: ectoderm, mesoderm, and endoderm. The cells of these layers are multipotent.

gastrulation The development of an inner cell mass into a gastrula.

gated channel A channel that opens in response to a stimulus. For example, a voltage-gated channel opens in response to a change in membrane potential.

GC box Transcriptional regulatory element with the sequence GGGCGG, used to increase the transcription level of a gene.

gel shift A phenomenon in electrophoresis whereby samples migrate more slowly when bound to another agent, such as DNA migrating more slowly when bound to a protein.

gene A stretch of DNA that functions as a unit to give rise to a polypeptide product via an RNA intermediate.

gene silencing Suppression of gene expression.

gene therapy The delivery of genetic material into cells for the purpose of altering cellular function.

germ cell A sperm or egg cell.

graft copolymer A segmented copolymer consisting of a linear backbone of one composition and branches of a different composition.

growth curve A classic plot of cell number (or mass) versus time. Characterized by lag, early log (exponential), late log (deceleration), plateau, and death phases.

guide RNA (gRNA) RNA fragment engineered to mimic CRISPR-RNA. It is bound by a Cas protein to target nuclease action during CRISPR-Cas-mediated gene editing.

GURT Acronym for "genetic use restriction technology," another name for *traitor technology*.

Hayflick limit A supposed limit to the number of times that cells can divide, based on the shortening of telomeres with each cell division.

heat-killing Exposure of enzymes to high temperatures to inactivate them.

helper lipid A neutral lipid used in conjunction with a cationic lipid for gene delivery. Although these neutral lipids do not bind to the carried polynucleotides, they help to improve transfection efficiency by aiding with escape from the late endosome or endolysosome.

heterocromatin Tightly packed eukaryotic DNA that is unavailable for transcription.

homology-directed repair A cellular method of repairing double-stranded DNA breaks that requires the presence of a homologous strand of DNA. This phenomenon is key to getting a donor template of DNA inserted into a genome via a CRISPR-Cas system.

homopolymer A polymer made of many copies of the same repeating unit.

housekeeping gene A gene that codes for proteins used in vital cellular processes. Because of the importance of these genes, they are constitutively expressed at nearly constant levels, independent of the cell cycle.

hybridization A molecular biology technique whereby a specific oligonucleotide sequence is located or analyzed via the binding of a labeled, complementary oligonucleotide probe.
hydrolysis The breaking of a chemical compound via the consumption of a water molecule.
hydrophilic Having an affinity for water; "water loving."
hydrophobic Immiscible with water; "water fearing."
hyperbranched Describes a highly branched polymer in which the bifurcations are irregularly spaced.
hypertonic A solution that has greater osmotic pressure versus another solution, such as that found on the inside of a cell.
hypotonic A solution that has lower osmotic pressure versus another solution, such as that found on the inside of a cell.
ice-minus A transgenic form of *Pseudomonas syringae* that underwent gene knockout to eliminate the expression of a membrane protein that served as a nucleation site for ice crystal formation.
induced pluripotent stem cell (iPSC) A differentiated adult cell that has been driven into an embryonic-like state.
inner cell mass The mass of cells on the interior of a blastocyst. These cells develop into the embryo.
insertion A type of DNA mutation whereby an extra base is inserted into the genome.
integrase The enzyme responsible for insertion (integration) of cDNA into genomic DNA.
integrin Member of a family of proteins used by cells to attach to extracellular surfaces.
intensity Similar to the brightness of a fluorophore, this descriptive parameter also takes into account pathlength, concentration, the intensity of the illuminating bulb, and the ability of the apparatus to gather and record photon emissions.
intercellular Located in the space between adjacent cells.
internal control In polymerase chain reaction, this is used to correct for pipettor errors and differences in initial cell concentrations when mRNA is isolated from cell cultures. Internal controls correspond to the amount of expression of a housekeeping gene within the cell; this level should be constant whether the cell is in a control or a test group.
intracellular Located within the interior of a cell.
intron A noncoding region used to separate exons within a gene (*c.f. exon*).
ionic detergent A detergent molecule that carries a charge in its head group. The charge can be positive or negative.
IR8 Also known as "IR8-288-3," or Miracle Rice, this plant was the result of extensive crossbreeding and cultivation. The plants were shorter but had higher yields and shorter cultivation times.
isoelectric point The pH at which the predominant net charge of a species of molecule in solution is zero.
isotonic A solution with the same osmotic pressure as another solution, such as that found on the inside of a cell.
kinetochore Structure that forms around the centromere on a chromosome undergoing mitosis. It is where fibers from the mitotic spindle will be attached.
Klenow fragment The large subunit of prokaryotic DNA polymerase I, used in molecular biology to make sticky ends blunt via $5' \rightarrow 3'$ polymerase activity and $3' \rightarrow 5'$ exonuclease activity.
knockdown A molecular biology technique whereby the expression of a gene is reduced. Knockdown can be achieved by RNA interference methods.
knockout A molecular biology technique whereby a gene is disrupted or removed from the genome, thus eliminating the expression of a functional form of the associated protein.
Kozak sequence The base sequence used as the final ribosomal assembly point in eukaryotic mRNA. It also serves to position the ribosome at the translational Start site (*c.f. Shine-Dalgarno sequence*).
lariat The loop structure formed as introns are spliced from a pre-mRNA molecule.
leak channel A channel that is always open.
leaky scanning A phenomenon whereby ribosomal subunits might ignore the first AUG codon in the mRNA and skip to a later instance, which allows the production of different protein products from a single mRNA.

ligand A molecule that can be bound by a receptor.

lipofection Gene delivery carried out via lipoplexes.

lipoplex A gene delivery complex formed by the interaction of DNA with lipids.

lipopolysaccharide A molecule class consisting of lipids conjugated to sugar polymers. A hallmark of the bacterial outer membrane.

locus The physical location of a gene within the genome.

lodging The falling over of a plant when the stalk is no longer able to support the weight of the head.

long-pass filter Light filter that allows wavelengths over a certain value to get through.

low-density lipoprotein (LDL) A construct of triglycerides, cholesterol, phospholipids, cholesterol esters, and apolipoprotein B. LDL is used to transport cholesterol and fatty acids from the liver to other body tissues.

luminal Referencing the interior side of a tube or pouch (e.g., an artery, the intestine, or the urinary bladder).

lysosome A vesicular organelle characterized as containing multiple degradative enzymes that are active at low pH.

lysosomotropic Able to penetrate lysosomes.

M phase The mitosis phase of the eukaryotic cell cycle, when the chromosomes condense and are separated into two complete genomes, each contained in its own nucleus.

maintenance (culture) Cells in a culture are kept alive, with a proportion being split into a new culture vessel when overpopulation is a concern, but the number of vessels is not increased (*c.f. expansion*).

malting Process of controlled germination by which complex carbohydrates within seeds are enzymatically broken down into simple sugars and disaccharides.

mediator (protein complex) Used in the building of the transcriptional machinery; joins together activator proteins, general transcription factors, and RNA polymerase II.

medical device patent Different from a composition of matter patent in that the device can be developed from existing technologies.

mesoderm Germ cell lineage originating from middle layer of the gastrula. These cells can develop into, for example, bone, cartilage, muscle, or blood cells.

messenger RNA (mRNA) A modified product of transcription, acted on by ribosomes during the process of translation.

method of use A type of patent that describes novel ways in which a molecule or material can be used. Composition of matter patents contain methods of use, but later additional uses can be discovered and claimed via separate patent applications.

Miracle Rice Also known as IR8 or IR8-288-3, this plant was the result of extensive crossbreeding and cultivation. The plants were shorter and had higher yields and shorter cultivation times than existing strains.

miRNA Micro RNA; a form of siRNA that is encoded in the genome and is produced by the cell itself.

mitosis A eukaryotic cell division process whereby a single nucleus is separated into two distinct nuclei, each containing its own complete copy of the genome.

mitotic spindle Formed during mitosis, primarily from microtubules; forms initially on opposite poles of the eukaryotic cell and pulls sister chromatids apart.

molecular weight marker In gel electrophoresis, a set of standards used as a reference to determine the size of samples.

monoculture The agricultural practice of relying on only a single genetic variant of a food crop.

monodisperse The special case where all polymer molecules in a sample have the same molecular weight. The polydispersity index of a monodisperse sample is 1.

mordant Used in the Gram stain to convert a dye into a form that is insoluble in water.

morula The solid, 32-cell structure that develops from a fertilized egg. It has a mulberry-like appearance.

multidrug resistance An ability carried by some cells (e.g., cancer cells) that affords them protection against certain drugs. The ability can be conferred via ABC transporters.

multinucleate The state of having more than one nucleus within a single cell.

multipass transmembrane protein A protein that spans the plasma membrane more than once.

multiple cloning site (MCS) A stretch of DNA that is rich in restriction enzyme recognition sequences; useful in genetic engineering.

multipotent Having the ability to differentiate into members of a single germ cell lineage (*c.f. pluripotent, totipotent*).

N:P ratio In gene delivery, the ratio of amines in the delivery vehicle to phosphates in the DNA (or RNA) to be delivered.

nonhomologous end joining A cellular method to repair DNA double-strand breaks. This method is prone to errors because any free ends of dsDNA can theoretically be used for the repair; a homologous DNA strand is not required for joining to occur.

nonionic detergent A detergent molecule with an uncharged (but polar) head group.

nonmetabolite A molecule that is not used by the cell during metabolism.

non-obvious In terms of patent awards, claims that are not a mere extension of existing technology or uses.

nonspecific binding The binding of a primer to a stretch of DNA that is not a 100% match.

nucleoside A DNA base attached to a ribose sugar that does not have a phosphate group attached (*c.f. nucleotide*).

nucleotide A DNA base attached to a phosphoribose (*c.f. nucleoside*).

number average molecular weight The sum of the molecular weights of all polymer molecules in a sample, divided by the total number of polymer molecules (*c.f. weight average molecular weight*).

Okazaki fragments The piecemeal fragments used to replicate the lagging strand in dsDNA replication, used to preserve replication of both leading and lagging strands in the $5' \rightarrow 3'$ direction.

open circle A plasmid architecture where each strand of the dsDNA is nicked but the nicks do not line up.

operator The prokaryotic version of the eukaryotic silencer, it is a noncoding portion of a gene that serves to inhibit transcription. See also *silencer*.

operon A set of prokaryotic genes under the control of a single promoter.

origin of replication The initial assembly site for DNA replication enzymes.

packed cell volume The volume of a cell pellet following centrifugation.

packing parameter A mathematical descriptor of a lipid that takes into account the effective volume of the lipid, the cross-sectional area of its polar head group, and the length of its hydrocarbon tail(s).

palindrome A dsDNA sequence that has the same identity for each strand when read in the $5' \rightarrow 3'$ direction.

PAM sequence See *protospacer-adjacent motif.*

passage Removal of a cell culture from one vessel and transfer into a new one. The process typically involves splitting the main culture into one or more subcultures, each with a significantly lower total number of cells. See also *split*.

passive transporter A membrane transporter that does not require energy (e.g., via adenosine triphosphate hydrolysis) to perform its function.

phagocytosis Endocytosis of large particles or entire cells. Often termed "cellular eating."

phosphatidylcholine A common membrane phospholipid, this molecule has a net neutral charge and typically resides on the extracellular face of the plasma membrane.

phosphatidylethanolamine A common membrane phospholipid, this molecule has a net neutral charge and typically resides on the cytoplasmic face of the plasma membrane.

phosphatidylserine A common membrane phospholipid, this molecule has a net negative charge and typically resides on the cytoplasmic face of the plasma membrane. When translocated to the extracellular face of the plasma membrane, it signals that the cell is undergoing apoptosis.

pinocytosis Endocytosis of dissolved substances. Often termed "cellular drinking."

pK_a The pH at which 50% of an ionizable group will exist in the ionized state in solution.

plasmid A circular DNA polymer.

pluripotent Having the ability to differentiate into members of any of the three germ cell lineages (*c.f. multipotent, totipotent*).

point of development plan A licensing strategy that identifies milestones the licensee must meet to ensure that an invention is being developed aggressively.

polycistronic A single mRNA that codes for multiple polypeptides. Typically found in prokaryotes.

polydispersity index A descriptor of the purity of polymer sizes in a sample; defined as the weight average molecular weight divided by the number average molecular weight.

polyplex A gene delivery complex formed by the interaction of DNA with a polymer.

porosity The percentage of a given volume of scaffold that is made up of pores.

pre-miRNA Abbreviation for "pre-micro RNA," the stem-loop structures cleaved from pri-miRNA transcripts.

Pribnow box The -10 binding site for the prokaryotic transcription factor σ^{70}. Although it has the consensus sequence TATAAT, it should not be confused with the eukaryotic TATA box.

primary cell Cell taken directly from a body or a tissue.

primary cell culture Laboratory-grown cell culture arising from primary cells. The cultured cells are not considered primary cells.

primary immune response Immune response mounted upon initial exposure to an antigen. Because the immune system is encountering the antigen for the first time, the patient may become sick while the body learns to create antibodies (and memory cells) to combat the infection.

primary sequence (primary structure) The sequence of amino acids in a polypeptide, starting with the N-terminal amino acid.

primer efficiency A measure of how close a primer set will come to allowing a doubling of the number of dsDNA amplicons in a single cycle of polymerase chain reaction under optimal conditions.

pri-miRNA Abbreviation for "primary micro RNA," the primary transcript of a gene encoding miRNA. It spontaneously forms stem-loop structures in the nucleus (see *pre-miRNA*).

probe A short DNA fragment designed to bind one strand of the amplicon at an interior site during quantitative polymerase chain reactions. The probe is labeled with both a fluorophore and a quencher. These are separated during the extension step, allowing for detection of the fluorophore.

professional phagocyte Cell that utilizes receptors to recognize objects, microbes, or other cells as foreign and then engulf or degrade them via phagocytosis. Examples of professional phagocytes include macrophages, neutrophils, and dendritic cells.

prokaryotic Adjective indicating that a cell is classified as a prokaryote. This means the cell does not have a membrane-bound nucleus, nor does it utilize membrane-bound organelles (*c.f. eukaryotic*).

promoter In eukaryotic genes, the site where RNA polymerase II binds. In general, promoters are characterized by their relatively close proximity to the transcriptional Start site.

promoter elements Distinct DNA sequences found within promoters. For example, the TATA box is a common promoter element.

proof of concept Experimental results demonstrating that a scientific idea is valid and can be built upon.

propidium iodide A dye that intercalates with dsDNA, fluorescing bright red upon stimulation.

protospacer-adjacent motif A specific sequence of DNA bases that must be present after a targeted DNA sequence for Cas proteins to cleave genomic DNA. Often called a PAM sequence.

purine A class of nucleotide bases (adenine and guanine) characterized by a bicyclic ring structure.

pyramided plant Plant containing genes for different forms of toxins to prevent consumption by insects that may have developed a resistance to one or more of the forms.

pyrimidine A class of nucleotide bases (thymine, cytosine, and uracil) characterized by a structure having a single, six-membered ring.

quantum yield (Φ) The probability that, once a photon has been absorbed, it will be emitted as a photon.

quaternary structure In protein folding, the interaction of two or more tertiary structures, such as when polypeptide subunits come together to form a functioning protein.

random copolymer A copolymer containing two or more different repeating units distributed unevenly throughout the molecule.

recognition sequence The specific DNA sequence that is bound by a particular restriction endonuclease.
recombinant DNA Genes or DNA sequences constructed by combining portions of DNA from more than one source.
recombinant protein A protein produced via the expression of recombinant DNA.
repeating unit The monomer used to create a polymer, drawn in its polymerized form. For example, the repeat unit of poly(ethylene) is $-[CH_2\text{-}CH_2]-$.
reporter gene A gene that encodes an easily detectable polypeptide, such as a fluorescent protein.
repressor In eukaryotes, a protein that acts to lower the transcription of a gene by binding to a silencer sequence. In prokaryotes, the repressor achieves the same effect by binding to an operator sequence.
respiratory acidosis Increased acidity (H^+ concentration) of the blood due to inadequate removal of CO_2 in the lungs.
respiratory alkalosis Increased alkalinity (decreased acidity) of the blood due to increased breathing rate or volume.
resting membrane potential The electrical potential across the cell membrane when the cell is at rest.
restriction endonuclease A bacterial enzyme that cuts DNA at defined locations. Originally named because of its ability to restrict the propagation of bacteriophage.
retrovirus A classification for viruses that deliver RNA into cells. The RNA is converted into cDNA inside the cell and then incorporated into the genome.
reverse transcriptase The enzyme responsible for carrying out the process of reverse transcription.
reverse transcription The process of converting a strand of RNA into DNA.
RFLPR Restriction fragment length polymorphism: a product of enzymatic genome degradation used to identify individuals at the genetic level.
RGD sequence The tripeptide sequence arginine-glycine-aspartate, used in a variety of proteins associated with cellular attachment.
ribosome A ribonuclear protein that carries out the act of translation in a cell.
RISC Acronym for RNA-induced silencing complex. This multimeric protein is the heart of RNA interference.
RNA interference (RNAi) The process whereby double-stranded RNA, perhaps formed by the delivery of specific antisense RNA molecules into the cytoplasm, is used to target and degrade sequence-specific sense RNA molecules.
RNA polymerase An enzyme that synthesizes RNA based on information provided by a DNA template.
Roundup Ready plant Plant engineered to be resistant to the herbicide glyphosate.
S phase The synthesis phase of the eukaryotic cell cycle, in which the genome is replicated.
sanitization The removal or destruction of microorganisms via mechanical means.
scramblase A type of phospholipid translocator that moves phospholipids back and forth between the two layers of the plasma membrane in a random fashion.
secondary cultures The result of passaging a primary cell culture into new culture vessels.
secondary immune response Immune response mounted when the patient has been exposed previously to a specific antigen. More antibodies will be produced and much more quickly (due to the presence of memory cells) than with the primary immune response.
secondary structure In protein folding, the spatial arrangement of the atoms in the protein backbone. It includes early-forming structures such as the α-helix, β-pleated sheet, and β-turn.
selective medium Medium that will restrict the growth of, or kill, all microorganisms except for the ones of specific interest.
semiconservative replication During cell division, half of the physical genome of the parent cell is passed to each of two daughter cells, along with a complementary copy of the DNA that is newly manufactured.
senescence The process of deterioration during old age.
sepsis A life-threatening condition brought about by the body's response to microorganisms in the blood and/or the tissues.

serotype Combination of exposed antigens that makes cells or agents, such as viruses, serologically distinguishable.
Shine-Dalgarno sequence In prokaryotic mRNA, the sequence used to locate the translational Start site. Having the sequence AGGAGG, it is typically found around four to seven bases upstream of a Start codon (*c.f. Kozak sequence*).
short-pass filter Light filter that allows wavelengths of less than a certain value to get through.
short tandem repeats (STRs) DNA repeats of two to five base pairs found throughout the genome. The number of repeats in a given locus is polymorphic but inherited. STRs are used in DNA fingerprinting to identify individuals.
shuttle breeding An agricultural technique whereby a crop is grown in one region during the winter and in another region during the summer, reducing the time needed to develop new varieties.
sigma (σ) factor A polypeptide used by prokaryotes to position RNA polymerase on a gene for transcription.
silencer A noncoding portion of a eukaryotic gene that serves to inhibit transcription.
silent mutation A base mutation that will have no effect on the sequence of a translated protein product.
single-pass transmembrane protein A protein that spans the plasma membrane once.
siRNA Abbreviation for "small interfering RNA," dsRNA fragments of 19 to 24 bp bound and processed by RISC complexes.
Slicer A protein involved with RNA interference. It acts to cleave single-stranded (sense) RNA molecules.
small nuclear ribonucleoprotein (snRNP) Protein that acts with snRNAs to form the spliceosome complex.
small nuclear RNA (snRNA) RNA molecule (e.g., U1, U2, U4, U5, U6) that participates in the recognition of intron–exon boundaries and helps to remodel phosphodiester bonds at these locations.
somatic cell A differentiated cell that helps to make up the body of an organism. Somatic cells are distinct from germ cells.
specificity A measure of how selective a receptor is for the associated ligand.
spliceosome The protein–RNA complex used to excise introns (in the form of lariats) from developing mRNA. It consists of snRNAs, five snRNPs, and many additional proteins.
splicing The removal of introns from pre-mRNA.
split (a culture) When the cells of a populated culture vessel are proportionally divided among new vessels. See also *passage*.
spontaneous A term from thermodynamics describing an event that lowers the free energy of a system. Lower free energy states are more thermodynamically stable, so spontaneous events are favorable due to an increase in thermodynamic stability.
static culture A culture of cells grown in an environment where the surrounding medium is not stirred or caused to flow across or through the cell support.
sterile Free of all viable life forms. The term also encompasses viruses.
sticky ends The result of a restriction endonuclease that cuts both strands of dsDNA, leaving single-stranded overhangs that are complementary to one another.
stock (of cells) A repository of frozen cells.
Stokes shift The difference between the optimal excitation and emission wavelengths of a fluorophore.
strong promoter A promoter sequence that has high affinity for RNA polymerase or transcription factors, thereby yielding a high rate of transcription of the exons it controls.
substrate Molecule that binds to an enzyme and serves as the reactant for the catalyzed reaction.
super-secondary structure In polypeptide folding, refers to modification of a secondary structure, as in the adjustment of a β sheet to form a β-barrel. Can also refer to combinations of secondary structures, such as in the helix–turn–helix structure.
surfactant A surface active agent: a molecule having a polar head group and a nonpolar tail. Examples include soaps and detergents.
symporter A membrane transporter that carries two (or more) molecules across the cell membrane at the same time and in the same direction.

T4 DNA polymerase Enzyme used in molecular biology to make sticky ends blunt via $5' \rightarrow 3'$ polymerase activity and $3' \rightarrow 5'$ exonuclease activity.

TALENS Acronym for transcription activator-like effector nucleases, which are used in a specific type of nuclease-mediated genome editing.

TATA-binding protein (TBP) The eukaryotic protein that binds to the TATA box and aids the binding and positioning of RNA polymerase.

TATA box Eukaryotic promoter element with the consensus sequence TATAAAT.

telomerase Enzyme responsible for lengthening telomeres through the addition of copies of the telomere repeat sequence.

telomere Short, repeating DNA sequence found at the ends of eukaryotic chromosomes.

template strand The strand of genomic DNA that is used directly by RNA polymerase for transcription. The sequence of the template strand is complementary to that of the coding strand (and the primary RNA transcript).

teratoma A tissue mass that contains differentiated cells from all three germ lineages.

terminator (sequence) An AT-rich sequence found at the end of a gene, used to signal where transcription can cease in prokaryotes.

terminator technology Used to produce plants having seeds that cannot be used for subsequent farming.

tertiary structure In polypeptide folding, the three-dimensional structure of a folded polypeptide.

titer A concentration, although not as precise as mass/volume. Titers are determined by a series of dilutions that are tested as being either positive or negative for an action or detection. The titer is the most dilute fraction that still yields a positive result. Used to quantify viral or antibody concentrations.

totipotent Having the ability to proliferate and differentiate into a complete organism (*c.f. multipotent, pluripotent*).

traitor technology Used to produce plants that require an engineered chemical switch to be able to grow. (See also *GURT*.)

transcription The first step of the "central dogma": the conversion of genomic DNA into an RNA copy. See also *translation*.

transcription factor Polypeptide involved with the regulation of gene transcription. Transcription factors bind to promoters, enhancers, and RNA polymerase.

transcriptome The complete set of genes being transcribed by a cell at a given moment.

transduction Gene delivery via a viral vector.

transfection Gene delivery via a nonviral method

transfer RNA (tRNA) A type of RNA molecule (73–93 nucleotides long) characterized by a specific folding pattern, an anticodon sequence, and an amino acid covalently attached to the 3′ terminus.

transferrin A unit consisting of apotransferrin bound with iron, used to transport iron into cells.

transformation a. The introduction of foreign genetic material into a (prokaryotic) cell. The conversion of a normal cell into a cancer cell.

transgenic A term that refers to cells or organisms that have received either engineered genetic material or genetic material from a different species.

translation The second step of the "central dogma": the conversion of messenger RNA into protein. *c.f. transcription*.

transmittance The ratio of light getting through a solution in the presence versus absence of a given solute.

transposon A genetic element that is able to translocate to different locations in the genome.

triskelion A shape having three lines emanating from a center. Used to describe the shape of clathrin molecules.

trophoblast The outer cell layer of a blastocyst.

tunica adventitia The outermost layer of an artery. It is made from collagen fibers and contains fibroblasts.

tunica intima The innermost layer of an artery. It is made from endothelial cells.

tunica media The middle layer of an artery. It is made from smooth muscle cells and elastic fibers.

uniporter A membrane transporter that carries a single molecule at a time, and always in the same direction.

unit (U) (of enzyme) The amount of enzyme needed to completely cut 1 μg of phage lambda DNA in 1 hour at optimal temperature.

upfront fee Money paid by a licensing company to an inventor upon the signing of a license agreement.

V-ATPase Vesicular ATPase, which pumps protons into a vesicle at the expense of adenosine triphosphate hydrolysis.

vitamin A molecule that is necessary for the health of an organism but cannot be synthesized by that organism.

weight average molecular weight The sum of the squares of the molecular weights of all polymer molecules in a sample, divided by the total number of polymer molecules times the sum of their molecular weights (*c.f. number average molecular weight*).

wet weight The weight of a cell pellet following centrifugation.

wild type The typical phenotype or genotype that predominates in natural populations.

wobble effect Used to explain the degeneracy of the genetic code, the wobble effect stems from the fact that only two of the three bases of the anticodon can come into close proximity to the codon, allowing for movement (wobble) of the tRNA molecule within the A site of the ribosome.

wort In brewing, a liquid nutrient broth produced by adding water to and mashing the product of malting, then filtering out the solid debris.

ZFNs Acronym for "zinc finger nucleases," which are used in a specific type of nuclease-mediated genome editing.

zwitterion A single molecule that contains both positive and negative charges. (From the German word *Zwitter*, which means "hermaphrodite.")

Index

Note: Page numbers followed by "f" indicate figures, "b" indicate boxes, and "t" indicate tables.

D

E

Q

R